D1711590

Concise Handbook of Mathematics *and* Physics

Alexander G. Alenitsyn
Eugene I. Butikov
Alexander S. Kondratyev

CRC Press
Boca Raton New York

Nauka Publishers
Fizmatlit, Moscow

Library of Congress Cataloging-in-Publication Data

Catalog record is available from the Library of Congress.

Deveoloped by Nauka Publishers, Fizmatlit of the Academy of Sciences of Russia, Moscow.

International Standard Book Number 0-8493-7745-5
Printed in the United States of America 1 2 3 4 5 6 7 8 9 0
Printed on acid-free paper

Preface

This handbook is addressed to a wide audience of readers: school and college students and their teachers, undergraduate students and professors of universities, engineering and pedagogical institutes. The handbook may be especially helpful for self-education.

The main purpose of the book is to help a reader to find or recollect a necessary information. The handbook envelopes all principal parts of contemporary college courses of physics and mathematics. One can find in it the definitions of fundamental notions and concepts, of physical and mathematical quantities. The handbook contains formulas that express the laws of physics, axioms and theorems of mathematics. There are many examples that help in understanding the material. All information both in physics and mathematics is arranged in a coherent system, so the reader gets additional conveniences, especially in the process of solving problems.

The systematic rather than the alphabetic arrangement of the material, with a lot of examples, makes the handbook similar to a concise textbook. Numerous problems and their solutions constitute an essential part of the book. The most important and sophisticated subjects are usually provided with detailed discussions. This quality of the book gives the reader a definite advantage in training for tests and examinations.

The authors are grateful to the executive director of Nauka Publishers Lyudmila Gladneva, and to the editorial staff of CRC Press, especially to Robert Stern, Tim Pletscher, and Susan Fox for their patience and helpful assistance.

Authors

Contents

Part I

MATHEMATICS

Chapter 1

Basic Notations, Formulas, and Concepts

This chapter contains commonly used elementary algebraic formulas, transformation rules for fractions, rules for rounding-off calculations, and a description of the mathematical induction method.

1.1 General Rules

1.1.1 Some notations

The following symbols, along with the usual arithmetic ones, are commonly used:

$\in$ means "belongs" $\notin$ means "does not belong"
$\Rightarrow$ means "implies" $\Leftrightarrow$ means "equivalent"
$\emptyset$ means "empty set"
$\subset$ means "proper subset" $\subseteq$ means "subset"
$\cup$ means "union" (of sets) $\cap$ means "intersection" (of sets)
N means the set of natural numbers
Z means the set of integral numbers
Q means the set of rational numbers
R means the set of real numbers
C means the set of complex numbers

1.1.2 The rules to remove brackets

1. $a(b+c) = ab + ac$, i.e., every summand is multiplied by a and the resulting products are added; in particular

   $-(a+b) = -a - b, \quad -(a-b) = -a + b;$

2. $(a+b)(c+d) = a(c+d) + b(c+d) = ac + ad + bc + bd$, or

   $(a+b)(c+d) = (a+b)c + (a+b)d = ac + bc + ad + bd$,

   that is, every summand from the first sum is multiplied by every summand from the second sum, and the resulting products are added.

1.1.3 Short methods of multiplication

$$(a+b)^2 = a^2 + 2ab + b^2, \qquad (a-b)^2 = a^2 - 2ab + b^2,$$

$$(a+b)^3 = a^3 + 3a^2b + 3ab^2 + b^3,$$

$$(a-b)^3 = a^3 - 3a^2b + 3ab^2 - b^3,$$

$$(a+b)(a-b) = a^2 - b^2,$$

$$(a+b)(a^2 - ab + b^2) = a^3 + b^3,$$

$$(a-b)(a^2 + ab + b^2) = a^3 - b^3,$$

$$a^3 + b^3 = (a+b)(a^2 - ab + b^2),$$

$$a^3 - b^3 = (a-b)(a^2 + ab + b^2),$$

$$(1-q)(1+q+q^2+\ldots+q^n) = 1 - q^{n+1},$$

$$(1+q)[1-q+q^2-\ldots(-1)^n q^n] = 1 + (-1)^n q^{n+1}.$$

The formulas given above are useful in simplification of algebraic expressions and can be read either from left to right or vice versa.

Examples

1. $$\frac{1-h^5}{1-h^2} = \frac{(1-h)(1+h+h^2+h^3+h^4)}{(1-h)(1+h)} = 1 + h^2 + \frac{h^4}{1+h};$$

2. $$\frac{x^2 - y^2}{(x-y)(\sqrt[3]{x} + \sqrt[3]{y})} + 3\sqrt[3]{xy} =$$

   $$\frac{(x-y)(\sqrt[3]{x} + \sqrt[3]{y})(x^{2/3} - x^{1/3}y^{1/3} + y^{2/3})}{(x-y)(\sqrt[3]{x} + \sqrt[3]{y})} + 3\sqrt[3]{xy} =$$

   $$x^{2/3} + 2x^{1/3}y^{1/3} + y^{2/3} = (\sqrt[3]{x} + \sqrt[3]{y})^2\,;$$

3. $(\sqrt{a} + \sqrt{b})^4 - (\sqrt{a} - \sqrt{b})^4 =$

   $$[(\sqrt{a}+\sqrt{b})^2 - (\sqrt{a}-\sqrt{b})^2]\cdot[(\sqrt{a}+\sqrt{b})^2 + (\sqrt{a}-\sqrt{b})^2] =$$

   $$(a+2\sqrt{ab}+b-a+2\sqrt{ab}-b)(a+2\sqrt{ab}+b+a-2\sqrt{ab}+b) = 8\sqrt{ab}(a+b).$$

1.1.4 Fractions

The expression of the form $\frac{m}{n}$ (or m/n) is called a *fractional number*, or a *fraction* (sometimes—a vulgar, or natural, fraction). It is assumed here that m and n are integers (see Section 2.2.1), $n \neq 0$, and m is not evenly divisible by n.

The fraction has the *numerator*, m, and the *denominator*, n. If $|m| < |n|$, the fraction is *proper*, and if $|m| > |n|$, it is *improper*. Any improper fraction may be represented as a mixed number, which consists of an integer and a proper fraction. For instance:

$$\frac{7}{3} = \frac{6+1}{3} = 2 + \frac{1}{3} = 2\frac{1}{3}$$

(read as "two and one-third"; don't confuse it with $2 \cdot \frac{1}{3} = \frac{2}{3}$).

1.1.5 The rules of handling with fractions

1. The value of the fraction is not altered after multiplying the numerator and the denominator of the fraction by the same number:

$$\frac{m}{n} = \frac{m \cdot b}{n \cdot b}.$$

This rule is often used in order to:

(i) simplify a fraction by canceling common factors, e.g.,

$$\frac{824}{515} = \frac{103 \cdot 8}{103 \cdot 5} = \frac{8}{5} = 1\frac{3}{5};$$

(ii) reduce fractions to the common denominator (for addition or subtraction of fractions), e.g.,

$$\frac{5}{63} - \frac{11}{42} = \frac{5}{3 \cdot 3 \cdot 7} - \frac{11}{2 \cdot 3 \cdot 7} = \frac{5 \cdot 2}{2 \cdot 3 \cdot 3 \cdot 7} - \frac{11 \cdot 3}{2 \cdot 3 \cdot 3 \cdot 7} = \frac{10}{126} - \frac{33}{126}.$$

Here the additional factor 2 was introduced to the numerator and the denominator in the first fraction and the additional factor 3 was introduced in the second fraction.

2. Addition (subtraction) of fractions having equal denominators:

$$\frac{m}{n} \pm \frac{l}{n} = \frac{m \pm l}{n},$$

i.e., in order to add (to subtract) fractions that have equal denominators, we add (subtract) the numerators and take the common denominator for the resulting fraction.

For example:

$$\frac{10}{126} - \frac{33}{126} = \frac{10-33}{126} = \frac{-23}{126} = -\frac{23}{126}.$$

3. In order to reduce given fractions to a common denominator we should fulfill the following steps:

(i) find the common multiple for all denominators (see Section 2.2.1), i.e., such an integer that it is divisible (without a remainder) by each denominator;

(ii) for every fraction, find the "additional factor", i.e., such a factor that complements the corresponding denominator to the common multiple;

(iii) multiply both the numerator and the denominator of each fraction by the corresponding additional factor; now the fractions have a common denominator.

After reducing fractions to a common denominator we can add or subtract them according to Rule 2.

Usually it is preferable to find the *minimal* common denominator (i.e., the minimal common multiple for denominators). To do this one has to factorize each denominator into prime factors, and then calculate the common denominator as a product of all prime factors taken in their highest powers. For instance,

$$\frac{7}{324} + \frac{11}{180} = \frac{7}{2^2 \cdot 3^4} + \frac{11}{2^2 \cdot 3^2 \cdot 5} = \frac{7 \cdot 5}{2^4 \cdot 3^4 \cdot 5} + \frac{11 \cdot 3^2}{2^4 \cdot 3^4 \cdot 5} =$$

$$\frac{35+99}{1620} = \frac{134}{1620} = \frac{67}{810}.$$

4. Multiplication of fractions:

$$\frac{m}{n} \cdot \frac{p}{k} = \frac{m \cdot p}{n \cdot k},$$

i.e., in order to multiply fractions we should multiply their numerators and do the same with their denominators.

5. Division of fractions:

$$\frac{m}{n} : \frac{p}{k} = \frac{m}{n} \cdot \frac{k}{p} = \frac{m \cdot k}{n \cdot p},$$

i.e., division of a fraction by another fraction is equivalent to multiplication of the first fraction by the inverted second fraction, for instance,

$$\frac{21}{5} : \frac{7}{15} = \frac{21}{5} \cdot \frac{15}{7} = \frac{21 \cdot 15}{5 \cdot 7} = 3 \cdot 3 = 9.$$

1.1.6 Fractional expressions

Fractional expressions (quotients), i.e., expressions of the form $\frac{A}{B}$ (or A/B) with A and B being any arbitrary numerical or algebraic expressions, have properties similar to those of fractions. In particular, multiplication or division is performed by using Rules 4 and 5; addition or subtraction is performed by means of reducing the fractional expressions to a common denominator:

$$\frac{A}{B} \pm \frac{C}{D} = \frac{A \cdot D}{B \cdot D} \pm \frac{C \cdot B}{D \cdot B} = \frac{A \cdot D \pm C \cdot B}{B \cdot D}.$$

For example,

$$\tan a + \cot a = \frac{\sin a}{\cos a} + \frac{\cos a}{\sin a} = \frac{\sin^2 a + \cos^2 a}{\sin a \cos a} = \frac{1}{\sin a \cos a}.$$

1.1.7 Proportions

Two quotients, related by the symbol of equality, form a *proportion*:

$$\frac{A}{B} = \frac{C}{D}.$$

Here A and D are called the *extreme terms*, and B and C are called the *middle* (or mean) *terms* of the proportion.

Properties of the proportion

For an arbitrary proportion

$$\frac{A}{B} = \frac{C}{D},$$

the following relations are correct:

$$A \cdot D = B \cdot C,$$

i.e., the product of the extreme terms is equal to the product of the middle terms;

$$\frac{A+B}{B} = \frac{C+D}{D}, \quad \frac{A-B}{B} = \frac{C-D}{D}, \quad \frac{A+B}{A-B} = \frac{C+D}{C-D},$$

or, more generally:

$$\frac{A+\alpha B}{B+\beta A} = \frac{C+\alpha D}{D+\beta C}$$

("derived proportions"), where α and β are arbitrary numbers; for instance,

$$\text{let} \quad \frac{x-3}{5} = \frac{y-2}{6}, \quad \text{then} \quad \frac{x+2}{5} = \frac{y+4}{6}, \quad \frac{x+2}{2x-1} = \frac{y+4}{2y+2}.$$

1.1.8 Percentage

One hundredth part of a given number is called one *percent* (1%) of this number; for instance, \$5 is 1% of \$500. Respectively, 2% is 0.02 of the given number, 10% is 0.1 of it, etc. In general, $p\%$ of a given number a corresponds to $p/100$ part of a. The number a is, therefore, 100% of a; 50% of a is half of a.

To calculate the percentage one can use the following rule: the number b is $p\%$ of the number a if:

$$b = \frac{p\%}{100\%} \cdot a, \quad \text{or} \quad p\% = 100\% \frac{b}{a}.$$

Examples

1. Suppose that John hit the target 32 times out of 50. The number of hits makes up $p\% = 100\% \cdot 32/50 = 64\%$.

2. 5% of a certain sum of money is \$7. What is the sum? The sum is $(\$7 \cdot 100)/5 = \$7 \cdot (100/5) = \$7 \cdot 20 = \$140$.

3. Suppose that over a period of one year the population of a city increased by 2% and now stands at 830,000. What was the population at the beginning of the year?

   Solution. 830,000 is 102% of the initial population, hence 1% corresponds to 830,000/102. So, there were originally 100%, or $(830,000/102) \cdot 100 \approx 813,700$.

1.2 Decimal Fractions

1.2.1 Introduction

A fraction whose denominator is a natural power of 10 can be written as a *decimal fraction*; for instance, $\frac{3}{100} = 0.03$ (read as: "zero point zero three"), $2\frac{812}{1000} = 2.812$. The decimal point is used to separate the integral (or entire) part from the fractional one. In some books, a comma is used instead of the decimal point.

1.2.2 Handling with decimal fractions

1. Adding or omitting zero figures from the right side of a decimal fraction does not change its value (see, however, Section 1.3):

$$3.1400 = 3.140 = 3.14\,.$$

2. In order to multiply a decimal by 10, one has to move the point over one digit (decimal figure) to the right:

$$2.812 \cdot 10 = 28.12; \quad 0.03 \cdot 100 = 0.3 \cdot 10 = 3.0 = 3\,.$$

3. Dividing a decimal by 10 is performed by moving the decimal point over one digit to the left:

$$28.12 : 10 = 2.812; \quad 2.812 : 10 = 0.2812; \quad 0.2812 : 10 = 0.02812;$$

$$30 : 1000 = 3.0 : 100 = 0.3 : 10 = 0.03\,.$$

4. Addition of positive decimals or subtraction of positive decimals (with a positive result!) may be performed by using the "column rule". We write the second number under the first one, with the decimal points vertically aligned, and perform the operation digit by digit (from the right to the left):

$$\begin{array}{r} 312.070 \\ + \quad \\ \underline{0.342} \\ 312.412 \end{array} \qquad \begin{array}{r} 312.070 \\ - \quad \\ \underline{0.342} \\ 311.728 \end{array}$$

5. Decimals may be multiplied simply as integers: for example, to multiply 312.07 by 0.342, simply multiply 31207 by 342 to obtain 10672794. Then count the total number of digits in fractional parts of the operands; in this example, we have two digits (07) in the fractional part of the first operand, and three digits (342) in the second one, so the total number is $2 + 3 = 5$. Now we insert the decimal point into the resulting number in order to show that the number of digits in the fractional part is equal to the above-mentioned total number:

$$312.07 \cdot 0.342 = 106.72794\,.$$

In addition (subtraction) or in multiplication (division) of decimals by using the approximation rules, it is customary to discard some "lower" digits (see Section 1.3). In the example above we could round the result up to 106.73.

6. In order to divide one decimal by another, we move the decimal point in each operand to the right side over the same number of digits, in order to make the divisor be an integer. Then we perform division by using the "corner rule", and keeping the proper number of digits (see the rounding-off rules in Section 1.3.1). The length of the integral part of the result can be found in the process of dividing the integral part of the dividend by the divisor (see Section 2.2.1):

$$5.35 : 2 = 2.675 \approx 2.68; \quad 5.35 : 0.2 = 53.5 : 2 = 26.75 \approx 26.8;$$

$$10 : 3.3 = 100 : 33 = 3.03030303\ldots \approx 3.03\,.$$

1.2.3 Normalized form of numbers

Integral numbers and decimal fractions can be written in the "normalized form" (with or without the exponential factor): the integral part of the number must consist of one significant digit, and the exponential factor (i.e., entire power of 10) must be placed after the number. For instance,

$$31.4 = 3.14 \cdot 10^1 .$$

The principal part (here: 3.14) is called the *mantissa*, the exponential factor is called the *exponent*. In some books the normalized form is presented as having a zero integral part:

$$3.14 = 0.314 \cdot 10^1, \quad 0.0018 = 0.18 \cdot 10^{-2} .$$

Some operations with normalized numbers

Before executing the addition or subtraction of the normalized numbers, we have to "match" their exponents:

$$3.14 \cdot 10^2 + 1.3 \cdot 10^{-2} = 3.14 \cdot 10^2 + 0.00013 \cdot 10^2 =$$
$$(3.14 + 0.00013) \cdot 10^2 = 3.14013 \cdot 10^2 \approx 3.14 \cdot 10^2 .$$

Attention: if the calculation is performed by using rounding, a specific phenomenon called "loss of precision" may occur (see Section 10.1.3).

To multiply (to divide) normalized numbers we should:

1. Multiply (divide) the decimals taking no notice of the exponents.
2. Add (subtract) the indexes of exponents.
3. If necessary, normalize and round off the result.

For example,

$$4.560 \cdot 10^4 \cdot 3.032 \cdot 10^{-3} = 13.82592 \cdot 10^1 = 1.382592 \cdot 10^2 \approx$$
$$1.383 \cdot 10^2 , \quad 3.14 \cdot 10^2 : 8.30 \cdot 10^3 = 0.37831325 \ldots \cdot 10^{-1} \approx$$
$$3.7831325 \cdot 10^{-2} \approx 3.78 \cdot 10^{-2} .$$

The normalized form of numbers is used, in particular, when handling with approximate numbers (see Section 1.3.2).

1.2.4 Repeating decimals

In some cases the process of dividing the numbers by using the "corner rule" becomes infinite, and the result appears in the form of a sequence of an infinitely repeating group of the same figures. Such an infinite decimal fraction is called a *repeating decimal*, and the group of repeating figures is called the *period* of the fraction; for instance, $101/33 = 3.06060606\ldots$ is the repeating fraction whose period is (06). A repeating fraction may be expressed in a written form by showing its period in parentheses, for example, $101/33 = 3.(06)$. The periodic (i.e., the repeating) part does not necessarily appear right after the decimal part, for instance, $47/30 = 1.5666666\ldots = 1.5(6)$; such fractions are called *mixed repeating* fractions.

The vulgar fraction m/n, where m and n are integers, is equal to a finite decimal fraction if and only if its denominator, being factorized into prime factors, contains (after reducing the fraction) no other factors but 2 and 5. For instance, $17/20 = 0.85$ and $51/60 = 0.85$ are finite decimals, but $17/21 = 0.809523809523\ldots = 0.(809523)$ is a repeating decimal.

Any repeating decimal may be transformed into a vulgar fraction of the form m/n by using the rules:

- If the periodic part immediately follows the decimal point, we add the integral part of the repeating decimal to a vulgar fraction whose numerator is equal to the period, and whose denominator is formed by the repeated figure 9, so that the number of "9s" is equal to the number of digits in the period:

$$16.(809523) = 16 + \frac{809523}{999999} = 16\frac{17}{21}.$$

- If there are some figures between the point and the periodic part, we must first move the point to the beginning of the periodic part (by using an exponent) and then apply the preceding rule:

$$1.5(6) = 15.(6) \cdot 10^{-1} = (15 + \frac{6}{9}) \cdot 10^{-1} = \frac{45+2}{3} : 10 = \frac{47}{30}.$$

Infinite *non-repeating* fractions (irrational numbers, see Section 2.2.3), cannot be transformed exactly into vulgar fractions.

1.3 Rounding off Numbers. Approximate Numbers

1.3.1 Rounding off

To round off a decimal fraction we omit the extreme right-hand figure, and after that we either keep the previous figure—in case we have omitted a figure 1, 2, 3, or 4 (shortage-rounding), or we increase the previous figure by 1—in case a figure 5, 6, 7, 8, or 9 has been omitted (excess-rounding):

$$41.32 \approx 41.3, \quad 41.36 \approx 41.4, \quad 41.35 \approx 41.4.$$

(the symbol $\approx$ is to be read as "approximately equal").

The special rule is ordinarily used in case of omitting the figure 5: we keep the previous figure if it is even, or we increase it by 1 if it is odd:

$$3.165 \approx 3.16, \quad \text{but} \quad 3.175 \approx 3.18\,.$$

After such an approximation the round-off error, i.e., the difference between the original number and its approximate representation, does not exceed (in absolute value) 5 units of the decimal place of the omitted figure:

$$|3.165 - 3.16| \leq 0.005, \quad |3.16 - 3.2| \leq 0.05\,.$$

Sometimes a less exact rounding off rule is used: the last right-side figure is simply omitted, then the round-off error does not exceed 10 units of this last digit, i.e., 1 unit of the previous digit:

$$|3.18 - 3.1| \leq 0.1\,.$$

Rounding off the integers is performed in a similar way; here the omitted figure is replaced by the zero figure:

$$38,074 \approx 38,070 \approx 38,100 \approx 38,000 \approx 40,000.$$

Rounding off the numbers is used in cases where an exact calculation is impossible, or where the exact representation of the result has no practical meaning, for instance, when we operate with approximate numbers (see Section 1.3.2).

1.3.2 Approximate numbers

As a result of rounding off, "approximate numbers" appear, i.e., numbers close to the original ones. Approximate numbers are obtained also as a result of the physical measurement of distances, time intervals, masses, temperatures, etc. because any measurement can be performed only with a limited accuracy, which is dependent upon the instrumental error.

Approximate numbers are to be written in the normalized form, and the length of the fractional part characterizes the accuracy of the approximate number; it is a practice to write only "true" figures whose correctness is known. For example, from the sentence "the diameter of the earth is equal to $1.27 \cdot 10^7$m" we know that the figures 1, 2, 7 have been reliably determined, but subsequent figures are unknown or not important, or make no sense (the shape of the earth is not spherical).

The 0-digit may be the last one in an approximate number; in such a case this 0-digit is considered to be true; so, the expressions like $3.50 \cdot 10^{-1}$ and $3.5 \cdot 10^{-1}$ have different meaning. It is customary to suppose that the error of an approximate number is not greater than a half of a unit of the last digit place.

When handling with approximate numbers it is worthwhile to keep in mind that the quantity of true figures in the resulting number cannot be greater than the minimal quantity of true figures in any operand. Nevertheless, it is necessary to keep one or two "reserve" digits in the process of intermediate calculations.

In round-off calculations (including computer calculations), it is recommended to avoid cases like "loss of precision", see Section 10.1.3.

1.4 The Mathematical Induction Method

Some formulas or statements (in particular, some equalities or inequalities), which contain an integral variable value $n \geq 1$, can be proved by the mathematical induction method. The proof consists of two parts:

1. We check the correctness of the given statement for $n = 1$ (the "induction base").
2. We assume that the given statement is correct for $n = k$, and then we prove, as a corollary, the correctness of the statement for $n = k + 1$ (the "induction step").

Now we can assert the correctness of the given statement for $n = 2$, $n = 3$, $n = 4, \ldots,$ i.e., for any value of n. In case the given statement is to be proved for $n \geq n_0$, the base should be taken with $n = n_0$.

Example

For all $n \geq 1$, prove the equality (see Section 5.1.5):

$$1 + 2 + \ldots + n = \tfrac{n(n+1)}{2}.$$

Proof. For $n = 1$ the equality is evident. Let us assume that

$$1 + 2 + \ldots + k = \tfrac{k(k+1)}{2}.$$

Consider the expression $1 + 2 + \ldots + k + (k + 1)$ and replace here $1 + 2 + \ldots + k$ by $k(k + 1)/2$ to obtain:

$$\tfrac{k(k+1)}{2} + (k + 1) = \tfrac{(k+1)(k+2)}{2}.$$

The inequality is proved.

Note. Using the mathematical induction method we cannot obtain new formulas or statements: this method is useful only in the proof of formulas (statements) that have been already supposed.

Chapter 2

Sets. Real Numbers. Functions

The concepts of sets, real numbers, and functions are fundamental ones in calculus. In this chapter we consider natural, integral, rational and irrational numbers, the general concept and essential characteristics of functional dependence, and also the graphs and main properties of the most important elementary functions.

2.1 Sets

2.1.1 Concept of a set

A *set* is a collection, or combination, of *elements* that share some common property, e.g., the set of students, the set of integers, the set of planets in the solar system, the set of points in a circle. If an element x belongs to set A, we write $x \in A$; if x does not belong to set A, we write $x \notin A$. The set that has no elements is said to be *empty* and is denoted by the symbol $\emptyset$. Examples of the empty set are the set of the obtuse angles of an equilateral triangle; the set of real roots of the equation $x^2 + 1 = 0$; a set of persons older than 300 years.

To indicate a set, in some cases it is convenient simply to write out all the elements of the given set: $\{ 1, 2, 3, 4 \}$ denotes a set that consists of the elements 1, 2, 3, 4. The symbol $\{ x \mid x < 1 \}$ denotes the set of all numbers x such that $x < 1$.

Two sets are said to be *equal* if they consist of the same elements; for instance,

$$\{x \mid x^2 + 3x + 2 = 0 \} = \{-2, \ -1 \} = \{-1, \ -2 \}.$$

The *union of sets* A and B (denoted by $A \cup B$) is the set composed of all elements that belong to at least one of these sets; for instance,

$$\{1, 3, 4\} \cup \{0, 1, 2\} = \{0, 1, 2, 3, 4\}.$$

The union of the set of rational numbers and the set of irrational numbers is the set of real numbers (see Section 2.2).

The *intersection* of sets A and B (denoted by $A \cap B$) is the set composed of all elements belonging to both the sets A and B; for instance,

$$\{1, 3, 4\} \cap \{0, 2\} = \emptyset, \qquad \{1, 3, 4\} \cap \{0, 1, 2, 3\} = \{1, 3\}.$$

The intersection of the set of integers and the set of natural numbers is the set of natural numbers (see Section 2.2.1).

2.1.2 Subsets

A *subset* P of a given set A (denoted by $P \subset A$) is the set composed of some elements of A, i.e., a subset is a part of the given set (including the cases $P = A$ and $P = \emptyset$).

Some examples of subsets: the set of natural numbers is a subset of the set of integers; the set of integral numbers is a subset of the set of real numbers; the set of real numbers is a subset of the set of complex numbers (see Section 7.1). The expression $\{x \in A \mid L\}$ means a subset of a set A, whose elements satisfy the indicated condition L. For instance, $\{x \in \mathbf{Z} \mid -2 < x < 9\}$ is a subset of set $\mathbf{Z}$ (the set of all integers) such that each integer x of the subset is greater than -2 and less than 9. If a set A consists of n elements, then the total quantity of its subsets (including $\emptyset$ and A) is 2^n.

2.1.3 Intervals

The following sets of numbers, called *intervals*, are often used:

1. Closed interval, or segment:

$$[a, b] = \{x \in \mathbf{R} \mid a \le x \le b\}.$$

2. Open interval, or interval:

$$(a, b) = \{x \in \mathbf{R} \mid a < x < b\}.$$

In some books open intervals are denoted as $]a, b[$.

3. Half-open intervals:

$$\begin{aligned}(a, b] &= \{x \in \mathbf{R} \mid a < x \le b\},\\ [a, b) &= \{x \in \mathbf{R} \mid a \le x < b\}.\end{aligned}$$

Alternative notation: $]a, b]$ and $[a, b[$, respectively.

4. Infinite intervals:

   - Semi-axes, or rays:

$$\begin{aligned}(-\infty,\ a) &= \{x \in \mathbf{R} \mid -\infty < x < a\};\\ (-\infty,\ a] &= \{x \in \mathbf{R} \mid -\infty < x \le a\};\\ (a,\ +\infty) &= \{x \in \mathbf{R} \mid a < x < +\infty\};\\ [a,\ +\infty) &= \{x \in \mathbf{R} \mid a \le x < +\infty\}.\end{aligned}$$

   - The real line (the real axis):

$$(-\infty, +\infty) = \mathbf{R}.$$

2.2 Real Numbers

2.2.1 Natural numbers

Natural numbers are the numbers used in counting: 1, 2, 3, ... The set of natural numbers is designated by the symbol **N**. In this set the operations of addition and multiplication are defined; the inverse operations—subtraction and division—are not applicable to all natural numbers.

A natural number m is said to be *divisible* by a natural number n if a natural number l exists such that $m = n \cdot l$; for instance, 12 is divisible by 3 but not by 5. Any natural number is divisible by 1 and by the number itself.

Natural numbers that divide a given natural number are called *divisors* of the given number; e.g., the numbers 1, 2, 3, 4, 6, 12 are divisors of the number 12; 1 and 7 are divisors of 7.

A *prime number* is a natural number $m > 1$ possessing only two divisors: 1 and m itself. Some first prime numbers: 2, 3, 5, 7, 11, 13, 17, 19, 23, 29, ... The set of prime numbers is infinite.

Some theorems about prime numbers

1. For any natural n there exists at least one prime number between n and $n!$ (see Section 6.1).

2. A natural number p greater than 2 is a prime number if and only if the number $(p-2)! - 1$ is divisible by p (*Leibnitz*).

3. Every prime number of the form $4n + 1$ is the sum of the squares of two natural numbers (*Fermat*).

Natural numbers that are not prime are called *composite numbers*. All natural even numbers, number 2 being the only exception, are composite. The principal theorem of arithmetic states that any natural number not equal to 1 may be factorized into its prime factors, and such factorization is unique regardless of the order of the prime factors.

Some tests for divisibility of natural numbers

1. A number is divisible by 2 if the last figure of the number is even or zero.
2. A number is divisible by 4 if its last two figures are zero or they represent together a number divisible by 4.
3. A number is divisible by 3 if the total sum of its figures is divisible by 3; the number is divisible by 9 if the sum of its figures is divisible by 9.
4. A number is divisible by 5 if its last figure is 5 or 0.

Integers (or *integral numbers*) are the following numbers: $\ldots, -3, -2, -1, 0, 1, 2, 3, \ldots$ The set of integers is denoted by the symbol $\mathbf{Z}$. Natural numbers are positive integers.

Let us introduce a special operation for natural numbers: "division with a remainder". This operation associates with a given pair of natural numbers, m and n, a pair of integral numbers, q and r, such that $m = q \cdot n + r$, where $0 \leq q$ and $0 \leq r < n$. The numbers m and n are called *dividend* and *divisor*, the numbers q and r are called *quotient* and *remainder*, respectively. The result is to be written as $\frac{m}{n} = q + \frac{r}{n}$; for instance,

$$\frac{7}{3} = 2 + \frac{1}{3}, \quad \frac{3}{7} = 0 + \frac{3}{7}, \quad \frac{12}{3} = 4 + \frac{0}{3}.$$

The *common divisor* for natural numbers m and n is a natural number p which is a divisor for both m and n. The *greatest common divisor* (G.C.D.) for the given numbers is the maximum of all common divisors for these numbers.

In order to find the G.C.D. for the numbers m and n, we can factorize these numbers into their prime factors and calculate the G.C.D. by multiplying all prime factors common to the given numbers.

> **Example:** $600 = 2^3 \cdot 3 \cdot 5^2$, $3780 = 2^2 \cdot 3^3 \cdot 5 \cdot 7$, hence
>
> the G.C.D. $= 2^2 \cdot 3 \cdot 5 = 60$.

To calculate the G.C.D. more rapidly, we can use the algorithm by *Euclid*. Let the natural numbers m and n be given, where $m > n$. At first we divide, with a remainder, the number m by the number n and obtain the remainder r_1; if $r_1 = 0$, then the G.C.D.$= n$. If $r_1 \neq 0$, we divide, with a remainder, n by r_1 and obtain the remainder r_2; if $r_2 = 0$, then the G.C.D. $= r_1$. If $r_2 \neq 0$, we divide, with a remainder, r_1 by r_2, and so on. For instance, $m = 1780$, $n = 600$, $r_1 = 180$, $r_2 = 60$, $r_3 = 0$, hence the G.C.D. equals $r_2 = 60$.

2.2.2 Rational numbers

Rational numbers are numbers of the form m/n where m and n are integers ($n \neq 0$). In the case of $n = 1$ (or if m is divisible by n) the rational number m/n is considered to be an integer. In the opposite case it is a fraction (see Section 1.1.4).

The set of rational numbers is designated by the symbol $\mathbf{Q}$. Evidently $\mathbf{N} \subset \mathbf{Z} \subset \mathbf{Q}$.

Every rational number can be written as a finite or an infinite repeating decimal fraction (see Section 1.2.4).

2.2.3 Irrational numbers

Irrational number can be defined as an infinite non-repeating decimal fraction. For example, the number

$$0.101001000100001\ldots$$

is an irrational one because the farther to the right, the more zeros we find between identity figures. The numbers

$$\sqrt{2} = 1.4142\ldots, \quad \sqrt{3} = 1.73205\ldots, \quad \pi = 3.1416\ldots, \quad e = 2.7183\ldots$$

are also examples of irrational numbers.

No irrational number can be represented as m/n where m and n are integers. Thus, if p is an irrational number, then the product $p \cdot n$ cannot be equal to an integer.

The union of the set of rational numbers and the set of irrational numbers is the set of *real numbers* (denoted by the symbol $\mathbf{R}$).

Real numbers can be depicted by dots on the x-axis in such a manner that the distance between the origin O and the dot depicting a given real number b is equal to $|b|$. The dots depicting all real numbers are distributed "everywhere densely" on the axis: between any two given real numbers there are infinitely many other real numbers. The set of irrational and the set of rational numbers have the same property of density. Any irrational number can be approximated by rational numbers with an arbitrarily high accuracy, in particular, by finite decimal fractions; for instance

$$\sqrt{2} \approx 1.4, \quad 1.41, \quad 1.414, \quad 1.4142, \quad 1.41421, \ldots$$

In practical calculations with a limited precision there is no difference between rational and irrational numbers.

2.2.4 Properties of arithmetic operations

Some properties of addition and multiplication operations with real numbers are as follows:

1. Commutativity: $a+b=b+a$; $a \cdot b = b \cdot a$.
2. Associativity: $(a+b)+c = a+(b+c)$; $(ab) \cdot c = a \cdot (bc)$.
3. Distributivity: $(a+b) \cdot c = ac + bc$.

2.3 Functions

2.3.1 Concept of a function

Let X be a numerical set, e.g., an interval. If a rule is given which puts every number $x \in X$ in a correspondence with some number y, then we call such a rule a *function* and write

$$y = f(x), \quad \text{or} \quad y = y(x).$$

The notation $y = f(x)$ (read as "y is equal to f of x") means that the function f establishes a correspondence of y to x. The number x is called the *argument value*, or the *argument*, of the function f; the number y is called the *function value* corresponding to the argument value x; y is also called the value of the function f at point x.

The set X of those numbers x for which the function f is defined is called the *domain of definition* of the function f. The domain of definition is usually denoted by $D(f)$. The set of numbers in the form $f(x)$, where $x \in D(f)$, is called the *range of values* for the function f. It is denoted by $R(f)$.

In practice, a function may be defined by various ways: by an analytic expression (i.e., by formulas), e.g., $y = x^2$, $y = \sin 3x$, $y = \ln x$; by a graph (see, for instance, Figure 2.1a); by a table; or by a verbal description. In physics and engineering the most important ways to define functions are analytic and graphic ones.

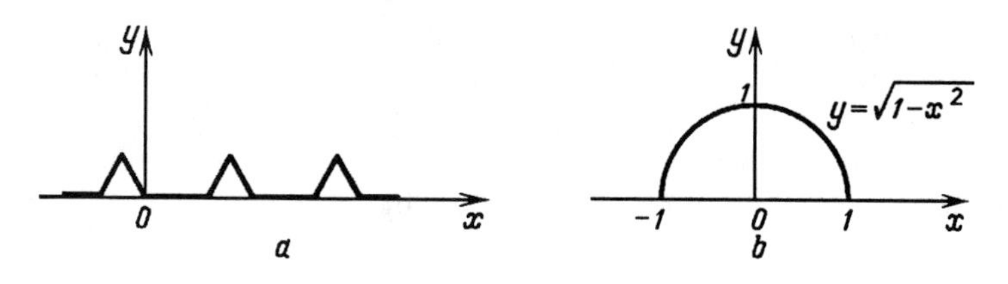

Figure 2.1: a – The function defined by a graph. b – A semicircle.

2.3.2 Variables

For $y = f(x)$, the *variables* x and y are said to be related by a *functional dependence*, x is called the *independent variable*, y the *dependent variable*. The concept of functional dependence is commonly used in physics.

For instance, when a point moves along a line its distance traveled is a function of time; for a gas in a closed container the pressure is a function of temperature; and for two electric charges the force of their interaction is a function of distance between the charges.

Remarks.

• The definition of the function given above is applicable to single-valued functions; in some special instances it is preferable to consider many-valued functions like $y = \pm\sqrt{x}$.

• In mathematics, physics, and engineering, functions of several independent variables, $y = f(x_1, x_2, \ldots, x_n)$, and also vector-valued functions are commonly used, e.g., the radius vector $\mathbf{r} = \mathbf{r}(t)$, see Section 13.1.1.

Examples

1. $y = x^2 + 1$, here $D(f) = (-\infty, +\infty)$, $R(f) = [1, +\infty)$.
2. $y = \sqrt{-x^2}$, $D(f) = R(f) = \{0\}$.
3. $y = (x^2 + 8)/(x - 1)$, $D(f) = (-\infty, 1) \cup (1, +\infty)$, $R(f) = (-\infty, -4] \cup [8, +\infty)$.
4. $y = \sin x$, $D(f) = (-\infty, +\infty)$, $R(f) = [-1,\ 1]$.

2.3.3 Graph

Let us introduce a rectangular coordinate system xy in a plane (see Section 8.3). The *graph* of a function $f(x)$ is the set of those points in the plane whose coordinates are (x, y), where $y = f(x)$. In other words, while the independent variable x passes through the domain of definition of $f(x)$, the point $(x,\ f(x))$ in the plane xy generates the graph of the function.

As a rule, the graph of a function is a "smooth" curve (see, however, Section 5.3.2). Figure 2.1*b* represents the graph of $y = \sqrt{1 - x^2}$; it is a semicircle. Graphic representation of functions is widely used in order to visually demonstrate the most essential characteristics of functional dependence; some practical methods to construct graphs are described in Sections 5.4.3 and 5.4.4.

2.4 Basic Characteristics of Functions

2.4.1 Monotonicity

A function $f(x)$ is said to be *increasing* over the interval $(a,\ b)$ if

$$f(x_2) \geq f(x_1) \quad \text{for} \quad x_2 > x_1, \tag{2.1}$$

i.e., an increase of x leads to an increase of $f(x)$. Here it is assumed that the interval (a, b) lies in the domain of definition of $f(x)$, and that x_1 and x_2 are arbitrary points in (a, b).

If, instead of (2.1), the following strict inequality is true:

$$f(x_2) > f(x_1) \quad \text{for} \quad x_2 > x_1, \tag{2.2}$$

the function $f(x)$ is said to be *strictly increasing*.

A function $f(x)$ is said to be *decreasing* over the interval (a, b) if

$$f(x_2) \leq f(x_1) \quad \text{for} \quad x_2 > x_1, \tag{2.3}$$

and it is said to be *strictly decreasing* if

$$f(x_2) < f(x_1) \quad \text{for} \quad x_2 > x_1. \tag{2.4}$$

Both increasing and decreasing functions are called *monotonic*; strictly increasing and strictly decreasing functions are called *strictly monotonic*.

While point x passes the interval (a, b) from a to b, the graph of an increasing function rises (for a decreasing—lowers). A function can be increasing over some intervals and decreasing over others, e.g., $y = \sqrt{1 - x^2}$ decreases over $x > 0$ and increases over $x < 0$ (Figure 2.1*b*).

A sufficient condition for a differentiable function to be strictly monotonic (see Section 5.4.3):

$$f'(x) \neq 0 \quad \text{for all} \quad x \in (a, b).$$

In the case $f'(x) > 0$ the function $f(x)$ is strictly increasing, and in the case $f'(x) < 0$ the function is strictly decreasing; in these inequalities we may admit that $f'(x) = 0$ at some isolated points.

Examples

1. $f(x) = \frac{x^2+8}{x-1}$, $f'(x) = \frac{x^2-2x-8}{(x-1)^2} = \frac{(x+2)(x-4)}{(x-1)^2}$.

   The sign of the last fractional expression can be found with the help of the interval method (see Section 3.7.5). The final result: $f'(x) > 0$ for $x < -2$ and $x > 4$, thus $f(x)$ is increasing here; $f'(x) < 0$ for $-2 < x < 4$, $x \neq -1$, thus $f(x)$ is decreasing here.

2. $y = x^3$. The function is strictly increasing for all $x \in \mathbf{R}$ in spite of the fact that $y'(0) = 0$.

2.4.2 Periodicity

A function $f(x)$ is said to be *periodic* if its values do not alter along with changing values of the argument by a certain constant number, i.e., $f(x)$ is periodic if for each x in the domain of definition, $D(f)$, the points $x+T$ and $x-T$ also belong to $D(f)$ and, moreover,

$$f(x+T) = f(x-T) = f(x).$$

The number T is called the *period* of the function. If T is a period, then kT for any integer $k \neq 0$ is also a period. It is customary to imply by the term "period" the minimal positive period; for instance, $\sin x$ and $\cos x$ are of the period 2π. The graph of a periodic function consists of infinitely repeating identical parts of some curve (Figure 2.2). In physics periodic functions are used to describe oscillatory processes (see Section 16.1).

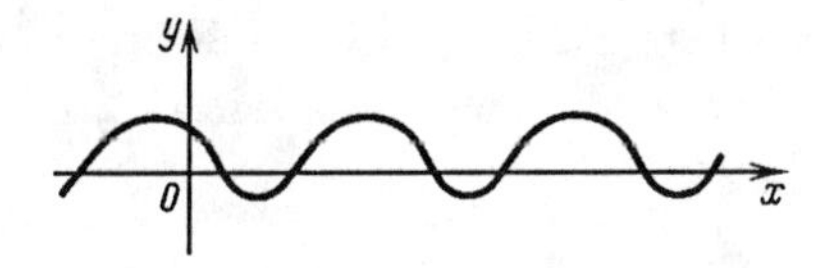

Figure 2.2: A periodic function.

2.4.3 Evenness and oddness

A function $f(x)$ is called an *even function* if the equality

$$f(-x) = f(x)$$

is true for any x in the domain of definition. Here the domain of definition of $f(x)$ is assumed to be symmetric with respect to point $x = 0$.

A function $f(x)$ is called an *odd function* if

$$f(-x) = -f(x).$$

For instance, the functions $\cos x$, x^{2n}, and $|x|$ are even, and the functions $\sin x$, x^{2n+1}, and $x/|x|$ are odd.

The graph of an even function is symmetric with respect to the y-axis, and the graph of an odd function is symmetric with respect to the origin of coordinates.

A function may possess no properties of evenness or oddness, for instance, $y = x^2 + x$.

Any function having a symmetric domain of definition may be represented as the sum of even and odd parts:

$$f(x) = f_1(x) + f_2(x),$$

where

$$f_1(x) = (f(x) + f(-x))/2, \quad f_2(x) = (f(x) - f(-x))/2.$$

The functions $f_1(x)$ and $f_2(x)$ are, respectively, the even and the odd parts of $f(x)$.

The sum and the difference of two even functions are even, the sum and the difference of two odd functions are odd, the product of two even or two odd functions is even, and the product of an even and an odd function is odd.

2.4.4 Boundedness

A function $f(x)$ is said to be *bounded* if for all x in its domain of definition, the inequality $|f(x)| \leq M$ (or the double inequality $M_1 \leq f(x) \leq M_2$) is true, where M, M_1, M_2 are some constant numbers.

Some examples of bounded functions: $|\sin x| \leq 1$, $|\arctan x| \leq \pi/2$, $0 \leq \exp(-|x|) \leq 1$.

2.5 Inverse Functions

2.5.1 Definition

A correspondence between the elements of a set A and the elements of a set B is said to be *one-to-one* if each element $a \in A$ corresponds exactly to one element $b \in B$ and, vice versa, each element $b \in B$ corresponds exactly to one element $a \in A$.

If a function f defines one-to-one correspondence between its domain of definition X and its range of values Y, then we say that the function f admits an *inverse function*, and that the function f is *invertible*.

The inverse function, by definition, is the rule which associates some number $x \in X$ with each number $y \in Y$, where $y = f(x)$. For the inverse function, its domain of definition is Y and its range of values is X. The given function and its inverse function are called together the "inverses". Examples of the inverses: sine and arcsine, exponent and logarithm, the second degree and square root.

It is a custom to use special notations for inverse functions, e.g., for $y = e^x$, we write $x = \ln y$ (see Section 2.8.3). In theoretical questions the inverse function to f is usually denoted by the symbol f^{-1}. Evidently, $(f^{-1})^{-1} = f$, i.e., the function which is inverse to the inverse of a given function is simply the original function f. For instance, $\ln(\exp(\ln x)) = \ln x$, because the exponential function and the natural logarithm are inverses.

2.5.2 Inverse function for a monotonic one

Every strictly monotonic function—either increasing or decreasing—has an inverse function, and this inverse function is also strictly monotonic. If the original function $f(x)$ is not monotonic, it takes the same values more than once, e.g., $y = x^2$ or $y = \sin x$. In order to define an inverse function at such a case, we can split the domain of definition of $f(x)$ into intervals of strict monotonicity of $f(x)$, and then define separate inverse functions for each monotonicity interval.

Examples

1. $y = x^3$. This function is monotonic over all of the real axis, that is, $x \in \mathbf{R}$. Hence the inverse function for $y = x^3$ is $\sqrt[3]{y}$. The domain of definition of the inverse function is also all of the real axis, $y \in \mathbf{R}$.

2. $y = x^2$. Here we can introduce two inverse functions: $x = \sqrt{y}$ and $x = -\sqrt{y}$, because there are two monotonicity intervals $(-\infty, 0)$ and $(0, +\infty)$ of the original function. Each of the inverse functions, $\sqrt{y}$ and $-\sqrt{y}$, give a single-valued mapping of the semi-axis $y \geq 0$ on its own part of the domain of definition $(-\infty, +\infty)$ of the original function $y = x^2$.

3. $y = \sin x$. For $x \in [-\pi/2, \pi/2]$ the inverse function is $x = \arcsin y$ (see Section 4.3.1).

2.5.3 Graph of the inverse function

It is customary to denote again by x the argument of the inverse function and by y its value: $y = f^{-1}(x)$. Then we can show the graphs of both the given function and its inverse on the same xy–plane. Evidently, these graphs are symmetric to each other with respect to the bisector of the I - III quadrants, hence the graph of the inverse function can be obtained by inverting (over the bisector) the plane where the graph of the original function is plotted (Figure 2.3).

2.6 Linear and Quadratic Functions. Modulus

2.6.1 Linear function

A *linear function* is of the form $y = ax + b$, where a and b are some constant *coefficients*. The domain of definition: $x \in \mathbf{R}$. The range of values: if $a \neq 0$, then $Y = \mathbf{R}$, i.e., the values of the linear function can be found anywhere on the axis of real numbers; in the case $a = 0$ the function degenerates into a constant $y = b$. The graph of a linear function

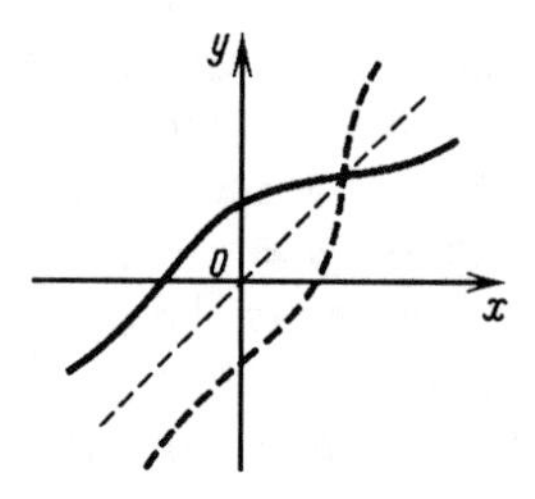

Figure 2.3: Inverses.

is a straight line (Figure 2.4). The number a defines the *slope* of the straight line: $a = \tan\alpha$, where α is the angle measured counterclockwise from the positive x-semi-axis to the straight line.

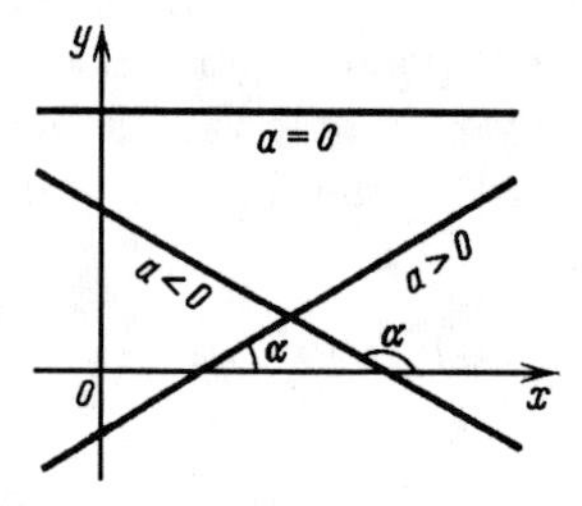

Figure 2.4: Graph of the linear function $y = ax + b$ for different values of a.

The linear function is monotonic for all x: it increases in the case $a > 0$ and it decreases in the case $a < 0$. For $a \neq 0$, the function is not periodic. In the case where $b = 0$, the dependence $y = ax$ is called *proportionality*, and the function is odd.

The inverse function exists if $a \neq 0$: $x = (y - b)/a$. It is a linear function, too.

Linear functional dependence is common in physics. For example, the distance $s = vt$ passed by a point moving uniformly along a straight line, or the velocity $v = v_0 + at$ given the motion is uniformly accelerated.

Sometimes the linear function is considered on the set $\mathbf{C}$ of complex numbers (see Section 7.1); for $a \neq 0$ it defines one-to-one mapping of $\mathbf{C}$ on $\mathbf{C}$.

2.6.2 Quadratic function

A *quadratic function* (or *quadratic trinomial*) is a function of the form $y = ax^2 + bx + c$, where $a \neq 0$. The domain of definition: $x \in \mathbf{R}$; the range of values: $[m, +\infty)$ for $a > 0$, or $(-\infty, M]$ for $a < 0$, where m is the minimal value of y, M is the maximal value of y (see below).

In the case $b = c = 0$ the graph of the quadratic function $y = ax^2$ is a parabola, whose vertex is at the origin of coordinates and whose axis of symmetry coincides with the y-axis (see Figure 2.5*a*).

In the general case, the quadratic trinomial $y = ax^2 + bx + c$ may be reduced to the canonical form:

$$ax^2 + bx + c = a\left[\left(x + \frac{b}{2a}\right)^2 - \frac{b^2}{4a^2}\right] + c = a(x - x_0)^2 + y_0,$$

where $x_0 = -b/(2a)$, $y_0 = -D/(4a)$, D is called the *discriminant*,

$$D = b^2 - 4ac.$$

Let us change the coordinate system by using the translation $x' = x - x_0$, $y' = y - y_0$, then the equation $y = ax^2 + bx + c$ alters its form to $y' = a(x')^2$. The coordinate system $x'y'$ is obtained from the xy-system by shifting to the right along the x-axis by x_0 and shifting up along the y-axis by y_0 (Figure 2.5*b*).

Hence, the graph of the quadratic function $y = ax^2 + bx + c$ is a parabola obtained by a parallel transfer of the parabola $y = ax^2$. The vertex P of the parabola has the coordinates $x_0 = -b/(2a)$, $y_0 = -D/(4a)$. At point x_0 the function $y = ax^2 + bx + c$ takes its extreme value $y_0 = -D/(4a)$: in the case $a > 0$ it is its minimal value m and in the case $a < 0$ it is its maximal value M. The quadratic function is monotonic to the left from x_0 and to the right from x_0. In the case $b = 0$ the function is even.

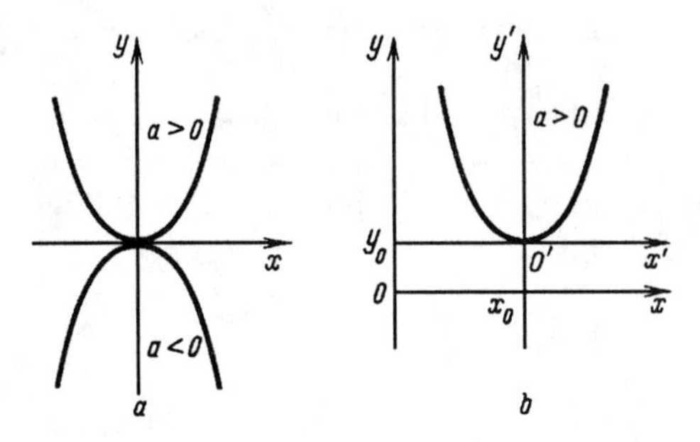

Figure 2.5: a – Parabola $y = ax^2$. It is infinite either upwards (for $a > 0$) or downwards (for $a < 0$). b – Translation of the coordinate system.

For any values of the coefficients a, b, c the graph of the quadratic function intersects the y-axis only once, at $x = 0$, $y = c$. The number of intersection points with the x-axis depends upon the value of the discriminant D:

If $D < 0$, then the parabola does not intersect the x-axis.

If $D = 0$, then there exists only one common point: contact at $x = -b/(2a)$.

If $D > 0$, then there are two points of intersection (see Section 3.2.2).

Some applications of the quadratic function in physics: any coordinate of a point which moves along a straight line with a constant acceleration is a quadratic function of time (see Section 13.1.4); the potential energy of a stretched spring is a quadratic function of the elongation (see Section 13.4.6).

Decomposition of a quadratic trinomial

If $D > 0$, then the quadratic trinomial may be decomposed into linear factors:

$$ax^2 + bx + c = a(x - x_1)(x - x_2), \qquad (2.5)$$

where x_1 and x_2 are the roots of the trinomial (see Section 3.2).

If $D = 0$, then $x_1 = x_2$ and hence $y = a(x - x_1)^2$.

If $D < 0$, it is impossible to factorize the trinomial in the way like (2.5) in the domain of real numbers, but in the domain of complex numbers it is always possible (see Sections 3.3 and 7.1).

2.6.3 Modulus

The *modulus* (or *absolute value*) of a real number x is defined as follows:

$$|x| = x \text{ for } x \geq 0, \quad |x| = -x \text{ for } x < 0.$$

Obviously $|x| \geq 0$ for any x. For the function $y = |x|$, the domain of definition is $x \in \mathbf{R}$, the range of values: $y \geq 0$. The function $|x|$ is even. The graph is the right angle (provided that the scales of both axes are the same, see Figure 2.6a).

Some properties of the function $|x|$

$$|ax| = |a| \cdot |x|, \quad |-x| = |x|, \quad \sqrt{x^2} = |x|,$$

$$|a + b| \leq |a| + |b|, \quad |a - b| \geq ||a| - |b||.$$

Let ε be a positive number. The inequality $|x - a| < \varepsilon$ is equivalent to the double inequality $-\varepsilon < x - a < \varepsilon$, or $a - \varepsilon < x < a + \varepsilon$. The inequality $|x - a| > \varepsilon$ is equivalent to a combination of two inequalities: $x - a > \varepsilon$, $x - a < -\varepsilon$, or (which is the same) $x < a - \varepsilon$, $x > a + \varepsilon$.

In other words, the inequality $|x-a|<\varepsilon$ is true for the interval $x \in (a-\varepsilon, a+\varepsilon)$, while the inequality $|x-a|>\varepsilon$ is true for the union of two half-lines (Figure 2.6*b*):

$$x \in (-\infty, a-\varepsilon) \cup (a+\varepsilon, +\infty).$$

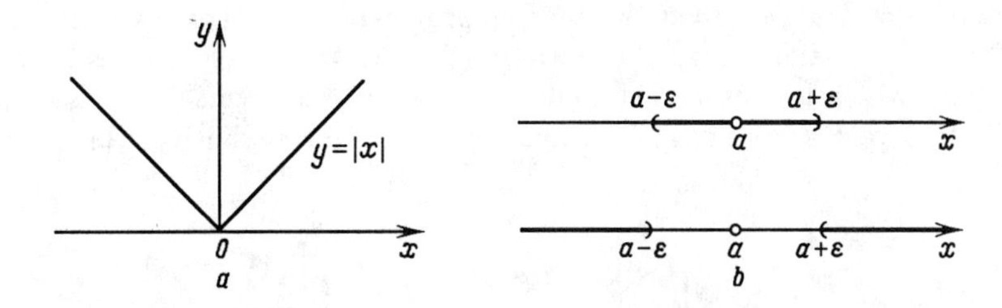

Figure 2.6: *a* – Graph of the function $y=|x|$. *b* – Geometric interpretation of solutions to inequalities $|x-a|<\varepsilon$ and $|x-a|>\varepsilon$.

2.7 Degree Function

The function $y = x^a$ is called a *degree function;* here a (*index*, or *exponent*) is an arbitrary constant real number. The domain of definition: $x > 0$, i.e., $x \in (0, +\infty)$; the range of values: $y > 0$, i.e., $y \in (0, +\infty)$. The degree function is monotonic: it is strictly increasing for $a > 0$, and strictly decreasing for $a < 0$. The shape of the graph depends upon the value of a (Figure 2.7).

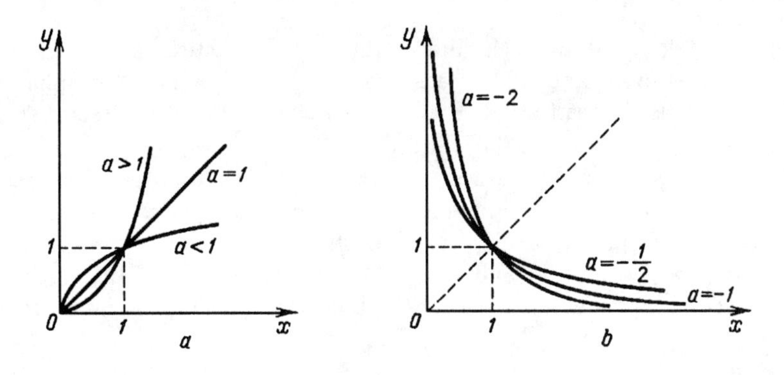

Figure 2.7: Graph of the degree function $y = x^a$ for $a > 0$ (a) and for $a < 0$ (b).

Extension for special values of the index

Let the index a be an integer, i.e., $a \in \mathbf{Z}$. For $a > 0$, we can consider the degree function on the whole axis of real numbers. For $a < 0$, we consider it on the same axis except for point $x = 0$ since the function has a discontinuity at this point (see Section 5.3.2).

In the special case $a = -1$, the dependence $y = 1/x$ (or, more generally, $y = k/x$) is called the *inverse proportionality*. For instance, the Ohm's law in the theory of electricity (Section 15.2.1) establishes that a current which is running through a given part of a circuit under a given constant voltage, is inversely proportional to the resistivity of this part.

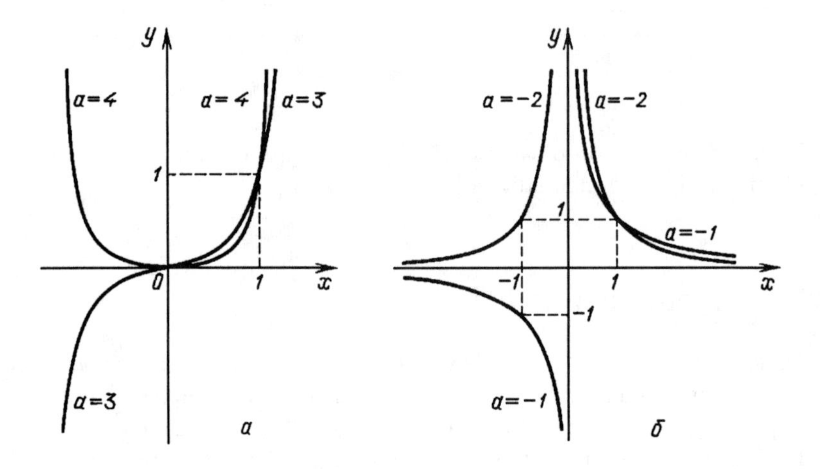

Figure 2.8: Graph of the degree function $y = x^a$ for a as integer.

For integral values of the index, the degree function is even in case the index is even, and odd in case it is odd. In Figure 2.8 graphs are plotted of the degree functions $y = x^3$, $y = x^4$, $y = x^{-1}$ ("rectangular hyperbola", see Section 9.3.3), and $y = x^{-2}$.

For some rational values of the index $a = m/n$, where m and n are integers, the degree function $y = x^a$ may also be considered over an extended domain of definition. If the index a is positive, that is, $a = m/n > 0$, where m and n are integers and n is odd, then $y = x^a$ is defined on the whole axis $\mathbf{R}$; if the index is negative, $a < 0$, we must delete point $x = 0$ from $\mathbf{R}$.

For instance, the function $x^{1/3}$ (or $\sqrt[3]{x}$, which is the same) is defined for the whole axis $x \in (-\infty, +\infty)$, while the function $x^{-1/3}$ is defined for this axis with the exception of the origin, i.e., $x \in (-\infty, 0) \cup (0, +\infty)$. For even values of n, and $m/n > 0$, the function $x^{m/n}$ is defined on the half-line $x \geq 0$. In particular, the function $\sqrt{x} = x^{1/2}$ if defined for $x \geq 0$.

2.8 Exponential and Logarithmic Functions

2.8.1 Exponential function

An *exponential function* is a function of the form $y = a^x$, where the *base* $a > 0$, $a \neq 1$. The domain of definition: $x \in \mathbf{R}$, the range of values: $y \in (0, +\infty)$. The main property of the exponential function:

$$a^{x_1+x_2} = a^{x_1} \cdot a^{x_2},$$

i.e., addition of values of the argument is equivalent to the multiplication of values of the function. Note that $a^0 = 1$, $a^1 = a$, $a^x > 0$.

The exponential function is strictly monotonic: it is increasing for $a > 1$ and decreasing for $a < 1$ (Figure 2.9*a*). For $a > 1$, the exponential function grows very fast when $x \to +\infty$ (so-called "exponential growth", it is faster than the growth of x^n for any $n > 0$).

The special number e (see Section 5.1.8) is often used as the base for the exponential function. The function e^x is denoted as well by $\exp(x)$.

The exponential dependence can be found widely in nature: for instance, in the process of radioactive decay the activity of radiation decreases exponentially in time (see Section 18.6.2).

2.8.2 Hyperbolic functions

The expressions e^x and e^{-x} can often be seen in the following standard combinations:

1. $\cosh x = (e^x + e^{-x})/2$ (*hyperbolic cosine*),

2. $\sinh x = (e^x - e^{-x})/2$ (*hyperbolic sine*),

3. $\tanh x = \frac{\sinh x}{\cosh x} = (e^x - e^{-x})/(e^x + e^{-x})$ (*hyperbolic tangent*).

Hyperbolic functions, in contrast to the trigonometric ones, are not periodic (Figure 2.9*b*). Properties of hyperbolic functions are the following: $\cosh x$ is even, $\sinh x$ and $\tanh x$ are odd. The functions $\cosh x$ and $\sinh x$ are unbounded, $\tanh x$ is bounded: $|\tanh x| < 1$. As $x \to \pm\infty$, $\tanh x \to \pm 1$.

The following formulas are valid:

$$\cosh^2 x - \sinh^2 x = 1, \quad \cosh 2x = \cosh^2 x + \sinh^2 x,$$

$$\sinh(x \pm y) = \sinh x \cdot \cosh x \pm \cosh x \cdot \sinh y,$$

$$\cosh(x \pm y) = \cosh x \cdot \cosh y \pm \sinh x \cdot \sinh y.$$

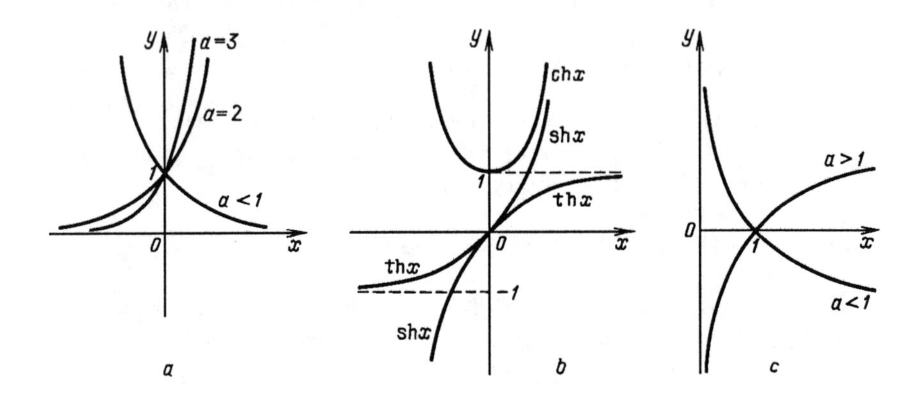

Figure 2.9: Graph of the degree function $y = x^a$ for a as integer.

2.8.3 Logarithmic function

The *logarithmic function* $y = \log_a x$ (read as "y is equal to the logarithm to base a of x"), with $a > 0$, $a \neq 1$, is the inverse function of the exponential one: if $y = a^x$, then $x = \log_a y$. The domain of definition: $x \in (0, +\infty)$, the range of values: $y \in (-\infty, +\infty)$. The logarithmic function is strictly monotonic: for $a > 1$, it is increasing, and for $a < 1$ it is decreasing (see Figure 2.9c).

The main logarithmic identity:

$$a^{\log_a x} = x \quad \text{for} \quad x \in (0, +\infty).$$

Properties of the logarithms (the arguments are assumed to be positive in all cases):

$$\log_a 1 = 0, \qquad \log_a a = 1,$$

$$\log_a(x \cdot y) = \log_a x + \log_a y, \quad \log_a(x/y) = \log_a x - \log_a y,$$

$$\log_a x^b = b \log_a x, \quad \log_a b = \frac{1}{\log_b a},$$

$$\log_a x = \log_b x \cdot \log_a b = \frac{\log_b x}{\log_b a}, \quad \log_{1/a} x = -\log_a x.$$

Examples

1. $\log_{1/3} 45 = -\log_3(5 \cdot 3^2) = -\log_3 5 - \log_3 3^2 = -\log_3 5 - 2.$

2. $4^{-\log_{\sqrt{2}} x} = [(\sqrt{2})^4]^{-\log_{\sqrt{2}} x} = (\sqrt{2})^{-4\log_{\sqrt{2}} x} = \sqrt{2}^{\log_{\sqrt{2}}(x^{-4})}.$

Logarithms to the base 10 (i.e., $\log_{10} x$) are called *decimal* (or common) logarithms and are denoted ordinarily by $\log x$; those to the base e (i.e., $\log_e x$) are called *natural* (or Napierian) logarithms and are denoted by $\ln x$. In various books (and countries), different notations are used, for instance, $\lg x$ for $\log_{10} x$, and $\operatorname{Log} x$ for $\log_e x$. In most books, the natural logarithms are designated as $\ln x$, and the same notation is used in this Handbook.

The natural and the decimal logarithms are related by the formula:

$$\log_{10} x = \log_e x \cdot \log_{10} e = \frac{\log_e x}{\log_{10} x},$$

where $\log_{10} e \approx 0.434$, $\log_e 10 \approx 2.30$.

Chapter 3

Equations and Systems of Equations. Inequalities

Many problems in physics, mathematics, engineering, and economics are related with investigation and solution of equations, inequalities, and systems of equations or inequalities. In this chapter we consider general properties of equations and inequalities, and discuss linear, quadratic and biquadratic equations, polynomials and algebraic equations, irrational and modulus equations, and linear and nonlinear systems of equations.

3.1 General Concepts

3.1.1 Concept of the equation

An *equation in one unknown* is a relation of the form

$$f(x) = 0, \tag{3.1}$$

where $f(x)$ is a given function of real or complex variable x. An alternative form of the equation: $f(x) = g(x)$, where $f(x)$ and $g(x)$ are two given functions, $f(x)$ is called the *left side* and $g(x)$ the *right side* of the equation. By transposing $g(x)$ to the left side we obtain $f(x) - g(x) = 0$, i.e., the equation of the type (3.1).

Any equation must be considered in its *domain of definition* (DD), i.e., in the domain of admissible values of x. This term means the numerical set on which all the functions occurred in the given equation are defined. For Equation (3.1), the DD coincides with the domain of definition of $f(x)$, for the equation $f(x) = g(x)$ it is the intersection of domains of definition of $f(x)$ and $g(x)$.

To solve Equation (3.1) means to find its *roots* (or *solutions*), i.e., such values of x that if we substitute any of these values into the equation, we obtain a true equality. In other words, to solve an equation $f(x) = 0$ means to find the *zeros* (or *roots*) of the function $f(x)$, i.e., the values of the argument x such that for these x the graph of $f(x)$ intersects the x-axis. The roots of equation are said to *satisfy* this equation. If there are several roots it is customary to index them: x_1, x_2, x_3, etc. For instance, $x_1 = 2$ and $x_2 = 3$ are the roots of the equation $x^2 - 5x + 6 = 0$; the equation $10^x = 0$ has no roots.

If we cannot solve the given equation exactly, we can seek *approximate solutions* (see Section 10.4).

3.1.2 Multiplicity of a root

Let the function $f(x)$ and all its derivatives (see Section 5.4) up to the order $(n-1)$ at a point x_0 be equal to zero, and the derivative of the n-th order be not equal to zero:

$$f(x_0) = f'(x_0) = \ldots = f^{n-1}(x_0) = 0, \quad f^n(x_0) \neq 0,$$

then we call x_0 the zero (or the root) of *multiplicity* n for the function $f(x)$. For example, $x = 0$ is a zero of multiplicity 2 for the function $y = \sin(3x^2)$ because $y(0) = 0$, $y'(0) = 0$, $y''(0) = 6$. The root x_0 of Equation (3.1) is said to be of *multiplicity* n if x is a zero of multiplicity n for $f(x)$.

3.1.3 Equivalent equations

Two equations $f_1(x) = g_1(x)$ and $f_2(x) = g_2(x)$ are said to be *equivalent* if each root of the first equation satisfies the second equation, and each root of the second equation satisfies the first one. In other words, the sets of roots of equivalent equations coincide. For example, the equations $3x - 6 = 0$ and $x = 2$ are equivalent (one root $x = 2$); the equations $x^2 = 4$ and $x = 2$ are not equivalent: the roots of the first equation are $x_1 = 2, x_2 = -2$, and the unique root of the second equation is $x = 2$. Equivalence of equations is denoted by the symbol $\Leftrightarrow$, e.g.:

$$3x - 6 = 0 \quad \Leftrightarrow \quad x - 2 = 0.$$

Equivalent transformations of an equation are the transformations that reduce the equation to an equivalent one:

- Adding an arbitrary number to both parts of the equation (in particular, transposing a summand from one side of the equation to another side and changing the sign).
- Multiplying (or dividing) both parts of the equation by an arbitrary nonzero number (in particular, by -1).

Moreover, for equations in the domain of real numbers, there are equivalent transformations as well:

- Raising both parts of the equation to any odd natural power (for instance, to the third power).
- Raising both parts of the equation to any even natural power (for instance, to the second power) provided that both parts be nonnegative.

3.1.4 Extraneous roots

If all roots of the equation $f_1(x) = g_1(x)$ satisfy the equation $f_2(x) = g_2(x)$, then we say that the second equation is the *consequence* (or *implication*) of the first equation, and write:

$$f_1(x) = g_1(x) \quad \Rightarrow \quad f_2(x) = g_2(x).$$

For instance, $x = 2 \Rightarrow x^2 = 4$, but the statement that $x^2 = 4 \Rightarrow x = 2$ would be false.

The operation of raising both parts of an equation to an even power is an important example of a nonequivalent transformation:

$$f(x) = g(x) \quad \Rightarrow \quad f^{2m}(x) = g^{2m}(x).$$

In general, after raising both parts of an equation to an even power we obtain so-called *outside* (or *extraneous*) roots; outside roots satisfy the equation $f(x) = -g(x)$. For example, the equation $x^2 = -1$ has no real roots; square both parts to obtain $x^4 = 1$, this equation has two real roots: $x = 1$, $x = -1$; both roots are outside for the given equation.

3.2 Linear and Quadratic Equations

3.2.1 Linear equation

Linear equation: $ax + b = 0$, where a and b are given real or complex numbers called *coefficients*. For any $a \neq 0$ the linear equation has a unique solution $x = -b/a$. For $a = 0$, $b \neq 0$, there are no solutions.

3.2.2 Quadratic equation

Quadratic equation: $ax^2 + bx + c = 0$, where a, b, c are given real or complex numbers called *coefficients*. If $a = 0$, the equation degenerates into a linear one; for $a \neq 0$ we can reduce the quadratic equation—dividing it by a—to the reduced form: $x^2 + px + q = 0$. In what follows we suppose that $a \neq 0$.

Any quadratic equation has roots, real or complex. For real values of a, b, c the roots are found as follows. First we calculate the *discriminant* $D = b^2 - 4ac$. Now three cases are possible:

1. If $D = 0$, then there exists only one root (of multiplicity two, see Section 3.1.2):

$$x = -\frac{b}{2a}. \tag{3.2}$$

2. If $D > 0$, then there are two real roots:

$$x_1 = \frac{-b + \sqrt{D}}{2a}, \quad x_2 = \frac{-b - \sqrt{D}}{2a}. \tag{3.3}$$

3. If $D < 0$, then there are no real roots, but two complex roots:

$$x_1 = \frac{-b + i\sqrt{-D}}{2a}, \quad x_2 = \frac{-b - i\sqrt{-D}}{2a}, \tag{3.4}$$

where i is the imaginary unit (see Section 7.1.1); in this case the roots are conjugate: $x_2 = \overline{x}_1$.

If the coefficients a, b, c are complex numbers, the formulas (3.2) through (3.4) are valid; the symbol $\sqrt{z}$ means any of two possible values for the complex square root (see Section 7.3.7).

Formula (3.3) for the reduced quadratic equation $x^2 + px + q = 0$ takes the form:

$$x_1 = -\frac{p}{2} + \sqrt{\left(\frac{p}{2}\right)^2 - q}, \quad x_2 = -\frac{p}{2} - \sqrt{\left(\frac{p}{2}\right)^2 - q},$$

which is convenient in the case of even value of p. For instance, the roots of the equation $x^2 - 14x - 576 = 0$ are

$$x = 7 \pm \sqrt{49 + 576} = 7 \pm \sqrt{625} = 7 \pm 25, \quad x_1 = 32, \quad x_2 = -18.$$

Theorem of Vieta: let x_1, x_2 be the roots of the reduced quadratic equation $x^2 + px + q = 0$, then $x_1 \cdot x_2 = q$, $x_1 + x_2 = -p$, i.e., the product of the roots is equal to the "free term" of the equation, and the sum of the roots is equal to the coefficient before x, taken with the opposite sign.

Vieta's theorem enables us to test quickly the roots, or —sometimes— to guess them. Thus, the numbers $x_1 = 50$, $x_2 = 7$ are easily seen to be the roots of the equation $x^2 - 57x + 350 = 0$ since $350 = 50 : 7$, $57 = 50 + 7$.

3.2.3 Biquadratic equation

Biquadratic equation $ax^4+bx^2+c=0$, by using the substitution $t=x^2$, is reduced to the quadratic one: $at^2+bt+c=0$. In the case of real-valued coefficients a, b, c, for $D=b^2-4ac<0$ there are no real roots; for $D>0$, there are two real values of $t:\ t_{1,2}=(-b\pm\sqrt{D})/(2a)$; for $D=0$ there is one real value of $t:\ t_1=-b/(2a)$. Therefore, the roots of the biquadratic equation are real or complex depending on the signs of t_1, t_2. In the complex plane $(x\in\mathbf{C})$ the roots of any biquadratic equation (with complex coefficients as well) exist.

3.3 Polynomials

3.3.1 Definitions

Polynomial of the degree n in the variable x is a function of the form

$$P(x)=a_0x^n+a_1x^{n-1}+\ldots+a_{n-1}x+a_n.$$

Here $a_0\neq 0$, n is integer, $n\geq 0$. The numbers $a_0, a_1, a_2,\ldots, a_n$ (real or complex) are called the *coefficients* of the polynomial. The polynomial of the zero degree $(n=0)$ is a constant.

Polynomials are considered either on the real x-axis $(x\in\mathbf{R})$ or in the complex plane $(x\in\mathbf{C})$.

Polynomials on the real x-axis: the variable x and the coefficients $a_0, a_1,\ldots, a_n$ are real numbers. The domain of definition is all of the real axis $\mathbf{R}$. The range of values:

for odd n it is $\mathbf{R}$;
for even n:
if $a_0<0$, then $(-\infty, M]$, where M is the greatest value of $P(x)$,
if $a_0>0$, then $[m,+\infty)$, where m is the smallest value of $P(x)$;
for $n=0$ the range of values consists of a unique point a.

A polynomial of nonzero degree is a nonperiodical, unbounded, and, in general, non-monotonic function. The graphs of some polynomials are represented in Figure 3.1.

Polynomials on the complex plane: the domain of definition and the range of values (if $n\neq 0$) are all of the complex plane $\mathbf{C}$.

3.3.2 Horner's method

In order to calculate the value of a polynomial at a given point x it is convenient to use the *Horner's method*: calculate consequently the numbers

$$p_0=a_0,\ p_1=p_0x+a_1,\ldots,\ p_n=p_{n-1}x+a_n,$$

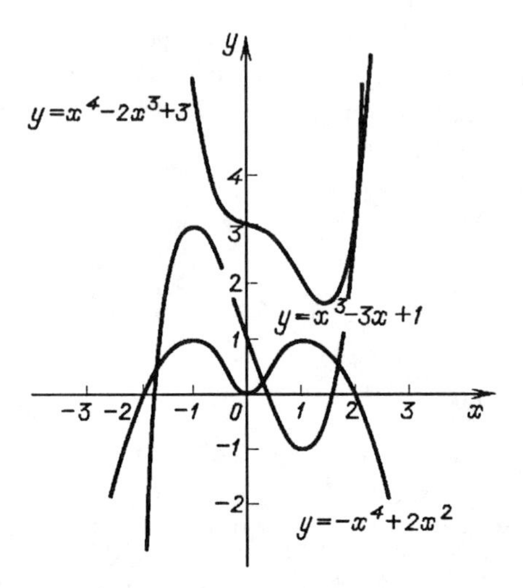

Figure 3.1: Graphs of some polynomials.

then $P(x) = p_n$. This algorithm is ordinarily used in computer calculations.

3.3.3 Polynomial algebra

1. Two polynomials, $P(x) = a_0x^m + \ldots + a_m$ and $Q(x) = b_0x^n + \ldots + b_n$ are said to be *equal* if they coincide identically, that is, $P(x) = Q(x)$ for all $x \in \mathbf{R}$ (or $x \in \mathbf{C}$). It is proved that two polynomials are equal if and only if their degrees are identical and their corresponding coefficients are equal to each other. However, two different polynomials may take common values at some separate points, for instance, $P(x) = x$ and $Q(x) = -x^2 + 2x$ coincide at $x = 0$ and $x = 1$.

Theorem: let two polynomials, $P(x)$ and $Q(x)$, both of the degree not greater than n, coincide at some $(n + 1)$ separate points $x_1, x_2, \ldots, x_{n+1}$, (i.e., $P(x_k) = Q(x_k)$ for $k = 1, 2, \ldots, n + 1$), then these polynomials are equal, that is, they coincide identically.

2. The degree of a sum of two polynomials is equal to the greatest degree of the summands (excepting the case when the degrees of summands are identical and $a_0 = -b_0$, here the degree of the sum is less than that of summands).

3. The product of two polynomials is a polynomial whose degree equals the sum of degrees of both co-factors. The coefficients of the resulted polynomial are obtained by removing the parentheses and gath-

ering the similar terms, e.g.,

$$(x+2)(x^2-2x+4)=x^3+8.$$

4. *Rational function* (or rational fraction) is an expression of the form:

$$R(x)=\frac{P(x)}{Q(x)},$$

where $P(x)$ and $Q(x)$ are polynomials. The fraction $P(x)/Q(x)$ is called *proper* if the degree of the numerator is less than the degree of the denominator.

Let m be the degree of the numerator $P(x)$ and n the degree of the denominator $Q(x)$. If $m \geq n$, then we can represent the rational function in the form

$$\frac{P(x)}{Q(x)}=p(x)+\frac{q(x)}{Q(x)}, \tag{3.5}$$

where $p(x)$ is an $(m-n)$-degree polynomial called the *integral part* (or *quotient*) of $P(x)/Q(x)$; $q(x)$ is a polynomial of the degree $k \leq (n-1)$ called the *remainder*. For example,

$$\frac{2x^2+1}{x-1}=2x+2+\frac{3}{x-1}.$$

Formula (3.5) may be rewritten as

$$P(x)=p(x)Q(x)+q(x).$$

If the remainder equals zero, the polynomial $P(x)$ is said to be *divisible* by the polynomial $Q(x)$, for instance, $(x^3+8)/(x^2-2x+4)=x+2$.

Theorem of Besout: a polynomial $P(x)$ being divided by $(x-b)$, the remainder is equal to $P(b)$, i.e., it equals the value of the polynomial at the point b.

Corollary: If $P(b)=0$, i.e., the number b is a root of the polynomial $P(x)$, then $P(x)$ is divisible by $(x-b)$. In other words, the n-degree polynomial $P(x)$ may be factorized as follows:

$$P(x)=(x-b)p(x),$$

where the degree of $p(x)$ is $(n-1)$.

In practice, the division of polynomials is performed by using a "corner rule". Previously we must order both the polynomials in descending powers of x, then write them in a line and separate with the help of a corner line:

$$2x^3+3x^2+x-1 \quad \left|\underline{x^2+2}\right. .$$

At first we divide the leading term of the dividend by the leading term of the divisor, here $2x^3/x^2$, and write the result under the horizontal line:

$$\begin{array}{l|l} 2x^3+3x^2+x-1 & \underline{x^2+2} \\ & 2x\ . \end{array}$$

Now we multiply the divisor (here: x^2+2) by the result (here: $2x$) and write this product (here: $2x^3+4x$) under the dividend, and then subtract the product from the dividend; the difference is to be written under the horizontal line:

$$\begin{array}{l|l} 2x^3+3x^2+x-1 & \underline{x^2+2} \\ \underline{2x^3+4x} & 2x \\ \quad 3x^2-3x-1\ . & \end{array}$$

Hence, the original problem is reduced to a similar problem of dividing a polynomial of a lower degree (here: $3x^2-3x-1$) by the original divisor. Subsequent operations are similar. In this example we come to the scheme:

$$\begin{array}{l|l} 2x^3+3x^2+x-1 & \underline{x^2+2} \\ \underline{2x^3+4x} & 2x+3 \\ \quad 3x^2-3x-1 & \\ \quad \underline{3x^2+6} & \\ \qquad -3x-7\ . & \end{array}$$

Further division is impossible; $2x+3$ is the integral part (quotient), $-3x-7$ the remainder.

3.3.4 Factorization of a polynomial

Each polynomial of the degree $n>0$ may be factorized into n linear co-factors:

$$P(x)=a_0(x-x_1)\dots(x-x_n), \tag{3.6}$$

where $x_1, x_2, \dots, x_n$ are the roots of the polynomial; in general the roots are complex. For example,

$$x^3-2x^2-x+2=(x-1)(x-2)(x+1),$$

$$x^3+8=(x+2)(x-1-i\sqrt{3})(x-1+i\sqrt{3}),$$

where $i=\sqrt{-1}$ (see Section 7.1). Some of the numbers $x_1, \dots, x_n$ may coincide and then Formula (3.6) should be rewritten as

$$P(x)=a_0(x-x_1)^{n_1}\dots(x-x_k)^{n_k}. \tag{3.7}$$

Here all the numbers $x_1, \dots, x_{n_k}$ are different, and the indices $n_1, \dots, n_k$ are multiplicities of the corresponding roots (see Section 3.1.2); $n_1+n_2+\dots+n_k=n$.

For instance,

$$x^7-3x^6+5x^5-7x^4+7x^3-5x^2+3x-1=(x-1)^3(x-i)^2(x+i)^2,$$

where $x_1 = 1$ is the root of multiplicity 3, $x_2 = i$ and $x_3 = -i$ are the roots of multiplicity 2.

Since the complex roots of a polynomial with real coefficients are pairwise conjugated (see Section 7.1.3), such a polynomial may be factorized into real factors:

$$P(x) = a_0(x - x_1)^{n_1} \dots (x - x_r)^{n_r}(x^2 + p_1 x + q)^{m_1} \dots (x^2 + p_k x + q_k)^{m_k},$$

where $x_1, \dots, x_r$ are the different real roots, and each quadratic trinomial $x^2 + p_j x + q_j$ $(j = 1, 2, \dots, k)$ has two mutually conjugated roots (see Section 3.2.2):

$$z_{1,2} = -p_j/2 \pm i\sqrt{q_j - (p_j/2)^2},$$

each root of multiplicity m_j. Here the relation holds:

$$n_1 + n_2 + \dots + n_r + 2(m_1 + \dots + m_k) = n.$$

For example, $x^3 + 8 = (x + 2)(x^2 - 2x + 4)$, and here $n_1 = 1$, $m_1 = 1$.

3.4 Algebraic Equations

3.4.1 Roots of an algebraic equation

An equation of the form

$$a_0 x^n + a_1 x^{n-1} + \dots + a_{n-1} x + a_n = 0, \tag{3.8}$$

where $a_0 \neq 0$, is called the *algebraic equation of the degree* n. Its roots are the roots of the corresponding polynomial $P(x) = a_0 x^n + \dots + a_n$. Linear, quadratic, and biquadratic equations are examples of algebraic equations.

The *fundamental theorem of algebra:* any algebraic equation of nonzero degree has at least one root (possibly, complex). From here and by using the Besout theorem (see Section 3.3.4) we obtain the corollary: any algebraic equation of the degree $n > 0$ has, in the complex plane, exactly n roots (every root is counted according its multiplicity). Thus, the equation $x^2 = 0$ has one double root $x_0 = 0$ or, in other words, two coinciding roots $x_1 = x_2 = 0$.

3.4.2 On general formulas for the roots

The problem of finding roots of algebraic equations is of great importance in theoretical and applied mathematics, physics, and engineering. There exist formulas which enable one to calculate the roots of algebraic equations of the degrees 1, 2, 3, and 4 in terms of fractional powers of some combinations of their coefficients. Such formulas for linear and quadratic equations are given in Section 3.2; *Cardano's* formulas are known for equations of the 3rd degree ("cubic equations"), and similar formulas exist also for the equations of the 4th degree. However, it has been proved (*Theorem of Abel*) that no formula for roots of arbitrary equations of the degree $n \geq 5$ may exist which could express the roots in terms of coefficients by using a finite number of operations such as addition, subtraction, multiplication, division, and raising to a fractional power.

3.4.3 Reduction of the degree of a polynomial

If we succeed in factorization of a given polynomial $P(x)$ into factors, the problem of solving the equation $P(x) = 0$ would be simplified. For instance,

$$x^3 - 6x^2 + 11x - 6 = (x-1)(x^2 - 5x + 6),$$

hence $x_1 = 1$, and two other roots satisfy the quadratic equation $x^2 - 5x + 6 = 0$.

In general, if we have somehow found (possibly by guess) a root x_1 of an algebraic n-degree equation $P(x) = 0$, the problem will be reduced, on dividing the polynomial $P(x)$ by $(x - x_1)$, to the problem of solving an equation of the degree $(n-1)$. Such a method is called "reduction of degree".

For example, the number $x_1 = 2$ satisfies the equation

$$x^4 - 2x^3 - 4x^2 + 11x - 6 = 0,$$

hence, dividing the left part by $(x-2)$ results in the cubic equation for two other roots:

$$x^3 - 4x + 3 = 0.$$

Further, $x_2 = 1$ is an evident root of this equation, etc.

3.4.4 Binomial algebraic equation

Binomial algebraic equation is of the form $ax^n + b = 0$, with a and b being real or complex numbers. They are solved in an explicit way: write the equation in the form $x^n = c$, where $c = -b/a$; if $b \neq 0$, we obtain n different roots, generally complex:

$$x_k = |c|^{1/n} \exp[i(\varphi + 2\pi k)/n], \quad \varphi = \arg c,$$

where $k = 0, 1, \ldots, n-1$ (see Section 7.3.7); if $b = 0$, the root $x = 0$ is unique. For real values of c, the number of real roots depends upon the sign of c and the evenness of n. Namely, if n is odd, there is only one real root $x_0 = \sqrt[n]{c}$; if n is even, then for positive c there are two real roots $x_1 = \sqrt[n]{c}$, $x_2 = -\sqrt[n]{c}$, and for negative c there are no real roots.

3.4.5 Rational equations

Rational equations are of the form $P(x)/Q(x) = 0$, with $P(x)$ and $Q(x)$ as polynomials. The domain of definition is all of the real x-axis (or complex plane) with the exception of the denominator's roots. In this domain, the rational equation is equivalent to the algebraic equation $P(x) = 0$.

3.5 Irrational and Modulus Equations

3.5.1 Irrational equations

Irrational equations contain expressions like $\sqrt[m]{P(x)}$, where $P(x)$ is a polynomial, e.g., the equation $x = \sqrt{2x^2 - 1}$. Usually the irrational equations are considered for real values of x. The domain of definition (DD) is the set of $x \in \mathbf{R}$ such that all the functions occurring in the equation are defined there. In the example above, the DD is the union of the half-lines:

$$(-\infty, -1/\sqrt{2}] \cup [1/\sqrt{2}, +\infty).$$

Some irrational equations are reducible to algebraic ones. Four simple types of such equations are described below.

1. The first type:

$$\sqrt[m]{P(x)} = Q(x), \tag{3.9}$$

where $P(x)$ and $Q(x)$ are polynomials. For odd m, the DD is all of the real axis, and for even m, the DD is the set of real x such that $P(x) \geq 0$ (see Section 3.7.5).

By raising both sides of Equation (3.9) to the degree m we obtain the algebraic equation

$$P(x) = Q^m(x). \tag{3.10}$$

If m is odd then (3.10) is equivalent to (3.9). If m is even, (3.10) is implied by (3.9): Equation (3.10) has outside roots, i.e., the roots of the equation $\sqrt[m]{P(x)} = -Q(x)$.

After we have found all the roots of (3.10) we must check each root by substitution into (3.9). However, there is an easier way: let the given root x_0 of (3.10) belong to the DD of the original equation (3.9), then it is sufficient to check the sign of $Q(x)$: if $Q(x_0) < 0$, then x is an outside root, and if $Q(x_0) \geq 0$, then x_0 is a root of the original equation.

Example

$\sqrt{2x^2 - 1} = x \Rightarrow 2x^2 - 1 = x^2 \Leftrightarrow x^2 = 1$. Both roots $x_1 = 1$ and $x_2 = -1$ are in the DD, but the root x_2 is outside, since $Q(x_2) = x_2 < 0$. Answer: $x = 1$.

2. The second type:

$$\sqrt[m]{P(x)} = \sqrt[n]{Q(x)},$$

where $P(x)$ and $Q(x)$ are polynomials. By raising to the degree $m \cdot n$, the equation is reduced to the algebraic one:

$$P^n(x) = Q^m(x).$$

Investigation of this equation is similar to that described above.

3. The third type:

$$P(x,\ \sqrt[m]{ax+b}) = 0,$$

in particular $P(x,\ \sqrt[m]{x}) = 0$. Here $P(x,y)$ is a polynomial in two variables x and y, i.e., a function of the form

$$P(x,y) = a_0 x^{j_0} y^{k_0} + \ldots + a_n x^{j_n} y^{k_n}.$$

The domain of definition for the equation $P(x,\ \sqrt[m]{ax+b}) = 0$ is either all of the real axis, or a half-line—that depends on the evenness of m. Substitution of $t = \sqrt[m]{ax+b}$ gives the algebraic equation

$$P\left(\frac{t^m - b}{a}, t\right) = 0.$$

Example

$x\sqrt{x} - 2\sqrt{x} + 1 = 0$. Substitute $t = \sqrt{x}$, where $t \geq 0$, and obtain $t^3 - 2t + 1 = 0$, or $(t-1)(t^2+t-1) = 0$. The roots are $t_1 = 1$, $t_2 = (-1+\sqrt{5})/2$, and $t_3 = (-1-\sqrt{5})/2$; we see that the root t_3 is outside the DD since $t_3 < 0$.

The answer: $x_1 = 1$, $x_2 = [(\sqrt{5}-1)/2]^2 = (3-\sqrt{5})/2$.

4. The fourth type:

$$\sqrt{P(x)} = \sqrt{Q(x)} + R(x),$$

where $P(x)$, $Q(x)$, $R(x)$ are polynomials. Here the the DD is the common part of the domains of definition of $\sqrt{P(x)}$ and $\sqrt{Q(x)}$, i.e.,

$$\{x | P(x) \geq 0\} \cap \{x | Q(x) \geq 0\}.$$

Square both sides of the equation: $\sqrt{P(x)} = \sqrt{Q(x)} + R(x) \Rightarrow P(x) = Q(x) + 2R(x)\sqrt{Q(x)} + R^2(x)$. Now transfer $2R(x)\sqrt{Q(x)}$ to the left and other terms to the right side, and once more square both sides:

$2R(x)\sqrt{Q(x)} = P(x) - Q(x) - R^2(x) \Rightarrow 4R^2(x)Q(x) = (P(x) - Q(x) - R^2(x))^2$.

Outside roots could appear twice in the process of these transformations. However, if, for instance, $R(x) \geq 0$ for all $x \in$ DD, then the first squaring of the equation is an equivalent transformation in the DD.

Example

$\sqrt{2x-1} = \sqrt{3x} - 1$. Here the DD is $[0.5, +\infty)$. Squaring gives only a consequence, but not an equivalent equation. Here it would be preferable to rewrite the equation in this way:

$\sqrt{3x} = \sqrt{2x-1} + 1 \Leftrightarrow 3x = 2x - 1 + 2\sqrt{2x-1} + 1 \Leftrightarrow 2\sqrt{2x-1} = x \Leftrightarrow x^2 - 8x + 4 = 0$. Both roots $x_{1,2} = 4 \pm 2\sqrt{3}$ are in the DD.

3.5.2 Modulus equations

Modulus equations contain expressions of the form $|f(x)|$. It is customary to consider such equations for real x. In order to simplify a modulus equation we first find the intervals of constant sign for all the functions that are situated in the modulus symbols. Then we rewrite the given equation for each interval using no modulus symbols ("interval method"). Thus the modulus equation will be reduced to a totality of usual equations over corresponding intervals, and the set of roots will be the union of corresponding sets of roots.

Example

$|x-1| = 2x + |x-2|$. The function $f_1 = x - 1$ changes its sign when x passes over the point $x_1 = 1$, and the function $f_2 = x - 2$ does the same at $x_2 = 2$. There are three intervals where the equation may be simplified; denote them as *I*, *II*, and *III* (Figure 3.2):

I. $1 - x = 2x + 2 - x$, whence $x_1 = -1/2$, $x_1 \in I$;

II. $x - 1 = 2x + 2 - x$, no roots here;

III. $x - 1 = 2x + x - 2$, whence $x_2 = 1/2$; it is an outside root since $x_2 \notin III$.

The answer: $x = -1/2$.

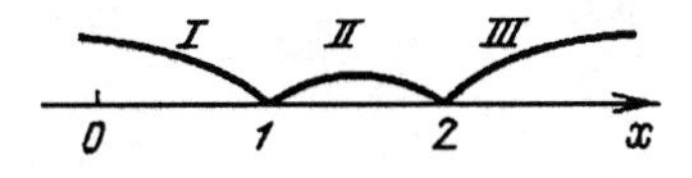

Figure 3.2: The solution of the equation $|x-1| = 2x + |x-2|$: three intervals.

3.6 Systems of Equations

3.6.1 Linear system, two equations

The *system of two linear equations in two unknowns* x *and* y is of the following form:

$$\begin{cases} a_1x + b_1y = c_1 \\ a_2x + b_2y = c_2 . \end{cases} \tag{3.11}$$

Here the *coefficients* a_1, a_2, b_1, b_2 and the *right-hand sides* c_1, c_2 are some given numbers (real or complex). To solve the system means to find all its *solutions*, i.e., all pairs of numbers (x, y) such that being substituted into Equations (3.11), they turn both equations into true equalities. For convenience, we shall sometimes write the system simply as $a_1x + b_1y = c_1$, $a_2x + b_2y = c_2$.

For example, the system

$$\begin{cases} x + y = 3 \\ x - y = 1 \end{cases} ,$$

or $x + y = 3$, $x - y = 1$, has the unique solution $x = 2$, $y = 1$, i.e., one pair $(2, 1)$; the system $x + y = 3$, $2x + 2y = 1$ has no solutions; the system $x + y = 3$, $2x + 2y = 6$ has an infinite set of solutions: every pair of numbers of the form $(a, 3 - a)$ with a as an arbitrary number, satisfies the equations.

The equations of the system (3.11) make simple geometric sense: each equation corresponds to a straight line in the xy-plane (see Section 9.1.1). The solution of (3.11) is the pair of coordinates of the intersection point for these straight lines. Therefore, the solution is unique if the lines intersect; no solution if the lines are parallel to each other; infinite quantity of solutions if the lines coincide.

Two systems of equations are said to be *equivalent* if all the solutions of the first system satisfy the second system and, vice versa, all the solutions of the second system satisfy the first system.

Such transformations as permutation of the equations, multiplication of both sides of an equation by a nonzero number, and adding to an equation of another equation multiplied by some number, are the equivalent transformations of the system because they transform the given system to an equivalent one.

In order to solve the system (3.11), we calculate the determinant $\Delta = a_1b_2 - a_2b_1$. If $\Delta \neq 0$ then there exists a unique solution and it is given by *Cramer's formulas:*

$$x = \frac{\Delta_x}{\Delta}, \quad y = \frac{\Delta_y}{\Delta},$$

where

$$\Delta_x = c_1 b_2 - c_2 b_1, \quad \Delta_y = a_1 c_2 - a_2 c_1.$$

If $\Delta = 0$, then there is either no solution or an infinite number of solutions.

A more convenient practical way of solving the system (3.11) is the *substitution method:* first "solve" one of the equations with respect to y or x (e.g., $y = (c_1 - a_1 x)/b_1$), and then substitute it into the other equation, thus we obtain a linear equation in one unknown (in the given case, in x.)

Example

Solve the system $2x+3y = 4$, $5x-2y = 3$. The first equation implies $x = 2 - 1.5y$; substitution of x into the second equation gives

$$5(2 - 1.5y) - 2y = 3, \quad \text{or} \quad 9.5y = 7, \quad y = 14/19.$$

Now x may be determined with the help of the substitution formula: $x = 2 - 1.5 \cdot 14/19 = 17/19$.

As a result of substituting $y = (c_1 - a_1 x)/b_1$, there may appear an equation of the form $0 \cdot x = q$. In such a case, for $q \neq 0$, the system has no solution; for $q = 0$, there are infinitely many solutions: x is arbitrary and y is expressed according to the substitution formula $y = (c_1 - a_1 x)/b_1$.

The *Gauss method* (or the *elimination method*) is essentially identical to the substitution method: multiply the first equation by b_2, the second one by b_1, and subtract the second equation from the first equation, thus we obtain a linear equation in one unknown x. In order to find y, we can operate now in a similar way: multiply the equations by a_2, a_1, and so on, or use any equation of the system (3.11) to express y in terms of x, which has already been found.

3.6.2 Linear system, three equations

The system of three linear equations in three unknowns x, y, z is of the form

$$\begin{cases} a_1 x + b_1 y + c_1 z = d_1 \\ a_2 x + b_2 y + c_2 z = d_2 \\ a_3 x + b_3 y + c_3 z = d_3. \end{cases} \tag{3.12}$$

The system (3.12) has the following geometric sense: each equation corresponds to some plane in the space (see Section 9.4.6), and the intersection set of two planes is a straight line. In the case where this line is not parallel to the third plane, there exists a unique intersection point for all three planes, so the system has a unique solution (x, y, z), where the numbers

x, y, z are the coordinates (see Section 8.3.3) of the intersection point. In the cases when some plane is parallel to another one, or when the intersection line of two planes is parallel to the third plane, the system has no solution.

The case is also possible that there exists an infinite number of solutions, e.g., all three planes intersect along one straight line.

Suppose the system (3.12) has a unique solution. The solution may be found with the help of the substitution method, or of the elimination method.

The substitution method: starting from the first equation, we express, for instance, the unknown x in terms of y and z : $x = (d_1 - b_1 y - c_1 z)/a_1$, then we replace x by this expression in the 2nd and the 3rd equations. Thus we obtain a system of two equations in two unknowns y and z; for how to solve it, see Section 3.6.1.

The elimination method: we divide the 1st equation by a_1 (if $a_1 \neq 0$) to obtain

$$x + b_1' y + c_1' z = d_1',$$

where $b_1' = b_1/a_1$, $c_1' = c_1/a_1$, $d_1' = a_1$. Then we multiply this equation by a_2 and subtract it from the 2nd equation, and as well we multiply it by a_3 and subtract it from the 3rd equation. Now the transformed system is of the form:

$$\begin{cases} x + b_1' y + c_1' z = & d_1' \\ \quad\;\; b_2' y + c_2' z = & d_2' \\ \quad\;\; b_3' y + c_3' z = & d_3'. \end{cases}$$

Two last equations may be subjected to the same procedure to get one equation of the form $\alpha z = \gamma$ with one unknown z. Now we find z and substitute it into the 2nd and 1st equations, further find y, and ultimately x. If a division is impossible (vanishing of a coefficient before unknown) we can replace equations.

In the process of eliminations, there may arise an equation of the form

$$0 \cdot x + 0 \cdot y + 0 \cdot z = g, \quad \text{where} \quad g \neq 0.$$

Since no values of x, y, z can satisfy such an equality, it means that there is no solution to the system.

If the system has an infinite number of solutions, then in the elimination process there must appear an equation of the kind

$$0 \cdot x + 0 \cdot y + 0 \cdot z = 0.$$

3.6.3 Nonlinear system

A general system of two equations in two unknowns x and y may be written as follows:

$$\begin{cases} f(x,y)=0 \\ g(x,y)=0, \end{cases} \tag{3.13}$$

where $f(x,y)$ and $g(x,y)$ are some given functions of two variables. Here the domain of definition (DD) is the set of all pairs (x,y) (or, what is equivalent, of points in the plane) such that both the functions $f(x,y)$ and $g(x,y)$ are defined. A solution of (3.13) is a pair of numbers (x,y) such that if we substitute them into both equations (3.13), we obtain true equalities.

For instance, the system $\begin{cases} x+y &= 3 \\ xy &= 2 \end{cases}$ admits two solutions: $(1,2)$ and $(2,1)$.

Some systems are solvable with the help of the substitution method. For instance, if the 1st equation (3.13) is solvable with respect to y : $y = h(x)$, where $h(x)$ is some known function, then the substitution of $h(x)$ into the 2nd equation gives the equation $g(x,h(x)) = 0$ in one unknown x.

Example

Solve the system $x+y=3$, $xy=2$. Starting from the 1st equation, find $y = 3-x$ and substitute it into the 2nd one: $x(3-x) = 2$, or $x^2-3x+2=0$, hence $x_1 = 1$, $x_2 = 2$ and, respectively, $y_1 = 3-1 = 2$, $y_2 = 3-2 = 1$.

In many cases there appear not one, but several functions of the kind $h(x)$, thus we generally get a corresponding totality of equations of the type $g(x,h(x)) = 0$.

Example

$$\begin{cases} x^2+y^2 &= 4 \\ xy &= 1. \end{cases}$$

From the first equation, $y = \sqrt{4-x^2}$ or $y = -\sqrt{4-x^2}$. Accordingly, two irrational equations, $x\sqrt{4-x^2} = 1$ and $x\sqrt{4-x^2} = -1$, arise, both are to be solved. In this example it is preferable to act in a different way: starting from the second equation, we express $y = 1/x$ and substitute it into the first equation to obtain $x^2+1/x^2 = 4$, or $x^4-4x^2+1=0$, i.e., a biquadratic equation (see Section 3.2.3).

Some systems are conveniently simplified by transforming to equivalent ones. Such transformations are: equivalent transformations of any equation (see Section 3.1.3), or replacement of the equations, or the addition of any equation (possibly multiplied by some number), to another equation.

Example

$$\begin{cases} x^2 + y^2 &= 4 \\ xy &= 1 \end{cases} \Longleftrightarrow \begin{cases} x^2 - 2xy + y^2 &= 2 \\ x^2 + 2xy + y^2 &= 6. \end{cases}$$

Here, we multiplied by 2 the second equation and subtracted it from the first equation, and also added it to the (original) first equation. Evidently,

$$\begin{cases} (x-y)^2 = 2 \\ (x+y)^2 = 6 \end{cases} \Longleftrightarrow \begin{cases} |x-y| = \sqrt{2} \\ |x-y| = \sqrt{6}, \end{cases}$$

and we obtained a set of four linear systems:

$$\begin{cases} x - y = \sqrt{2} \\ x + y = \sqrt{6} \end{cases} \begin{cases} x - y = -\sqrt{2} \\ x + y = \sqrt{6} \end{cases} \begin{cases} x - y = -\sqrt{2} \\ x + y = -\sqrt{6} \end{cases} \begin{cases} x - y = \sqrt{2} \\ x + y = -\sqrt{6}. \end{cases}$$

As a result, we have four solutions of the given system:

$$(\frac{\sqrt{6}+\sqrt{2}}{2}, \frac{\sqrt{6}-\sqrt{2}}{2}), \ (\frac{\sqrt{6}-\sqrt{2}}{2}, \frac{\sqrt{6}+\sqrt{2}}{2}),$$

$$(\frac{-\sqrt{6}-\sqrt{2}}{2}, \frac{\sqrt{2}-\sqrt{6}}{2}), \ (\frac{\sqrt{2}-\sqrt{6}}{2}, \frac{-\sqrt{6}-\sqrt{2}}{2}).$$

3.7 Inequalities

3.7.1 Definitions

The relations such as

$$A < B, \quad A > B, \quad A \leq B, \quad A \geq B, \quad A \neq B,$$

(to read: "A is less than B, A is greater than B, A is less than or equal to B, A is greater than or equal to B, A is not equal to B") are called *inequalities*.

Here A and B must be real numbers or real-valued functions. The inequality consists of the *left-hand* side and the *right-hand side* separated by the *symbol of inequality*. For the symbols $>$ or $<$ the inequality is said to be *strict*.

An inequality may be either *true* (correct, valid) or *false* (incorrect, invalid). For instance, $7 \geq -2$ and $1-\sqrt{2} < 0$ are true inequalities, but $7 \leq -2$ is a false one; $x^2 - 1 < 0$ is true for $x \in (-1, 1)$ and false for $x \notin (-1, 1)$; inequality $x^2 + 1 > 0$ is true for all $x \in \mathbf{R}$.

Consider two inequalities. If, from the fact that the first inequality is true, it follows that the second one is true, then the second inequality is said to be *implied* by the first one. We show this property by using the symbol $\Rightarrow$, for example, $x > 1 \Rightarrow x \geq 0$. If the two given inequalities are implied by each other, we call them *equivalent* and designate this with the help of the symbol $\Leftrightarrow$. For example, $2^a < 2^b \Leftrightarrow a < b$ for any real a and b.

3.7.2 Basic properties of inequalities

1. An arbitrary number may be added to both sides of an inequality:

$$A < B \Leftrightarrow A + \gamma < B + \gamma.$$

2. Both sides of an inequality may be multiplied by an arbitrary positive number:

$$A < B \Leftrightarrow \gamma A < \gamma B \quad \text{for} \quad \gamma > 0.$$

3. Both sides of an inequality may be multiplied by an arbitrary negative number, in this case the symbol of inequality must be changed to its opposite:

$$A < B \Leftrightarrow \gamma A > \gamma B \quad \text{for} \quad \gamma < 0.$$

Non-strict inequalities have similar properties.

A *system of inequalities* (two or more inequalities) is said to be *true* (or *satisfied*) if all its inequalities are true.

For instance, the system

$$\begin{cases} 7 > 2 \\ -3 \leq 1 \end{cases} \text{ is true, and the system } \begin{cases} 7 > 2 \\ 3 \leq 1 \end{cases} \text{ is not true;}$$

$$\text{the system } \begin{cases} 7 > 2 \\ x - 1 > 0 \end{cases} \text{ is true only for } x \in (1, +\infty).$$

True inequalities whose symbols are identical may be added term by term to obtain a true inequality:

$$\begin{cases} A < B \\ C < D \end{cases} \Rightarrow A + C < B + D.$$

If the numbers on both sides of two inequalities are positive and the symbols identical, we may multiply such inequalities term by term:

$$\begin{cases} A > B \\ C > D \end{cases} \Rightarrow AC > BD.$$

Any inequality, both sides of which are positive, may be raised to a natural power:

$$A < B \Leftrightarrow A^m < B^m \quad \text{for} \quad A > 0,\ B > 0,\quad m \in \mathbf{N}.$$

Any inequality (regardless the sign of its sides) may be raised to an odd natural power:

$$A < B \Leftrightarrow A^{2n+1} < B^{2n+1}.$$

Double inequalities are of the form:

$$A < B < C, \quad A < B \leq C, \quad \text{etc.}$$

A double inequality means a system of inequalities, respectively:

$$\begin{cases} A < B \\ B < C, \end{cases} \quad \begin{cases} A < B \\ B \leq C, \end{cases} \quad \text{etc.}$$

3.7.3 Problems related with inequalities

Two kinds of problems are related with inequalities:

(a) to prove validity (or invalidity) of a given inequality;

(b) to solve an equality, i.e., to find those values of variables occurring in the inequality, which make it true.

Examples

1. Prove that $a + 1/a \geq 2$ for any $a > 0$.

2. Solve the inequality $\sqrt{3x-1} \leq x$.

Some useful true inequalities

1. $|a+b| \leq |a| + |b|$, $|a-b| \geq ||a| - |b||$, so called the *triangle inequalities;*

2. $a + 1/a \geq 2$ for $a \in (0, +\infty)$;

3. $|ab| \leq (a^2 + b^2)/2$;

4. $\sqrt{ab} \leq (a+b)/2$ for $a > 0$, $b > 0$, i.e., the geometric mean of two numbers is less than or equal to their arithmetical mean;

5. $(a_1 b_1 + \ldots + a_n b_n)^2 \leq (a_1^2 + \ldots + a_n^2)(b_1^2 + \ldots + b_n^2)$, the *Cauchy-Bunyakowskii* (or *Schwartz*) inequality; in particular:

$$(ab + cd)^2 \leq (a^2 + c^2)(b^2 + d^2),$$

$$(a_1 + \ldots + a_n)^2 \leq n(a_1^2 + \ldots + a_n^2);$$

6. $|\sin x| \leq |x|$ for $x \in \mathbf{R}$, $0 < \sin x < x$ for $x \in (0, \pi)$,
$\sin x > (2/\pi)x$ for $x \in (0, \pi/2)$;

7. $\tan x > x$ for $x \in (0, \pi/2)$.

Attention: in inequalities 6 and 7 the angle x is measured in radians (see Section 9.1.5).

3.7.4 Domain of definition

In order to *solve* an inequality of the form $f(x) < g(x)$ or $f(x) \leq g(x)$, we must first find the domain of definition (DD) of the inequality, i.e., the set of those x for which both the functions $f(x)$ and $g(x)$ are defined. In some cases we are able, by using equivalent transformations (see Section 3.7.2), to simplify the given inequality and then solve it. The answer is ordinarily represented as an interval or a union of intervals.

3.7.5 Algebraic inequalities

We consider the following five kinds of algebraic inequalities:

1. $P(x) > 0$, or $P(x) < 0$, where $P(x)$ is a polynomial. If we factor the polynomial (see Section 3.3.4):

$$P(x) = a_0(x-x_1)^{n_1} \dots (x-x_r)^{n_r}(x^2+p_1x+q_1)^{m_1} \dots (x^2+p_kx+q_k)^{m_k},$$

then the intervals where $P(x) > 0$ and, as well, where $P(x) < 0$, can be easily determined with the help of the *interval method.*

According to the interval method we mark on the x-axis the real roots of the polynomial $P(x)$, i.e., the points $x_1, x_2, \dots, x_r$. Such points split the real axis into some intervals; in each interval the sign of $P(x)$ is constant because the sign may change only when x passes one of the points $\{x_j\}$, for instance, x_1. The sign really changes in the case where the corresponding power index n is odd, but it does not change if n is even. All the quadratic factors, e.g., $x^2 + p_1x + q_1$, are strictly positive over all of x-axis. In order to find the sign of $P(x)$ in each interval, it is sufficient to find it in one of them. Thus the inequality is solved.

Remark. The sign of $P(x)$ for x greater than the greatest of $x_1, x_2, \dots, x_r$ is constant and coincides with the sign of a_0.

Example. Consider $P(x) = -3(x+1)x^3(x-1)^2(x^2+x+1)$. The intervals of constant sign of $P(x)$ are depicted by arcs in Figure 3.3. The solution of the inequality $P(x) > 0$ is the interval $(-1, 0)$, the solution of $P(x) < 0$ is the union of three intervals: $(-\infty, -1) \cup (0, 1) \cup (1, +\infty)$.

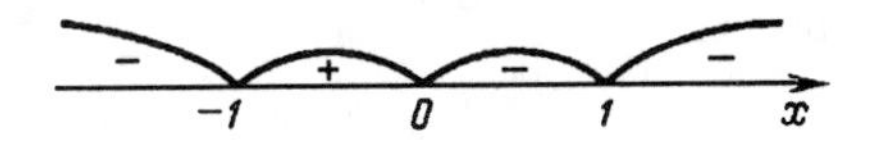

Figure 3.3: The intervals, where the polynomial $-3(x+1)x^3(x-1)^2(x^2+x+1)$ is of constant sign.

2. Non-strict inequalities $P(x) \geq 0$ or $P(x) \leq 0$ are solved in the same way, but the end-points of intervals must now be included. For

example, the inequality

$$(x - 0.5)(x - 2)^2(x - 4) \geq 0$$

has the solution $(-\infty, 0.5) \cup \{2\} \cup (4, +\infty)$, the end-point $x = 2$ of the interval (2, 4) is included because $x = 2$ satisfies the inequality.

3. Inequality $P(x) < Q(x)$, where $P(x)$ and $Q(x)$ are polynomials, may be simplified by carrying $Q(x)$ to the left-hand side, and then it may be solved by the interval method.

4. $P(x)/Q(x) < 0$, where $P(x)$ and $Q(x)$ are polynomials. The domain of definition is all of the real axis except for the roots of the denominator. Multiplying both sides of the inequality by $Q^2(x)$ gives an equivalent (in the DD) inequality $P(x)Q(x) < 0$.

5. An inequality of the fractional type is simply reduced to the preceding one:

$$\frac{P_1(x)}{Q_1(x)} < \frac{P_2}{Q_2} \Leftrightarrow \frac{P_1Q_2 - Q_1P_2}{Q_1Q_2} < 0\,.$$

3.7.6 Irrational inequalities

Irrational inequalities contain expressions like $\sqrt[m]{P(x)}$, where $P(x)$ is a polynomial. Methods for reduction of irrational inequalities to algebraic ones are analogous to the methods for corresponding equations (see Section 3.5.1), but here some specific distinctions appear concerning the raising of inequalities to an even power.

As an example, consider the inequality $\sqrt{P(x)} < Q(x)$; two possibilities must be examined:

(a) $Q(x) \geq 0$, then the given inequality is equivalent to $P(x) < Q^2(x)$ (for x in the DD);

(b) $Q(x) < 0$, in this case, squaring of the given inequality is impossible; the given inequality has no solution since $\sqrt{P(x)} \geq 0$.

Consider now the inequality $\sqrt{P(x)} \geq Q(x)$; for x such that $Q(x) \geq 0$, it may be squared: $P(x) = Q^2(x)$, and for x such that $Q(x) < 0$, it is true for all x in the DD.

Examples

1. Solve the inequality:

$$\sqrt{3x} \geq 1 + \sqrt{2x - 1}. \tag{3.14}$$

Here the domain of definition is $[0.5, +\infty)$. Since both sides are non-negative, square them:

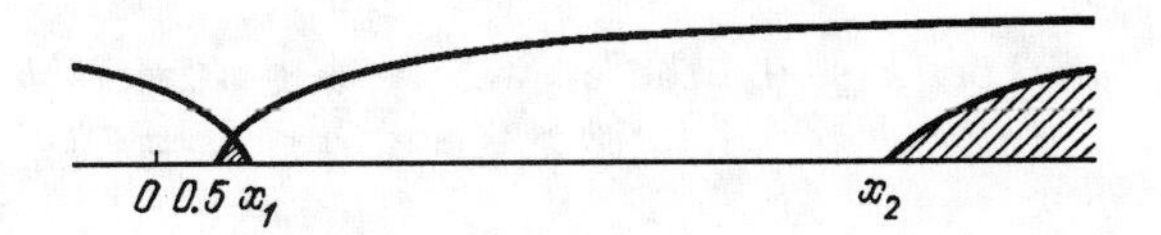

Figure 3.4: The solution of inequality (3.14).

$$3x \geq 1 + 2x - 1 + 2\sqrt{2x-1} \Leftrightarrow 2\sqrt{2x-1} \leq x \Leftrightarrow 4(2x-1) \leq x^2 \Leftrightarrow$$
$$x^2 - 8x + 4 \geq 0 \Leftrightarrow (x - x_1)(x - x_2) \geq 0,$$

where x_1 and x_2 are the roots of the equation $x^2 - 8x + 4 = 0$, $x_{1,2} = 4 \pm 2\sqrt{3}$. The inequality obtained is true outside the interval (x_1, x_2). Taking into account the DD, we get the final answer:

$$x \in [0.5,\ 4 - 2\sqrt{3}] \cup [4 + 2\sqrt{3},\ +\infty),$$

see Figure 3.4.

Remark: the assertion that $0.5 < x_1$ may be proved in this way:

$$0.5 < 4 - 2\sqrt{3} \Leftrightarrow 4\sqrt{3} < 7 \Leftrightarrow 48 < 49.$$

Since the resulting inequality is true and the transformations above are equivalent, we conclude that $0.5 < 4 - 2\sqrt{3}$.

2. $\sqrt{x^2 - 2x} > x - 4$; the DD: $(-\infty,\ 0] \cup [2,\ +\infty)$.

Two cases must be considered:

(a) $x \geq 4$. Square the original inequality: $x^2 - 2x > x^2 - 8x + 16 \Leftrightarrow 3x > 8$; this case leads to $x \in [4,\ +\infty)$.

(b) $x < 4$. The left-hand side of the original inequality is positive, the right-hand side is negative, so any value $x \in [2,\ 4)$ or $x \in (-\infty,\ 0]$ is suitable. Answer:

$$x \in (-\infty,\ 0] \cup [2,\ +\infty).$$

3. $\sqrt{x^2 - 2x} < x - 4$. Two cases must be considered:

(a) $x \geq 4$. Square the inequality to obtain $3x < 8$; the last inequality is not true for $x \geq 4$;

(b) $x < 4$. The left-hand side is non-negative, the right-hand one negative, hence no solution exists.

The answer: the given inequality has no solution.

3.7.7 Transcendental inequalities

Transcendental inequalities may contain exponential, logarithmic, and trigonometric functions. In some cases such inequalities can be solved utilizing special properties of those functions.

Examples

1. $a^x > b$. For $b \leq 0$, the answer is $x \in \mathbf{R}$; for $b > 0$ and $a > 1$, $x \in (\log_a b,\ +\infty)$; for $b > 0$ and $a < 1$, $x \in (-\infty,\ \log_a b)$.

2. $4^x - 5 \cdot 2^x + 6 \geq 0$. Let $t = 2^x$, $t > 0$. The given inequality now takes the form $t^2 - 5t + 6 \geq 0$, whence $t \in (0,\ 2] \cup [3,\ +\infty)$, consequently, $x \in (-\infty,\ 1] \cup [\log_2 3,\ +\infty)$.

3. $-3\log_{0.7}(x+1) < 6$. Here, the DD is: $x > -1$. On dividing the inequality by (-3) we obtain: $\log_{0.7}(x+1) > -2 \Leftrightarrow \log_{0.7}(x+1) > \log_{0.7}((0.7)^{-2}) \Leftrightarrow x + 1 < (0.7)^{-2} \Leftrightarrow x < -1 + (0.7)^{-2}$.
   The answer: $-1 < x < 51/49$.

3.7.8 Inequalities with modulus symbol

Inequalities that contain the modulus function may be investigated in a similar way to the corresponding equations, i.e., by the interval method (see Section 3.5.2). As a result we obtain a set of simpler inequalities, each inequality on its own interval, and then give the answer as a union of the corresponding intervals.

Example

$|x+1| + |2x-1| > |x-1| - 3$. Transition points: -1, 0.5, 1. Respectively, there are four intervals:

$$(-\infty,\ -1),\quad (-1,\ 0.5),\quad (0.5,\ 1),\quad (1,\ +\infty).$$

For instance, on the interval $(-1,\ 0.5)$ the given inequality is equivalent to the following one:

$$x + 1 - (2x - 1) > 1 - x - 3, \quad \text{or} \quad 1 > -3,$$

and we see that this relation is true regardless of x. This means that all of the interval $(-1,\ 0.5)$ satisfies the given inequality. Other intervals may be considered similarly.

Chapter 4

Trigonometry

Trigonometry was developed under investigation of correlations between the sides and the angles of triangles. Trigonometric functions, as well as polynomials, are met practically in all kinds of formulas and calculations. In trigonometry, both the radian and the degree measures for angles are used (see Section 9.1). Throughout this chapter we assume the angle α to be measured by radians (if the opposite is not specified), so the value α is a dimensionless number (see Section 12.2), and $\alpha \in (-\infty, +\infty)$.

4.1 Trigonometric Functions

4.1.1 Trigonometric functions of an acute angle

The simplest way to introduce the trigonometric functions is to use a right (i.e., rectangular) triangle. In a right triangle $\triangle ABC$ (see Figure 4.1) the side AB is called the *hypotenuse*, the side AC the *leg* (or the *cathetus*), the side BC also the leg (or cathetus).

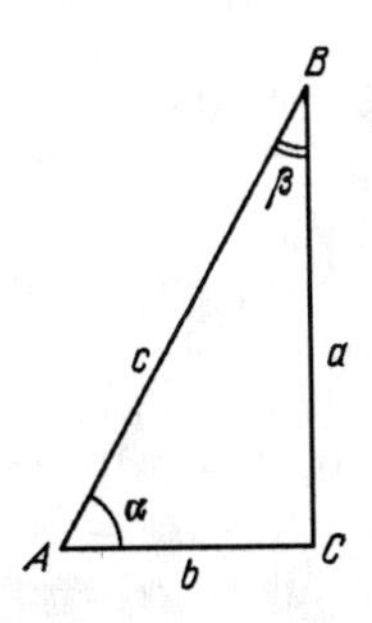

Figure 4.1: The introduction of the trigonometric functions.

The ratio of the leg BC to the hypotenuse AB is called the *sine of the angle* α *lying opposite* to the leg BC:

$$\sin\alpha = \frac{|BC|}{|AB|}, \quad \text{or} \quad \sin\alpha = \frac{a}{c},$$

where a, b, c are the corresponding lengths of the triangle sides.

The ratio $|AC|/|AB|$ is called the *cosine of the angle* α *adjacent* to the leg AC:

$$\cos\alpha = \frac{b}{c}.$$

The ratio of the leg $|BC|$ to the leg $|AC|$ is called the *tangent*, and the inverse ratio is called the *cotangent* of the angle α:

$$\tan\alpha = \frac{a}{b}, \quad \cot\alpha = \frac{b}{a}.$$

These equalities imply that

$$\tan\alpha = \frac{\sin\alpha}{\cos\alpha}, \quad \cot\alpha = \frac{1}{\tan\alpha} = \frac{\cos\alpha}{\sin\alpha}.$$

The following formulas are evident:

$$\sin\alpha = \cos\beta, \quad \tan\alpha = \cot\beta,$$

where α and β are the acute angles of a right triangle. The acute angles are "complements": $\alpha + \beta = \pi/2$.

Two other functions are used comparatively seldom:

$$\sec\alpha = \frac{1}{\cos\alpha}, \quad \csc\alpha = \frac{1}{\sin\alpha}.$$

Remark. In some books the symbols tg and ctg are used instead of tan and cot, respectively.

4.1.2 Trigonometric functions of arbitrary values of argument

Let us take a ray OM_0 and rotate it counterclockwise about the point O (Figure 4.2a). As a result of the rotation by an angle α, the ray OM_0 will be mapped onto the ray OM_α, and the point M_0, having passed an arc whose center is O, will coincide with the point M_0. If the rotation angle is right, i.e., $\alpha = \pi/2$, the arc $\smile M_0M_\alpha$ is a quarter of the circumference; if the angle is straight ($\alpha = \pi$), the arc $\smile M_0M_\alpha$ is a half of the circumference; if the angle is total ($\alpha = 2\pi$), the arc $\smile M_0M_\alpha$ is the total circumference, and then $M_\alpha = M_0$.

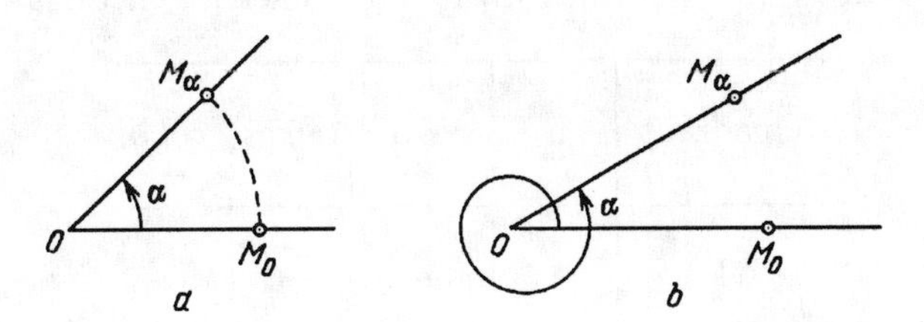

Figure 4.2: The result of rotation by angle α: the ray OM_0 is mapped onto the ray OM_α.

If we continue to rotate the ray counterclockwise, the point M_0 will continue its motion along the circumference, but now the rotation angle becomes greater than the total one (see Figure 4.2*b*, where the angle $\alpha = 2\pi + \pi/6$).

It is conventional to call the counterclockwise direction *positive* ($\alpha > 0$), and the clockwise *negative* ($\alpha < 0$).

Consider a Cartesian (i.e., rectangular) coordinate system xy in the plane (see Section 8.3.2) and such a circle that its center is at the origin of the coordinate system and the radius is unit (Figure 4.3). Such a circle is called the *unit circle*.

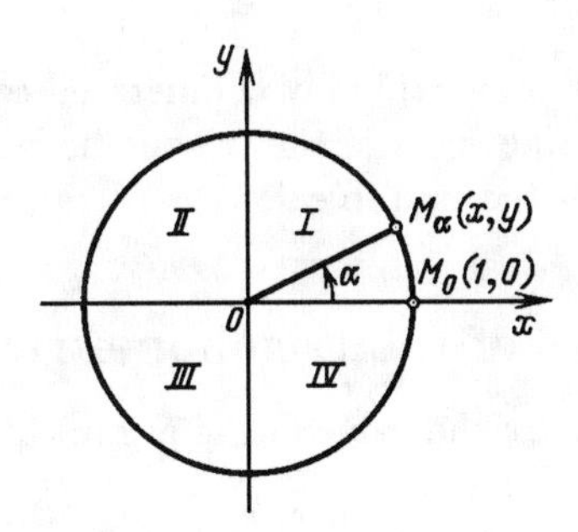

Figure 4.3: Unit circle.

The point M_α of the circumference is obtained from the point $M_0(1,0)$ as a result of rotation of the ray OM_0 by the angle α. Note that if the angle α differs from the angle β by an integer number of total revolutions, i.e., by $360° \cdot m$ $(m \in \mathbf{Z})$, then the points M_α and M_0 coincide.

Denote by x and y the coordinates of the point M_α; evidently,

$$x^2 + y^2 = 1$$

Table 4.1: Sign rule for the trigonometric functions

Quadrant	$\sin\alpha$	$\cos\alpha$	$\tan\alpha$	$\cot\alpha$
I	+	+	+	+
II	+	−	−	−
III	−	−	+	+
IV	−	+	−	−

(the circumference equation).

The trigonometric functions are defined as follows:

$$\sin\alpha = y, \quad \cos\alpha = x, \quad \tan\alpha = \frac{\sin\alpha}{\cos\alpha}, \quad \cot\alpha = \frac{\cos\alpha}{\sin\alpha}.$$

For positive acute angles these definitions are identical to ones given in Section 4.1.1.

The *quadrants* are conventionally enumerated counterclockwise from I to IV (Figure 4.3). The signs of the trigonometric functions according to the quadrants are listed in Table 4.1.

4.1.3 Properties of the trigonometric functions

As usual, we denote by x the value of the argument and by y the corresponding value of the function.

1. $y = \sin x$. The graph is called the *sine curve* (see the continuous line in Figure 4.4). The domain of definition: $x \in (-\infty, +\infty)$; the range of values: $y \in [-1, 1]$. The function is 2π-periodic, odd, bounded, and continuous. The intervals of monotonicity are: $(-\pi/2 + n\pi, \ \pi/2 + n\pi)$; the roots: $x_n = n\pi$ $(n \in \mathbf{Z})$.

2. $y = \cos x$. The graph is called the *cosine curve* (see the dashed line in Figure 4.4). The cosine curve is obtained from the sine curve as a result of the shifting of it to the left by $\pi/2$. The domain of definition: $x \in (-\infty, +\infty)$; the range of values: $y \in [-1, 1]$. The function is 2π-periodic, even, bounded, and continuous. The intervals of monotonicity: $(n\pi, \ \pi + n\pi)$; the roots: $x_n = \pi/2 + n\pi$ $(n \in \mathbf{Z})$.

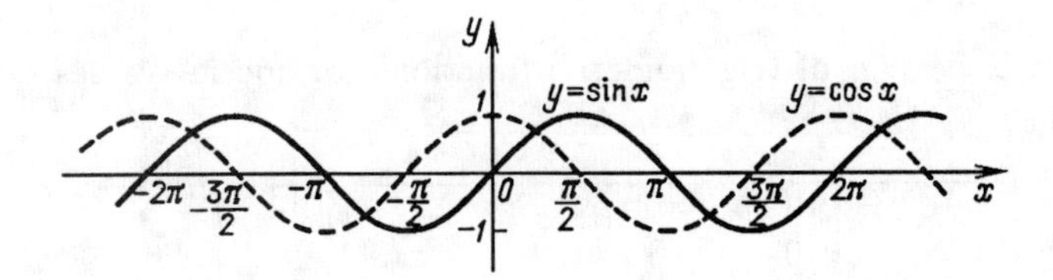

Figure 4.4: Graphs of the sine and the cosine functions.

3. $y = \tan x$. The graph is the *tangent curve* (see the continuous line in Figure 4.5). The domain of definition: all of the real axis $(-\infty, +\infty)$ with the exception of the points $x_k = \pi/2 + k\pi$ $(k \in \mathbf{Z})$; the range of values: $(-\infty, +\infty)$. The function is π-periodic, odd, unbounded, and continuous for $x \neq \pi/2 + k\pi$, and has the infinite discontinuities at points $x_k = \pi/2 + k\pi$ $(k \in \mathbf{Z})$. Every interval of continuity of $\tan x$ is an interval of increase.

4. $y = \cot x$. The graph is the *cotangent curve* (see the dashed line in Figure 4.5). The domain of definition: all of the real axis $(-\infty, +\infty)$ with the exception of the points $x_m = m\pi$ $(m \in \mathbf{Z})$; the range of values: $(-\infty, +\infty)$. The function is π-periodic, odd, unbounded, and continuous for $x \neq m\pi$, and has infinite discontinuities at points $x_m = m\pi$ $(m \in \mathbf{Z})$. Every interval of continuity of $\cot x$ is an interval of decrease.

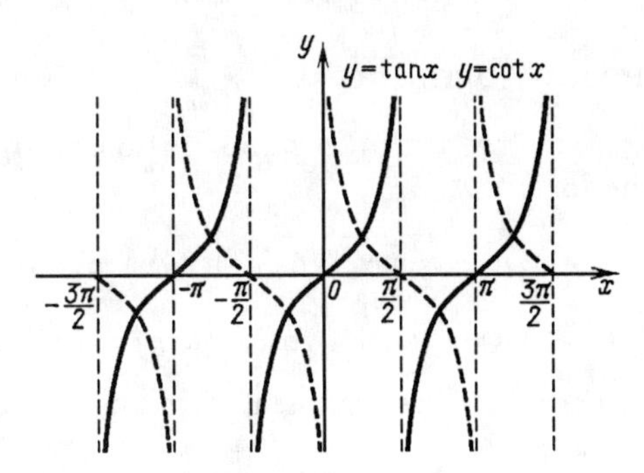

Figure 4.5: Graphs of the tangent and cotangent functions (vertical dashed lines are the asymptotes).

Trigonometric functions for special values of argument

Some values of trigonometric functions are given in Table 4.2. Notice that the symbol ∞ is used there in the cases where the function is not

Table 4.2: Values of trigonometric functions for special values of their argument

	0 $0°$	$\pi/6$ $30°$	$\pi/4$ $45°$	$\pi/3$ $60°$	$\pi/2$ $90°$	π $180°$
$\sin\alpha$	0	$1/2$	$1/\sqrt{2}$	$\sqrt{3}/2$	1	0
$\cos\alpha$	1	$\sqrt{3}/2$	$1/\sqrt{2}$	$1/2$	0	-1
$\tan\alpha$	0	$1/\sqrt{3}$	1	$\sqrt{3}$	∞	0
$\cot\alpha$	∞	$\sqrt{3}$	1	$1/\sqrt{3}$	0	∞

defined at the given point α_0; in such cases there exists an infinite limit (see Section 5.2.3):

$$\lim_{\alpha\to\alpha_0} f(\alpha) = \infty.$$

4.2 Formulas of Trigonometry

4.2.1 Reduction formulas

The *reduction formulas* are useful in simplification of arguments of the trigonometric functions:

$$\sin(\alpha + 2\pi n) = \sin\alpha, \quad \cos(\alpha + 2\pi n) = \cos\alpha,$$

$$\tan(\alpha + \pi n) = \tan\alpha, \quad \cot(\alpha + \pi n) = \cot\alpha,$$

i.e., addition of a period to the argument does not change the value of the function;

$$\sin(\alpha + \pi) = -\sin\alpha, \quad \cos(\alpha + \pi) = -\cos\alpha,$$

i.e., addition of a half-period to the argument changes the sign of sine and cosine;

$$\sin(\alpha \pm \pi/2) = \pm\cos\alpha, \quad \cos(\alpha \pm \pi/2) = \mp\sin\alpha,$$

$$\tan(\alpha \pm \pi/2) = -\cot\alpha, \quad \cot(\alpha \pm \pi/2) = -\tan\alpha,$$

$$\sin(\alpha + \pi n) = (-1)^n \sin\alpha, \quad \cos(\alpha + \pi n) = (-1)^n \cos\alpha,$$

$$\sin(\alpha + \pi/2 + \pi n) = (-1)^n \cos\alpha, \quad \cos(\alpha + \pi/2 + \pi n) = (-1)^{n+1} \sin\alpha.$$

In the formulas above, n is an arbitrary integer, $n \in \mathbf{Z}$.

The oddness and evenness properties:

$$\sin(-\alpha) = -\sin\alpha, \quad \cos(-\alpha) = \cos\alpha,$$

$$\tan(-\alpha) = -\tan\alpha, \quad \cot(-\alpha) = -\cot\alpha.$$

The functions of complementary angles:

$$\sin(\pi - \alpha) = \sin\alpha, \quad \cos(\pi - \alpha) = -\cos\alpha,$$

$$\sin(\pi/2 - \alpha) = \cos\alpha, \quad \cos(\pi/2 - \alpha) = \sin\alpha,$$

$$\tan(\pi/2 - \alpha) = \cot\alpha, \quad \cot(\pi/2 - \alpha) = \tan\alpha.$$

Examples

1. $\sin 1873^\circ = \sin(73^\circ + 360^\circ \cdot 5) = \sin 73^\circ = \sin(90^\circ - 17^\circ) = \cos 17^\circ$.
2. $\cos(-27\frac{3}{8} \cdot \pi) = \cos(-28\pi + \frac{5\pi}{8}) = \cos\frac{5\pi}{8} = \cos(\pi - \frac{3\pi}{8}) = -\cos\frac{3\pi}{8}$.
3. $\cot 539^\circ = \cot(539^\circ - 180^\circ \cdot 3) = -\cot 1^\circ$.

Reduction formulas may be derived from the addition (subtraction) formulas listed below, or by using some geometric speculations.

4.2.2 The basic formulas of trigonometry

$$\tan\alpha \cdot \cot\alpha = 1,$$

$$\sin^2\alpha + \cos^2\alpha = 1 \quad (\textit{Pythagoras' theorem}),$$

$$\sin(\alpha \pm \beta) = \sin\alpha\cos\beta \pm \cos\alpha\sin\beta,$$

$$\cos(\alpha \pm \beta) = \cos\alpha\cos\beta \mp \sin\alpha\sin\beta.$$

Consequence formulas:

$$\sin 2\alpha = 2\sin\alpha\,\cos\alpha, \quad \cos 2\alpha = \cos^2\alpha - \sin^2\alpha,$$

$$\cos 2\alpha = 2\cos^2\alpha - 1, \quad \cos 2\alpha = 1 - 2\sin^2\alpha,$$

$$1 + \cos\alpha = 2\cos^2\frac{\alpha}{2}, \quad 1 - \cos\alpha = 2\sin^2\frac{\alpha}{2},$$

$$\tan 2\alpha = \frac{2\tan\alpha}{1 - \tan^2\alpha}, \quad \cot 2\alpha = \frac{\cot^2\alpha - 1}{2\cot\alpha},$$

$$\tan(\alpha \pm \beta) = \frac{\tan\alpha \pm \tan\beta}{1 \mp \tan\alpha\,\tan\beta}, \quad \cot(\alpha \pm \beta) = \frac{\cot\alpha\,\cot\beta \mp 1}{\cot\beta \pm \cot\alpha},$$

$$1+\tan^2\alpha = \frac{1}{\cos^2\alpha}, \quad 1+\cot^2\alpha = \frac{1}{\sin^2\alpha},$$

$$\sin\alpha + \sin\beta = 2\sin\frac{\alpha+\beta}{2}\cos\frac{\alpha-\beta}{2},$$

$$\sin\alpha - \sin\beta = 2\sin\frac{\alpha-\beta}{2}\cos\frac{\alpha+\beta}{2},$$

$$\cos\alpha + \cos\beta = 2\cos\frac{\alpha+\beta}{2}\cos\frac{\alpha-\beta}{2},$$

$$\cos\alpha - \cos\beta = -2\sin\frac{\sin\alpha+\beta}{2}\sin\frac{\alpha-\beta}{2},$$

$$\sin\alpha\,\sin\beta = \frac{1}{2}\left[\cos(\alpha-\beta) - \cos(\alpha+\beta)\right],$$

$$\cos\alpha\,\cos\beta = \frac{1}{2}\left[\cos(\alpha-\beta) + \cos(\alpha+\beta)\right],$$

$$\sin\alpha\,\cos\beta = \frac{1}{2}\left[\sin(\alpha-\beta) + \sin(\alpha+\beta)\right],$$

$$\sin\alpha = \frac{2t}{1+t^2}, \quad \cos\alpha = \frac{1-t^2}{1+t^2}, \quad \tan\alpha = \frac{2t}{1-t^2}, \quad t = \tan\frac{\alpha}{2},$$

$$\tan\frac{\alpha}{2} = \frac{1-\cos\alpha}{\sin\alpha}, \quad \cot\frac{\alpha}{2} = \frac{1+\cos\alpha}{\sin\alpha},$$

$$\sin^2\frac{\alpha}{2} = \frac{1}{2}(1-\cos\alpha), \quad \cos^2\frac{\alpha}{2} = \frac{1}{2}(1+\cos\alpha),$$

$$\sin 3\alpha = 3\sin\alpha - 4\sin^3\alpha, \quad \cos 3\alpha = 4\cos^3\alpha - 3\cos\alpha.$$

For trigonometric functions of multiple angles (i.e., of the angles such as $n\alpha$) see Section 7.3.6.

4.3 Inverse Trigonometric Functions

4.3.1 Arcsine

Since the sine function is non-monotonic, in order to define its inverse function we have to choose some interval of monotonicity for the sine. The following convention is customary: the *arcsine* of a number $x \in [-1,\ 1]$ is the angle $y \in [-\pi/2,\ \pi/2]$ such that its sine equals x:

$$y = \arcsin x, \quad \sin y = x.$$

Some examples: $\arcsin 0.5 = \pi/6$, $\arcsin 0 = 0$, $\arcsin 1 = \pi/2$, $\arcsin(-1/\sqrt{2}) = -\pi/4$, but $\arcsin 2$ does not exist as a real number (however, in the complex domain, the equation $\sin z = 2$ is solvable: see Section 7.5).

The graph of the function $y = \arcsin x$ is plotted in Figure 4.6 (continuous line). The domain of definition: $x \in [-1, 1]$; the range of values: $y \in [-\pi/2, \pi/2]$. The function is increasing, odd, bounded, and continuous.

4.3.2 Arccosine

The *arccosine* is the inverse function for the cosine:

$$y = \arccos x, \quad \cos y = x, \quad x \in [-1, 1], \quad y \in [0, \pi].$$

The graph is plotted in Figure 4.6 (dashed line). The domain of definition: $x \in [-1, 1]$; the range of values: $y \in [0, \pi]$. The function is decreasing, bounded, continuous, and neither odd nor even.

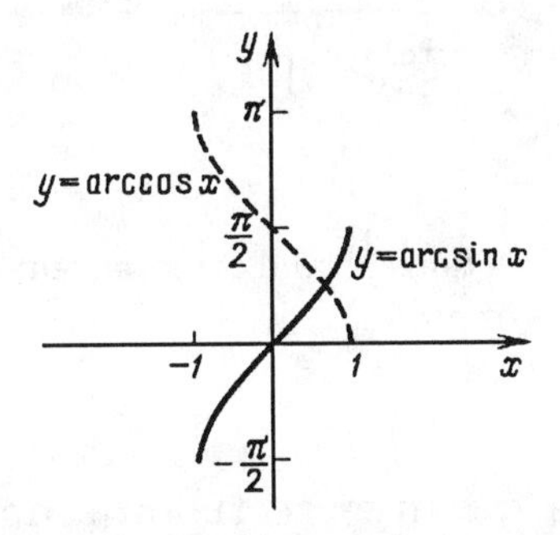

Figure 4.6: Graphs of arcsine and arccosine functions.

4.3.3 Arctangent

The *arctangent* is the inverse function for the tangent:

$$y = \arctan x, \quad \tan y = x, \quad x \in (-\infty, +\infty), \quad y \in (-\pi/2, \pi/2).$$

The graph is plotted in Figure 4.7 (continuous line). The domain of definition: $x \in (-\infty, +\infty)$; the range of values: $y \in (-\pi/2, \pi/2)$. The function is increasing, bounded, continuous, and odd; there are finite limits:

$$\lim_{x \to +\infty} \arctan x = +\pi/2, \qquad \lim_{x \to -\infty} \arctan x = -\pi/2.$$

4.3.4 Arccotangent

The *arccotangent* is the inverse function for the cotangent:

$$y = \operatorname{arccot} x, \quad \cot y = x, \quad x \in (-\infty, +\infty), \quad y \in (0, \pi).$$

The graph is plotted in Figure 4.7 (thick dashed line). The domain of definition: $x \in (-\infty, +\infty)$; the range of values: $y \in (0, \pi)$. The function is decreasing, bounded, continuous, and neither odd nor even; there are finite limits:

$$\lim_{x \to -\infty} \operatorname{arccot} x = \pi, \quad \lim_{x \to +\infty} \operatorname{arccot} x = 0.$$

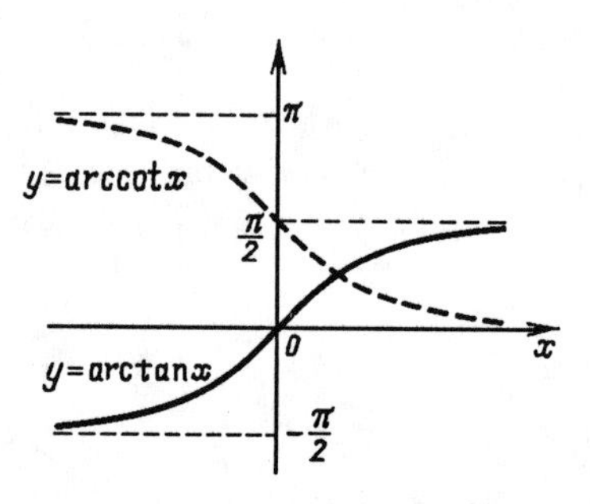

Figure 4.7: Graphs of arctangent and arccotangent functions (horizontal dashed lines are the asymptotes).

4.3.5 Formulas for inverse trigonometric functions

$$\sin(\arcsin x) = x, \quad \cos(\arccos x) = x,$$

$$\arcsin(\sin x) = x \quad \text{for} \quad x \in [-\pi/2,\ \pi/2],$$

$$\arccos(\cos x) = x \quad \text{for} \quad x \in [0, \pi],$$

$$\arcsin(-x) = -\arcsin x, \quad \arccos(-x) = \pi - \arccos x,$$

$$\arctan(-x) = -\arctan x, \quad \arctan(-x) = \pi - \arctan x,$$

$$\arcsin x + \arccos x = \pi/2, \quad \arctan x + \operatorname{arccot} x = \pi/2,$$

$$\arcsin x = \arccos \sqrt{1 - x^2} \quad \text{for} \quad x \in [0,\ 1],$$

$$\arccos x = \arcsin \sqrt{1 - x^2} \quad \text{for} \quad x \in [0,\ 1],$$

$$\operatorname{arccot} x = \arctan(1/x) \quad \text{for} \quad x \in (0,\ +\infty).$$

4.4 Trigonometric Equations and Inequalities

4.4.1 Simplest trigonometric equations

The simplest trigonometric equations and their solutions:

$$\sin x = 0, \quad x = n\pi; \quad \tan x = 0, \quad x = n\pi;$$

$$\cos x = 0, \quad x = \pi/2 + n\pi; \quad \cot x = 0, \quad x = \pi/2 + n\pi;$$

$$\sin x = 1, \quad x = \pi/2 + 2n\pi; \quad \sin x = -1, \quad x = -\pi/2 + 2n\pi;$$

$$\cos x = 1, \quad x = 2n\pi; \quad \cos x = -1, \quad x = (2n+1)\pi;$$

$$\sin x = a, \quad x = (-1)^n \arcsin a + n\pi, \quad a \in [-1, 1];$$

$$\cos x = a, \quad x = \pm \arccos a + 2n\pi, \quad a \in [-1, 1];$$

$$\tan x = a, \quad x = \arctan a + n\pi;$$

$$\cot x = a, \quad x = \operatorname{arccot} a + n\pi.$$

In the formulas above, n is an arbitrary integer ($n \in \mathbf{Z}$).

4.4.2 Equation reducible to the simplest one

In order to solve an equation of the type $f(h(x)) = 0$, where $h(x)$ is one of the trigonometric functions, we introduce a new unknown, $t = h(x)$, and then find the roots $t_1, t_2, \ldots, t_n$ of the resulting equation $f(t) = 0$. For each of these roots we write and solve the corresponding equation:

$$h(x) = t_1, \quad h(x) = t_2, \ldots, \quad h(x) = t_n. \tag{4.1}$$

The total answer is a union of all solution sets of (4.1).

Examples

1. $2\cos^2 2x - 9\cos 2x + 4 = 0$.

   Putting $t = \cos 2x$, where $t \in [-1, 1]$, we get a quadratic equation $2t^2 - 9t + 4 = 0$, whose roots are $t_1 = 4$, $t_2 = 0.5$. The root t_1 brings no solution to the original equation because $|\cos 2x| \le 1$. The root t_2 corresponds to the equation $\cos 2x = 1/2$ whence $2x = \pm\pi/3 + 2n\pi$, or

   $$x = \pm\pi/6 + n\pi, \quad n \in \mathbf{Z}.$$

2. $2\cos^2 x - 4\sin x \cos x + 1 = 0$.

   This equation is reducible to a form that contains only one trigonometric function. At first we write $2\cos^2 x - 4\sin x\ \cos x + \sin^2 x + \cos^2 x = 0$, then divide both sides by $\cos^2 x$ and obtain

$$\tan^2 x - 4\tan x + 3 = 0.$$

   This operation is correct because for the values of x such that $\cos^2 x = 0$, the original equation is obviously not satisfied. Now the substitution $t = \tan x$ leads to the equation $t^2 - 4t + 3 = 0$, whose roots are $t_1 = 1$, $t_2 = 3$. Thus we obtained two equations, $\tan x = 1$ and $\tan x = 3$, whence

$$x = \pi/4 + n\pi, \quad x = \arctan 3 + n\pi, \quad n \in \mathbf{Z}.$$

4.4.3 Equation $a\sin x + b\cos x = c$

Some trigonometric equations may be reduced to the form

$$a\sin x + b\cos x = c, \quad \text{where} \quad a \neq 0, \quad b \neq 0. \tag{4.2}$$

The left-hand expression in (4.2) may be rewritten as sine (or cosine) of a shifted argument:

$$a\sin x + b\cos x = D\left(\frac{a}{D}\sin x + \frac{b}{D}\cos x\right),$$

where $D = \pm\sqrt{a^2 + b^2}$, and the sign before the root symbol must coincide with the sign of a. The coefficients a/D and b/D can be taken as the cosine and the sine of an auxiliary angle γ:

$$a/D = \cos\gamma, \quad b/D = \sin\gamma.$$

The angle γ is determined as

$$\gamma = \arctan(b/a), \quad \gamma \in (-\pi/2, \pi/2).$$

Therefore,

$$a\sin x + b\cos x = D\sin(x + \gamma), \tag{4.3}$$

and Equation (4.2) is now reduced to the simplest form

$$\sin(x + \gamma) = d, \quad \text{with} \quad d = c/D.$$

Note the physical meaning of (4.3): the sum of two harmonic oscillations having identical frequency is a harmonic oscillation of the same frequency (see Section 16.2).

Example: $\cos x + \sin x + 3\cos 2x = 0$. Since $\cos 2x = \cos^2 x - \sin^2 x$, the given equation is equivalent to the following one:

$$(\cos x + \sin x)[1 + 3(\cos x - \sin x)] = 0.$$

Thus we have a set of two equations:

(a) $\cos x + \sin x = 0 \Leftrightarrow \tan x = -1, \quad x = -\pi/4 + n\pi, \quad n \in \mathbf{Z}$;

(b) $\sin x - \cos x = \frac{1}{3}$, or $\sqrt{2}(\frac{1}{\sqrt{2}}\sin x - \frac{1}{\sqrt{2}}\cos x) = \frac{1}{3}$.

Here the angle $\gamma = \arctan(-1) = -\pi/4$, so the equation turns to $\sin(x - \pi/4) = 1/(3\sqrt{2})$, and finally:

$$x = \pi/4 + (-1)^n \arcsin\frac{1}{3\sqrt{2}} + \pi n, \quad n \in \mathbf{Z}.$$

Another way of solving Equation (4.2) is to substitute $t = \tan(x/2)$ and apply the formulas from Section 4.2.2:

$$\sin x = \frac{2t}{1+t^2}, \qquad \cos x = \frac{1-t^2}{1+t^2}.$$

Such a substitution transforms (4.2) into

$$\frac{2at}{1+t^2} + \frac{b(1-t^2)}{1+t^2} = c.$$

After a simplification, we come to a quadratic equation whose roots t_1 and t_2 correspond to the equations

$$\tan\frac{x}{2} = t_1, \quad \tan\frac{x}{2} = t_2.$$

The third way of solving Equation (4.2) is to transfer the term $b\cos x$ to the right-hand side and then square both sides; it brings a quadratic equation for $\cos x$:

$$2(1 - \cos^2 x) = b^2\cos^2 x - 2bc\cos x + c^2.$$

This method has a deficiency: the squaring gives rise to outside roots, so it is necessary to test the roots.

4.4.4 Trigonometric inequalities

Trigonometric inequalities are to be reduced to the simplest ones:

$$\sin x < a, \quad \cos x < a, \quad \tan x < a, \quad \cot x < a,$$

or to analogous opposite (strict or not) inequalities. In order to solve a simplest trigonometric inequality it is profitable to imagine the corresponding graph or the unit circle.

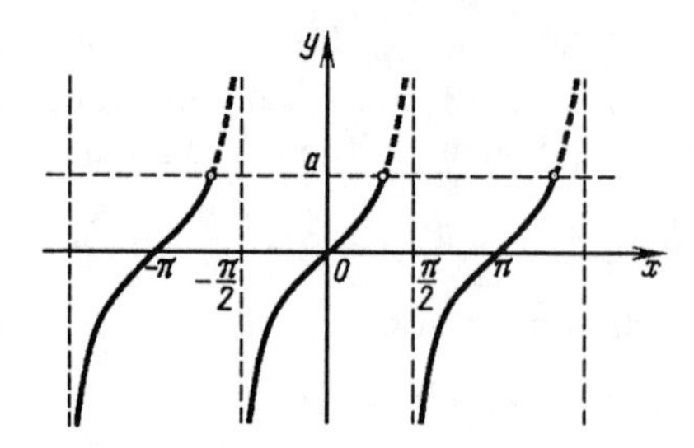

Figure 4.8: The solution of inequality $\tan x < a$.

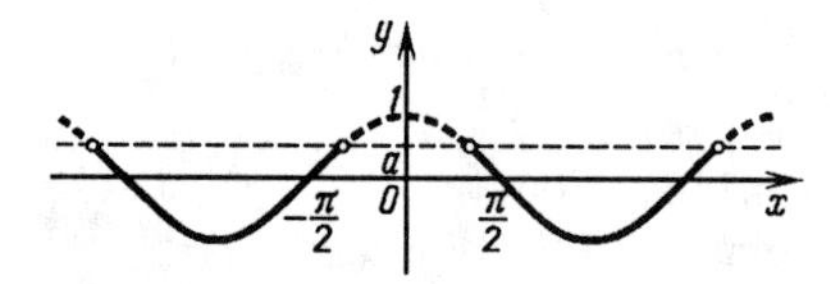

Figure 4.9: The solution of inequality $\cos x \le a$.

Examples

1. $\tan x < a$.

   For arbitrary $a \in \mathbf{R}$, the solution is the union of intervals such as $(-\pi/2+\pi n, \arctan a+\pi n)$, $n \in \mathbf{Z}$ (corresponding to the continuous lines in Figure 4.8).

2. $\cos x \le a$.

   For $a < -1$, no solution; for $a = -1$, $x = \pi + 2\pi n$, $n \in \mathbf{Z}$; for $-1 < a \le 1$, $x \in [\arccos a + 2\pi n, 2\pi - \arccos a + 2\pi n]$, $n \in \mathbf{Z}$ (corresponding to the continuous lines in Figure 4.9); finally, for $a > 1$, $x \in \mathbf{R}$.

Chapter 5

Elements of Calculus

The concept of the limit of a sequence or a function is in the base of calculus. The notion of the derivative is widely used in differential calculus for investigation of functions, while the integral calculus works at the problem of reconstructing the function whose derivative is given. Differential equations and also series are often to be met in the setting and solving of many mathematical and physical problems.

5.1 Sequences

5.1.1 Concept of a sequence

A *numerical sequence* is a mapping of the set $\mathbf{N}$ of natural numbers (or of a subset of $\mathbf{N}$) into the set $\mathbf{R}$ of real numbers. In other words, a sequence is a numerical set that is ordered by numbers of its elements.

A sequence may be *finite:* $x_1, x_2, \ldots, x_n$, or *infinite:* $x_1, x_2, \ldots,$ $x_n, \ldots$. The numbers x_1, $x_2, \ldots, x_n, \ldots$ are called the *terms* of the sequence. Examples: the sequence of positive integers $1, 2, 3, \ldots$ (here $x_n = n$); the finite sequence $1, 2, 3, 4, 5$; the sequence of decimal approximations to the ratio $1/3$: $x_1 = 0.3$, $x_2 = 0.33$, $x_3 = 0.333, \ldots$ (see Section 1.2.4). For brevity, we denote the sequence $x_1, x_2, \ldots, x_n, \ldots$ by $\{x_n\}$; the value x_n is called the *general term* of the sequence.

5.1.2 Limit of a sequence

It may happen that the terms of an infinite sequence approximate a constant number, and the more exact such an approximation becomes, the greater the number of the term becomes.

Definition: A number a is called the *limit* of the sequence x_1, $x_2, \ldots, x_n, \ldots$ if, for every given number $\varepsilon > 0$, there exists a natural number N such that for all $n > N$ the following inequality holds:

$$|x_n - a| < \varepsilon.$$

In this case we write:

$$\lim_{n\to\infty} x_n = a, \quad \text{or} \quad \lim_{n\to\infty} = a, \quad \text{or} \quad x_n \longrightarrow a,$$

and say: "a is the limit of x_n as n tends to infinity", or "x_n tends to a as n tends to infinity", or "x_n converges to the limit a". The sequence having a limit is said to be *convergent.*

The fact that the sequence $\{x_n\}$ converges to limit a means geometrically that the corresponding points x_1,x_2,x_3,... on the real axis **R** come arbitrarily close to point a as the number of x_n increases infinitely. In other words, for every given (arbitrary small) neighborhood of point a all terms of the sequence $\{x_n\}$, after some number N, belong to this neighborhood (see Figure 5.1).

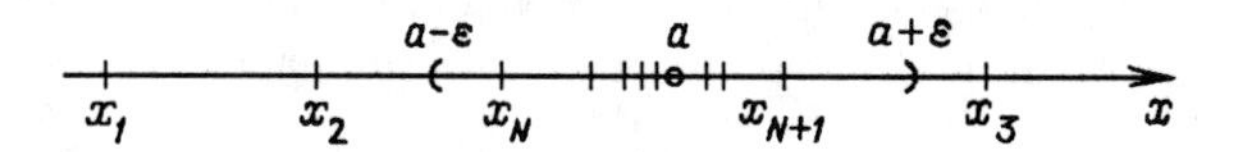

Figure 5.1: All terms of the sequence x_n, after some number N, belong to the neighborhood $(a - \varepsilon, a + \varepsilon)$ of point a.

Examples

1. $\lim_{n\to\infty} 1/n = 0$ since $|1/n| < \varepsilon \Longleftrightarrow n > 1/\varepsilon$, so N can be chosen as an arbitrary positive integer greater than $1/\varepsilon$. Thus, if $\varepsilon = 0.1$, then $N > 10$; if $\varepsilon = 0.001$, then $N > 1000$, etc.

2. $\lim_{n\to\infty} \frac{n}{2n+1} = \frac{1}{2}$.

Theorems about convergent sequences

1. Every convergent sequence admits a unique limit.

2. The limit of a constant sequence is equal to this constant:

$$\lim_{n\to\infty} C = C.$$

3. The limit of a sum is equal to the sum of the limits:

$$\lim_{n\to\infty} (x_n + y_n) = \lim_{n\to\infty} x_n + \lim_{n\to\infty} y_n.$$

4. A constant multiplier may be taken outside of the limit symbol:

$$\lim_{n\to\infty} (\gamma x_n) = \gamma \lim_{n\to\infty} x_n.$$

5. The limit of a product is equal to the product of the limits:

$$\lim_{n\to\infty} (x_n \cdot y_n) = \lim_{n\to\infty} x_n \cdot \lim_{n\to\infty} y_n.$$

6. The limit of a quotient is equal to the quotient of limits:

$$\lim_{n\to\infty} \frac{x_n}{y_n} = \frac{\lim_{n\to\infty} x_n}{\lim_{n\to\infty} y_n}, \quad \text{if} \quad \lim_{n\to\infty} y_n \neq 0,$$

7. Any convergent sequence is bounded.

8. Any bounded monotonic sequence has a limit.

9. If $a_n \leq x_n \leq b_n$ and the sequences a_n and b_n both tend to a common limit C, then

$$\lim_{n\to\infty} x_n = C.$$

Remark. In Theorems 3 – 6, all the limits occurring in the right-hand side of the equalities are assumed to be existing.

5.1.3 Monotonic and bounded sequences. Infinitesimal sequences

A sequence $\{x_n\}$ is said to be *increasing* if, for every n, the inequality $x_{n+1} \geq x_n$ is true. The sequence is said to be *decreasing* if $x_{n+1} \leq x_n$, *strictly increasing* if $x_{n+1} > x_n$, and *strictly decreasing* if $x_{n+1} < x_n$ for every n.

Both increasing and decreasing sequences are called *monotonic*; both strictly increasing and strictly decreasing sequences are called *strictly monotonic*.

A sequence is said to be *bounded* if $|x_n| \leq C$, or if $C_1 \leq x_n \leq C_2$, where C, C_1, C_2 are some constants.

Examples: calculation of limits with the help of Theorems 2 – 6.

1. $\lim_{n\to\infty}(1+\frac{1}{n}+\frac{2}{n^2}) = \lim_{n\to\infty} 1+\lim_{n\to\infty} \frac{1}{n} + 2\lim_{n\to\infty} \frac{1}{n^2} = 1$, since $\lim_{n\to\infty} 1 = 1$, $\lim_{n\to\infty} \frac{1}{n} = 0$, $\lim_{n\to\infty} \frac{1}{n^2} = (\lim_{n\to\infty} \frac{1}{n})^2 = 0$;

2. $\lim_{n\to\infty} \frac{2n+1}{3n-5} = \lim_{n\to\infty} \frac{n(2+1/n)}{n(3-5/n)} = \frac{\lim_{n\to\infty}(2+1/n)}{\lim_{n\to\infty}(3-5/n)} = \frac{2}{3}$.

If a sequence has the zero limit it is called the *infinitesimal* sequence.

Theorem. If $\{x_n\}$ is an infinitesimal sequence and $\{y_n\}$ is a bounded one, then the sequence $\{x_n \cdot y_n\}$ is infinitesimal.

For instance, let $x_n = 1/n$ (infinitesimal sequence) and $y_n = \sin n$ (bounded sequence). We conclude that the sequence $z_n = \frac{\sin n}{n}$ is infinitesimal, i.e., that $\lim_{n\to\infty} \frac{\sin n}{n} = 0$.

5.1.4 Infinite limit

If the general term x_n of a sequence $\{x_n\}$ increases infinitely when $n \longrightarrow \infty$, then we may speak about an "infinite limit" of the sequence.

Definition. The limit of the sequence $x_1, x_2, \ldots, x_n, \ldots$ is *equal to infinity* if, for every given (arbitrary large) number $E > 0$, there exists a number N such that for all $n > N$ the inequality holds:

$$|x_n| > E. \tag{5.1}$$

In this case we write

$$\lim_{n\to\infty} x_n = \infty, \qquad \text{or} \qquad x_n \to \infty \text{ as } n \to \infty.$$

For example, $\lim_{n\to\infty} \sqrt{n} = \infty$ because $\sqrt{n} > E \Leftrightarrow n > E^2$, hence any natural number $N > E^2$ is appropriate.

If, instead of the inequality (5.1), the other inequality $x_n > E$ is correct, then we write:

$$\lim_{n\to\infty} x_n = +\infty.$$

If, instead of (5.1), the inequality $x_n < -E$ is correct, then we write:

$$\lim_{n\to\infty} x_n = -\infty.$$

For instance,

$$\lim_{n\to\infty} \ln n = +\infty, \qquad \lim_{n\to\infty} \ln 1/n = -\infty.$$

Theorem:

(i) if $\lim_{n\to\infty} x_n = \infty$, then $\lim_{n\to\infty} \frac{1}{x_n} = 0$;

(ii) if $\lim_{n\to\infty} x_n = 0$, then $\lim_{n\to\infty} \frac{1}{x_n} = \infty$.

Here it is assumed that $x_n \neq 0$ for $n > n_0$.

5.1.5 Arithmetic progression and series

A sequence $\{a_n\}$ (finite or infinite) is called the *arithmetic progression* if every next term of it is equal to the previous one plus a constant number d, which is called the *common difference* of the progression:

$$a_{n+1} = a_n + d, \quad n = 1, 2, 3, \ldots$$

The general term formula:

$$a_n = a_1 + (n-1)d.$$

Remark. Often, the term "arithmetic progression" is used for the finite case and "arithmetic series" for the infinite case.

Examples

1. The numbers 3, 5, 7, 9 form an arithmetic progression, where $a_1 = 3,\ d = 2.$

2. The numbers $20.5,\ 15.5,\ 10.5,\ 5.5,\ 0.5,\ -4.5,\ -9.5, \ldots$ form an arithmetic progression, where $a_1 = 20.5,\ \ d = -5.$

The main property of the arithmetic progression:

$$a_n = \frac{1}{2}(a_{n-1} + a_{n+1}),$$

i.e., every term is equal to the arithmetical mean of the previous and the next terms.

The sum $s_n = a_1 + a_2 + \ldots + a_n$ of n terms of the arithmetic progression:

$$s_n = n\frac{a_1 + a_n}{2}, \quad \text{or} \quad s_n = n\left[a_1 + \frac{(n-1)d}{2}\right]. \tag{5.3}$$

In particular, the sum of n first natural numbers:

$$1 + 2 + 3 + \ldots + n = \frac{n(n+1)}{2}.$$

Example. A factory makes instruments. It is known that during the first year the production was 10,000, and it increased by 500 instruments every year. How many instruments had been made over the period of 10 years?

Solution. The annual production forms an arithmetic progression whose first term $a_1 = 10,000$, the common difference $d = 500$, and the number of terms $n = 10$. Using the second formula in (5.3) we find:

$$s_{10} = 10 \cdot (10,000 + 9 \cdot 500/2) = 122,500.$$

5.1.6 Geometric progression and series

A sequence b_n (finite or infinite) is called the *geometric progression* if each of its next terms is equal to the previous one multiplied by a constant number q, which is called the *ratio of the progression*:

$$b_{n+1} = b_n \cdot q, \quad n = 1, 2, 3, \ldots. \tag{5.4}$$

The general term formula:

$$b_n = b_1 \cdot q^{n-1}.$$

Examples

1. The numbers 3, 6, 12, 24, 48 form a geometric progression where $b_1 = 3, \quad q = 2$.

2. The numbers $40, -20, 10, -5, 5/2, -5/4, \ldots$ form a geometric series, where $b_1 = 40, \quad q = -1/2$.

3. Assume that the activity (i.e., the decay speed) of a radioactive isotope of some chemical element decreases continuously in time in such a way that it becomes twice smaller every 60 seconds. The values of activity at the initial moment and $1, 2, 3, \ldots$ minutes later form a geometric progression $b_1, b_2, b_3, \ldots$, whose ratio $q = 0.5$. After 30 minutes the activity will be $b_{31} = b_1 q^{30} = 2^{-30} \cdot b_1$, i.e., approximately 10^9 times less than the initial one.

The main property of the geometric progression:

$$b_n^2 = b_{n-1} \cdot b_{n+1}, \tag{5.5}$$

i.e., every term is equal (in absolute value) to the geometrical mean of the previous and the next terms.

The sum $s_n = b_1 + b_2 + \ldots + b_n$ of n terms of the geometric progression:

$$s_n = \frac{b_1 - q b_n}{1 - q}, \quad \text{or} \quad s_n = b_1 \frac{1 - q^n}{1 - q}. \tag{5.6}$$

Example. A factory makes instruments. During the first year the production was 10,000 instruments, and it increased every year by 5%. How many instruments had been made over the period of 10 years?

Solution. The annual production is a geometric progression with $b_1 = 10{,}000$, $q = 1.05$ and $n = 10$. Using the second formula (5.6) we find:

$$s_{10} = 10{,}000 \cdot \frac{1.05^{10} - 1}{0.05} \approx 10{,}000 \cdot 12.578 = 125{,}780.$$

5.1.7 Infinitely decreasing geometric progression

Infinitely decreasing geometric progression, or *geometric series*, is the infinite geometric progression whose ratio is less (in absolute value) than 1:

$$|q| < 1.$$

In this case Formulas (5.4) – (5.6) are valid and, moreover, there exists a limit for the sum S_n of n terms of the progression:

$$S = \lim_{n\to\infty} S_n = b_1 \frac{1}{1-q}. \tag{5.7}$$

Examples

1. Find the sum of the series (see Section 5.8):

$$S = 1 + \frac{1}{2} + \frac{1}{2^2} + \frac{1}{2^3} + \dots .$$

Using (5.7) we obtain

$$S = 1 \cdot \frac{1}{1-0.5} = 2.$$

2. The *Zeno's paradox*. Can Achilles overtake the tortoise? Achilles, whose speed is V, tries to overtake the tortoise, which is creeping away from him at the speed $v < V$. The initial distance between them is d_1. While Achilles is passing this distance d_1, the tortoise will creep the distance d_2; while Achilles is passing this distance d_2, the tortoise creeps further to the distance d_3, and so on. It looks like Achilles never overtake the tortoise(?).

Explanation. The time necessary for Achilles to pass the distance d_1 is $t_1 = d_1/V$; then $d_2 = t_1 \cdot v = d_1 \cdot v/V$. The time necessary for Achilles to pass the distance d_2 is $t_2 = d_2/V = (d_1/V) \cdot v/V$; then $d_3 = d_1 \cdot (v/V)^2$, and so on.

After n steps we obtain:

$$d_n = d_1 \cdot (v/V)^{n-1}, \quad t_n = (d_1/V) \cdot (v/V)^{n-1}.$$

Now we see that the time intervals $t_1, t_2, t_3, \dots$ form an infinitely decreasing geometric progression whose first term is $b_1 = d_1/V$, and the ratio $q = v/V < 1$. The total time of Achilles's movement is equal to the sum of all time intervals:

$$t = t_1 + t_2 + t_3 + \dots = \frac{d_1}{V} \cdot \frac{1}{1 - v/V} = \frac{d_1}{V - v}.$$

So, Achilles will overtake the tortoise over the finite time t. This result is evident because one can find this time immediately, not using the geometrical progression—but simply with the help of dividing the initial distance d_1 by the relative speed $V - v$.

5.1.8 The number "e"

The sequence

$$x_n = \left(1 + \frac{1}{n}\right)^n$$

has a finite limit $e = 2.7182818\ldots$. It is proved that e is an irrational number (see Section 2.2.3).

5.2 Limit of Function

5.2.1 Definition and theorems

The concept of limit of a function is fundamental in calculus. If values of a given function approximate some number b while the values of the argument approximate a number a, then b is said to be the limit of the function at the point a.

Definition (in "epsilon-delta" language). The number b is called the *limit of the function $f(x)$ at point a* if for any positive (arbitrary small) number ε, there exists a number $\delta > 0$ such that for all x satisfying the inequality

$$0 < |x - a| < \delta, \tag{5.8}$$

the following inequality holds:

$$|f(x) - b| < \varepsilon. \tag{5.9}$$

It is conventional to write

$$\lim_{x \to a} f(x) = b, \quad \text{or} \quad f(x) \to b \quad \text{as} \quad x \to a.$$

The function $f(x)$ is also said to "tend to the limit b as x tends to a".

Remark. In this definition, it is assumed that the function $f(x)$ is defined in a neighborhood of the point a, though possibly not at the point a, and that the point x, which occurs in (5.8) and (5.9), belongs to this neighborhood.

The fact that $f(x) \to b$ as $x \to a$ makes the following geometrical sense: the points of the graph $y = f(x)$ in the xy-plane approximate the point (a, b) as x approximates the point a in the real x-axis (Figure 5.2a). The inequalities (5.8) and (5.9) emphasize the fact that for all x close enough to a, the points of the graph of $f(x)$ are situated inside an arbitraryly chosen narrow strip like $b - \varepsilon < y < b + \varepsilon$ (see Figure 5.2b).

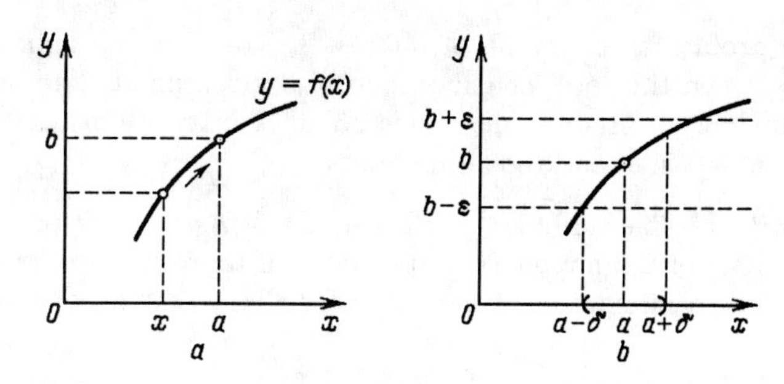

Figure 5.2: Points of the graph of $f(x)$: (a) – approximate the point (a, b) as $x \to a$, (b) – are in strip $b - \varepsilon < y < b + \varepsilon$ for $a - \delta < x < a + \delta$.

Theorems about limits of functions

1. The limit of a given function at a given point, if it exists, is unique.

2. The limit of a constant function is equal to this constant:

$$\lim_{x \to a} C = C.$$

3. The limit of a sum is equal to the sum of limits:

$$\lim_{x \to a}[f(x) + g(x)] = \lim_{x \to a} f(x) + \lim_{x \to a} g(x).$$

4. A constant multiplier may be taken outside of the limit symbol:

$$\lim_{x \to a}[\gamma f(x)] = \gamma \lim_{x \to a} f(x).$$

5. The limit of a product is equal to the product of limits:

$$\lim_{x \to a}[f(x) \cdot g(x)] = \lim_{x \to a} f(x) \cdot \lim_{x \to a} g(x).$$

6. The limit of a quotient is equal to the quotient of limits:

$$\lim_{x \to a} \frac{f(x)}{g(x)} = \frac{\lim_{x \to a} f(x)}{\lim_{x \to a} g(x)}, \quad \text{if} \quad \lim_{x \to a} g(x) \neq 0.$$

Remark. In Theorems 3 – 6, it is assumed that all the limits occurring in the right-hand side of the formulas exist.

Examples: calculation of limits with the help of Theorems 2 – 6.

1. $\lim_{x \to 0} \frac{3+x^2}{7x-4} = \frac{\lim_{x \to 0}(3+x^2)}{\lim_{x \to 0}(7x-4)} = \frac{\lim_{x \to 0} 3 + \lim_{x \to 0} x \cdot \lim_{x \to 0} x}{7 \lim_{x \to 0} x - \lim_{x \to 0} 4} = -0.75;$

2. $\lim_{x \to 1} \frac{x^2-1}{x-1} = \lim_{x \to 1} \frac{(x-1)(x+1)}{x-1} = \lim_{x \to 1}(x+1) = 2.$

5.2.2 One-side limits

In some problems it may be of interest to consider the behavior of a function not in the total neighborhood of a given point (like in Section 5.2.1), but only from one side of this point. Thus there arises a concept of so-called left-side and right-side limits.

Definition. The number b_1 is called the *left-side limit* (or the limit from the left) of a function $f(x)$ at point a, if to every number $\varepsilon > 0$ we can associate a number $\delta > 0$ such that for all x satisfying the inequality

$$a - \delta < x < a \tag{5.10}$$

the following inequality holds:

$$|f(x) - b_1| < \varepsilon. \tag{5.11}$$

It is conventional to write

$$\lim_{x \to a-0} f(x) = b_1, \quad \text{or} \quad f(x) \to b_1 \quad \text{as} \quad x \to a\text{–}0, \quad \text{or} \quad f(a-0) = b_1.$$

In the definition above, it is assumed that the function $f(x)$ is defined in a left half-neighborhood of the point a, i.e., for all $x \in (a - \gamma, a)$ with some $\gamma > 0$, and that x in (5.10) and (5.11) belongs to this half-neighborhood.

The right-side limit (or the limit from the right) is defined in a similar way, and we write

$$\lim_{x \to a+0} f(x) = b_2, \quad \text{or} \quad f(x) \to b_2 \quad \text{as} \quad x \to a+0, \quad \text{or} \quad f(a+0) = b_2.$$

In this case the inequality (5.10) must be replaced by the following one:

$$a < x < a + \delta.$$

Example

$\lim_{x \to 0-0} \frac{|x|}{x} = -1$; $\lim_{x \to 0+0} \frac{|x|}{x} = +1$; hence, $\lim_{x \to 0} \frac{|x|}{x}$ does not exist.

The ordinary limit $\lim_{x \to a} f(x)$ exists if and only if both the left-side and the right-side limits at the given point are existing and equal to each other.

5.2.3 Infinite limit

If the values of a function infinitely increase while the values of its argument approach some point a, we speak about an infinite limit of the function.

Definition. A function $f(x)$ has the *infinite limit* at the point a, if to every number $E > 0$ (arbitrary large), we can associate a number $\delta > 0$ such that for all x satisfying the inequality

$$0 < |x - a| < \delta \tag{5.12}$$

the following inequality holds:

$$|f(x)| > E. \tag{5.13}$$

It is conventional to write

$$\lim_{x \to a} f(x) = \infty \quad \text{or} \quad f(x) \to \infty \quad \text{as} \quad x \to a$$

(read: "the limit of $f(x)$ is equal to infinity" or "$f(x)$ tends to infinity as x tends to a").

In case the function tends to infinity while $x \to a$, the graph $y = f(x)$ is receding infinitely from the x-axis and approaching the vertical straight line $x = a$ (the *vertical asymptote*, see Figure 5.3).

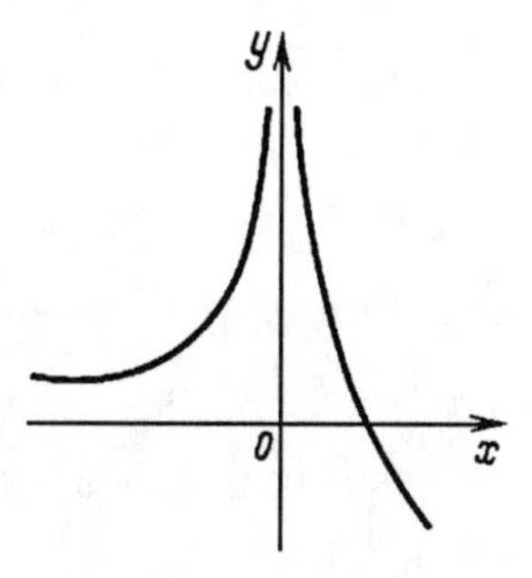

Figure 5.3: Infinite limit at point $x = 0$.

If for those x which satisfy (5.12), the other inequality $f(x) > E$ holds instead of (5.13), then we say that

$$\lim_{x \to a} f(x) = +\infty.$$

Similarly, if the inequality $f(x) < -E$ holds instead of (5.13), we say that

$$\lim_{x \to a} f(x) = -\infty.$$

Left-side and right-side infinite limits are defined by an analogy.

Remark. It may happen that the left-side limit is finite but the right-side limit is infinite (or vice versa), for instance:

$$\lim_{x \to -0} \exp(1/x) = 0, \quad \lim_{x \to +0} \exp(1/x) = +\infty.$$

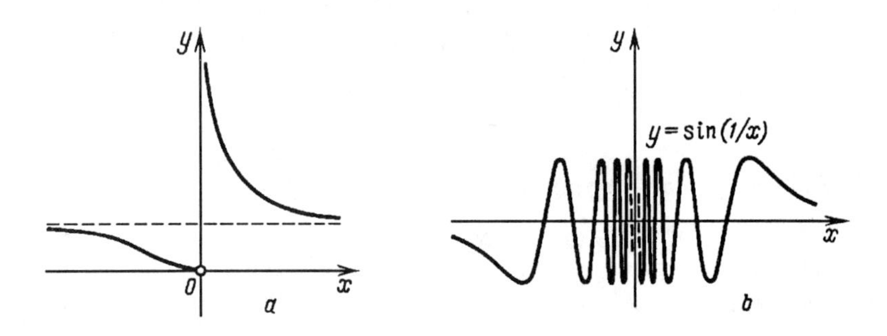

Figure 5.4: (a) – The left limit is 0 and the right one is $+\infty$, (b) – no limit at $x = 0$.

The corresponding graph is plotted in Figure 5.4a. It may also happen that as $x \longrightarrow a$, there are neither finite nor infinite limits. For instance,

$$\lim_{x\to 0} \sin \frac{1}{x}$$

does not exist: while x is approximating zero, the values of the function are oscillating from -1 to $+1$ with an increasing frequency (Figure 5.4b).

Examples: one-side and infinite limits.

1. $\lim_{x\to 0-0} \frac{|x|}{x^3} = -\infty$, $\lim_{x\to 0+0} \frac{|x|}{x^3} = +\infty$;
2. $\lim_{x\to 2-0} \arctan \frac{1}{2-x} = +\frac{\pi}{2}$, $\lim_{x\to 2+0} \arctan \frac{1}{2-x} = -\frac{\pi}{2}$;
3. $\lim_{x\to \pi/2-0} \tan x = +\infty$, $\lim_{x\to \pi/2+0} \tan x = -\infty$.

5.2.4 Limit at infinity

If the values of a function approximate some number while the argument x increases infinitely, we speak about a limit of the function at infinity.

Definition. The number b_1 is called the *limit of a function* $f(x)$ *at positive infinity* if to every number $\varepsilon > 0$ we can associate a number $\Delta > 0$ such that for all x satisfying the inequality

$$x > \Delta, \tag{5.14}$$

the following inequality holds:

$$|f(x) - b_1| < \varepsilon\,.$$

It is conventional to write

$$\lim_{x\to +\infty} f(x) = b_1, \quad \text{or} \quad f(x) \longrightarrow b_1 \ \text{ as } \ x \to +\infty.$$

In this definition it is assumed that $f(x)$ is defined in the vicinity of $+\infty$, i.e., for all $x > \gamma$, where γ is some positive number.

The geometric sense of the fact that $\lim_{x\to+\infty} f(x) = b_1$ is the following one: while x is infinitely receding from the origin of coordinates to the right, the graph of the function is approaching the straight line $y = b_1$ (the ***horizontal asymptote***, see Figure 5.5).

To define the limit of function at negative infinity we replace (5.14) by the inequality $x < -\Delta$. Geometrically, the fact that $f(x) \to b_2$ as $x \to -\infty$, denotes that the graph of $f(x)$ is approaching the straight line $y = b_2$ as x is receding infinitely from the coordinate origin to the left (Figure 5.5).

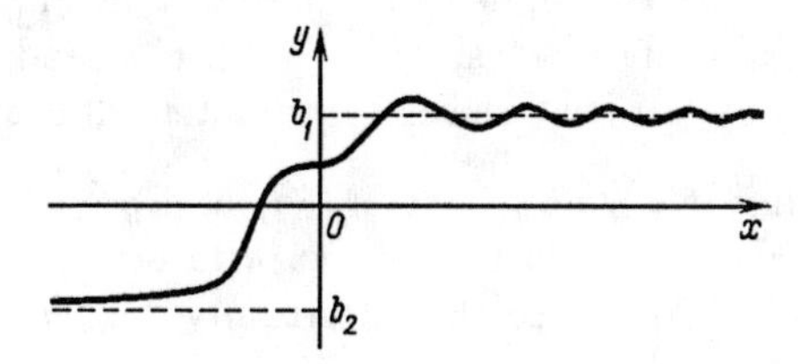

Figure 5.5: Finite limits as $x \to -\infty$ and $x \to +\infty$

Examples

1. $\lim_{x\to\pm\infty} \arctan x = \pm\pi/2$ (see Section 4.3).

2. $\lim_{x\to\pm\infty} x^2 = +\infty$.

3. $\lim_{x\to-\infty} e^x = 0, \quad \lim_{x\to+\infty} = +\infty$.

5.2.5 Rules to calculate the limits of functions

For a continuous function (see Section 5.3), it is easy to calculate its limit at any point which is in the domain of definition: it is enough to substitute the corresponding value of the argument, i.e.,

$$\lim_{x\to x_0} f(x) = f(x_0).$$

In particular, this rule is valid for elementary functions such as x^a, a^x, $\log_a x$, $\sin x$, $\cos x$, and their compositions.

Examples

1. $\lim_{x\to\pi/4} \sin x = \sin(\pi/4) = \sqrt{2}/2$.

2. $\lim_{x\to1-0} \sqrt{1-x^2} = \sqrt{1-1} = 0$.

3. $\lim_{x\to 0}\frac{\sin x}{x} \neq \frac{\sin 0}{0}$, because zero is outside the domain of definition.

If $\lim_{x\to a} f(x) = 0$, then $\lim_{x\to a}\frac{1}{f(x)} = \infty$, i.e., a reciprocal to an infinitesimal value is an infinitely large value; and vice versa, if $\lim_{x\to a} f(x) = \infty$, then $\lim_{x\to a}\frac{1}{f(x)} = 0$, i.e., a value which is reciprocal to an infinitely large value is an infinitesimal one. Here a denotes a finite point or the symbol ∞.

Example

Consider the electrostatic field generated by a point charge Q, and measure its potential $\varphi(r)$ at a distance r from the charge, then $\varphi = Q/r$ (see Section 12.2). Let $r \to 0$, i.e., the observer tends to the charge, then the potential is infinitely increasing: $\varphi \to \infty$. If r tends to infinity (i.e., the observer recedes from the charge), the potential decreases: $\varphi \to 0$.

Expressions like $f(x)/g(x)$, where $f(x)$ and $g(x)$ both tend to zero or to infinity as $x \to a$, are called the *uncertainties* of the kinds $0/0$ or ∞/∞, respectively. To reduce the uncertainty $f(x)/g(x)$ means to find $\lim_{x\to a}\frac{f(x)}{g(x)}$.

Rules to reduce the uncertainties of $0/0$ or ∞/∞ kinds

1. Identical transformations of the expression.

2. Using the "basic limits":

$$\lim_{x\to 0}\frac{\sin x}{x} = 1, \quad \lim_{x\to 0}\frac{\tan x}{x} = 1,$$

$$\lim_{x\to 0}\frac{e^x - 1}{x} = 1, \quad \lim_{x\to 0}\frac{\log_e 1 + x}{x} = 1,$$

$$\lim_{x\to 0}(1+x)^{1/x} = e, \quad \lim_{x\to 0}\frac{(1+x)^a - 1}{x} = a,$$

where "e" is the base of natural logarithms. To read more about e-number see Section 5.1.8.

3. Using the following *L'Hospital's rule:* if there exists a limit (finite or infinite) of the quotient of the derivatives, i.e., $\lim f'(x)/g'(x)$, then there exists a limit of the quotient $f(x)/g(x)$, and these limits are equal to each other:

$$\lim_{x\to a}\frac{f(x)}{g(x)} = \lim_{x\to a}\frac{f'(x)}{g'(x)}.$$

In some cases one must apply L'Hospital's rule several times in succession.

Examples

1. $\lim_{x\to 1} \frac{x^2-1}{\sqrt{x}-1} = \lim_{x\to 1} \frac{(\sqrt{x}-1)(\sqrt{x}+1)(x+1)}{\sqrt{x}-1} =$

$\lim_{x\to 1} (\sqrt{x}+1)(x+1) = 2 \cdot 2 = 4;$

2. $\lim_{x\to\infty} \frac{x^2-x+1}{3x^2-\sqrt{x}-2} = \lim_{x\to\infty} \frac{x^2(1-x^{-1}+x^{-2})}{x^2(3-x^{-3/2}-2x^{-2})} = \frac{1}{3};$

3. $\lim_{x\to 0} \frac{\sin 5x}{\sin 8x} = \lim_{x\to 0} \frac{\sin 5x}{5x} \cdot \frac{8x}{\sin 8x} \cdot \frac{5}{8} =$

$\frac{5}{8} (\lim_{x\to 0} \frac{\sin 5x}{5x})/(\lim_{x\to 0} \frac{\sin 8x}{8x}) = \frac{5}{8};$

4. $\lim_{x\to 0} \frac{\sin 5x}{\sin 8x} = \lim_{x\to 0} \frac{(\sin 5x)'}{(\sin 8x)'} = \lim_{x\to 0} \frac{5\cos 5x}{8\cos 8x} = \frac{5}{8};$

5. $\lim_{x\to\infty} \frac{\ln x}{x} = \lim_{x\to\infty} \frac{1/x}{1} = 0;$

6. $\lim_{x\to 0} \frac{\sin x - x}{x^3} = \lim_{x\to 0} \frac{\cos x - 1}{3x^2} = \lim_{x\to 0} \frac{-\sin x}{6x} = -\frac{1}{6}.$

In the last example, we had to apply twice L'Hospital's rule, since a new uncertainty of the type 0/0 appeared as a result of the first application of the rule.

7. $\lim_{x\to 1} \frac{\sin(x-1)}{(x^2-1)^2} = \lim_{x\to 1} \frac{\cos(x-1)}{4x(x^2-1)} = \frac{1}{4} \lim_{x\to 1} \frac{1}{x^2-1} = \infty.$

The uncertainties of the type $0\cdot\infty$ or $\infty-\infty$ may be first transformed to the type $0/0$ or ∞/∞, and then considered using the rules given above.

Examples

1. $\lim_{x\to 0} (x \ln x) = \lim_{x\to 0} \frac{\ln x}{1/x} = \lim_{x\to 0} (-x) = 0.$

2. $\lim_{x\to 0} [(\frac{1}{\sin x} - \frac{1}{x})\frac{1}{x}] = \lim_{x\to 0} \frac{x-\sin x}{x^2 \sin x} = \lim_{x\to 0} (\frac{x}{\sin x} \cdot \frac{x-\sin x}{x^3}) =$
$\lim_{x\to 0} \frac{x}{\sin x} \cdot \lim_{x\to 0} \frac{x-\sin x}{x^3} = 1 \cdot \lim_{x\to 0} \frac{1-\cos x}{3x^2} = \lim_{x\to 0} (\frac{\sin x}{6x}) = \frac{1}{6}.$

Here we applied L'Hospital's rule twice.

The uncertainty 1^∞ means an expression like $(u(x))^{v(x)}$, where $u(x) \to 1$, $v(x) \to \infty$, as $x \to a$. For instance, $\lim_{x\to 0} (1+x)^{1/x} = e$. One possible way to reduce such an uncertainty is by the application of logarithms:

$$\lim_{x\to a} u(x)^{v(x)} = \exp[\lim_{x\to a} (v(x) \ln u(x))]. \tag{5.15}$$

The uncertainty of the kind 0^0 is also reduced by using logarithms according to Formula (5.15).

Examples

1. Let $P = \lim_{x\to 0} (1+2x)^{\cot x}$. By formula (5.15),

$P = \exp[\lim_{x\to 0} \cot x \cdot \ln(1+2x)],$

$\lim_{x\to 0} \cot x \cdot \ln(1+2x) = \lim_{x\to 0} [\cos x \cdot \frac{\ln(1+2x)}{\sin x}] =$

$(\lim_{x\to 0} \cos x) \cdot \lim_{x\to 0} \frac{\ln(1+2x)}{\sin x} = 1 \cdot \lim_{x\to 0} \frac{2}{(1+2x)\cos x} = 2$

(L'Hospital's rule was applied twice); thus, $P = e^2$.

2. $\lim_{x\to 0+0} x^x = \exp[\lim_{x\to 0+0} (x \ln x)] = e^0 = 1.$

Here we used the fact that $\lim_{x\to 0+0} (x \ln x) = 0$ (see above).

5.3 Continuity of Function. Discontinuities

5.3.1 Continuity

A function $f(x)$ is said to be *continuous at point* a if its limit at this point is equal to its value at this point:

$$\lim_{x\to a} f(x) = f(a). \qquad (5.16)$$

In this definition it is assumed that the function is defined at the point a and also in a neighborhood of this point.

A function $f(x)$ is said to be *continuous from the left* at point a if

$$\lim_{x\to a-0} f(x) = f(a),$$

i.e., the left-side limit (see Section 5.2.2) exists and equals to the value of the function at the point a. Similarly, $f(x)$ is said to be *continuous from the right* at point a if

$$\lim_{x\to a+0} f(x) = f(a).$$

We say that:

• A function is continuous in the interval (a, b) if it is continuous at every point of this interval;

• A function is continuous on the closed interval $[a, b]$ if it is continuous on the (opened) interval (a, b) and continuous from the left at the point b and from the right at the point a.

The graph of a continuous function looks like a continuous line. Any elementary function is continuous in its domain of definition. For instance, polynomials, sine, cosine, and exponential function are continuous in all of the real axis, while the tangent function is continuous for $x \neq \pi/2 + n\pi$ $(n \in \mathbf{Z})$. Any rational function (see Section 3.3.4) is continuous at all points such that denominator of the ratio is not equal to zero.

Properties of the continuous functions

1. The sum and the product of two continuous functions are continuous.

2. A quotient of two continuous functions is continuous at all points where the denominator is not equal to zero.

3. Let $g(x)$ be continuous at a point $x = a$ and $f(u)$ be continuous at the point $u = g(a)$, then the *composition function* $f(g(x))$ is continuous at the point $x = a$. For example, $\sin(x^3+1)$ is continuous for $x \in \mathbf{R}$ since the polynomial $x^3 + 1$ is continuous for $x \in \mathbf{R}$ and $\sin u$ is also continuous for $u \in \mathbf{R}$; here $u = x^3 + 1$.

4. A function that is continuous on a closed interval is bounded.

5. A function that is continuous on a closed interval takes its maximal and minimal values.

6. Let $f(x)$ be continuous on a closed interval $[a, b]$ and $f(a) \cdot f(b) < 0$, i.e., inside the interval the function changes the sign, then the function admits at least one root inside the interval (a, b) (*theorem of Cauchy*).

5.3.2 Discontinuities

If the equality (5.16) is invalid at some point x_0, then we say that the function $f(x)$ has a *discontinuity* (or a gap) at this point. Here it is assumed that the function is defined from the left and the right sides of x_0, i.e., for $a < x < x_0$ and $x_0 < x < b$. The point x_0 is called the *discontinuity point* of the function.

Classification of discontinuities

1. *First type of discontinuity:*

   (i) A function $f(x)$ has a *removable discontinuity* at point x_0 if there exists a finite $\lim_{x\to x_0} f(x)$, but either $f(x_0) \neq \lim_{x\to x_0} f(x)$ or the function $f(x)$ is not defined at the point x_0 (e.g., the points x_1 and x_2 in Figure 5.6*a*). In this case we can remove the discontinuity and obtain a continuous function by putting $f(x_0) = \lim_{x\to x_0} f(x)$; in practice, all removable discontinuities are usually assumed to be removed;

   (ii) A function $f(x)$ has an *unremovable discontinuity* (or a *jump*) at point x_0 if there exist both one-side limits

$$b_1 = \lim_{x\to x_0-0} f(x) \text{ and } b_2 = \lim_{x\to x_0-0} f(x),$$

but $b_1 \neq b_2$ (as the points x_3 and x_4 in Figure 5.6*a*).

2. *Second type of discontinuity:*

   At least one of the two one-side limits is either infinite or does not exist (as the points x_1 and x_2 in Figure 5.6*b*).

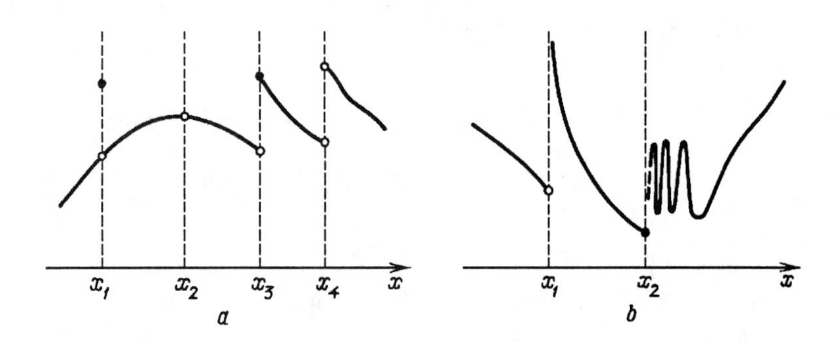

Figure 5.6: Discontinuities of the first type (*a*) and the second type (*b*).

Examples

Discontinuities of the first type:

1. $f(x) = \frac{\sin x}{x}$, a removable discontinuity at $x_0 = 0$ (shown by a continuous line in Figure 5.7).

2. $f(x) = \frac{|x|}{x}$, a jump at $x_0 = 0$ (accented line on Figure 5.7).

Discontinuities of the second type:

1. $f(x) = e^{1/x}$, $x_0 = 0$, $\lim_{x \to 0-0} f(x) = 0$, $\lim_{x \to 0+0} f(x) = +\infty$ (Figure 5.4*a*).

2. $f(x) = \sin(1/x)$, $x_0 = 0$, $\lim_{x \to 0 \pm 0} f(x)$ do not exist (Figure 5.4*b*).

3. $f(x) = \tan x$, $x_0 = \pi/2$, $\lim_{x \to \pi/2 \pm 0} f(x) = \mp\infty$.

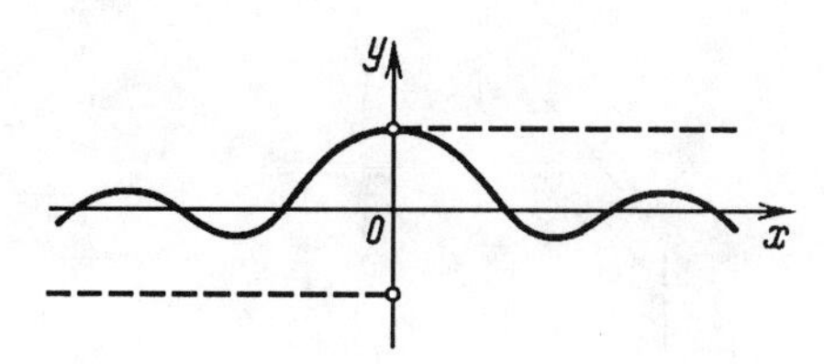

Figure 5.7: Graphs of $y = \frac{\sin x}{x}$ (continuous line) and $y = \frac{|x|}{x}$ (dashed line).

5.4 Derivative. Differentiation Rules

5.4.1 Derivative

The derivative of a function characterizes the variation speed of this function under variation of its argument. *Definition:* the *derivative* $f'(x_0)$ of the function $f(x)$ at point x_0 is the limit, as $x \to x_0$, of the difference quotient of the function and its argument:

$$f'(x_0) = \lim_{\Delta x \to 0} \frac{f(x + \Delta x) - f(x)}{\Delta x}.$$

In this definition the function $f(x)$ is assumed to be defined in a neighborhood of x_0.

Alternative notations are also used for the derivative:

$$\frac{df(x_0)}{dx}, \quad \frac{df(x)}{dx}|_{x_0}, \quad Df(x_0).$$

The geometrical sense of the derivative: $f'(x_0)$ is equal to $\tan\alpha$, where α is the angle between the positive x-semiaxis and the straight line tangential to the graph of $f(x)$ at $x = x_0$. (Figure 5.8*a*).

The physical sense of the derivative: let a variable φ depend on time t, i.e., $\varphi = \varphi(t)$, then the speed of variation of this variable at the moment t_0 is equal to the derivative $\varphi'(t_0)$. For example, let the x-coordinate of a moving point depend on the time as $x = x(t)$, then the instantaneous velocity of the point is $v_x = x'(t)$, and the acceleration is equal to the derivative of the velocity: $a_x(t_0) = v_0'(t_0)$.

A function that has the derivative at a given point is said to be *differentiable* at this point. A function having the derivative at every point of an interval (a, b) is said to be differentiable in this interval. A function differentiable at a point x_0 is necessarily continuous at this point. However, a function may be continuous and have no derivative, for instance, $y = x^{1/3}$ is continuous at $x = 0$, but $y'(0)$ does not exist.

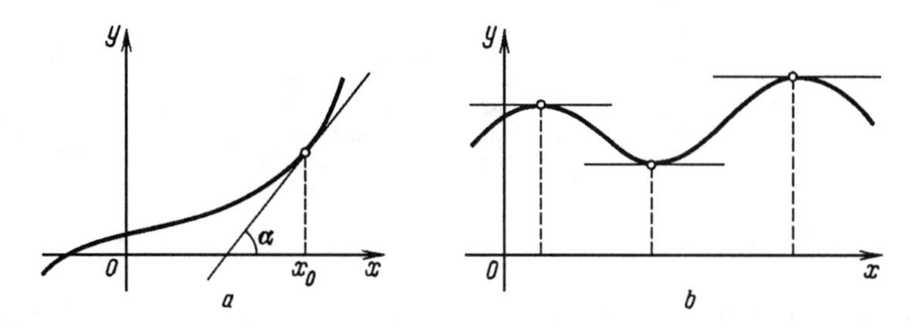

Figure 5.8: (*a*) – Geometrical sense of the derivative. (*b*) – The tangent line is horizontal.

5.4.2 Derivatives of higher order

The value of the derivative depends on the point where it is calculated, hence we can consider the derivative as a function of x and write: $f'(x)$. The derivative of a derivative (if it exists) is called the *second derivative*, or the *derivative of the second order* for the function $f(x)$:

$$f''(x) = (f'(x))'.$$

In a similar way we introduce the third and the fourth derivatives, and so on:

$$f'''(x) = (f''(x))', \quad f^{IV}(x) = (f'''(x))', \ldots.$$

For derivatives of order n, the following symbols are used:

$$f^{(n)}(x), \quad \text{or} \quad \frac{d^n f(x)}{dx^n}, \quad \text{or} \quad D^n f(x).$$

5.4.3 Application of derivatives

In order to investigate extrema (maxima or minima) of a continuous function by the help of derivative, we apply the *Fermat theorem*: if function $f(x)$ at the extremal point x_0 admits a derivative, then the derivative equals zero:

$$f'(x_0) = 0. \tag{5.17}$$

A point x_0 is called the point of (local) *maximun* of a function $f(x)$ if its value at this point is greater than its values in a vicinity of this point:

$$f(x_0) > f(x) \quad \text{for} \quad x \neq x_0.$$

A point x_0 is called the point of (local) *minimum* of $f(x)$ if its value at this point is less than its values in a vicinity of this point:

$$f(x_0) < f(x) \quad \text{for} \quad x \neq x_0.$$

Table 5.1: Investigation of the function near a critical point

Sign of $f'(x)$ for $x < x_0$	Sign of $f'(x)$ for $x > x_0$	Variation of $f(x)$ for $x < x_0$	Variation of $f(x)$ for $x > x_0$	Behavior of $f(x)$ at x_0
$+$	$-$	$\nearrow$	$\searrow$	maximum
$-$	$+$	$\searrow$	$\nearrow$	minimum
$+$	$+$	$\nearrow$	$\nearrow$	inflection
$-$	$-$	$\searrow$	$\searrow$	inflection

The geometrical sense of (5.17): a straight line tangential to the graph of $f(x)$ at the point of maximum or minimum, is parallel to the x-axis (Figure 5.8*b*).

The points where the derivative is equal to zero or does not exist, are called the *critical points* of the function. In order to find extrema of a given function, we find all its critical points and determine the sign of the derivative in a vicinity of such a point (Table 5.1). In some cases, however, it may be more convenient to use the following rule applying the second derivative (of course, if it exists at the critical point and is nonzero):

- If $f''(x_0) > 0$, then x_0 is a minimum point for $f(x)$;
- If $f''(x_0) < 0$, then x_0 is a maximum point for $f(x)$.

5.4.4 Inflection

Consider a curve line in the xy-plane and a straight line tangent to the curve at a given point P. The point P is called the *point of inflection* for the curve if, in a vicinity of this point, the curve is split by P into two parts: one part is situated from one side of the tangent line while the other part is situated from the other side (see Figure 5.9). At the inflection point, the convexity of the curve is changed by concavity.

Inflections of the graph of a function are called inflections of this function, the x-coordinate of a point of inflection is also called the *inflection point* of the function. In order to find inflections of a given function we

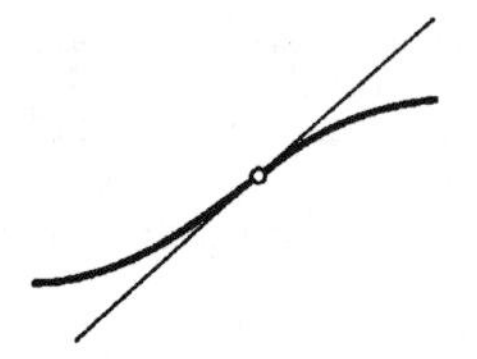

Figure 5.9: Point of inflection.

search the points "suspicious to admit inflection": at such points, $f''(x)$ equals zero or does not exist. To answer the question "is there an inflection at the suspicious point x_*, or not?", we check the sign of $f''(x)$ in a vicinity of x_*: if $f''(x)$ has different sign from the left and the right sides of x, then x_* is the inflection point, but if $f''(x)$ is of the same sign from both sides of x_*, there is no inflection.

Example: $f(x) = x^3 - 3x + 1$. Here $f'(x) = 3x^2 - 3$, (see Section 5.4.5 below). Critical points are determined as the roots of the equation $3x^2 - 3 = 0$, whence $x_1 = -1$, $x_2 = 1$. Further, $f''(x) = 6x$. Since $f''(-1) = -6 < 0$ and $f''(1) = 6 > 0$, x_1 is a maximum point, x_2 is a minimum point. There is a point $x_3 = 0$, suspicious to admit inflection; from the fact that $f''(x) < 0$ for $x < 0$ and $f''(x) > 0$ for $x > 0$, it follows that the point $x_3 = 0$ is an inflection point indeed. The function is graphically represented in Figure 5.10.

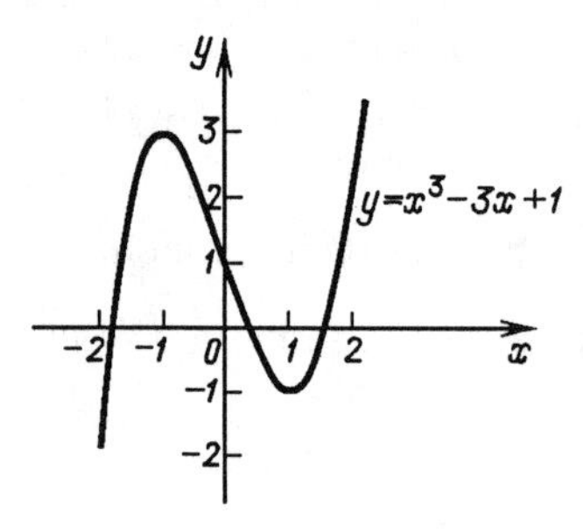

Figure 5.10: The graph of $y = x^3 - 3x + 1$.

5.4.5 Differentiation rules

Calculation of a derivative, or differentiation, is performed by using the following rules:

1. $C' = 0$ —the derivative of a constant is zero.

2. $(f(x) + g(x))' = f'(x) + g'(x)$ —the derivative of a sum is equal to the sum of derivatives.

3. $(\gamma f(x))' = \gamma f'(x)$ —a constant multiplier may be taken outside of the differential symbol.

4. $(f(x) \cdot g(x))' = f'(x)\, g(x) + f(x)\, g'(x)$ —to find the derivative of a product we differentiate the factors in turn.

5. $(\frac{f(x)}{g(x)})' = \frac{f'(x)g(x)-f(x)g'(x)}{g^2(x)}$, in particular:

$$(\tfrac{1}{g(x)})' = -\tfrac{1}{g^2(x)} \cdot g'(x).$$

6. $[f(g(x))]' = f'(g) \cdot g'(x)$ —the "chain rule" for differentiation of a composite function.

Here it is assumed that the derivatives on the right side of the equalities in Rules 2 through 6 exist.

Derivatives of the elementary functions are represented in Table 5.2.

Examples

1. $f(x) = \sin\sqrt{3x+1}$,

$$f'(x) = \cos\sqrt{3x+1} \cdot (\sqrt{3x+1})' = \tfrac{\cos\sqrt{3x+1}}{2\sqrt{3x+1}} \cdot (3x+1)' = \tfrac{3\cos\sqrt{3x+1}}{2\sqrt{3x+1}}.$$

2. $f(x) = x\ln x + \arctan(e^x/x)$,

$$f'(x) = \ln x + x \cdot \tfrac{1}{x} + \tfrac{1}{1+e^{2x}/x^2} \cdot \tfrac{xe^x - e^x}{x^2} = \ln x + 1 + \tfrac{(x-1)\,e^x}{x^2+e^{2x}}.$$

3. $f(x) = \sin\sin\sin x$,

$$f'(x) = (\cos\sin\sin x) \cdot (\cos\sin x) \cdot \cos x.$$

Along with Rules 1 through 6, we applied here the table of derivatives.

5.5 Some Differential Equations

5.5.1 Introduction

An equation that contains derivatives of the unknown function is called a *differential equation*. Many problems in theoretical and mathematical physics, mechanics, electromagnetic theory, wave processes, control theory, biology, and so on may be set in terms of differential equations. In what follows, we consider only the *ordinary* differential equations, i.e., equations for a function of one variable.

Table 5.2: Table of derivatives

$f(x)$	$f'(x)$	$f(x)$	$f'(x)$	$f(x)$	$f'(x)$
x^n	nx^{n-1}	$\log_e x$	$\frac{1}{x}$	$\sinh x$	$\cosh x$
$\sqrt{x}$	$\frac{1}{2\sqrt{x}}$	$\log_a x$	$\frac{1}{x \log_e a}$	$\cosh x$	$\sinh x$
$\frac{1}{x}$	$-\frac{1}{x^2}$	$\sin x$	$\cos x$	$\arcsin x$	$\frac{1}{\sqrt{1-x^2}}$
e^x	e^x	$\cos x$	$-\sin x$	$\arccos x$	$-\frac{1}{\sqrt{1-x^2}}$
a^x	$a^x \log_e a$	$\tan x$	$\frac{1}{\cos^2 x}$	$\arctan x$	$\frac{1}{1+x^2}$
C	0	$\cot x$	$-\frac{1}{\sin^2 x}$	$\text{arccot}\ x$	$-\frac{1}{1+x^2}$

A first-order differential equation is of the general form $F(x, y, y') = 0$. Here F is a given function of three variables x, y, y'. The variable x is the *independent variable*, $y = y(x)$ is the *unknown* (desired) *function* of x; $y' = y'(x)$ is its derivative.

The *solution* of the differential equation is a function $y(x)$ such that if we substitute it into the equation, we obtain a true equality $F(x, y(x), y'(x)) = 0$ for all x over a given interval. A *general solution* is a family (a set) of solutions that contain an arbitrary constant as a parameter, so that any solution can be obtained by choosing a correspondent value of this constant.

A differential equation of the n-th order is of general form $F(x, y, y', y'', \ldots, y^{(n)}) = 0$, and its general solution depends on n independent arbitrary constants.

5.5.2 Some simple differential equations

1. $y' = 0$.

General solution: $y(x) = C$, where C is an arbitrary constant, i.e., for any choice of the number C, the equation is satisfied by $y = C$. As an example, consider a point moving in such a way that its velocity is $v(t)$, t is the time. Let $v(t)$ satisfy the differential equation $v'(t) = 0$, then $v(t) = C$, that is, the point is moving with a constant velocity.

2. $y' = g(x)$, where $g(x)$ is a given function.

The general solution is of the form $y(x) = \int g(x)\,dx + C$, where C is an arbitrary constant (see Section 5.6). For example, let $v_x(t)$ be the x-projection of velocity of a moving point, then its x-coordinate satisfies the differential equation $x'(t) = v_x(t)$, whence:

$$x(t) = \int v_x(t)\,dt + C.$$

The question of how to determine the constant C is answered in Section 5.5.3.

3. $y' = ay$, where a is a given constant.

Such a differential equation physically means that the speed of variation of the variable $y(t)$ is proportional to its value at instant t. General solution: $y(x) = C\exp(ax)$, where C is an arbitrary constant. For $a < 0$, $y(x) \to 0$ as $x \to +\infty$ (damping process), and for $a > 0$, $y(x) \to +\infty$ as $x \to +\infty$ (avalanche process).

4. $y'' + \omega^2 y = 0$, where ω is a given constant.

This equation is met in the oscillation theory (see Section 16.3.1). For instance, small oscillations of a mathematical pendulum are well described by such an equation.

The general solution may be written in alternative forms:

$$y(x) = A\cos\omega x + B\sin\omega x, \quad \text{or}$$
$$y(x) = C\exp(i\omega x) + D\exp(-i\omega x), \quad \text{or}$$
$$y(x) = E\sin(\omega x + \phi).$$

Here $A,\ B,\ C,\ D,\ E,\ \phi$ are arbitrary constants.

5. $y'' - a^2 y = 0$, where a is a given constant.

General solution:

$$y(x) = A\exp(ax) + B\exp(-ax), \quad \text{or}$$
$$y(x) = C\cosh(ax) + D\sinh(ax),$$

where $A,\ B,\ C,\ D$ are arbitrary constants. Solutions are of monotonic character: damping or avalanche increasing as x tends to infinity.

6. $y'' = 0$.

General solution: $y = Ax + B$, where A and B are arbitrary constants.

5.5.3 Some problems related to differential equations

• *Cauchy problem*. For first-order differential equation $F(x, y, y') = 0$, an initial condition $y(x_0) = y_0$ is prescribed; for second-order equation $F(x, y, y', y'') = 0$, two initial conditions $y(x_0) = y_0$ and $y'(x_0) = y_1$ are prescribed. Here x_0 is a given point on the x-axis, y_0 and y_1 are given numbers.

Example. The process of radioactive decay is described by the differential equation $N'(t) = -\beta N(t)$, where β is a constant characterizing the decay speed, $N(t)$ is the total quantity of radioactive nuclei at the instant t. Let N_0 be the quantity of radioactive nuclei at the moment $t_0 = 0$. At the time t, we have $N(t) = C\exp(-\beta t)$, the constant C may be found from the condition $N(0) = N_0$: $N_0 = C\exp(0)$, i.e., $C = N_0$. Hence, the answer is $N(t) = N_0\exp(-\beta t)$ (see Section 18.6.2).

• *Boundary value problem.* For second-order differential equation $F(x, y, y', y'') = 0$, two boundary value conditions at the end points of a given interval $[a, b]$ are prescribed: $y(a) = A$, $y(b) = B$, where A and B are given numbers.

Example. A problem of determinating the equilibrium form of a homogeneous heavy chain whose ends are fixed at the same altitude H, may be reduced to a problem for the equation $y'' - a^2y = 0$ with the boundary value conditions $y(-b) = H$, $y(b) = H$. Here the distance between the end points of the chain is assumed to be given and equal to $2b$, the parameter a is a given constant number.

Solution. Use the general solution $y(x) = C_1\cosh(ax) + C_2\sinh(ax)$; the conditions at $x = -b$ and $x = b$ give the system for C_1 and C_2:

$$H = C_1\cosh(-ab) + C_2\sinh(-ab),$$
$$H = C_1\cosh(ab) + C_2\sinh(ab).$$

Since $\sinh(-p) = -\sinh(p)$, $\cosh(-p) = \cosh(p)$ (see Section 2.8.2), we obtain $C_2 = 0$, $C_1 = H/\cosh(ab)$.

The answer: $y = H\cdot\cosh(ax)/\cosh(ab)$, this curve is called the "catenary". The "deflection" of the chain (Figure 5.11) is equal to

$$\Delta h = h - y(0) = h(1 - 1/\cosh(ab)).$$

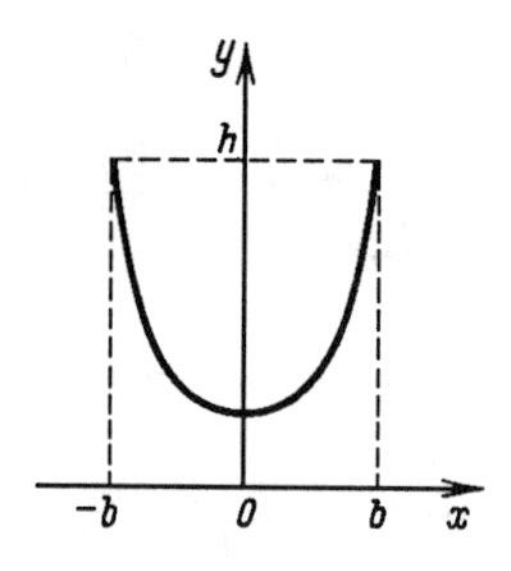

Figure 5.11: Catenary.

5.6 Antiderivative. Indefinite Integral

5.6.1 Antiderivative

In many problems of physics and mathematics we meet the following question: given the derivative of some function, how do we find this function? In connection with this question, we introduce the concepts of an antiderivative and an indefinite integral.

Definition. If two functions, $f(x)$ and $F(x)$, are related by the relation $F'(x) = f(x)$ for $x \in (a, b)$, we call $F(x)$ the *antiderivative*, or *primitive*, of $f(x)$ over (a, b). For instance, $\sin x$ is an antiderivative of $\cos x$ for all $x \in \mathbf{R}$, $\sqrt{x}$ has $(2/3)x^{3/2}$ as an antiderivative for $x > 0$. Thus, calculation of an antiderivative is the inverse operation with respect to differentiation.

Any continuous function admits an antiderivative. The antiderivative is always continuous. In this section, we assume continuity of all functions to be met.

Basic properties of antiderivatives

1. If $F(x)$ is an antiderivative of $f(x)$, then for every constant C, the function $F(x) + C$ is also an antiderivative of $f(x)$.
2. Any two antiderivatives $F_1(x)$ and $F_2(x)$ of a given function differ from each other by a constant:

$$F_1(x) - F_2(x) = C.$$

5.6.2 Indefinite integral

The *indefinite integral* of a function $f(x)$ is, by definition, the general expression $F(x) + C$ for all antiderivatives of the given function $f(x)$. Here $F(x)$ is some antiderivative of $f(x)$ and C is an arbitrary constant. The indefinite integral is denoted by the symbol $\int f(x)\,dx$ (read "integral of the function $f(x)$ differential x"). Obviously,

$$\int f(x)\,dx = F(x) + C, \quad (\int f(x)\,dx)' = f(x).$$

The function $f(x)$ is called the *integrand* (or the function which is to be integrated), x is called the *integration variable*. Indefinite integrals of some elementary functions are represented by the Table 5.3.

Properties of indefinite integrals

1. $\int 0\,dx = C$.
2. $\int (f(x) + g(x))\,dx = \int f(x)\,dx + \int g(x)\,dx$ —the integral of a sum of functions is equal to the sum of their integrals.
3. $\int \gamma f(x)\,dx = \gamma f(x)\,dx$ —a constant multiplier may be taken outside of the integral symbol.

Table 5.3: Table of indefinite integrals

$f(x)$	$\int f(x)\,dx$	$f(x)$	$\int f(x)\,dx$
1	$x+C$	$\cos x$	$\sin x + C$
$x^n\ (n \neq -1)$	$\frac{1}{n+1}x^{n+1}+C$	$\frac{1}{a^2+x^2}$	$\frac{1}{a}\arctan\frac{x}{a}+C$
$\frac{1}{x}$	$\log_e\lvert x\rvert+C$	$\frac{1}{a^2-x^2}$	$\frac{1}{2a}\log_e\lvert\frac{a+x}{a-x}\rvert+C$
e^x	e^x+C	$\frac{1}{\sqrt{a^2-x^2}}$	$\arcsin\frac{x}{a}+C$
a^x	$a^x/\log_e a+C$	$\frac{1}{\sqrt{x^2+b}}$	$\log_e\lvert x+\sqrt{x^2+b}\rvert+C$
$\sin x$	$-\cos x+C$		

The following formulas may be of use in integration of the functions:

• Integration by parts formula:

$$\int u'(x)v(x)\,dx = u(x)v(x) - \int u(x)v'(x)\,dx,$$

it may be also written as

$$\int f(x)g(x)\,dx = F(x)g(x) - \int F(x)g'(x)\,dx, \tag{5.18}$$

where $F(x)$ is some antiderivative of $f(x)$;

• change of variable formula: if $x = q(t)$, then

$$\int f(x)\,dx = \int f(q(t))q'(t)\,dt. \tag{5.19}$$

Examples

1. $\int x \ln x\,dx$. Let us put $f(x) = x$, $g(x) = \ln x$,

   then $F(x) = 0.5x^2$, $g'(x) = 1/x$. Using (5.18), we obtain

   $\int x\ln x\,dx = 0.5x^2\ln x - \int 0.5x\,dx = 0.5x^2\ln x - 0.25x^2 + C.$

2. $\int f(ax+b)\,dx$, where a and b are constants, $a \neq 0$. Let us change the integration variable: $t = ax + b$, or $x = (t-b)/a$. Here $q(t) = (t-b)/a$, $q'(t) = 1/a$. From (5.19):

$$\int f(ax+b)\,dx = \frac{1}{a}\int f(t)\,dt.$$

For instance, $\int \sin(3x+7)\,dx = \frac{1}{3}\int \sin t\,dt = -\frac{1}{3}\cos t + C = -\frac{1}{3}\cos(3x+7) + C$.

3. $\int x^3 \exp(x^4)\,dx = \frac{1}{4}\int \exp t\,dt = \frac{1}{4}e^t + C, \quad t = x^4$.

Generally speaking, the integration is a much more difficult problem than the differentiation. The derivative of any elementary function may be determined in the form of elementary function, but integrals of elementary functions are expressible in terms of elementary functions for some special kinds of integrands only. For example, the integral

$$\int \frac{\sin x}{x}\,dx$$

is inexpressible in terms of elementary functions.

5.7 Definite Integral and its Applications

5.7.1 Definition

We come to the concept of definite integral when trying to measure the area of some plane domain, e.g., the area under the graph of a function (see Figure 5.12a, the domain $ABCD$). Namely, let us split the domain $ABCD$ (Figure 5.12a) into a set of small zones, replace each zone (approximately) by a rectangle, and calculate a sum of the areas of all the rectangles. Thus we obtain an approximate value for the area of the domain (see also Section 10.3).

Definition. Let a function $f(x)$ be defined on the interval $[a, b]$. Split $[a, b]$ arbitrarily into n partial intervals $\Delta_1, \Delta_2, \ldots, \Delta_n$ (Figure 5.12b) and take, in each partial interval, an arbitrary point $t_i \in \Delta_i$ $(i = 1, 2, \ldots, n)$, then make up the *integral sum*:

$$\sum_{i=1}^{n} f(t_i)\,\Delta x_i,$$

where Δx_i denotes the length of the partial interval Δ_i. Now let n increase infinitely, $n \to \infty$, and the maximal length of A tend to zero:

$$\max_{1<i<n} \Delta x_i \to 0.$$

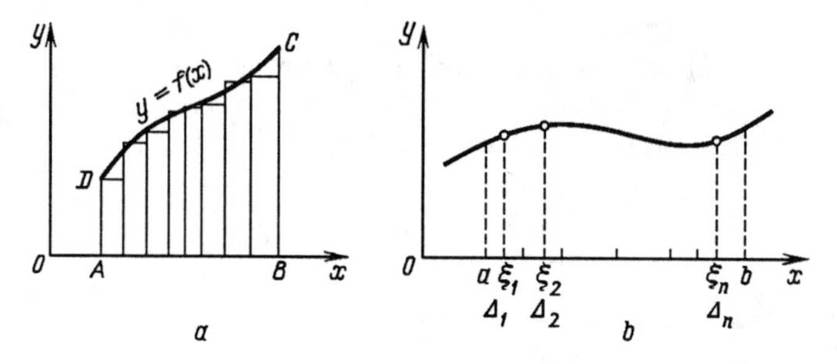

Figure 5.12: a – The figure $ABCD$ is approximated by a sum of small rectangles. b – The definition of the definite integral.

The limit of the integral sum (if this limit exists) is called the *definite integral* of the function $f(x)$ over the interval $[a, b]$:

$$\int_a^b f(x)\,dx = \lim_{n\to\infty} \sum_{i=1}^{n} f(t_i)\Delta x_i .$$

The symbol $\int_a^b f(x)\,dx$ is to be read as "summation from a to b of $f(x)$ differential x". The function $f(x)$ is the *integrand*, x the *variable of integration*, a the *lower bound*, b the *upper bound*, and $[a, b]$ the *interval of integration* (summation).

Remarks.

• In the definition above we imply that the integral sum has the same limit whatever are the way of splitting the interval $[a, b]$ and of choosing the points $t_i \in \Delta_i$.

• The definite integral depends on a, b , and $f(x)$, and it does not depend upon the variable x, so we can take any other letter to denote the variable of integration:

$$\int_a^b f(x)\,dx = \int_a^b f(y)\,dy = \int_a^b f(s)\,ds, \quad \text{etc.}$$

The definite integral of any continuous function exists. In the sequel, we assume all the integrands to be continuous (or to have only a finite number of discontinuity points of the 1st type, see Section 5.3.2).

The geometrical meaning of the definite integral: if $f(x) \geq 0$, then $\int_a^b f(x)\,dx$ is numerically equal to the area of a curvilinear trapezoid (Figure 5.13a); if $f(x)$ changes the sign in (a, b), then $\int_a^b f(x)\,dx$ is equal to the "algebraic sum" of the areas of corresponding curvilinear trapezia: we assign the sign (+) to those parts of the graph that are above the x-axis, and the sign (−) to those parts that are below the x-axis (Figure 5.13b).

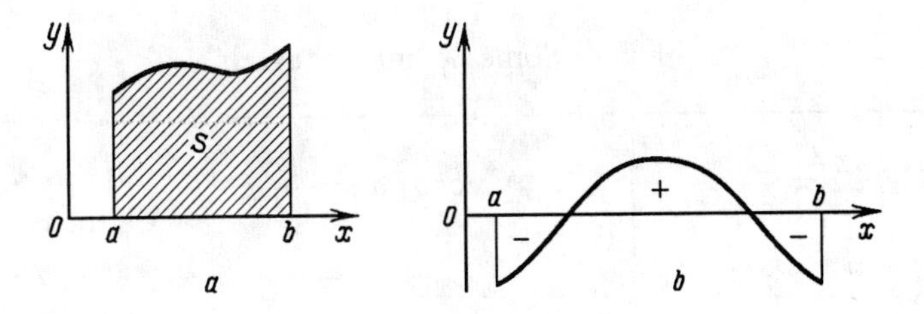

Figure 5.13: The definite integral is equal to the area of a curvilinear trapezoid (a), or to the algebraic sum of areas of such trapezoids (b).

Some physical applications of the definite integral: let $f(x)$ denote the x-projection of a force, then

$$\int_a^b f(x)\,dx = A$$

is the work done by this force in displacement of a point along the x-axis over the distance $[a, b]$; let $v(t)$ be the velocity of a moving point and $v_x(t)$ its x-projection, then the value

$$\int_{t_1}^{t_2} v_x(t)\,dt = x_2 - x_1$$

is equal to the x-projection of the displacement during the time interval from t_1 to t_2 (Section 13.1.1).

5.7.2 Properties of the definite integral

1. $\int_a^b 0 \cdot dx = 0, \quad \int_a^b 1 \cdot dx = b - a;$

2. $\int_a^b f(x)\,dx = -\int_b^a f(x)\,dx, \quad \int_a^a f(x)\,dx = 0;$

3. $\int_a^b (f(x) + g(x))\,dx = \int_a^b f(x)\,dx + \int_a^b g(x)\,dx$ —the integral of a sum of functions is equal to the sum of their integrals;

4. $\int_a^b \gamma f(x)\,dx = \gamma \int_a^b f(x)\,dx$ —a constant multiplier is taken outside of the integral symbol;

5. $\int_a^b f(x)\,dx = \int_a^c f(x)\,dx + \int_c^b f(x)\,dx$ —the definite integral is additive with respect to the interval;

6. $\left|\int_a^b f(x)\,dx\right| \le (b - a) \cdot \max_{a \le x \le b} |f(x)|.$

Definite integrals may be calculated applying the following *Newton–Leibnitz formula*:

Table 5.4: Some definite integrals

$\int_0^\infty \frac{\sin x}{x}\,dx = \frac{\pi}{2}$	$\int_0^\infty e^{-x^2}\cos bx\,dx = \frac{\sqrt{\pi}}{2}e^{-b^2/4}$
$\int_0^\infty \frac{\sin x}{\sqrt{x}}\,dx = \sqrt{\frac{\pi}{2}}$	$\int_0^{\pi/2}\log_e \tan x\,dx = 0$
$\int_0^\infty \frac{\cos x}{\sqrt{x}}\,dx = \sqrt{\frac{\pi}{2}}$	$\int_0^\infty e^{-x^2}\log_e x\,dx = -C \approx -0.5772$, where C is Euler's constant
$\int_{-\infty}^\infty \sin x^2\,dx = \sqrt{\frac{\pi}{2}}$	$\int_0^{\pi/2}(\sin x)^{2k}\,dx = \frac{(2k-1)!!}{(2k)!!}\cdot\frac{\pi}{2}$
$\int_{-\infty}^\infty \cos x^2\,dx = \sqrt{\frac{\pi}{2}}$	$\int_0^{\pi/2}(\cos x)^{2k}\,dx = \frac{(2k-1)!!}{(2k)!!}\cdot\frac{\pi}{2}$
$\int_0^\infty e^{-a^2x^2}\,dx = \frac{\sqrt{\pi}}{2a}$	$\int_0^{\pi/2}(\sin x)^{2k+1}\,dx = \frac{(2k)!!}{(2k+1)!!}$
$\int_0^\infty x^n e^{-ax}\,dx = \frac{n!}{a^{n+1}}$, $a > 0,\ n \in \mathbf{N}$	$\int_0^{\pi/2}(\cos x)^{2k+1}\,dx = \frac{(2k)!!}{(2k+1)!!}$, $k \in \mathbf{N}$, about $n!$ and $n!!$ see Section 6.1.4

$$\int_a^b f(x)\,dx = F(b) - F(a), \quad \text{or} \quad \int_a^b f(x)\,dx = [F(x)]_a^b,$$

where $F(x)$ is any antiderivative of $f(x)$. The symbol $[F(x)]_a^b$ (in some books: $F(x)|_a^b$ or $|_a^b F(x)$) is to be read as "the double substitution $F(x)$ from a to b", or "the substitution $F(x)$ in the bounds a and b".

In some cases, definite integrals can be calculated without the application of the Newton–Leibnitz formula. Various special methods (such as complex variable theory, differentiation or integration with respect to a parameter, and so on) are known to give exact values of integrals in such cases. Some definite integrals are given in Table 5.4. About the approximate calculation of definite integrals see Section 10.3.

5.7.3 Improper integrals: infinite interval

- Improper integral of a continuous function over a semi-axis.

Definition:

$$\int_a^\infty f(x)\,dx = \lim_{b\to+\infty}\int_a^b f(x)\,dx.$$

In the case where a finite limit exists, we say that $\int_a^\infty f(x)\,dx$ is convergent; in the opposite case, it is divergent. Instead of the symbol $\int_a^\infty f(x)\,dx$ it is also conventional to write $\int_a^{+\infty} f(x)\,dx$.

Examples

1. $\int_0^\infty e^{-x}\,dx = \lim_{b\to\infty} \int_0^b e^{-x}\,dx = \lim_{b\to\infty}(e^0 - e^{-b}) = 1$, the integral is convergent.

2. $\int_2^\infty \frac{dx}{x} = \lim_{b\to\infty} \int_2^b \frac{dx}{x} = \lim_{b\to\infty}(\ln b - \ln 2) = \infty$, the integral is divergent.

The geometric meaning of the improper integral of a positive function: the integral is equal to the area of an infinite curvilinear trapezoid (Figure 5.14). The integral is convergent in the case that the curve line $y = f(x)$ is rapidly enough coming close to the x-axis, as $x \to \infty$, and divergent in the opposite case.

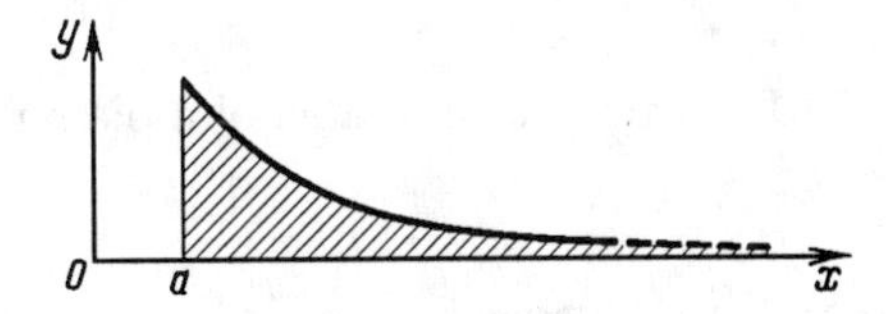

Figure 5.14: Finite area of an infinite figure.

A condition sufficient for convergence: let $|f(x)| \le M/x^\alpha$ for $x \ge x_0$, with $\alpha > 1$ and some $x_0 > 0$, then the integral is convergent.

• Improper integral over all the x-axis is defined similarly:

$$\int_{-\infty}^{+\infty} f(x)\,dx = \lim_{b\to+\infty} \left(\lim_{a\to-\infty} \int_a^b f(x)\,dx \right).$$

Improper integrals are met in some physical problems. For instance, let a point electric charge q be transferred from a distance r_0 to infinity by action of the electric field due to a point charge Q, then the work of the electric field (Section 15.1.2) may be found as

$$\frac{Qq}{4\pi\epsilon_0} \int_{r_0}^{\infty} \frac{dr}{r^2} = \frac{Qq}{4\pi\epsilon_0 r_0}.$$

5.7.4 Improper integrals: discontinuous function

Let a be a point of discontinuity of $f(x)$, then the improper integral of this function over the interval (a, b) is defined as follows:

$$\int_a^b f(x)\,dx = \lim_{c\to a+0} \int_c^b f(x)\,dx.$$

Example

$\int_0^1 \ln x\,dx = \lim_{c\to 0+0} \int_c^1 \ln x\,dx = \lim_{c\to 0+0}([x\ln x]_c^1 - \int_c^1 x\cdot \frac{1}{x}\,dx) =$ $\lim_{c\to 0+0}(-(c\ln c) - 1 + c) = -1 - \lim_{c\to 0+0}(c\ln c) = -1.$

Here we used the integration by parts formula and the well-known limit: $\lim_{x\to 0+0}(x\ln x) = 0$ (see Section 5.2.5).

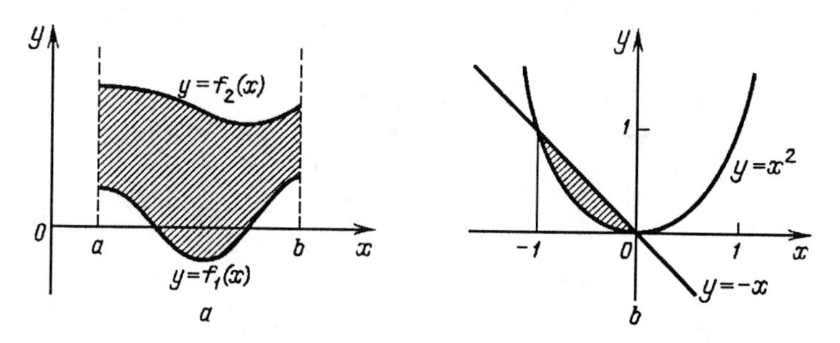

Figure 5.15: The calculation of area of a planar figure.

5.7.5 Some applications of the definite integral

1. Let S denote the area of a planar domain bounded by the curve lines $y = f_1(x)$ and $y = f_2(x)$, where $f_1(x) \le f_2(x)$, $a \le x \le b$, and also by the straight lines $x = a$ and $x = b$ (Figure 5.15*a*), then:

$$S = \int_a^b [\,f_2(x) - f_1(x)\,]\,dx.$$

For example, the parabola $y = x^2$ and the straight line $y = -x$ bound together a domain (Figure 5.15*b*) whose area is:

$$\int_{-1}^{0} (-x - x^2)\,dx = -\left[\frac{x^2}{2} + \frac{x^3}{3}\right]_{-1}^{0} = \frac{1}{2} - \frac{1}{3} = \frac{1}{6}.$$

2. The length L of a curve $y = f(x)$ over the interval (a, b):

$$L = \int_a^b \sqrt{1 + [\,f'(x)\,]^2}\,dx.$$

For example, the length of the "catenary" $y = H\cdot\frac{\cosh(ax)}{\cosh(ab)}$ (see Section 5.5.3) in the case $aH = \cosh(ab)$ is equal to

$$\int_{-b}^{b} \sqrt{1+\sinh^2 ax}\,dx = \int_{-b}^{b} \cosh(ax)\,dx = \frac{1}{a}[\sinh ax\,]_{-b}^{b} = \frac{2}{a}\sinh(ab).$$

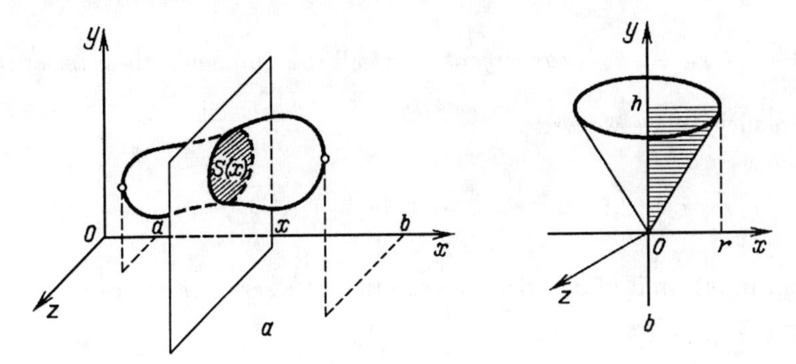

Figure 5.16: The calculation of volume of a spatial body.

3. Let V denote the volume of a spatial body whose cross-sectional area $S(x)$ is known for every $x \in [a,b]$, see Figure 5.16*a*. The following formula is valid:

$$V = \int_a^b S(x)\, dx.$$

In particular, the volume of a body of revolution, obtained by a rotation around the x-axis of a curvilinear trapezoid $0 \leq y \leq f(x)$, $a \leq x \leq b$, is equal to

$$V = \pi \int_a^b [f(x)]^2\, dx.$$

For example, the volume of a cone (Figure 5.16*b*) obtained by a rotation around the y-axis of a triangle whose vertexes are $(0,0)$, $(0,h)$, (r,h), is equal to

$$V = \pi \int_0^h (\frac{r}{h}y)^2\, dy = \frac{\pi r^2}{h^2} \cdot \frac{1}{3}h^3 = \frac{1}{3}\pi r^2 h.$$

4. The area of the surface obtained by a rotation around the x-axis of a planar curve line $y = f(x)$ over $x \in (a,b)$ is equal to

$$F = 2\pi \int_a^b |f(x)|\sqrt{1 + [f'(x)]^2}\, dx.$$

5.8 Some Information about Series

5.8.1 Convergence of a series

Let x_n be an infinite numerical sequence (see Section 5.1). Consider sums of the form

$$s_n = x_1 + x_2 + \ldots + x_n,$$

we call them the *partial sums*. If a finite limit of the sequence of the partial sums exists, $S = \lim_{n\to\infty} s_n$, then we say that the *series* $x_1 +$

$x_2 + \ldots + x_n + \ldots$ *is convergent*, and call the number S the *sum of the series.*

In such a case we write:

$$S = x_1 + x_2 + \ldots + x_n + \ldots = \sum_{n=1}^{\infty} x_n.$$

If no (finite) limit of s_n exists, we say that the *series is divergent.*

Examples

1. The *infinitely decreasing geometrical progression:* (see Section 5.1)

$$1 + q + q^2 + q^3 + \ldots$$

is convergent for $|q| < 1$, and its sum equals $S = \frac{1}{1-q}$.

2. The *harmonic series*

$$1 + \frac{1}{2} + \frac{1}{3} + \frac{1}{4} + \ldots$$

is divergent, and $\lim_{n\to\infty} s_n = +\infty$.

A necessary condition for convergence: if a series $\sum_{n=1}^{\infty} x_n$ is convergent, then $x \to 0$; this condition is not sufficient: it may happen that $x \to 0$, as $n \to \infty$, but the series $\sum_{n=1}^{\infty} x_n$ is divergent.

5.8.2 Tests for convergence of series

1. *Comparison test*: let $0 \leq a_n \leq b_n$ and the series $\sum_{n=1}^{\infty} b_n$ be convergent, then the series $\sum_{n=1}^{\infty} a_n$ is also convergent.

Example

The series

$$1 + \frac{1}{1!} + \frac{1}{2!} + \frac{1}{3!} + \ldots$$

is convergent. This proposition is a consequence of the following evident inequality:

$$n! = 1 \cdot 2 \cdot 3 \cdot \ldots n \geq 2^{n-1},$$

whence $1/n! \leq 1/2^{n-1}$, and of the fact that the geometric progression $1 + \frac{1}{2} + \frac{1}{4} + \frac{1}{8} + \ldots$ is convergent (see Section 5.1.7).

Remark: $1 + \frac{1}{1!} + \frac{1}{2!} + \frac{1}{3!} + \ldots = e$ (see Section 5.1.8).

2. *D'Alembert test*: assume that $a_n > 0$ and that there exists

$$\lim_{n\to\infty} \frac{a_{n+1}}{a_n} = q.$$

Then in the case $0 \leq q < 1$ the series $\sum_{n=1}^{\infty} a_n$ is convergent, and in the case $q > 1$ it is divergent.

Example

Consider the series $\sum_{n=1}^{\infty} \frac{n}{2^n}$. Here $a_n = \frac{n}{2^n}$, $q = \lim_{n\to\infty} \frac{n+1}{2n} = \frac{1}{2}$, thus the series is convergent.

3. *Cauchy integral test*: if $f(x)$ is a monotone decreasing continuous positive function, then the series $\sum_{n=1}^{\infty} f(n)$ and the improper integral $\int_1^{\infty} f(x)\,dx$ are either both convergent or both divergent.

Remark: To read more about improper integrals see Section 5.7.3.

Examples

1. The series $\sum_{n=1}^{\infty} \frac{1}{n^2}$ is convergent since $\int_1^{\infty} \frac{1}{x^2}\,dx = [-\frac{1}{x}]_1^{\infty} = 1$.

2. The series $\sum_{n=1}^{\infty} \frac{1}{n}$ is divergent since $\int_1^{\infty} \frac{dx}{x} = [\ln x]_1^{\infty} = \infty$.

5.8.3 Series whose members are of arbitrary sign

We say that the series $\sum_{n=1}^{\infty} a_n$ is *absolutely convergent* if the series $\sum_{n=1}^{\infty} |a_n|$ is convergent. In the case where the series $\sum_{n=1}^{\infty} a_n$ is convergent, but $\sum_{n=1}^{\infty} |a_n|$ is not, we say that the series $\sum_{n=1}^{\infty} a_n$ is *conventionally* (or *non-absolute*) *convergent.* For example, $\sum_{n=1}^{\infty} \frac{(-1)^n}{n^2}$ is absolutely convergent, while the series $\sum_{n=1}^{\infty} \frac{(-1)^n}{n}$ is conventionally convergent.

Properties of absolutely convergent series remind us of ordinary finite sums: an absolutely convergent series admits arbitrary transposition of members, and this does not alter the sum. On the contrary, the sum of a conventionally convergent series depends upon the order of summands: by a suitable transposition of its members, we can make the sum to be equal to any prescribed number (*Riemann's theorem*).

5.8.4 Power series

Functional series is a series whose members are functions:

$$f_1(x) + f_2(x) + \ldots + f_n(x) + \ldots.$$

In practice, the most important functional series are the *power series*:

$$\sum_{n=0}^{\infty} b_n (x-a)^n.$$

Properties of power series

• A power series is convergent over a *convergency interval*, i.e., for $-R < x - a < R$, the number R is called the *convergence radius.*

• A power series admits differentiation and integration term by term at any point inside its interval of convergence: given

$$S(x) = \sum_{n=1}^{\infty} b_n (x-a)^n,$$

we obtain

$$S'(x) = \sum_{n=0}^{\infty} n b_n (x-a)^{n-1},$$

$$\int S(x)\, dx = \sum_{n=0}^{\infty} b_n \frac{(x-a)^{n+1}}{n+1} + C,$$

and both operations may be applied repeatedly.

Any power series over the complex plane is convergent inside its *circle of convergence*, i.e., for $|z-a| < R, \quad z \in \mathbf{C}$ (see Section 7.1).

The intervals of convergence for the most important power series (for $x \in \mathbf{R}$):

1. $-\infty < x < +\infty$:

$$e^x = 1 + x/1! + x^2/2! + x^3/3! + x^4/4! + \dots$$
$$\sin x = x - x^3/3! + x^5/5! - \dots$$
$$\cos x = 1 - x^2/2! + x^4/4! - \dots,$$

2. $-1 < x < 1$:

$$\frac{1}{1-x} = 1 + x + x^2 + x^3 + \dots,$$

3. $-1 < x \leq 1$:

$$\ln(1+x) = x - x^2/2 + x^3/3 - x^4/4 + \dots.$$

Chapter 6

Combinatorics

Combinatorics considers problems such as to choosing some subset in a given finite set and arranging the chosen elements in a certain order. Two examples: How many various four-digit numbers may be written by using the digits 1, 2, 3, 4? How many various teams consisting of 11 football players may be formed amongst the given 15 football players?

6.1 Permutations. Arrangements. Combinations

6.1.1 Permutations

Consider n different elements $x_1, x_2, \ldots, x_n$. A *permutation of n given elements* is their sequence taken in a certain order. For instance, 3, 1, 4 and 4, 3, 1 are two different permutations of the digits 1, 3, 4; STOP, POST, OPTS, and TOPS are four different permutations of the letters O, P, S, T. Two permutations distinguish from each other by order of their elements only.

6.1.2 Factorial

The total number P_n of all permutations of n given elements is equal to the product of all natural numbers from 1 to n : $P_n = 1 \cdot 2 \cdot 3 \cdot \ldots \cdot n$. Such a product is denoted by the symbol $n!$ (read as "n factorial").

In addition, we define $0! = 1! = 1$.

Example. Six volleyball players may be placed in the playground by $6! = 1 \cdot 2 \cdot 3 \cdot 4 \cdot 5 \cdot 6 = 720$ fashions.

For factorials, the following recursive formula is true:

$$(n+1)! = n! \cdot (n+1).$$

Table 6.1: Values of Factorial

n	1	2	3	4	5	6	7	8	9	10
$n!$	1	2	6	24	120	720	5040	40320	362880	3628800

6.1.3 Stirling formula

The variable $n!$ grows very fast as $n \to \infty$. Table 6.1 demonstrates the behavior of $n!$ in dependence on n. For large values of n (in practice, for $n \geq 3$), we may use the approximate *Stirling formula:*

$$n! \approx \left(\frac{n}{e}\right)^n \cdot \sqrt{2\pi n}.$$

A more precise approximation is given by the modified Stirling formula:

$$n! \approx \left(\frac{n}{e}\right)^n \cdot \sqrt{2\pi n}\left(1 + \frac{1}{12n}\right).$$

6.1.4 Semi-factorial

In some formulas one meets the symbol $n!!$ ("semi-factorial"):

$$n!! = 2 \cdot 4 \cdot 6 \cdot \ldots \cdot (2k) \quad \text{for even } n = 2k;$$

$$n!! = 1 \cdot 3 \cdot 5 \cdot \ldots \cdot (2k-1) \quad \text{for odd } n = 2k-1.$$

Evidently, $n! = n!!\,(n-1)!!$, $(2k)!! = 2^k\,k!$.

6.1.5 Arrangements

Consider n different elements $x_1, x_2, \ldots, x_n$. Let us choose any m elements among them and arrange the chosen elements in some order. Such ordered sequences are called the *arrangements* of m elements chosen among n given elements. Two different arrangements are distinguished from each other by their elements and, as well, by the order of the elements.

Examples

1. Choose 4 pupils among the given 9 pupils and invite them to take 4 given chairs (one pupil to one chair).

2. Put 7 given enumerated balls into 7 boxes chosen among 20 given enumerated boxes (one ball into one box).

For both tasks above, each choice of elements and engaging places form an arrangement.

The total number of arrangements of m elements chosen among n given elements is denoted by A_n^m. There is a formula:

$$A_n^m = \frac{n!}{(n-m)!} = n(n-1)\dots(n-m+1).$$

Thus, for the first task above, the solution is $A_9^4 = 9 \cdot 8 \cdot 7 \cdot 6 = 3024$.

6.1.6 Combinations

Consider n different elements $x_1, x_2, \dots, x_n$. Let us choose any m elements among them; such a subset (see Section 2.1.2) is called the *combination* of m elements (chosen) from n given elements. Two different combinations distinguish from each other by their elements only, while the mutual order of the elements is indifferent. For instance, each team containing 11 participants chosen amongst 15 given sportsmen is a combination of 11 from 15.

The total number of combinations of m elements from n ones is denoted by C_n^m or by $\binom{n}{m}$. The following formulas are true:

$$C_n^m = \frac{n!}{m!(n-m)!}, \quad C_n^m = C_n^{n-m}, \quad C_n^m = \frac{A_n^m}{m!}.$$

In addition, we define $C_n^0 = 1$.

Examples

$$C_n^n = 1, \quad C_3^2 = \frac{3!}{2! \cdot 1!} = 3, \quad C_{15}^{10} = \frac{15 \cdot 14 \cdot 13 \cdot 12}{1 \cdot 2 \cdot 3 \cdot 4} = 1365.$$

6.2 Binomial Formula and Applications

6.2.1 Binomial formula

Consider the expression $(a+b)^n$, where n is a natural number. The following *binomial formula* is valid:

$$(a+b)^n = \sum_{k=0}^{n} C_n^k a^{n-k} b^k, \tag{6.1}$$

where C_n^k is the number of combinations of k from n (see 6.1.6). The numbers C_n^k are also called the *binomial coefficients*. Formula (6.1) may be written in an explicit form as

$$(a+b)^n = a^n + C_n^1 a^{n-1} b + C_n^2 a^{n-2} b^2 + \dots + C_n^{n-1} a b^{n-1} + b^n. \tag{6.2}$$

Replace b in (6.2) by $(-b)$ to obtain the formula

$$(a-b)^n = \sum_{k=0}^{n}(-1)^k C_n^k a^{n-k} b^k$$
$$= a^n - C_n^1 a^{n-1} b + C_n^2 a^{n-2} b^2 - \ldots + (-1)^n b^n. \quad (6.3)$$

Special cases for $n = 2$, $n = 3$ are represented in Section 1.1.

Example

$(a-b)^4 = a^4 - 4a^3b + 6a^2b^2 - 4ab^3 + b^4.$

6.2.2 Properties of the binomial coefficients

1. $C_n^n = C_n^0 = 1, \ \ C_n^1 = n.$
2. $C_n^k = C_n^{n-k}$ –the symmetry.
3. $\sum_{k=0}^{n} C_n^k = 2^n, \ \ \sum_{k=0}^{n}(-1)^k C_n^k = 0.$
4. $C_{n+1}^k = C_n^{k-1} + C_n^k$ –recursive formula. (6.4)
5. $C_{m+n}^k = C_m^0 C_n^k + C_m^1 C_n^{k-1} + \ldots + C_m^p C_n^{k-p} + \ldots + C_m^k C_n^0.$ (6.5)
6. $(C_n^0)^2 + (C_n^1)^2 + \ldots + (C_n^n)^2 = C_{2n}^n$–corollary to (6.5).

6.2.3 Pascal triangle

Formula (6.4) enables us to calculate the binomial coefficients in a recursive way (i.e., the next coefficient using the previous ones). Such a calculation may be conveniently performed by using the *Pascal triangle rule:*

1

1 1

1 2 1

1 3 3 1

1 4 6 4 1

1 5 10 10 5 1

....

Every horizontal line in this triangle diagram consists of binomial coefficients for the corresponding power of the binomial: first, second, third, and so on. Each coefficient (except the extreme ones) is obtained by the addition of two adjacent coefficients above it, for instance: $4 = 1 + 3$, $6 = 3 + 3$, $10 = 4 + 6$, and so on.

6.2.4 Sum of natural numbers in a certain power

Consider a problem to determine the sum of the first n natural numbers taken in the m-th power. Denote this sum by $S^m(n)$:

$$S^m(n) = \sum_{k=1}^{n} k^m = 1^m + 2^m + \ldots + n^m.$$

For $m = 1$, we have already found such a sum (see Section 1.4):

$$S^1(n) = \frac{n(n+1)}{2}.$$

For $m \geq 2$, the following recursive relation is valid:

$$(m+1)S^m(n) = n(n+1)^m - S^1(n) - C_m^1 S^2(n)$$
$$-C_m^2 S^3(n) - \ldots - C_m^{m-2} S^{m-1}(n). \tag{6.6}$$

In particular, for $m = 2$, formula (6.6) gives

$$3S^2(n) = n(n+1)^2 - S^1(n), \quad \text{or}$$

$$S^2(n) = \sum_{k=1}^{n} k^2 = \frac{n(n+1)(2n+1)}{6} \tag{6.7}$$

For $m = 3$, we obtain from (6.6)

$$4S^3(n) = n(n+1)^3 - S^1(n) - C_3^1 S^2(n).$$

By using (6.7) we get, finally,

$$S^3(n) = \frac{n^2(n+1)^2}{4}, \quad \text{or} \quad S^3(n) = [S^1(n)]^2,$$

or, in the explicit form

$$1^3 + 2^3 + 3^3 + \ldots + n^3 = (1 + 2 + 3 + \ldots + n)^2.$$

Chapter 7

Complex Numbers

Complex numbers had been invented as a generalization of real numbers in order to provide solvability of arbitrary algebraic equations: some algebraic equations have no roots in real domain but they become solvable in complex numbers. This extension of the concept of numbers proved to be very fruitful in various areas of physics, mathematics, and engineering.

7.1 Basic Concepts

7.1.1 Notations

Complex numbers are the pairs of real numbers (a, b) for which certain operations are defined. The set of complex numbers is denoted by the symbol **C**.

A complex number $z = (a, b)$ consists of the *real part*, a, and the *imaginary part*, b, designated respectively as $\operatorname{Re} z$ and $\operatorname{Im} z$. For instance,

$$\operatorname{Re}(3, -2) = 3, \quad \operatorname{Im}(3, -2) = -2.$$

A number of the form $(a, 0)$ is identified with the real number a, i.e., $(a, 0) = a$, so the concept of complex numbers is a generalization of the concept of real numbers. In particular, the number $(0, 0)$ coincides with the real number 0 and is called the zero. Complex numbers of the form $(0, b)$ are called *imaginary* (pure imaginary); specifically, the number $(0, 1)$ is called the *imaginary unit* and is denoted by the letter i (in some books by the letter j).

7.1.2 Rules to operate with complex numbers

1. The equality $(a_1, b_1) = (a_2, b_2)$ is equivalent to the system of two equalities: $a_1 = a_2$, $b_1 = b_2$, i.e., two complex numbers are equal

when their real and imaginary parts are respectively equal.

2. Addition: $(a_1, b_1) + (a_2, b_2) = (a_1 + a_2, b_1 + b_2)$, i.e., in order to add complex numbers we add their real parts and, separately, their imaginary parts.

3. Multiplication: $(a_1, b_1) \cdot (a_2, b_2) = (a_1a_2 - b_1b_2, a_1b_2 + a_2b_1)$ (see also Section 7.2); in particular, $i^2 = i \cdot i = -1$.

4. Subtraction and division are defined as operations inverse to addition and multiplication, respectively; division by zero is undefined.

The operations of addition and multiplication are commutative, associative, and distributive (see Section 2.2.4).

Inequalities $>$, $<$, $\geq$, and $\leq$ are not defined for complex numbers.

7.1.3 Conjugate numbers

For $z = (a, b)$, we call the number $(a, -b)$ the *conjugate* (or complex conjugate) of z, and denote this number by the symbol $\overline{z}$ or z^*. For instance,

$$(2, -3) = \overline{(2, 3)} = (2, 3)^*.$$

It is evident that

$$\overline{\overline{z}} = z, \quad z \cdot \overline{z} = a^2 + b^2 \geq 0,$$

$$\operatorname{Re} z = (z + \overline{z})/2, \quad \operatorname{Im} z = (z - \overline{z})/(2i),$$

$$\overline{(z_1 + z_2)} = \overline{z_1} + \overline{z_2}, \quad \overline{(z_1 \cdot z_2)} = \overline{z_1} \cdot \overline{z_2}.$$

7.2 Algebraic Form of Complex Number

A complex number $z = (a, b)$ may be written in the *algebraic form* (or algebraic representation):

$$z = a + ib, \quad \text{or} \quad z = a + bi,$$

for instance, $(-3, 4) = -3 + 4i$. The conjugate number is written as

$$\overline{z} = a - ib.$$

The algebraic form is convenient to perform the operations with complex numbers (such as addition, multiplication, removing the parentheses, and so on) as with usual real binomials, where we must replace, after all reductions, the expression i^2 by (-1). As a consequence we obtain the multiplication formula from Section 7.1.2:

$$(a_1 + ib_1)(a_2 + ib_2) = (a_1a_2 - b_1b_2) + i(a_1b_2 + a_2b_1).$$

Examples

1. $i^3 = -i, \quad i^4 = 1, \quad i^5 = i, \quad i^{n+4} = i^n.$

2. $(-2+3i)\cdot 4 + (1+i)\cdot(5+2i) = -8+12i+5+5i+2i+2i^2 =$

   $-5+19i.$

Division of complex numbers is easily performed by multiplication of both the numerator and the denominator by a number conjugate to the denominator:

$$\frac{2+3i}{-1+2i} = \frac{(2+3i)(-1-2i)}{(-1+2i)(-1-2i)} = \frac{-2+6+(-3-4)i}{(-1)^2+2^2} = \frac{4-7i}{5}.$$

Complex number $z = x+iy$ may be depicted in the plane by a point whose coordinates are (x, y), e.g., the numbers $2+i$, $2-i$, $1-i$ shown in Figure 7.1. The abscissa axis is called the *real axis*, the ordinate axis is called the *imaginary axis*, and the plane is called the *complex plane*. A pair of mutual conjugate numbers is depicted by a pair of points symmetric with respect to the real axis.

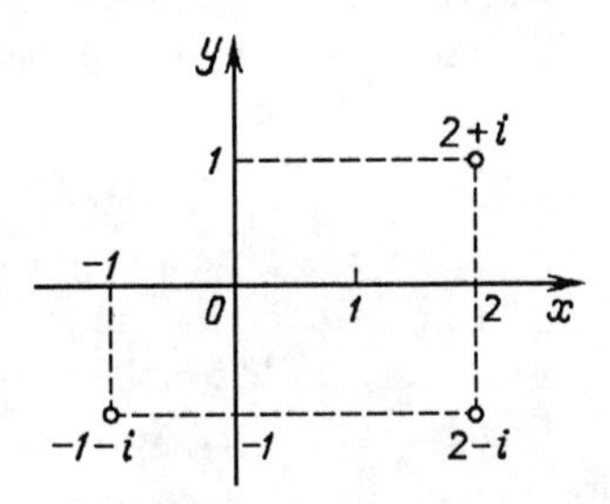

Figure 7.1: The complex number is depicted by a dot in the plane.

7.3 Trigonometric and Exponential Forms

7.3.1 Vector interpretation

Complex number $z = x + iy$ will be identified with a vector in the xy-plane if we locate the beginning point of the vector at point O (coordinate origin) and define the projections of the vector to the coordinate axes as x and y (Figure 7.2). For complex numbers, the addition and subtraction rules coincide with the corresponding rules for vectors (see Section 8.1). Vector interpretation is widely used with a view of representation of harmonic oscillations, alternating sine-shaped currents, or voltages (see Sections 15.5.4 and 16.4.2).

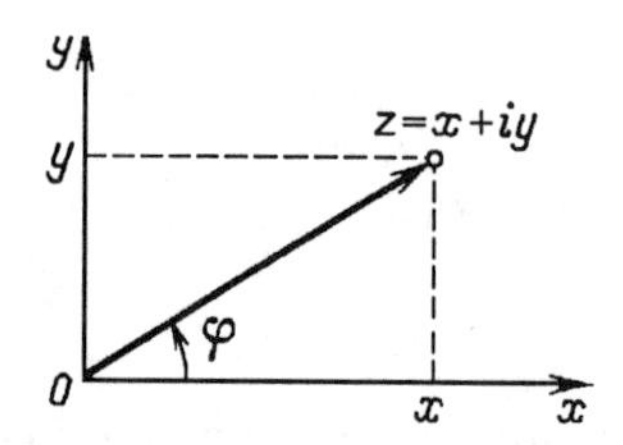

Figure 7.2: Vector interpretation of the complex number.

7.3.2 Modulus and argument of a complex number

The *modulus* (or absolute value) of a complex number z is defined as the length of the corresponding vector:

$$|z| = \sqrt{x^2 + y^2}, \quad |z| \geq 0.$$

The *argument* $\arg z = \varphi$ is defined as the angle (measured in radians) between the vector and the positive semiaxis x; the positive direction for counting φ is, conventionally, counterclockwise. For any $z \neq 0$, there exists an infinite set of values for φ which differ from each other by $2\pi k \; (k = 0, 1, 2, \dots)$.

The value of the argument chosen in the interval $-\pi < \varphi \leq \pi$ is called usually the *principal value*; sometimes the interval for the principal value is taken as $0 \leq \varphi < 2\pi$. The principal value of the argument can be calculated according to Table 7.1, while for the numbers having zero real or imaginary part according to Table 7.2. Note to Table 7.1: the symbol ρ denotes there the modulus of the number $z = x + iy$, $\rho = \sqrt{x^2 + y^2}$.

The modulus and the argument of a complex number z are the same as the polar coordinates of the point z on the xy-plane (see Section 8.3.4).

7.3.3 Trigonometric form

The *trigonometric form* (or trigonometric representation) of a complex number:

$$z = \rho(\cos\varphi + i\sin\varphi).$$

Examples

1. $1 + 0 \cdot i = 1 \cdot (\cos 0 + i \sin 0)$.
2. $i = 1 \cdot (\cos \frac{\pi}{2} + i \sin \frac{\pi}{2})$.
3. $1 + i = \sqrt{2}(\cos \frac{\pi}{4} + i \sin \frac{\pi}{4})$.

Table 7.1: Formulas for the argument of a complex number

Quadrant	Sign of x, y	Formula for the argument
I	$x > 0,\ y > 0$	$\varphi = \arctan \frac{y}{x} = \arcsin \frac{y}{\rho} = \arccos \frac{x}{\rho}$
II	$x < 0,\ y > 0$	$\varphi = \arccos \frac{x}{\rho} = \arctan \frac{y}{x} + \pi = \pi - \arcsin \frac{y}{\rho}$
III	$x < 0,\ y < 0$	$\varphi = \arctan \frac{y}{x} - \pi = -\arccos \frac{x}{\rho} = -\pi - \arcsin \frac{y}{\rho}$
IV	$x > 0,\ y < 0$	$\varphi = \arctan \frac{y}{x} = \arcsin \frac{y}{\rho} = -\arccos \frac{x}{\rho}$

Table 7.2: Arguments of real and imaginary numbers

Sign of x, y	Argument of z	Sign of x, y	Argument of z
$x > 0,\ y = 0$	$\varphi = 0$	$x < 0,\ y = 0$	$\varphi = \pi$
$x = 0,\ y > 0$	$\varphi = \pi/2$	$x = 0,\ y < 0$	$\varphi = -\pi/2$

4. $-1-i=\sqrt{2}(\cos\frac{5\pi}{4}+i\sin\frac{5\pi}{4})$.

5. $3+4i=5(\cos\varphi+i\sin\varphi),\quad \varphi=\arctan(4/3)$.

7.3.4 Exponential form

Exponential form (or exponential representation) of a complex number:

$$z=|z|e^{i\varphi},\quad \text{or}\quad z=\rho e^{i\varphi},\quad \text{or}\quad z=\rho\exp(i\varphi),$$

where the number $e\approx 2.718$ (see Section 5.1.8). For instance, $e^{i\pi/2}=i$, $e^{2k\pi i}=1$. A number having the exponential form $e^{i\varphi}$ is of unit modulus for any real value of φ.

7.3.5 Euler formulas

$$e^{i\varphi}=\cos\varphi+i\sin\varphi,$$

$$\cos\varphi=\frac{e^{i\varphi}+e^{-i\varphi}}{2},$$

$$\sin\varphi=\frac{e^{i\varphi}-e^{-i\varphi}}{2i}.$$

Corollary:

$$\cos(ix)=(e^{-x}+e^{x})/2=\cosh x,$$

$$\sin(ix)=(e^{-x}-e^{x})/(2i)=i\sinh x,$$

where $\cosh x$ and $\sinh x$ are the hyperbolic functions (see Section 2.8.2).

An example of using Euler formulas:

$$1+i\sqrt{3}=2(\cos\pi/3+i\sin\pi/3)=2e^{i\pi/3}.$$

Trigonometric and exponential forms of complex numbers are especially convenient in physical applications and also in multiplication or division, and as well in raising to a power. Namely:

$$z_1\cdot z_2=\rho_1\cdot\rho_2 e^{i(\varphi_1+\varphi_2)},\qquad \frac{z_1}{z_2}=\frac{\rho_1}{\rho_2}e^{i(\varphi_1-\varphi_2)}.$$

In other words, for multiplication of complex numbers we multiply their moduli and add their arguments, while for division we divide the modulus of the numerator by the modulus of the denominator and subtract the denominator's argument from the numerator's argument.

Example

$$3\,(\cos\tfrac{\pi}{3}+i\sin\tfrac{\pi}{3})\cdot 4(\cos\tfrac{\pi}{6}+i\sin\tfrac{\pi}{6})=12(\cos\tfrac{\pi}{2}+i\sin\tfrac{\pi}{2})=12i,$$

or

$$3e^{i\pi/3}\cdot 4e^{i\pi/6}=12e^{i\pi/2}=12i.$$

Multiplication of complex numbers may be illustrated geometrically as follows:

1. Multiplying a complex number z by a real positive factor p enlarges the length of a vector corresponding to z exactly p times and conserves the direction of the vector.

2. Multiplying a complex number z by a real negative factor q enlarges the length of the vector $|q|$ times and also changes the direction of the vector to its opposite.

3. Multiplying a complex number z_1 by a complex number z_2 enlarges the length of the vector z_1 exactly $|z_2|$ times and also rotates the obtained vector by the angle $\arg z_2$ with respect to the vector z_1.

7.3.6 De Moivre formula

The following formula (*de Moivre formula*) is known for raising complex numbers to an integer power:

$$z^n = \rho^n(\cos n\varphi + i \sin n\varphi), \quad \text{or} \quad z^n = \rho^n e^{i\varphi}.$$

Example

$$(1-i)^8 = (\sqrt{2})^8 e^{8i(-\pi/4)} = 16e^{-2\pi i} = 16.$$

The De Moivre formula can be used in the calculation of trigonometric functions with multiple arguments, for instance:

1. $\sin 2x = \operatorname{Im} e^{2ix} = \operatorname{Im}(\cos x + i \sin x)^2 =$
   $\operatorname{Im}(\cos^2 x + 2i \cos x \sin x - \sin^2 x) = 2 \sin x \cos x.$

2. $\cos 3x = \operatorname{Re} e^{3ix} = \operatorname{Re}(\cos x + i \sin x)^3 =$
   $\operatorname{Re}(\cos^3 x + 3i \cos^2 x \sin x - 3 \cos x \sin^2 x - i \sin^3 x) =$
   $\cos^3 x - 3 \cos x \sin^2 x.$

3. $\sin 5x = \operatorname{Im} e^{5ix} = \operatorname{Im}(\cos x + i \sin x)^5 =$
   $\operatorname{Im} \sum_{k=0}^{5} C_5^k (\cos x)^{5-k} (i \sin x)^5 = C_5^1 \cos^4 x \sin x -$
   $C_5^3 \cos^2 x \sin^3 x + C_5^5 \sin^5 x = 5 \cos^4 x \sin x - 10 \cos^2 x \sin^3 x + \sin^5 x.$
   (see the Binomial formula in Section 6.2.1).

De Moivre formula is useful as well in calculation of sums like

$$A = \sum_{k=0}^{n} q^k \cos kx \quad \text{and} \quad B = \sum_{k=1}^{n} q^k \sin kx.$$

For instance, find

$$A = \text{Re} \sum_{k=0}^{n} z^k, \quad \text{where } z = qe^{ix}.$$

We know from Section 1.1 that the last sum equals $(1-z^{n+1})/(1-z)$, hence

$$A = \text{Re}\,\frac{1 - q^{n+1}e^{ix(n+1)}}{1 - qe^{ix}} =$$

$$\text{Re}\,\frac{(1 - q^{n+1}e^{ix(n+1)})(1 - qe^{-ix})}{(1 - qe^{ix})(1 - qe^{-ix})} =$$

$$\frac{1}{1 - 2q\cos x + q^2}\text{Re}\,(1 - q^{n+1}e^{ix(n+1)} - qe^{-ix} + q^{n+2}e^{inx}) =$$

$$\frac{1 - q\cos x - q^{n+1}\cos(n+1)x + q^{n+2}\cos nx}{1 - 2q\cos x + q^2}.$$

7.3.7 Roots of complex numbers

Let m be a natural number. The root of order m of number zero is equal to zero. Raising a complex number $z \neq 0$ to the power $1/m$ (i.e., taking the root of order m) leads to m various values:

$$z^{1/m} = \sqrt[m]{z} = \rho^{1/m}(\cos\frac{\varphi + 2\pi k}{m} + i\sin\frac{\varphi + 2\pi k}{m}), \tag{7.1}$$

or

$$z^{1/m} = \sqrt[m]{z} = \rho^{1/m}\exp\frac{\varphi + 2\pi k}{m}, \tag{7.2}$$

where $k = 0, 1, 2, \ldots, m-1$; $\rho^{1/m} > 0$. Putting here $-\pi < \varphi \leq \pi$ and $k = 0$, we obtain so-called "principal value" of the root.

Example

$\sqrt[3]{-8} = \sqrt[3]{8}\,e^{i(\pi + 2\pi k)/3}$, and that brings three values:

$z_1 = 2e^{i\pi/3} = 2(\cos\frac{\pi}{3} + i\sin\frac{\pi}{3}) = 1 + i\sqrt{3}$ —the principal value,

$z_2 = 2e^{i\pi} = 2(\cos\pi + i\sin\pi) = -2$,

$z_3 = 2e^{5i\pi/3} = 2(\cos\frac{5\pi}{3} + i\sin\frac{5\pi}{3}) = 1 - i\sqrt{3}$.

The roots of order m of a complex number z are depicted in the complex plane by points $z_1, z_2, \ldots, z_m$, located at vertices of a regular m-side polygon, the points belong to a circumference whose radius is $\rho^{1/m}$ and the center $z_0 = 0$ (Figure 7.3, the case $m = 6$). In particular, the square root has two values distinguished by the sign:

$$\sqrt{1} = \pm 1, \ \sqrt{-1} = \pm i, \ \sqrt{i} = \pm(1+i)/\sqrt{2}.$$

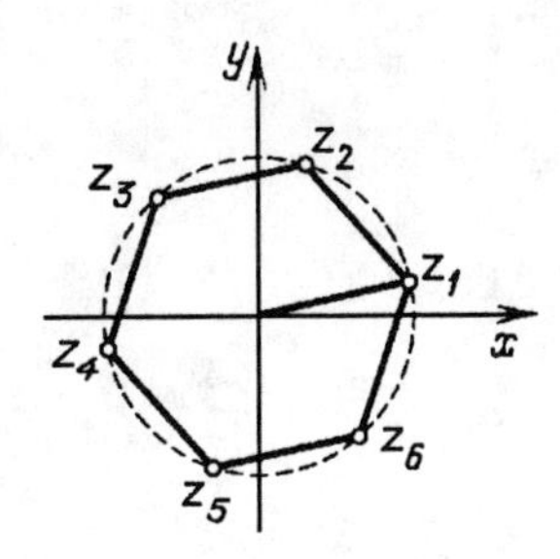

Figure 7.3: The root of order m of a complex number has m different values.

7.4 Logarithms of Complex Numbers

The logarithm (natural logarithm) of a complex number $z \neq 0$ has an infinite set of values:

$$\text{Log}\, z = \log_e \rho + i(\varphi + 2\pi k), \quad k = 0, \pm 1, \pm 2, \ldots .$$

Here $\log_e \rho$ means the real logarithm to the base e (natural logarithm, see Section 2.8.3). The relation $\exp(\text{Log}\, z) = z$ is true for complex numbers.

Examples

1. $\text{Log } 1 = \log_e 1 + 2i\pi k = 2i\pi k$.
2. $\text{Log } (-1) = \log_e 1 + i(\pi + 2\pi k) = i\pi(2k+1)$.
3. $\text{Log } (1 + i\sqrt{3}) = \log_e 2 + i(\pi/3 + 2\pi k)$.
4. $\text{Log } i = \log_e 1 + i(\pi/2 + 2\pi k) = i\pi(2k + 1/2)$.

The principal value of logarithm: $\log z = \log_e \rho + i\varphi$, where $-\pi < \varphi \leq \pi$.

7.5 Complex Roots of Equations

For algebraic equations in complex domain, see Section 3.4. Consider now the "bynomial" equation $z^a = b$, where z is an unknown complex number, $b \neq 0$ a given complex number, and $a = m/n$ a given rational real number. There are m solutions:

$$z_k = |b|^{\frac{1}{a}} \exp[\frac{i}{a}(\varphi + 2\pi k)], \tag{7.3}$$

where $\varphi = \arg z, \quad k = 0, 1, 2, \ldots, m-1$.

If a is a real irrational number (see Section 2.2.3) then there exists an infinite set of solutions, and also Formula (7.3) is valid with $k = 0, \pm 1, \pm 2, \ldots$

Examples

1. $z^{2/3} = 1 + i$, $z_k = 2^{\frac{3}{4}} \exp[\frac{3i}{2}(\frac{\pi}{4} + 2\pi k)]$, $k = 0$, $k = 1$.

2. $z^{\pi} = 1 + i$, $z_k = 2^{\frac{1}{2\pi}} \exp[\frac{i}{\pi}(\frac{\pi}{4} + 2\pi k)] = 2^{\frac{1}{2\pi}} \exp(\frac{i}{4} + 2ik)$, $k = 0, \pm 1, \pm 2, \ldots$

The equation $\sin z = b$ may have complex solutions if b is complex or if it is real with $|b| > 1$. To solve the equation, use the Euler formula:

$$\sin z = (e^{iz} - e^{-iz})/(2i),$$

denote $w = \exp(iz)$, and obtain

$$w - w^{-1} = 2ib, \quad \text{or} \quad w^2 - 2ibw - 1 = 0,$$

whence $w_{1,2} = ib \pm \sqrt{1 - b^2}$; here the symbol $\sqrt{q}$ means any of two admissible values of the complex square root of q. Now z is found by taking the logarithm:

$$iz = \operatorname{Log} w, \quad z = -i[\log_e |w| + i(\arg w + 2\pi k)], \tag{7.4}$$

with $k = 0, 1, 2, \ldots$. Thus we obtained two infinite series of solutions that correspond to $w = w_1$ and $w = w_2$.

Example: Find z such that $\sin z = 2$.
Solution. Here $w_{1,2} = 2i \pm i\sqrt{3} = i\,(2 \pm \sqrt{3})$, $\arg w_{1,2} = \pi/2$, hence

$$z = -i\operatorname{Log}(2 \pm \sqrt{3}) + (\pi/2 + 2\pi k).$$

Equations like $\cos z = b$, $\tan z = b$ with complex values of b are to be solved in a similar way.

Chapter 8

Vectors. Coordinates. Displacements and Symmetries

Vectors and vector values are commonly used in many problems of physics and mathematics. Thus, the basic laws of mechanics and of electromagnetism may be formulated most conveniently in the vector form. In practical calculations it is a custom to rewrite vector formulas to their coordinate representation. Vectorial and coordinate techniques proved to be very useful in geometry as well. This chapter, along with a description of operations above vectors and coordinates, also contains concepts and examples of symmetry and similarity of plane and spatial figures.

8.1 Vectors. Projections

8.1.1 Vectors

A pair of ordered points A and B defines a *vector* $\overrightarrow{AB}$. Point A is the *beginning*, or the *origin*, while point B is the *end*, or the *extremity*. A vector may be thought about as a directed straight line segment (see Figure 8.1a, where the direction of the vector $\overrightarrow{AB}$ is shown by an arrow). In the case $A = B$ (coincidence of the beginning and the end) the vector $\overrightarrow{AB}$ is said to be the *zero-vector*. The zero-vector is depicted as a dot, and its direction is undefined.

Examples of vector values in physics: translation, velocity, acceleration, force, and momentum (Sections 13.1.2 and 13.2.1).

The *modulus* (or *magnitude*, or *length*, or *absolute value*) of $\overrightarrow{AB}$ is, by definition, the length of the segment AB; the modulus of $\overrightarrow{AB}$ is denoted by $|\overrightarrow{AB}|$, or AB.

Two vectors AB and CD are considered as *equal* if the following two conditions are satisfied:

(i) $|\overrightarrow{AB}|=|\overrightarrow{CD}|$, i.e., moduli of the vectors are equal to each other;

(ii) The rays AB and CD are parallel and have the same direction.

For instance,

$\overrightarrow{A_1B_1}=\overrightarrow{A_2B_2}$, but $\overrightarrow{A_1B_1}\neq\overrightarrow{A_3B_3}$, $\overrightarrow{A_1B_1}\neq\overrightarrow{A_1B_4}$ (Figure 8.1*b*).

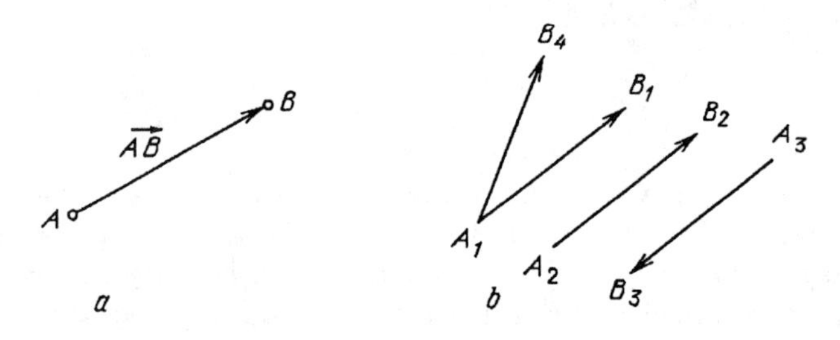

Figure 8.1: Oriented segments.

We see that any given vector $\overrightarrow{AB}$ corresponds to an infinite set of vectors that are equal to $\overrightarrow{AB}$: they are obtained from $\overrightarrow{AB}$ by all possible translations (i.e., parallel shifts). Under such a definition, the vectors are called *free* (or unfixed) because the position of the beginning point is indifferent. However, in certain applications, it may be preferable to use vectors having a fixed beginning point. For instance, in physics we often meet the "radius-vector" of a point, its beginning is fixed at the origin of the coordinate system (see Section 13.1.1).

A vector may be either denoted by one letter with an arrow above it: $\vec{a}$, or printed using boldface: $\mathbf{a}$. The zero-vector is denoted by $\vec{0}$ or $\mathbf{0}$.

Operations with vectors

1. The sum $\mathbf{a}+\mathbf{b}$ of the vectors $\mathbf{a}$ and $\mathbf{b}$ can be found by using the *triangle rule* (Figure 8.2*a*) or the *parallelogram rule* (Figure 8.2*b*). Both the rules are equivalent to each other. The triangle rule is preferable in practice when we add several vectors (Figure 8.2*c*).

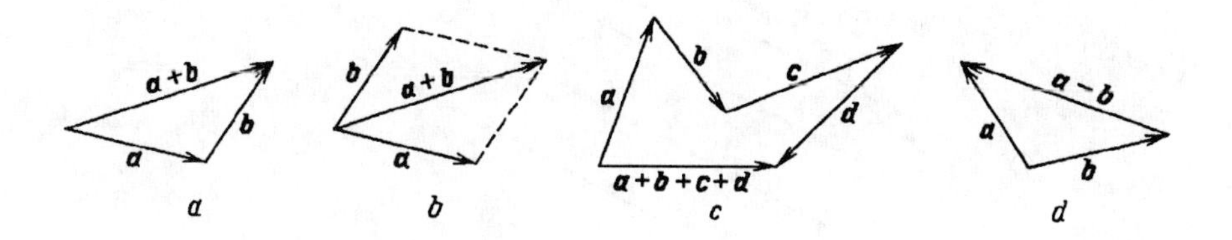

Figure 8.2: Sum and difference of vectors.

Addition of vectors is commutative and associative:

$$\mathbf{a}+\mathbf{b}=\mathbf{b}+\mathbf{a}, \qquad \mathbf{a}+(\mathbf{b}+\mathbf{c})=(\mathbf{a}+\mathbf{b})+\mathbf{c}.$$

Note that $\mathbf{a}+\mathbf{0}=\mathbf{a}$.

If $\mathbf{a}+\mathbf{b}=\mathbf{0}$, then $\mathbf{b}$ is called the *opposite vector* to $\mathbf{a}$, and it is denoted by -$\mathbf{a}$. The vector -$\mathbf{a}$ has same length as $\mathbf{a}$, but its direction is opposite. The difference $\mathbf{a}-\mathbf{b}$ is defined as the sum $\mathbf{a}+(-\mathbf{b})$, i.e., the subtraction of a vector may be replaced by the addition of its opposite vector. The triangle rule is applicable here in the following way: the vectors $\mathbf{a}$ and $\mathbf{b}$ are to be situated as having a common beginning point, then $\mathbf{a}-\mathbf{b}$ will be the vector whose beginning is the end of $\mathbf{b}$ and its end is the end of $\mathbf{a}$ (Figure 8.2*d*).

2. The product of a vector $\mathbf{a}$ and a real number q is, by definition, such a vector (denoted by $q\mathbf{a}$ or $\mathbf{a}q$) that its modulus is $|q|\cdot|\mathbf{a}|$ and the direction coincides with the direction of $\mathbf{a}$ for $q>0$, while it is opposite to the direction of $\mathbf{a}$ for $q<0$.

Hence, the multiplication of a vector by a positive number is simply its extension (or contraction), while multiplication by a negative number is an extension (contraction) accompanied by the changing of its direction (Figure 8.3). Note that

$$1\cdot\mathbf{a}=\mathbf{a}, \qquad 0\cdot\mathbf{a}=\mathbf{0}.$$

Collinear vectors are the vectors that lie in one straight line or in parallel straight lines (Figure 8.3). Collinear vectors differ from each other only by a numerical multiplier. A zero-vector is collinear to any vector.

Coplanar vectors are three or more vectors that lie in one plane or in parallel planes. Any three coplanar vectors are connected by a relation:

$$\alpha\mathbf{a}+\beta\mathbf{b}+\gamma\mathbf{c}=0,$$

where at least one of the numbers α, β, or γ is not 0.

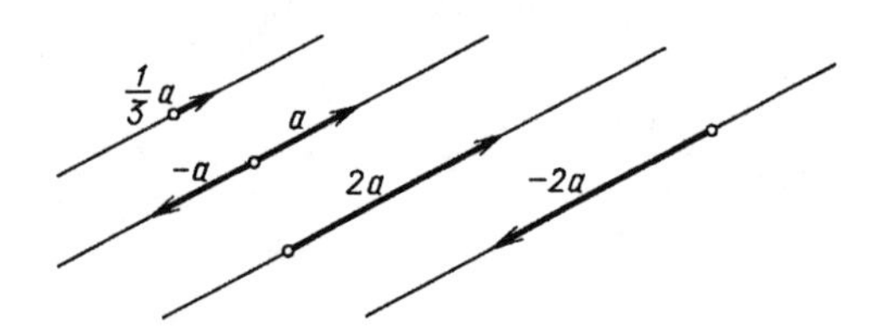

Figure 8.3: Collinear vectors.

8.1.2 Projections

Axis l is a straight line on which a positive direction is chosen; an arbitrary chosen point of the axis separates it into positive and negative *semiaxes*.

Projection a_l, or $(\mathbf{a})_l$, of a vector $\mathbf{a}$ on an axis l is, by definition, the product of the vector's modulus and the cosine of the angle φ between the vector and the positive direction of l:

$$(\mathbf{a})_l = |\mathbf{a}| \cos\varphi.$$

The projection of a vector is positive when φ is an acute angle (Figure

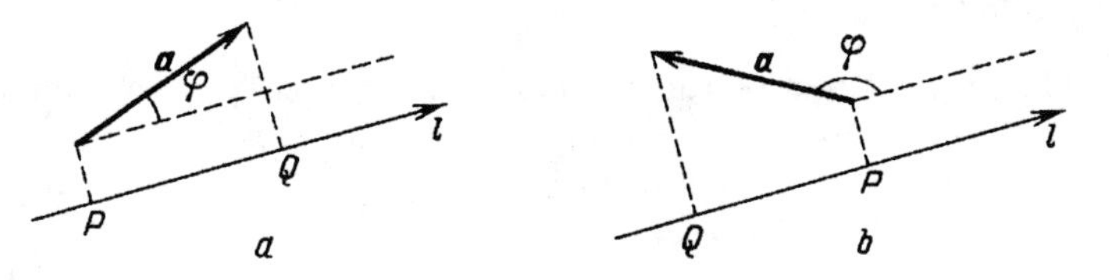

Figure 8.4: Projection of a vector on an axis.

8.4*a*), negative when the angle is obtuse (Figure 8.4*b*), and zero when the angle is a right angle. Evidently,

$$(\mathbf{a})_l = \pm|PQ|,$$

where P and Q are the feet of perpendiculars dropped from the beginning and the end points of the vector $\mathbf{a}$ towards the axis l ; the sign "+" corresponds to an acute angle, and "−" to an obtuse one.

Coordinates, or *scalar components*, a_x, a_y, a_z of a vector $\mathbf{a}$ with respect to a Cartesian coordinate system in the space (see Section 8.3) are, by definition, the projections of $\mathbf{a}$ on the coordinate axes x, y, z:

$$\begin{aligned} a_x &= (\mathbf{a})_x = |\mathbf{a}| \cos\alpha, \\ a_y &= (\mathbf{a})_y = |\mathbf{a}| \cos\beta, \\ a_z &= (\mathbf{a})_z = |\mathbf{a}| \cos\gamma. \end{aligned}$$

Here α, β, γ are the angles between $\mathbf{a}$ and the corresponding positive semiaxes (Figure 8.5). The coordinates of a vector are also called its *projections* (on the coordinate axes).

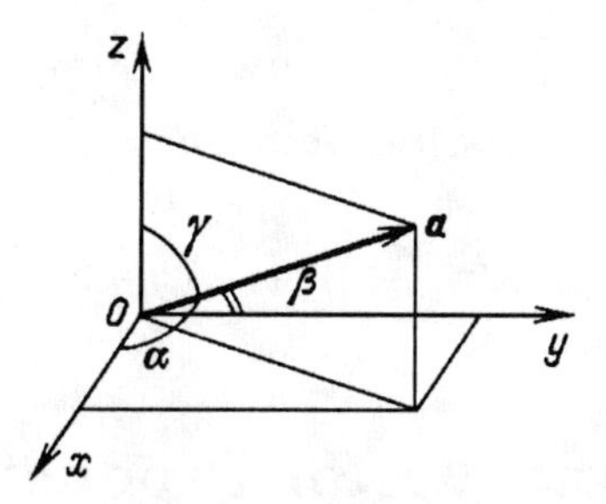

Figure 8.5: The angles α, β, γ between the vector $\mathbf{a}$ and the coordinate axes.

If a vector is given by two points $A(x_1, y_1, z_1)$ and $B(x_2, y_2, z_2)$, then its projections are equal to the differences of the corresponding coordinates of these points (see Section 8.3):

$$(\overrightarrow{AB})_x = x_2 - x_1, \quad (\overrightarrow{AB})_y = y_2 - y_1, \quad (\overrightarrow{AB})_z = z_2 - z_1. \tag{8.2}$$

In the addition of vectors, their projections are respectively added, in multiplication by a number, they are multiplied by this number:

$$\begin{array}{ll} (\mathbf{a}+\mathbf{b})_x = a_x + b_x, & (\gamma\,\mathbf{a})_x = \gamma\,(\mathbf{a})_x \\ (\mathbf{a}+\mathbf{b})_y = a_y + b_y, & (\gamma\,\mathbf{a})_y = \gamma\,(\mathbf{a})_y \\ (\mathbf{a}+\mathbf{b})_z = a_z + b_z, & (\gamma\,\mathbf{a})_z = \gamma\,(\mathbf{a})_z \end{array}$$

The modulus of a vector $\mathbf{a}(a_x, a_y, a_z)$ is calculated according to the formula:

$$|\mathbf{a}| = \sqrt{(a_x)^2 + (a_y)^2 + (a_z)^2}. \tag{8.3}$$

In the case where all vectors are considered in the plane xy, all preceding formulas hold, but the third coordinate is to be omitted:

$$|\mathbf{a}| = \sqrt{(a_x)^2 + (a_y)^2}.$$

8.1.3 Expansion in unit vectors

Any vector $\mathbf{a}$ in the plane may be expanded in unit vectors $\mathbf{i}, \mathbf{j}$:

$$\mathbf{a} = a_x\mathbf{i} + a_y\mathbf{j}.$$

In space, the expansion in unit vectors **i**, **j**, **k** (see Section 8.3) is the following one:

$$\mathbf{a} = a_x\mathbf{i} + a_y\mathbf{j} + a_z\mathbf{k}. \tag{8.4}$$

The vectorial summands $a_x\mathbf{i}$, $a_y\mathbf{j}$, $a_z\mathbf{k}$ are called the *vector components* of **a** respect to the axes x, y, z.

Example

Expand, in unit vectors, the vector $\mathbf{c} = 3\mathbf{a} + 2\mathbf{b}$, where $\mathbf{a}(1, 0, 2)$, $\mathbf{b}(-2, 1, 0)$.

Solution: $\mathbf{c} = (3 \cdot 1 + 2 \cdot (-2))\mathbf{i} + (3 \cdot 0 + 2 \cdot 1)\mathbf{j} + (3 \cdot 2 + 2 \cdot 0)\mathbf{k} = -\mathbf{i} + 2\mathbf{j} + 6\mathbf{k}$.

8.2 Scalar and Vector Products

8.2.1 Scalar product

The *scalar* (or *inner*, or dot) *product* $\mathbf{a} \cdot \mathbf{b}$ of two vectors **a** and **b** is a number which is equal to the product of moduli of the vectors and the cosine of the angle between them:

$$\mathbf{a} \cdot \mathbf{b} = |\mathbf{a}| \cdot |\mathbf{b}| \cos(\mathbf{a};\ \mathbf{b}).$$

A scalar product is alternatively denoted as $\mathbf{a}\mathbf{b}$ or $(\mathbf{a} \cdot \mathbf{b})$, and sometimes as $(\mathbf{a}, \mathbf{b})$.

Evidently, the scalar product is positive for the acute angle between the vectors, negative for the obtuse angle and zero for the right angle. In physics, the scalar product is met, for instance, in the definition of work (Section 13.4.5).

A scalar product may be expressed in terms of the projections of the vectors:

$$\mathbf{a} \cdot \mathbf{b} = a_x b_x + a_y b_y + a_z b_z.$$

Examples

1. $\mathbf{a}(1, 2, -1)$, $\mathbf{b}(0, 3, 4)$; $\mathbf{a} \cdot \mathbf{b} = 1 \cdot 0 + 2 \cdot 3 + (-1) \cdot 4 = 2$ (acute angle).
2. $\mathbf{a}(3, -2)$, $\mathbf{b}(-1, 2)$; $\mathbf{a} \cdot \mathbf{b} = -3 - 4 = -7$ (obtuse angle).
3. $\mathbf{a}(2, 1, -1)$, $\mathbf{b}(-1, 1, -1)$; $\mathbf{a} \cdot \mathbf{b} = -2 + 1 + 1 = 0$ (right angle).

The angle φ between two vectors **a** and **b** is calculated by using the formula:

$$\cos\varphi = \frac{\mathbf{a} \cdot \mathbf{b}}{|\mathbf{a}| \cdot |\mathbf{b}|} = \frac{a_x b_x + a_y b_y + a_z b_z}{\sqrt{a_x^2 + a_y^2 + a_z^2} \cdot \sqrt{b_x^2 + b_y^2 + b_z^2}}.$$

Example

$\mathbf{a}(3,-2)$, $\mathbf{b}(-1,2)$; $\cos\varphi = \frac{-7}{\sqrt{(3^2+2^2)(1^2+2^2)}} = -\frac{7}{\sqrt{65}} \approx -0.682$,

whence $\varphi = \arccos(-\frac{7}{\sqrt{65}}) \approx 2.62\ rad \approx 150°$.

8.2.2 Vector product

The *vector product* $\mathbf{a} \times \mathbf{b}$ of two vectors $\mathbf{a}$ and $\mathbf{b}$ is a vector $\mathbf{c}$, which is defined as follows:

(i) $|\mathbf{c}| = |\mathbf{a}| \cdot |\mathbf{b}| \sin(\mathbf{a};\mathbf{b})$, i.e., the modulus of the vector product is numerically equal to the area of a parallelogram constructed with the vectors $\mathbf{a}$ and $\mathbf{b}$;

(ii) $\mathbf{c} \perp \mathbf{a}$, $\mathbf{c} \perp \mathbf{b}$, i.e., the vector $\mathbf{c}$ is perpendicular to the plane in which lie both vectors $\mathbf{a}$ and $\mathbf{b}$;

(iii) Three vectors $\mathbf{a}$, $\mathbf{b}$, $\mathbf{c}$ form a *right trihedral*, that is, the shortest rotation of $\mathbf{a}$ towards $\mathbf{b}$ is seen, from the end point of $\mathbf{c}$, as a counter-clockwise rotation (see Figure 8.6*a*).

If the vectors $\mathbf{a}$ and $\mathbf{b}$ are collinear, then $\mathbf{a} \times \mathbf{b} = \mathbf{0}$.

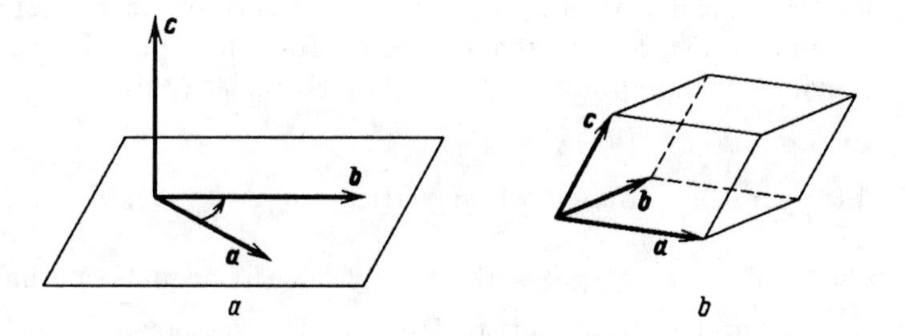

Figure 8.6: (*a*) – Vector product of vectors. (*b*) – Right trihedral.

The vector product is alternatively denoted as $[\mathbf{a}\,\mathbf{b}]$, sometimes as $[\mathbf{a},\mathbf{b}]$. It is also called the *cross* or *outer* product.

Remark. In some books, the right trihedral is called the "direct trihedral". The definition above is given for a right (or "direct") coordinate system (see Section 8.3.3); for a left (or "retrograde") system, the right trihedral mentioned in (iii) should be replaced by the left, or retrograde, trihedral.

The concept of vector product is used in physics, for instance, in the definition of moment of force (Section 13.6.1) and of Lorentz' force (Section 15.3.2).

Properties of a vector product

1. $\mathbf{a} \times \mathbf{a} = \mathbf{0}$.
2. $\mathbf{a} \times \mathbf{b} = -\mathbf{b} \times \mathbf{a}$ –the vector product is anticommutative.
3. $(\mathbf{a} + \mathbf{b}) \times \mathbf{c} = \mathbf{a} \times \mathbf{c} + \mathbf{b} \times \mathbf{c}, \quad (\gamma\mathbf{a} \times \mathbf{b}) = \gamma(\mathbf{a} \times \mathbf{b})$.

The coordinates of the vector product with respect to a Cartesian coordinate system (see Section 8.3.3):

$$\begin{aligned} (\mathbf{a} \times \mathbf{b})_x &= a_y b_z - a_z b_y, \\ (\mathbf{a} \times \mathbf{b})_y &= a_z b_x - a_x b_z, \\ (\mathbf{a} \times \mathbf{b})_z &= a_x b_y - a_y b_x. \end{aligned} \tag{8.5}$$

A vector product may be briefly written by using a determinant of the 3rd order:

$$\mathbf{a} \times \mathbf{b} = \begin{vmatrix} \mathbf{i} & \mathbf{j} & \mathbf{k} \\ a_x & a_y & a_z \\ b_x & b_y & b_z \end{vmatrix}.$$

The vector product of unit vectors $\mathbf{i}$, $\mathbf{j}$, $\mathbf{k}$:

$$\mathbf{i} \times \mathbf{j} = \mathbf{k}, \quad \mathbf{j} \times \mathbf{k} = \mathbf{i}, \quad \mathbf{k} \times \mathbf{i} = \mathbf{j}. \tag{8.6}$$

In practice, calculation of the vector product is conveniently done by using Formulas (8.5), or by expanding both co-factors in unit vectors, then removing the parentheses and applying Formulas (8.6).

Example

Find the area of a triangle whose vertices are $A(1, 0, 1)$, $B(-1, 1, 2)$, $C(2, 3, 0)$.

Solution. The area S_{ABC} of the triangle ΔABC equals one half of the area of a parallelogram constructed with the vectors $\overrightarrow{AB}$ and $\overrightarrow{AC}$. The coordinates of vectors are found using (8.2): $\overrightarrow{AB}$ $(-2, 1, 1)$, $\overrightarrow{AC}$ $(1, 3, -1)$. The area of the parallelogram may be calculated as a modulus of the vector product $\overrightarrow{AB} \times \overrightarrow{AC}$:

$$\overrightarrow{AB} \times \overrightarrow{AC} = (-2\mathbf{i} + \mathbf{j} + \mathbf{k}) \times (\mathbf{i} + 3\mathbf{j} - \mathbf{k}) = -4\mathbf{i} - \mathbf{j} - 7\mathbf{k}.$$

The modulus is found by Formula (8.3):

$$|\overrightarrow{AB} \times \overrightarrow{AC}| = \sqrt{4^2 + 1^2 + 7^2} = \sqrt{66}.$$

The answer: $S_{ABC} = \sqrt{66}/2$.

Remark. The volume of a parallelepiped constructed with three vectors $\mathbf{a}$, $\mathbf{b}$, $\mathbf{c}$ (Figure 8.6*b*) may be calculated applying the formula

$$V = |\mathbf{a} \cdot (\mathbf{b} \times \mathbf{c})|.$$

8.3 Coordinate Systems

8.3.1 Coordinate axis

Let us take a straight line, mark some point O (the *origin*) in it, fix a *scale* (i.e., choose a unit interval to measure distances), and assign a *positive direction* to one of the half-lines. Thus the straight line becomes a *coordinate axis.*

In Figure 8.7a, the unit interval for the coordinate axis x is denoted by OE, the direction from O towards E is taken to be positive (shown by an arrow). The origin O splits the axis into two rays: *positive half-line* (the point E is in it) and *negative half-line.*

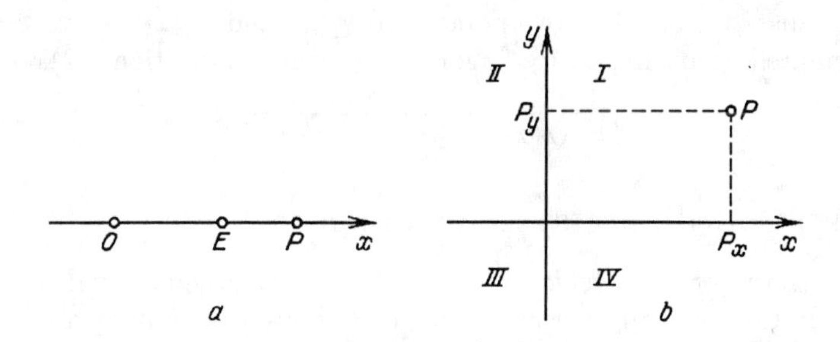

Figure 8.7: (a) – Coordinate axis, (b) – Cartesian coordinate system.

The *coordinate* of a point P belonging to the axis x is, by definition, the number $x = \pm|OP|$, where $|OP|$ denotes the length of the interval OP, and the sign "+" is taken if P belongs to the positive half-line, while the sign "−" if P belongs to the negative one, and we put $x = 0$ if $P = O$. The distance between two points P_1 and P_2 which lie in the axis x, is equal to $|x_1 - x_2|$, where x_1, x_2 are, respectively, the coordinates of the two points P_1, P_2.

The *unit vector* of an axis x (Figure 8.7a) is the vector OE, its length is 1, and its direction coincides with the positive direction of the axis.

8.3.2 Coordinate system in a plane

A *Cartesian* (or *rectangular*) *coordinate system* in a plane is defined by a pair of mutually perpendicular coordinate axes having a common origin O and identical scales (Figure 8.7b). The coordinate axes are usually denoted by letters x and y, and are called the *abscissa* and *ordinate* axes, respectively.

Unit vectors are usually denoted by **i** and **j**, for the x- and y-axes, respectively. The coordinate plane is denoted as xy, or xOy. In some special cases, the scales may be taken different for x and y.

The coordinate axes split the plane xy into four *quadrants:* I, II, III, IV (Figure 8.7*b*).

Take a point P in the plane xy, drop perpendiculars from P to the coordinate axes, and denote by P_x, P_y the feet of the corresponding perpendiculars (Figure 8.7*b*). The coordinate of point P_x in the x-axis is called the abscissa of point P, the coordinate of P_y in the y-axis is called the ordinate of P. The coordinates x, y of a point P are traditionally written near the name of the point: $P(x, y)$. The set of points in a coordinate plane and the set of their coordinate pairs are in one-to-one correspondence (see Section 2.5.1).

The distance between two points $P_1(x_1, y_1)$ and $P_2(x_2, y_2)$ in a plane is calculated according to Pythagoras' theorem (see Section 9.2):

$$d = \sqrt{(x_1 - x_2)^2 + (y_1 - y_2)^2}.$$

8.3.3 Coordinate systems in space

A *Cartesian* (or *rectangular*) *coordinate system* in space is defined by a trihedral of mutually perpendicular coordinate axes having a common origin O and identical scales (Figure 8.8). The coordinate axes in space are usually denoted by letters x, y, z, and the coordinate system by xyz. The space coordinate system can be *right* (or "direct"), see Figure 8.8*a*, and *left* (or "retrograde"), see Figure 8.8*b*; they are equivalent, but in practice the right systems are used more often than the left ones.

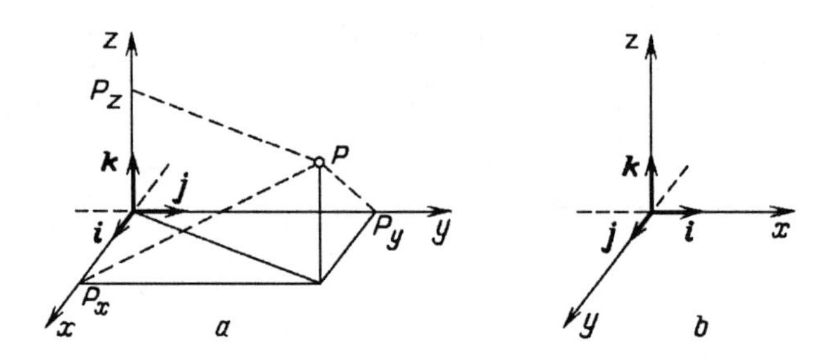

Figure 8.8: Cartesian coordinate systems in space: *a* – right system, *b* – left system.

Unit vectors on the axes are denoted by **i**, **j**, **k**, their directions coincide with positive directions of the axes (Figure 8.8). Unit vectors of a right coordinate system form a *right trihedral* (see Section 8.2.2), and those of a left system—a *left trihedral.*

The *coordinate planes* xy, yz, xz split the space into 8 *octants.*

The coordinates x, y, z of a point P in space are determined analogous to the plane case. Let us drop perpendiculars from the point P to the axes x, y, z, and let the feet of the perpendiculars be P_x, P_y, P_z; the coordinates x, y, z of P are the corresponding coordinates of P_x at x-axis, P_y at y-axis, and P_z at z-axis. The coordinates are traditionally written near the name of a point: $P(x, y, z)$. The points of space and their coordinate triples are in one-to-one correspondence.

Example

Let a cube $ABCDA'B'C'D'$ whose dimension is of length $d = 3$ be inscribed into the 1st octant (where $x > 0$, $y > 0$, $z > 0$) in such a way that the vertex A is located at the origin and the face $ABCD$ is inscribed into the 1st quadrant of the plane xy (Figure 8.9). The coordinates of the cube's vertices are:

$$A(0,0,0),\ B(3,0,3),\ C(3,3,0),\ D(0,3,0),$$

$$A'(0,0,3),\ B'(3,0,3),\ C'(3,3,3),\ D'(0,3,3).$$

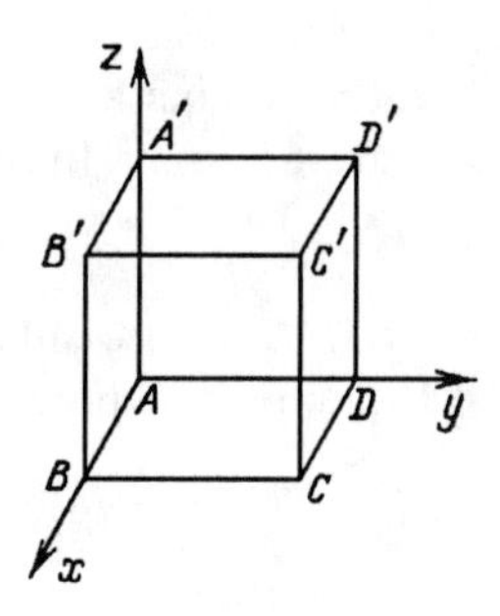

Figure 8.9: A cube is inscribed into the I octant.

The distance between two points $P_1(x_1, y_1, z_1)$ and $P_2(x_2, y_2, z_2)$ in space is calculated according to Pythagoras' theorem (Section 9.5.3):

$$d = \sqrt{(x_1 - x_2)^2 + (y_1 - y_2)^2 + (z_1 - z_2)^2}.$$

In particular, the distance between a point $P(x, y, z)$ and the origin O equals

$$d_0 = \sqrt{x^2 + y^2 + z^2}.$$

8.3.4 Polar coordinates

The *polar coordinate system* in the plane is defined by a point O (the *pole*) and a *polar axis* (i.e., a ray emitted from the pole—as the ray OE in Figure 8.10a). A unit interval is chosen as a scale to measure the distances (the unit interval OE in Figure 8.10a).

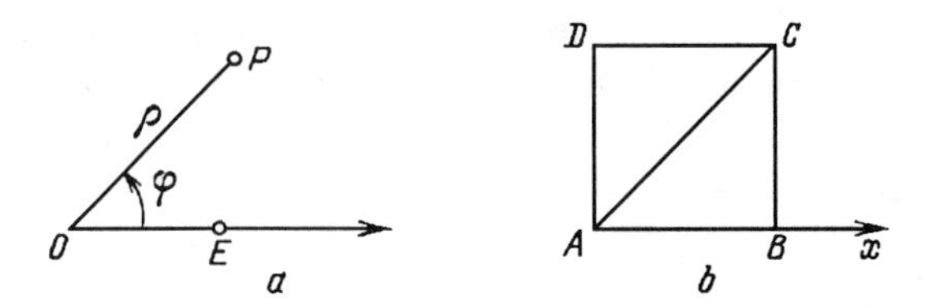

Figure 8.10: Polar coordinates.

The polar coordinates (ρ, φ) of a point P in a plane are, by definition, the distance ρ from point P to the pole O, and the angle φ between the polar axis and the vector $\overrightarrow{OP}$. It is a custom to consider the angle as positive if it is counted counterclockwise, and negative in the opposite case. In the case $P = O$ we put $\rho = 0$, and the value φ is undefined.

If we agree to take the values of the angle in the range $0 \le \varphi < 2\pi$, then the points in the plane and the pairs of their polar coordinates are in one-to-one correspondence (except the point $P = O$). Alternatively, the range for φ could be chosen, for instance, as $-\pi < \varphi \le \pi$, or, in general, $-\pi + \alpha < \varphi \le \pi + \alpha$, where α is a fixed arbitrary chosen angle.

Example

The vertices of a square $ABCD$ whose side is of length $d = 3$ (the square is shown in Figure 8.10b) have the polar coordinates

$$A(0, \varphi_0),\ B(3, 0),\ C(3\sqrt{2}, \pi/4),\ D(3, \pi/2),$$

where φ_0 is arbitrary.

Place the pole of the polar system at the origin of a Cartesian coordinate system and superpose the polar axis with the positive x-semi-axis, then we obtain the connection formulas

$$x = \rho \cos \varphi, \quad y = \rho \sin \varphi,$$

whence

$$\rho = \sqrt{x^2 + y^2}, \quad \tan \varphi = y/x, \quad \sin \varphi = y/\rho, \quad \cos \varphi = x/\rho.$$

The formulas to emphasize the angle φ in terms of x and y are found in Table 7.1.

Let us complement the polar coordinates ρ, φ by the z-axis that is perpendicular to the xy-plane, then we get the cylindrical coordinates ρ, φ, z in space (Figure 8.11).

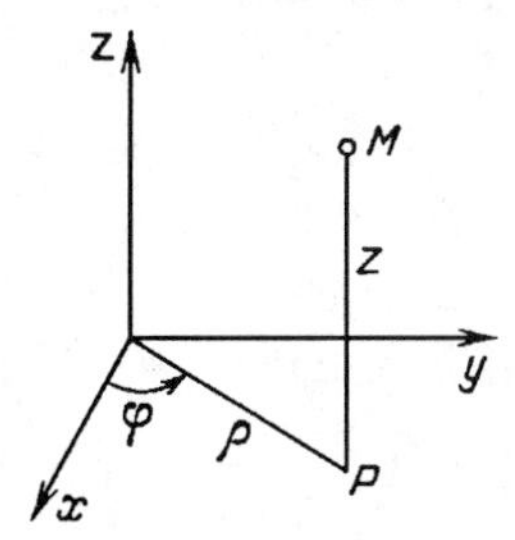

Figure 8.11: Cylindrical coordinates.

8.4 Displacement. Symmetry. Similarity

8.4.1 Displacement

Displacement (or moving) *of a plane* is a one-to-one transformation over points of the plane such that the distances are preserved: let a point A be mapped on a point A', B on B', then $|A'B'| = |AB|$. The angles are also preserved under displacements. Some examples of displacements: *translation* (Figure 8.12*a*), *rotation about a point* (Figure 8.12*b*), *reflection with respect to an axis* (or *axial symmetry*, Figure 8.12*c*). Any displacement of a plane may be represented as a superposition of displacements of the three kinds mentioned above. In some cases one speaks also about displacements of geometrical figures in a plane. Two figures are said to be *equal* (or congruent) to each other if there exists a displacement which causes the figures to coincide (Figure 8.12—all the triangles are equal to each other).

Displacement of space is defined similarly: it is one-to-one transformation over points of space such that the distances are preserved. The angles are also preserved, and the geometrical figures transform into equal figures.

Some examples of displacements in space: *translation* (Figure 8.13*a*), *rotation about an axis* (Figure 8.13*b*), and *reflection with respect to a plane* (specular reflection, Figure 8.13*c*). Translation and/or rotation of figures in space can be considered as real displacements, but to perform a reflection in space it would be necessary to leave our three-dimensional space. So, no real displacement of space can compel, for example, a

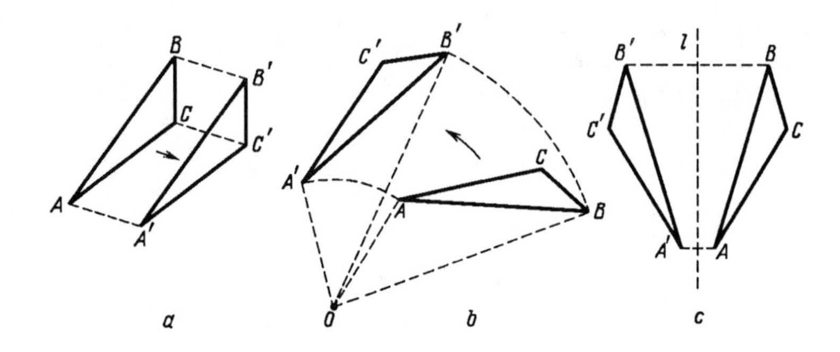

Figure 8.12: Displacements in a plane: a – translation, b – rotation, c – reflection with respect to an axis.

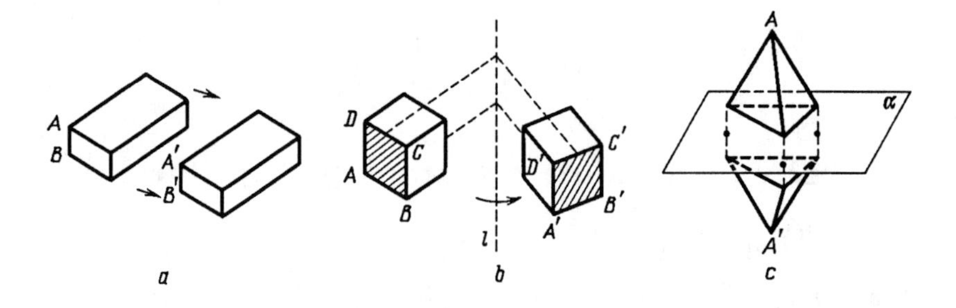

Figure 8.13: Displacements in space: a – translation, b – rotation, c – specular reflection.

right and a left gloves to coincide; nevertheless they are considered (in geometry!) to be equal. The same note concerns, as well, the right and the left trihedrals of vectors (see Section 8.2.2).

8.4.2 Symmetry

Two points M and M' in a plane or in space are said to be *symmetric* (to each other) *with respect to a center* O, if O is the midpoint of the segment MM' (Figure 8.14a). Two points M and M' are said to be *symmetric with respect to an axis* l if the segment MM' is perpendicular to the axis l, and MM' intersects the axis at the midpoint of MM' (Figure 8.14b).

A figure F is *symmetric* to a figure F' *with respect to a center* O (or with respect to an axis l) if all points of F are symmetric to corresponding points of F' with respect to the center O (or to the axis l).

In a plane, a figure is said to possess the *central symmetry* if it is symmetric to itself with respect to some center. Such a figure is said to

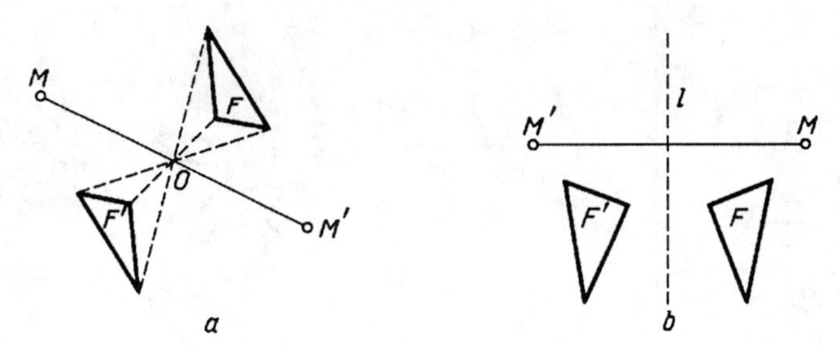

Figure 8.14: Points M, M' and figures F, F' are symmetric with respect to a center (a), or to an axis (b).

be a centrally symmetric one, and it may be matched with itself by using a rotation of it in the plane about the angle 180° (around the center of symmetry).

Some examples of planar centrally symmetric figures: a circle, an ellipse (see Section 9.3.2), a parallelogram, stars with even numbers of rays, regular polygons with even numbers of sides, and as well some flowers (approximately); see Figure 8.15.

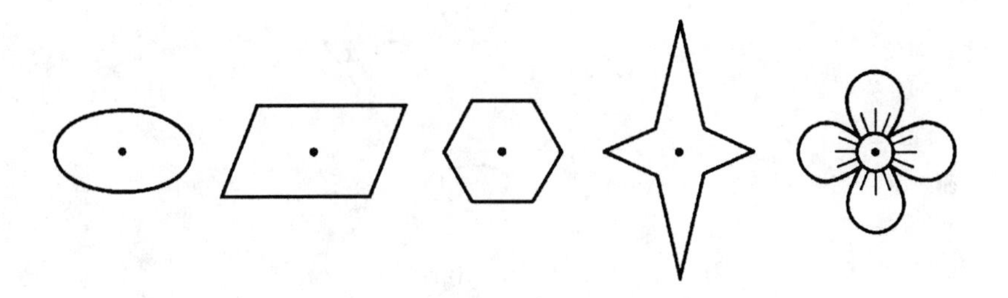

Figure 8.15: Centrally symmetric figures.

A plane figure is said to possess the *axial symmetry* (an axially symmetric figure) if it is symmetric to itself with respect to an axis belonging to the plane. Such a figure may be matched with itself by using a rotation of the plane (in space) around the axis about the angle 180°. Some examples of axially symmetric figures: isosceles trapezoid (Section 9.2.12), parabola (Sections 2.2.6 and 9.3.4), stars, regular polygons, and, as well, many flowers and insects (approximately), see Figure 8.16. A plane figure may possess both kinds of symmetry, e.g., circle, ellipse, hyperbola (see Section 9.3), rhombus, and so on.

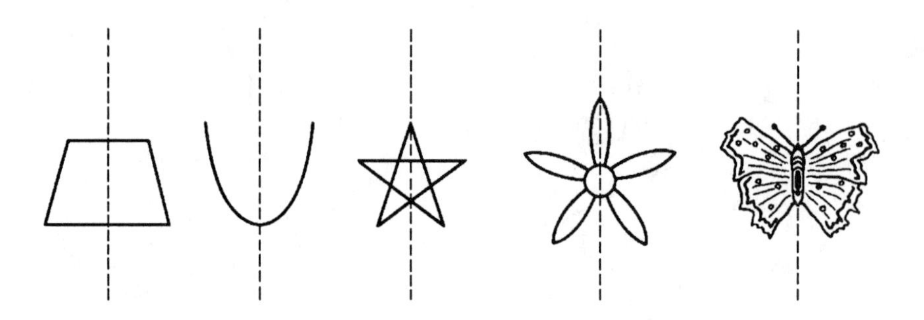

Figure 8.16: Axially symmetric figures.

8.4.3 Spatial symmetry

In space, a special kind of symmetry is considered: two points M and M' are said to be *symmetric* (to each other) *with respect to a plane* if the segment MM' is perpendicular to the plane and it intersects with the plane at the midpoint of MM' (Figure 8.17). We can say that point M' is a *specular reflection*, or a mirror image, of M by the symmetry plane.

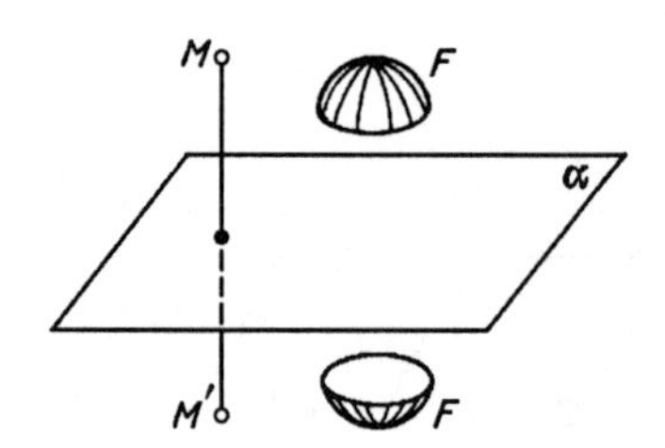

Figure 8.17: Specular reflection.

A figure F is said to be *symmetric to a figure* F' *with respect to the plane* if all points of F are symmetric to corresponding points of F' with respect to the plane (Figure 8.17). A figure possesses a *mirror* (or specular) *symmetry* if it is symmetric to itself with respect to a plane. Examples of mirror symmetric figures: ball, cube, cylinder, rectangular prism (see Section 9.5.3), as well as most animals and many engineering tools (approximately).

In some fields (in particular, in crystallography), one more kind of spatial symmetry is considered, the *axial symmetry of the n-th order*. We say that a figure possesses the n-th order symmetry with respect to an axis if it can be matched with itself by using a rotation around the axis about a minimal angle $\varphi = 360°/n$, where n is an integer. Thus a cube has symmetries of the 3rd order with respect to each internal diagonal, and of a 4th order with respect to a straight line passing through the centers of any two opposite faces, and it has also a 2nd-order symmetry with respect to a straight line passing through the midpoints of any two opposite ribs. In brief, a cube possesses the axes of the 4th order (three axes), the 3rd order (four axes), and the 2nd order (six axes). A snowflake has one symmetry axis of the 6th order. Any crystal (see Section 14.6.1) may possess only the axes of the 2nd, 3rd, 4th, and 6th order, because only such axial symmetries are compatible to the *translation symmetry*, i.e., to the spatial periodicity of a crystal. As examples of figures that have plane translational symmetry, we mention the parquet or the wallpaper.

8.4.4 Similarity

Similitude (or dilation) of space is defined as a one-to-one transformation of points in space under which all distances are multiplied by a certain positive number:

$$|A'B'| = \gamma|AB|,$$

the number γ is called the *similarity coefficient*. Any similitude may be represented as a composition of a displacement and a dilation (or compressing); see Figure 8.18. Two figures F and F' are called *similar* (to each other) if F' is a result of similitude over F. Some examples of similar figures: similar triangles (see Section 9.2.4), any segments, circles, squares, balls, regular tetrahedrons (see Section 9.5.7), reduced models of engineering constructions.

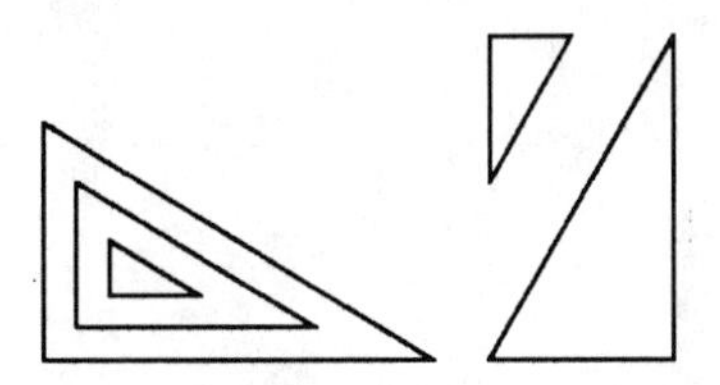

Figure 8.18: Similar figures.

For similar figures F and F', their linear sizes are in ratio $\gamma : 1$, the areas of plane figures and the surfaces of spatial figures are in ratio $\gamma^2 : 1$, and volumes of spatial figures are in ratio $\gamma^3 : 1$. These facts result in an interesting phenomenon: two bodies similar in geometrical meaning are, in general, not similar in physical nature.

For example, the mass of an animal is proportional to the cubic power of its linear size, but stability of its bones is proportional only to the second power—we find here an explanation why giant organisms cannot exist: in the process of continuous growing, the animal's body would become so heavy that at a certain moment its bones should be necessarily broken down. Another example of such a phenomenon: let two steel balls, a big and a small one, be heated to the same temperature and then be allowed to become cold, both in open air; then we can discover that the small ball becomes cold faster than the big one. The explanation is that the process of heat exchange goes through the surface (which is proportional to the second power of the diameter), while, on the other hand, the store of intrinsic energy in a ball is proportional to its volume and hence to the third power of the diameter. One more example: big droplets of mist go down faster than little ones because their weight is proportional to the third power and the resistive force—to the first power of the linear size (see Section 13.9.3).

Chapter 9

Geometry. Stereometry

Various figures in planes or in space can be investigated by considering triangles, as far as the figures may by treated as ones constructed of triangles. In this chapter we consider angles, triangles, and polygons in a plane, polyhedrons and bodies of revolution in space, and also the concept of the curvature of a surface.

9.1 Points, Straight Lines, and Angles in a Plane

9.1.1 Points and straight lines

A straight line lying in a plane splits (divides, separates) the plane into two *half-planes*. Arbitrary point O situated in a straight line splits the line into two *rays*, or *half-lines*. The point O is the common origin of the two rays; the rays are said to be *emitted* from the origin. To denote a ray we use two letters, the first one for the beginning and the second one for some other point of the ray (e.g., rays OA and OB in Figure 9.1).

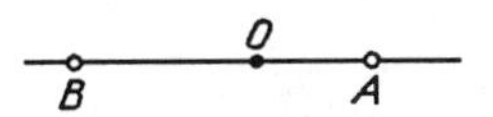

Figure 9.1: The straight line is split by point O into two rays.

For any two (different) points in a plane, it is possible to draw exactly one straight line passing through the points. Two different straight lines in a plane either *intersect* at a point, or they are *parallel* (in the last case they do not intersect). Parallelism is commonly denoted by the symbol $\|$, for instance, $a \| b$. If a given point is not in the given straight line, then there exists exactly one other straight line passing the point and being parallel to the given straight line.

A straight line in a plane may be introduced by its equation related to a Cartesian xy coordinate system:

$$ax + by + c = 0,$$

where a, b, c are constants, and x and y are the coordinates of a variable point $M(x, y)$ in the straight line. The vector $\vec{N}$ with the coordinates a, b (see Section 8.1) is called the "normal" to the line; this vector is perpendicular to the line. In the case $b \neq 0$ the equation may be rewritten as $y = \alpha x + \beta$ (see Section 2.6.1).

9.1.2 Segment

Two points A and B in a straight line define together a *segment* denoted as AB or BA; the points A and B are the *endpoints* (or ends) of the segment; the length of AB is denoted as $|AB|$ or simply AB. Two segments are considered as equal (to each other) if their lengths coincide. By using a displacement (see Section 8.4), a segment may be matched with any other segment equal to the given one.

If a point O is situated in a segment AB between the points A and B, then $|AB| = |AO| + |OB|$ (Figure 9.1).

9.1.3 Angle

Two rays emitted from a common origin form an *angle*, the rays are called the *sides* of the angle, and the common origin of the rays is called the *vertex*. An angle formed by the rays OA_1 and OA_2 is denoted as $\angle A_1OA_2$ or $\angle A_2OA_1$ (Figure 9.2a). In practical problems it is more convenient to regard the angle as a measure of rotation of a ray about its origin up to a prescribed position (Figure 9.2b). In geometry, all the angles are conventionally considered as positive ones.

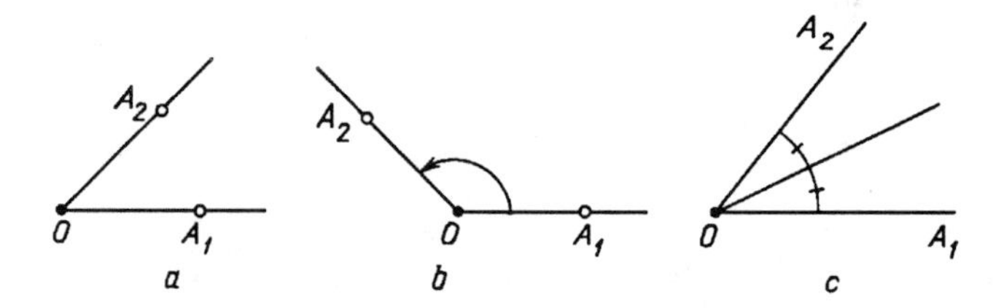

Figure 9.2: An angle as a pair of rays (a), or as a measure of rotation of a ray (b).

Two angles are considered as *equal* (to each other) if they admit matching by a suitable displacement (see Section 8.4.1). A *bisector* (or *bisectrix*) *of the angle* is a ray emitted from the vertex and splitting the angle into two equal parts (Figure 9.2*c*).

Let a ray OA_1 rotate about a point O and perform a complete revolution (then $A_2 = A_1$). The resulted angle is called *complete* (or full), see Figure 9.3*a*. A half of a complete angle is the *straight* angle; Figure 9.3*b*. Two sides of a straight angle lie on the same straight line. A half of a straight angle is called the *right* angle; Figure 9.3*c*.

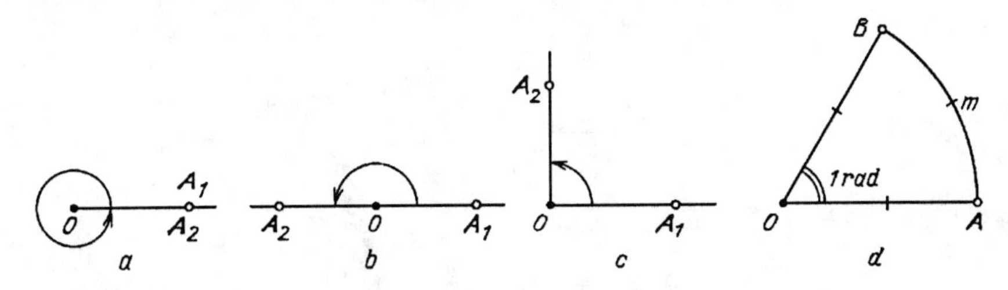

Figure 9.3: Angles: *a* – complete, *b* – straight, *c* – right, *d* – equal to 1 radian.

9.1.4 Degree

A *degree* (denoted by the symbol $^\circ$) is defined as 1/360 part of the complete angle. The right angle contains 90°, the straight angle 180°, the complete angle 360°. An angle less than 90° is called *acute* (Figure 9.2*a*, *c*), while an angle greater than 90° but smaller than 180° is called *obtuse* (Figure 9.2*b*). An *angle minute* is 1/60 part of a degree, and an *angle second* is 1/60 part of a minute. For example, the record 57°17′45″ is read as "57 degrees 17 minutes 45 seconds". One can measure the angles with the help of a protractor, and with a greater precision—by a goniometer; in geodesy the angles are measured by a theodolite.

9.1.5 Radian

A *radian* is defined as a central angle (in a circle, see Section 9.3.1) corresponding to an arc whose length is equal to the radius of the circle (Figure 9.3*d*). Being emphasized in degrees, 1 radian is near 57°17′45″. A right angle is equal to $\pi/2$ radians, a straight angle π radians, a complete angle 2π radians.

Let α be the value of an angle in degrees and β the same in radians, then

$$\alpha = \beta \cdot 180^\circ/\pi, \quad \text{or} \quad \beta = \alpha\pi/180^\circ.$$

For instance, an angle 30° is, in radians, $30^\circ\pi/180^\circ = \pi/6$.

9.1.6 Intersection of straight lines

When two straight lines intersect, they form four angles (Figure 9.4*a*). There are conventional names: the angles 1 and 2 (as well as 2 and 3, 3 and 4, 4 and 1) are *adjacent*; the angles 1 and 3 (as well as 2 and 4) are *vertical*. Vertical angles are equal to each other, and adjacent angles together give 180°.

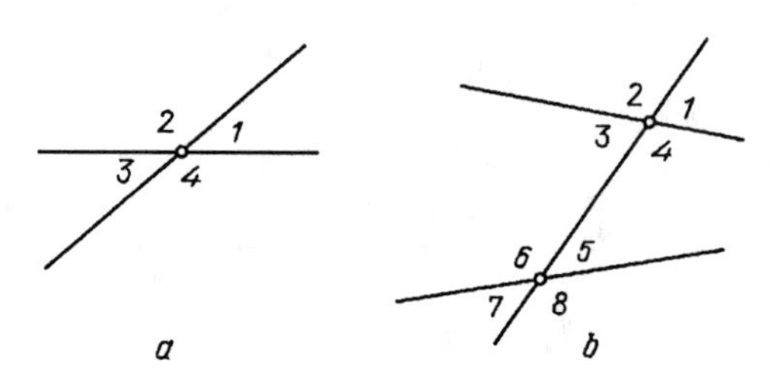

Figure 9.4: Angles: *a* – vertical (1 and 3, 2 and 4) and adjacent (1 and 2, 2 and 3, 3 and 4, 4 and 1), *b* – one-sided interior (4 and 5, 3 and 6), *c* – opposite interior (4 and 6, 3 and 5).

Two straight lines are called *perpendicular*, or *orthogonal* (to each other), if they intersect and form right angles. Perpendicularity of straight lines is denoted by the symbol $\perp$, e.g., $a \perp b$.

A straight line that intersects two other straight lines makes—together with these lines—eight angles (Figure 9.4*b*). There are conventional names: the angles 4 and 5 (as well as 3 and 6) are *one-sided interior*, while the angles 4 and 6 (as well as 3 and 5) are *opposite interior.*

Criteria for parallelism of straight lines

Two given straight lines are parallel if:

- There is another straight line parallel to each of the given lines.
- The given straight lines are intersected by another straight line and the opposite interior angles are equal.
- The given straight lines are intersected by another straight line and the one-sided interior angles, being added, give 180^0.
- There is another straight line perpendicular to each of the given lines.

Theorem of Thales

Let several mutually parallel straight lines intersect the sides of an angle (Figure 9.5, $n = 3$). Then the segments of the parallel lines situated inside the angle, are proportional:

$$\frac{|OA_1|}{|OB_1|} = \frac{|A_1A_2|}{|B_1B_2|} = \ldots = \frac{|A_{n-1}A_n|}{|B_{n-1}B_n|}.$$

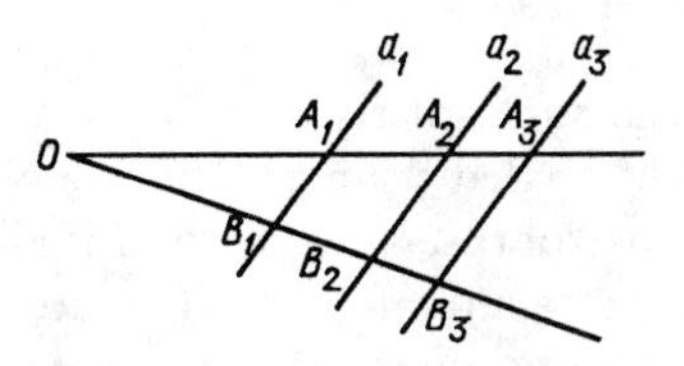

Figure 9.5: Parallel straight lines intersecting two sides of an angle.

Corollary: if several mutually parallel straight lines intersect the sides of an angle, and the segments cut from one side are equal to each other, then the segments cut from the second side are also equal to each other (the *theorem of Thales*).

9.2 Triangles. Polygons

9.2.1 Triangle

Let A, B, C be arbitrary points that do not lie in one straight line. The figure consisting of three segments AB, BC, AC is called a *triangle* ABC (the notation: ΔABC). Sometimes by the term "triangle" one means a part of the plane bounded by the segments AB, BC, AC. Each of the points A, B, C is the *vertex*, each of the segments AB, BC, AC is the *side* of the triangle (Figure 9.6).

Each of the angles $\angle CAB$, $\angle ABC$, $\angle BCA$ of the triangle ΔABC is usually denoted by one of Latin letters A, B, C, or by one of Greek letters α, β, γ; a small arc depicted inside the angle can help us in identification the angle (Figure 9.6). The angle $\angle A$ is called *adjacent* to the side AB and as well to AC, and it is called *opposite* (or alternative) to the side BC (also, the side BC is called the opposite side to the angle $\angle A$). Also, the angle $\angle B$ is adjacent to AB and BC, and $\angle C$ is adjacent to AC and BC, while $\angle B$ and AC, $\angle C$ and AB are mutually opposite.

The sum of the angles in a triangle is $180°$ ($= \pi$ radians):

$$\alpha + \beta + \gamma = 180°.$$

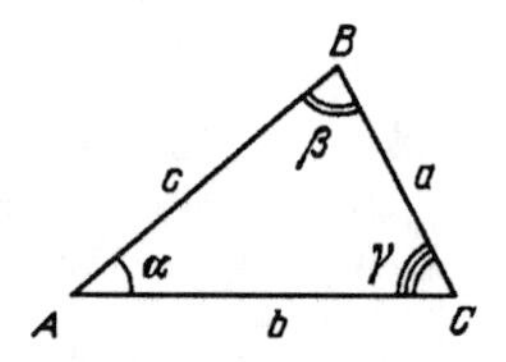

Figure 9.6: Angles and sides of a triangle.

In case where one of the angles is right, we have a *right triangle*; in some books it is called "right-angled triangle". If one of the angles is obtuse, the triangle is *obtuse*, or "obtuse-angled", and if all the angles are acute, the triangle is *acute*, or "acute-angled". (Figure 9.7). In an arbitrary triangle, opposite the greater angle there is situated the greater side, and opposite the equal angles there are situated the equal sides.

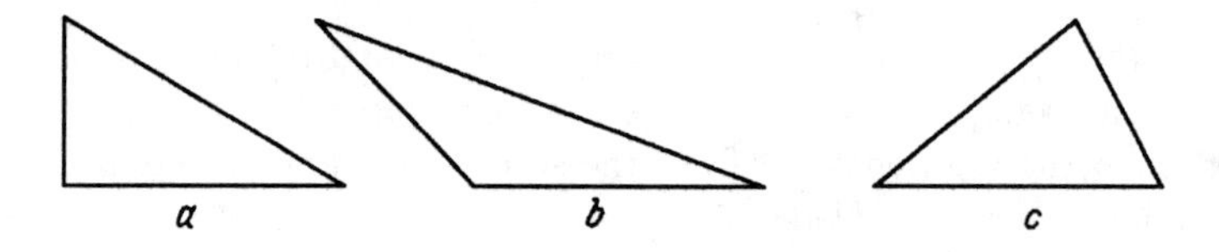

Figure 9.7: Triangles: a – right, b – obtuse, c – acute.

9.2.2 Elements of a triangle

In a triangle ABC let the point M be the midpoint of AB, the ray CS be the bisector of the angle $\angle C$, and $CH \perp AB$, $MP \perp AB$ (Figure 9.8a). The following terms are conventional: CM is the *median of* AB, CS the *bisector of angle* C, CH the *altitude from* C, and the straight line MP is the *middle perpendicular to*median AB. For an obtuse triangle, two of its altitudes are situated outside the triangle (Figure 9.8b).

Properties of a bisector of a triangle

- Each point of the bisector is equidistant to the sides of the angle.
- The bisector splits the opposite side into parts proportional to the adjacent sides:

$$|AS|/|BS| = |AC|/|BC|.$$

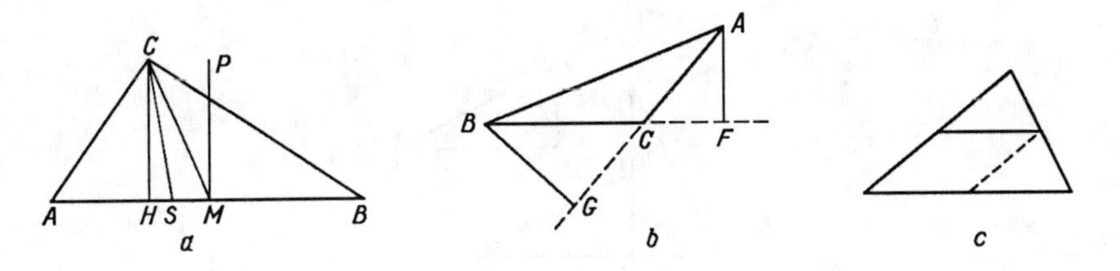

Figure 9.8: (a) – Elements of a triangle: CH – the altitude, CS – the bisector, CM – the median, MP – the middle perpendicular; (b) – two of the altitudes of an obtuse triangle are outside the triangle; (c) – the middle line of a triangle.

Four remarkable points in a triangle

1. Three medians intersect at one point (the "baricenter" of a homogeneous triangular plate, see Section 13.4.2).
2. Three bisectors intersect at one point (the center of an inscribed circle, see Section 9.2.5).
3. Three altitudes (or their continuations) intersect at one point (the "orthocenter").
4. Three middle perpendiculars intersect at one point (the center of a circumscribed circle, see Section 9.2.5; the barycenter of a homogeneous triangle frame, see Section 13.4.2).

The intersection point of medians splits every median to the ratio $2:1$ (starting from the vertex).

The middle line of a triangle (that is, the line joining the midpoints of two sides of a triangle) is parallel to the third side and is a half of it (Figure 9.8c).

9.2.3 Equal triangles

Two triangles are called *equal* (to each other) if they may be matched by using a suitable displacement (see Section 8.4.1). For example, $\triangle ABC$ on the Figure 9.9 may be matched with $\triangle A_1B_1C_1$ by a translation, or with $\triangle AB_2C_2$ by a rotation, or with $\triangle ABC_3$ by a symmetry (reflection) with respect to the axis AB.

Criteria for equality of triangles

Two triangles are equal if:

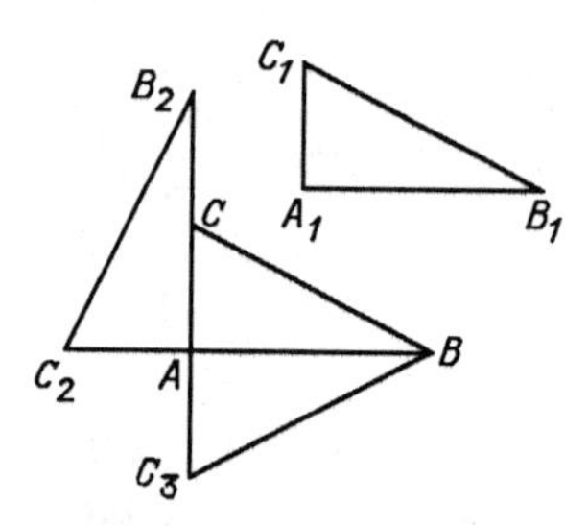

Figure 9.9: Equal triangles: they may be matched by a suitable displacement.

- Two sides and the angle between them in one triangle are equal, respectively, to two sides and the angle between them, in another triangle (the criterion "by two sides and the angle between them").
- A side and two angles adjacent to this side in one triangle are equal, respectively, to a side and two adjacent sides in another triangle (the criterion "by a side and adjacent angles").
- Three sides of one triangle are equal, respectively, to three sides of another triangle (the criterion "by three sides").

In a right triangle, the side opposite to the right angle is called the *hypotenuse* and each of two other sides is called the *leg* (or the *cathetus*.)

Two right triangles are equal by two elements:

- By two legs.
- By a leg and the hypotenuse.
- By the hypotenuse and an acute angle.
- By a leg and an adjacent acute angle.
- By a leg and an opposite acute angle.

A triangle whose two sides are equal to each other is said to be *isosceles* ($AC = BC$ in Figure 9.10a). In this case, the third side is the *base* of a triangle, the equal sides are the *lateral* sides. An isosceles triangle ΔABC is symmetrical with respect to the axis, which is perpendicular to the base and passes through the opposite vertex (Figure 9.10a). The segment CH is at the same time the altitude, the median, and the bisector, and it splits ΔABC into two equal right triangles.

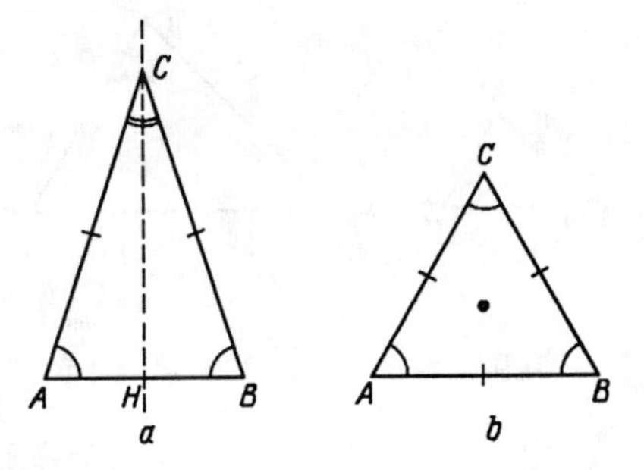

Figure 9.10: Triangles: isosceles (a) and equilateral (b).

In an *equilateral* (or *regular*) triangle all the sides are equal to each other, and all the angles are 60° (Figure 9.10b). An equilateral triangle has a symmetry axis of the 3rd order (see Section 8.4.3) which is perpendicular to the plane of the triangle and passing through the center of the triangle (where all of the four remarkable points coincide).

9.2.4 Similar triangles

Two triangles are said to be *similar* if one of them may be transformed into another by a suitable similitude (see Section 4.4). We can say that similar triangles are of the same form but of different size. Corresponding sides of similar triangles are proportional and the corresponding angles are equal (Figure 9.11):

$$\angle A = \angle A_1, \quad \angle B = \angle B_1, \quad \angle C = \angle C_1,$$

$$\frac{|A_1B_1|}{|AB|} = \frac{|B_1C_1|}{|BC|} = \frac{|A_1C_1|}{|AC|}.$$

The ratio $\gamma = |A_1B_1|/|AB|$ is called the *similarity coefficient*. The similarity coefficient characterizes the expansion of the plane that converts ΔABC into $\Delta A_1B_1C_1$. In similar triangles, the ratios of all corresponding elements (e.g., medians, bisectors, altitudes, middle perpendiculars) are equal to γ and the ratio of areas equals γ^2.

Criteria for similarity of triangles

Two triangles are similar if:

- Two angles of one triangle are equal, respectively, to two angles of the second triangle.

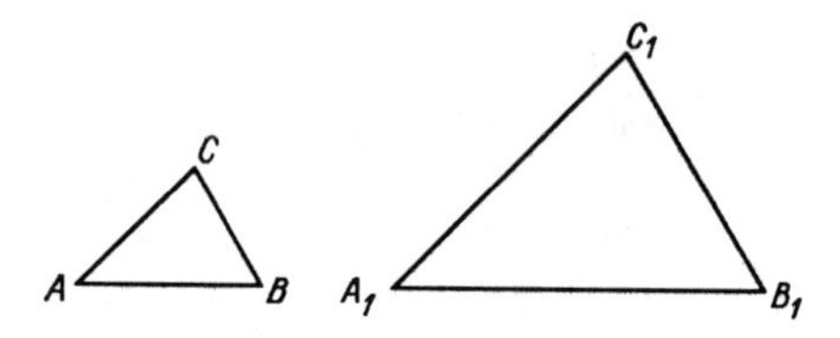

Figure 9.11: Similar triangles.

- Two sides of one triangle are proportional to two sides of the second triangle and the angles between the corresponding sides are equal.
- Three sides of one triangle are proportional to three sides of the second triangle.

9.2.5 Formulas for a triangle

Any element of a triangle may be calculated in terms of several other elements. Let S denote the area; a, b, c the sides; α, β, γ the angles; p the semi-perimeter: $p = (a + b + c)/2$; R the radius of the circumscribed circle, r the radius of the inscribed circle; h_c the altitude, m_c the median, l_c the bisector emitted from the vertex C.

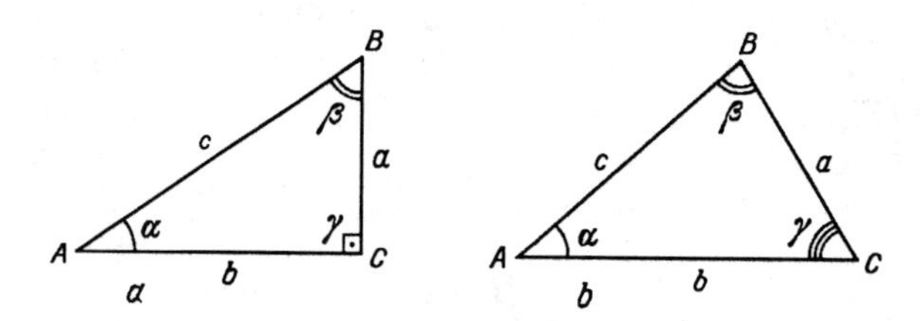

Figure 9.12: Triangles: a – right, b – general.

1. *Formulas for a right triangle* ($\gamma = 90°$, Figure 9.12a):

$$\alpha + \beta = 90°,$$

$$c^2 = a^2 + b^2 \quad \text{(the Pythagoras theorem)},$$

$$a = c \sin\alpha = b \tan\alpha, \quad \sin\alpha = \cos\beta, \quad \tan\alpha = \cot\beta,$$

$$S = \frac{1}{2}ab, \quad r = \frac{ab}{a + b + c}, \quad h_c = abc,$$

$$R = m_c = \frac{1}{2}c, \quad l_c = \frac{bc\sqrt{2}}{b+c}.$$

2. *Formulas for a general triangle* (Figurc 9.12*b*):

$$\alpha + \beta + \gamma = 180°,$$

$$c^2 = a^2 + b^2 - 2ab\cos\gamma \quad \text{(the cosine theorem)},$$

$$\frac{a}{\sin\alpha} = \frac{b}{\sin\beta} = \frac{c}{\sin\gamma} = 2R \quad \text{(the sine theorem)},$$

$$S = \frac{1}{2}ah_a, \quad S = \frac{1}{2}ab\sin\gamma, \quad S = rp,$$

$$S = \sqrt{p(p-a)(p-b)(p-c)} \quad \text{(the Heron formula)},$$

$$h_a = b\sin\gamma = c\sin\beta, \quad m_a = \frac{1}{2}\sqrt{b^2 + c^2 + 2bc\cos\alpha},$$

$$l_a = \frac{2bc}{b+c}\cos\frac{\alpha}{2}, \quad r = \frac{2S}{a+b+c},$$

$$r = p\tan\frac{\alpha}{2}\tan\frac{\beta}{2}\tan\frac{\gamma}{2} = 4R\sin\frac{\alpha}{2}\sin\frac{\beta}{2}\sin\frac{\gamma}{2}.$$

9.2.6 Polygons

Let several segments be situated in one plane; for each segment, we choose one of its endpoints and call it the *initial point* or the *origin*, and the other endpoint—simply the *end*. If the initial point of the second segment coincides with the end of the first segment, the initial point of the third segment coincides with the end of the second segment, and so on, then the totality (union) of the segments is called the *polygonal line* (no two of adjacent segments are supposed to lie in one straight line).

The segments that form the polygonal line are the *links* (or *components*), the endpoints of the segments are the *vertices* (or *nodes*) of the line (e.g., the polygonal line $A_1A_2...A_8$ in Figure 9.13*a*). A polygonal line is *simple* if it does not intersect itself; it is *closed* if the end of the last segment coincides with the initial point of the first segment (Figure 9.13*b*).

Polygon is a simple closed polygonal line. The links of the line are the *sides*, the vertices of the line are the *vertices* of the polygon. A polygon having n sides is called n-polygon (or n-angle polygon). Sometimes by the term "polygon" one means a part of the plane bounded by a simple closed polygonal line. The *perimeter* of the polygon is the sum of lenghts of all sides.

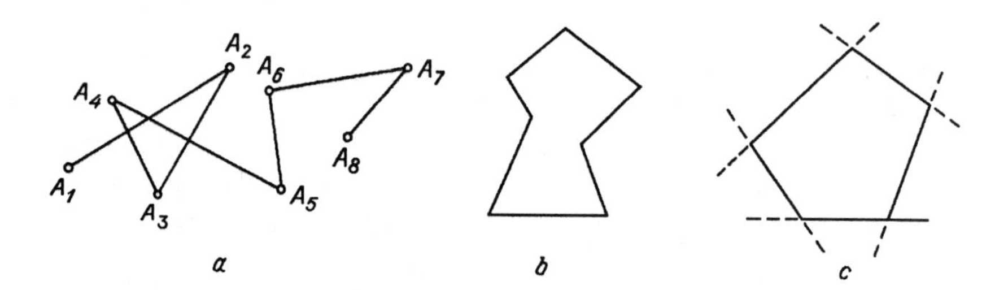

Figure 9.13: Polygonal line (a), polygon (b), and convex polygon (c).

Let us draw a straight line along a side of a polygon; this line will split the plane into two half-planes. Let the polygon be contained as a whole in one of the half-planes. If the same property is valid with respect to each side of the polygon, we say that the polygon is *convex* (Figure 9.13c).

The angles of a convex polygon are the interior angles formed by the pairs of adjacent sides. Each angle of a convex polygon is less than 180°. The sum Q of all angles is equal to $(n-2)\cdot 180°$, or

$$Q = (n-2)\cdot \pi,$$

where n is the number of the sides. A nonconvex polygon has at least one (interior) angle greater than 180°.

If a polygon is circumscribed around a circle (here the circumference is assumed to be tangent to each side of the polygon), then it is necessarily convex, and its area $S = pr$, where r is the radius of the circle, p the semi-perimeter, that is, p is half of the sum of all sides.

A polygon is said to be *regular* if all its sides are equal to each other and all its angles are equal to each other. Any regular polygon is convex. Examples of regular polygons: the equilateral triangle, the square, the regular pentagon or hexagon, etc. The plane may be completely covered either by equilateral triangles, or squares, or regular hexagones.

For any regular polygon, a circle may be inscribed into and another circle circumscribed around the polygon. The following formulas are true:

$$R = \frac{a}{2\sin(\pi/n)}, \quad r = \frac{a}{2\tan(\pi/n)},$$

where a is the side, n the number of polygon sides, R the radius of the circumscribed circle, r the radius of the inscribed one; r is called the *apothem* of the regular polygon. The area of a regular polygon:

$$S = \frac{1}{2}nar = \frac{1}{2}na^2\cot\frac{\pi}{n}.$$

9.2.7 Parallelogram

A *parallelogram* is the quadrangle whose opposite sides are mutually parallel (Figure 9.14a).

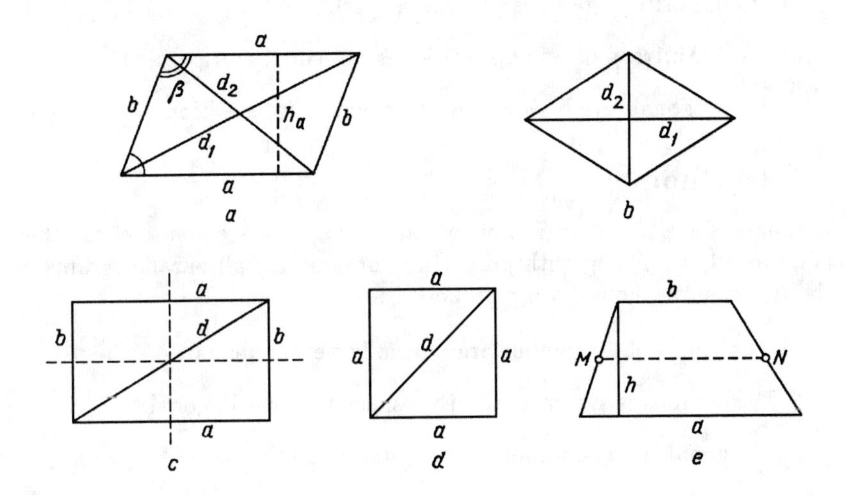

Figure 9.14: Parallelogram (a), rhombus (b), rectangle (c), square (d), and trapezoid (e).

Along with properties common for all convex polygons, the parallelogram shows the following special properties:

1. The opposite sides are equal.

2. The opposite angles are equal.

3. The diagonals are bisected by the point of intersection.

4. The intersection point of diagonals is the center of symmetry of the parallelogram.

5. The sides and the diagonals are related by the formula

$$d_1^2 + d_2^2 = 2(a^2 + b^2).$$

The area of a parallelogram (see Section 8.2.2):

$$S = ah_a = bh_b, \quad S = ab\sin\alpha = ab\sin\beta, \quad S = |\mathbf{a} \times \mathbf{b}|.$$

Criteria for parallelogram

A quadrangle is a parallelogram if:

- Its opposite sides are mutually equal.
- Its opposite angles are mutually equal.
- Two arbitrary of its opposite sides are equal and parallel.
- Its diagonals are bisected by the point of intersection.

9.2.8 Rhombus

A *rhombus* is a parallelogram whose all four sides are equal to each other (Figure 9.14*b*). Along with properties common for all parallelograms, a rhombus shows the following properties:

1. Diagonals of a rhombus are mutually perpendicular.
2. A rhombus is symmetric with respect to each diagonal.
3. Diagonals of a rhombus are also its bisectors.
4. The area of a rhombus is related with the diagonals by the formula

$$S = \frac{1}{2} d_1 d_2.$$

9.2.9 Rectangle

A *rectangle* is a parallelogram such that all four of its angles are right ones (Figure 9.14*c*). Along with properties common for all parallelograms, the rectangle shows the following properties:

1. Its diagonals are mutually equal.
2. A rectangle has two axes of symmetry (as shown by dashed lines in Figure 9.14*c*).
3. The area of a rectangle: $S = ab$.
4. The diagonal: $d = \sqrt{a^2 + b^2}$.

9.2.10 Square

A *square* is a rectangle whose sides are all equal to each other (Figure 9.14*d*). The square shows all properties, both of the rectangle and the rhombus, and also the following formulas are true:

$$S = a^2, \quad d = a\sqrt{2}\,.$$

9.2.11 Trapezoid

A *trapezoid*, or *trapezium*, is a quadrangle such that two of its sides are parallel and two other sides are not parallel to each other (Figure 9.14*e*). The parallel sides are called the *bases*, and the two other sides are called the *lateral sides*. The segment MN connecting midpoints of the lateral sides is called the *middle line*. The middle line is parallel to the bases and equal to a half of their sum.

The area of the trapezoid:

$$S = \frac{a+b}{2} \cdot h,$$

where a and b are the bases, h is the altitude (Figure 9.14*e*).

A trapezoid is said to be *isosceles* if its lateral sides are equal to each other. For an isosceles trapezoid, a straight line passing through the midpoints of the bases is the axis of symmetry.

9.3 Circle, Ellipse, Hyperbola, Parabola

9.3.1 Circle

The *circumference* is the set of points in a plane that are at a constant distance from a fixed point named the *center*. The *radius* of a circumference is the segment that connects the center and some point of the circumference. The part of the plane that is bounded by a circumference is called the *circle*; sometimes, the term "circle" is understood as a circumference.

The equation describing a circumference with the center $A_0(x_0, y_0, z_0)$ and the radius r:

$$(x - x_0)^2 + (y - y_0)^2 = r^2.$$

Here x and y are the Cartesian coordinates (see Section 8.3.2) of the variable point on the circumference.

A straight line and a circumference may have either two generic (i.e., common) points, or a single generic point—the point of *tangency*, or no generic points (Figure 9.15*b*). A straight line that has exactly one point in common with a circumference is called the *tangent line*. The tangent line is perpendicular to a radius drawn from the center to the point of tangency (Figure 9.15*c*).

A segment connecting two points of a circumference is called the *chord* or the *span*. A chord passing through the center is called the *diameter*, it is the maximal chord and equals two radii.

The *arc* is a part of the circumference bounded by two points, say A and B; it is denoted as $\smile AmB$ or $\overset{\frown}{AB}$ (Figure 9.16*a*). The *central angle*

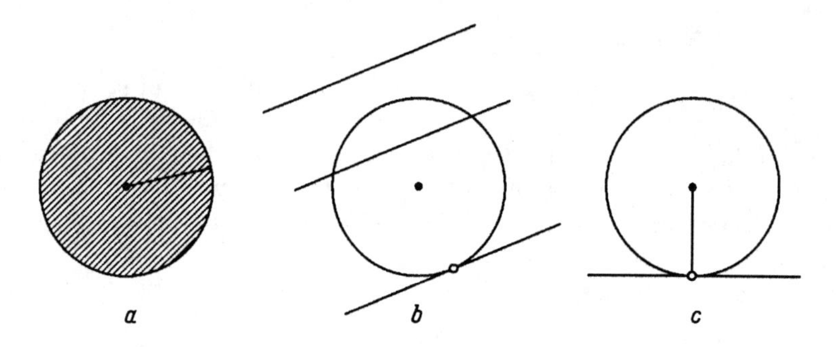

Figure 9.15: Circle (a), circumference and a straight line (b), tangent line to the circumference (c).

α and the *inscribed angle* β are basing, or resting, on the corresponding arc $\smile AmB$ (Figure 9.16a).

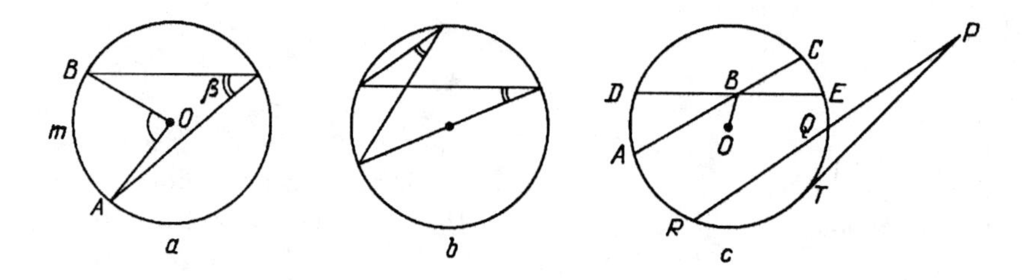

Figure 9.16: a – Central and inscribed angles basing on the same arc, b – equal inscribed angles, c – secant and tangent lines, and two intersecting chords.

Theorem: if an inscribed and a central angle are based on the same arc, then the inscribed angle equals half of the central angle: $\beta = \alpha/2$.

All inscribed angles based on the same arc are equal to each other (Figure 9.16b). An inscribed angle basing on a diameter is a right one.
Corollary: a triangle inscribed in a circle is a right triangle in the case where one of its sides is the diameter.

If two chords intersect inside the circle (Figure 9.16c), then the products of the corresponding parts of the chords are equal:

$$|DB| \cdot |BE| = |AB| \cdot |BC| = r^2 - |OB|^2.$$

$$|PQ| \cdot |PR| = |PT|^2.$$

Two straight lines tangent to a circle are symmetric to each other with respect to the axis passing through the center and the intersection point of the tangent lines (Figure 9.17*a*). In the case where the tangent lines are parallel, the symmetry axis is parallel to the tangent lines and it passes through the center.

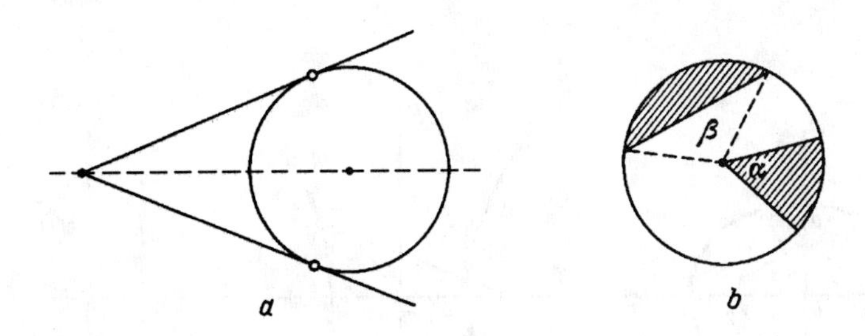

Figure 9.17: *a* – Two tangent lines to a circumference, *b* – sector and segment.

The length l of the circumference and the area S of the circle:

$$l = 2\pi r = \pi d, \qquad S = \pi r^2,$$

where $\pi \approx 3.1415926536 \approx 3.14$, r is the radius, d the diameter.

An arc is measured by the corresponding central angle:

$$l_\alpha = \alpha r,$$

where l_α is the length of the arc, α the central angle (in radians).

A *sector* is a part of a circle bounded by a central angle and a corresponding arc (Figure 9.17*b*). The area S_α of a sector whose central angle is α (in radians):

$$S_\alpha = \frac{1}{2}\alpha r^2.$$

A *segment* is a part of a circle bounded by an arc and its chord (which is called "the span ") (Figure 9.17*b*). The area S_{seg} of a segment whose central angle is β (in radians):

$$S_{seg} = \frac{1}{2}r^2(\beta - \sin\beta).$$

9.3.2 Ellipse

Let us fix in a plane two arbitrary points F_1 and F_2 (*focuses*) and consider the points of the plane such that for each point M, the distances from

M to the focuses, being added, give a constant number:

$$|F_1M| + |F_2M| = 2a.$$

The set of all such points is an *ellipse* with the focuses F_1 and F_2 (Figure 9.18a).

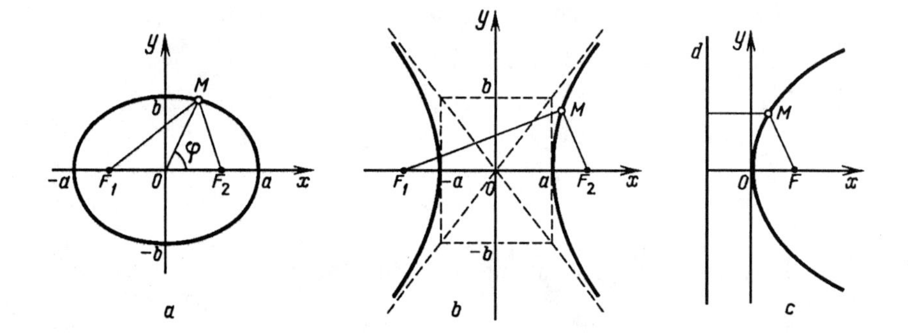

Figure 9.18: Ellipse (a), hyperbola (b), and parabola (c).

Draw the x-axis through the focuses and locate the origin of a Cartesian coordinate system x, y at the midpoint of F_1F_2; then

$$\frac{x^2}{a^2} + \frac{y^2}{b^2} = 1$$

is the equation of ellipse. Here x and y are the coordinates of a variable point of the ellipse, a is the *major semiaxis*, b is the *minor semiaxis*, $b = \sqrt{a^2 - c^2}$, $2c = |F_1F_2|$ is the distance between focuses, $c < a$. There are two axes of symmetry and a center of symmetry.

An ellipse may be described by the parametric equations:

$$x = a\cos\varphi, \quad y = b\sin\varphi,$$

where φ is the angle in the polar coordinate system (8.3.2), the pole is the center of the ellipse, $\varphi = \angle MOF_2$ (Figure 9.18a; see also Section 13.1.1).

Each planet of the solar system moves along an ellipse (approximately), the sun located at one of its focuses (see Section 13.5.1). An elliptic mirror has an interesting property: if we locate a point source of light at one focus, then the rays of light emitted from the source and then

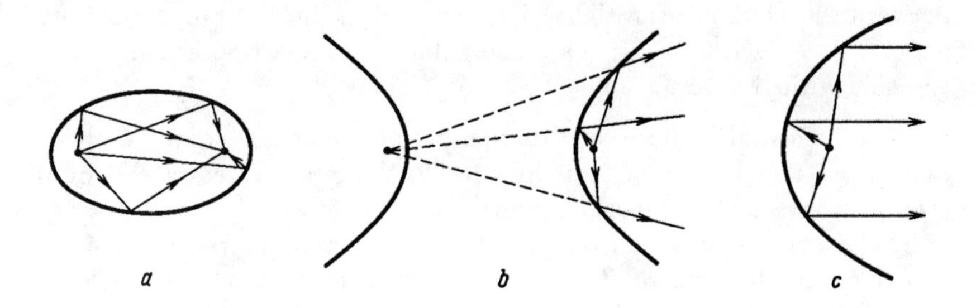

Figure 9.19: Optical properties of the mirrors: a – elliptic, b – hyperbolic, c – parabolic.

reflected by the mirror will all intersect each other in the second focus (Figure 9.19a).

The area S of the figure bounded by ellipse:

$$S = \pi ab.$$

9.3.3 Hyperbola

Let us fix two arbitrary points F_1 and F_2 (*focuses*) in a plane, and consider the set of points in the plane such that for each point M, the distances from the point to the first and to the second focus differ by a constant number:

$$|F_1M| - |F_2M| = \pm 2a.$$

The set of all such points is called the ***hyperbola*** with the focuses F_1 and F_2 (Figure 9.18b). A hyperbola consists of two ***branches*** which are infinite and symmetric to each other.

Draw the x-axis through the focuses and locate the origin of a Cartesian coordinate system at the midpoint of F_1F_2; then

$$\frac{x^2}{a^2} - \frac{y^2}{b^2} = 1$$

is the equation of hyperbola. Here a is the *real semiaxis*, b is the *imaginary semiaxis*, $b = \sqrt{c^2 - a^2}$, $2c = |F_1F_2|$ is the distance between focuses, $c > a$. There are two axes of symmetry and a center of symmetry.

The straight lines $y = \pm\frac{b}{a}x$ are called the *asymptotes* of the hyperbola; a point of hyperbola approaches infinitely close to an asymptote while $|x| \to \infty$.

In the case $a = b$ the hyperbola is said to be *rectangular*, the angle between the asymptotes is right. The graph of the inverse proportionality $y = k/x$ (see Section 2.7.2) is a rectangular hyperbola whose asymptotes coincide with the axes x and y.

Some celestial bodies—such as non-periodic comets—move, approximately, along a branch of some hyperbola, the Sun located at the focus (see Section 13.5.1). Hyperbolic mirror has the following property: rays of light which are emitted from one focus of a hyperbola, reflect from the corresponding branch, and then go to infinity in such directions that they seem to be emitted by an imaginary source located at the second focus (Figure 9.19*b*).

9.3.4 Parabola

A *parabola* is the set of points in a plane that are equidistant from a fixed straight line d (the *directrix*) and from a fixed point F (the *focus*); see Figure 9.18*c*.

Let us draw, through the focus, the x-axis perpendicular to the directrix, and locate the origin of a Cartesian coordinate system at the midpoint between the focus and the directrix; then

$$y^2 = 2px$$

will be the equation of the parabola. Here p is the distance from the focus to the directrix.

The x-axis is the symmetry axis for the parabola. No asymptotes. The property of a parabolic mirror: the light rays emitted by a point source located at the focus, after reflection by the mirror, go to infinity parallel to the symmetry axis (Figure 9.19*c*). This property is used in engineering to construct projectors.

9.3.5 Curvature of a curve

Consider a smooth (i.e., having no angle points) curve in a plane (Figure 9.20). Let us take three points in the curve, A_1, A_2, and A_3, close to each other, and draw a circumference through the points; it is known that for three arbitrary points—which do not lie in one straight line—there exists exactly one circumference passing through the points.

The center O of the circumference (Figure 9.20) can be found as the point of intersection of the middle perpendiculars to the segments A_1A_0 and A_0A_2. Let us move A_1 and A_2 towards A_0; the limiting position O_0 of the point O is called the *center of curvature* of the curve at point A_0, and the limiting radius—the *radius of curvature* at A_0. An arc of a circumference with such a radius is said to *approximate* the curve in the neighborhood of the point A_0.

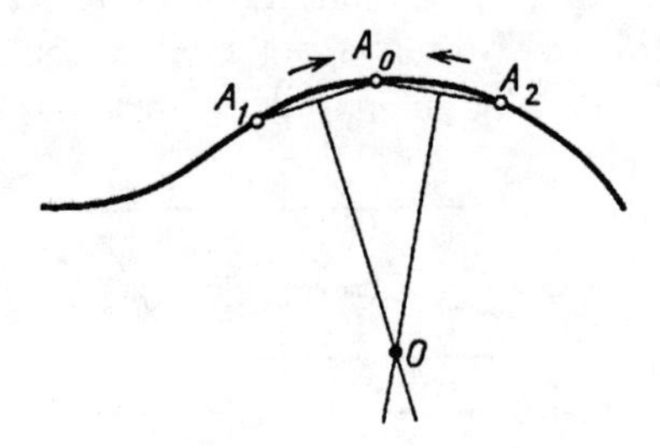

Figure 9.20: Center and radius of a planar curve.

Let a curve in the xy-plane be given by the equation $y = f(x)$, then the radius of curvature at point (x_0, y_0) is equal to $R = 1/K$, where

$$K = \frac{|f''(x_0)|}{[1 + (f'(x_0))^2]^{3/2}}.$$

The number K is called the *curvature of the curve.*

9.4 Planes and Straight Lines in Space

9.4.1 Parallelism of planes and lines

A plane splits the space into two *half-spaces*. Two different planes in space are either *parallel* (i.e., have no points in common) or *intersect* along a straight line. A straight line is either *parallel* to a plane (i.e., has no points in common) or *intersects* with it at one point, or as a whole, *lies* in the plane.

A criterion for parallelism of a straight line and a plane: if a given straight line is parallel to another straight line lying in the plane, then the given line is parallel to the plane.

A criterion for parallelism of two straight lines: if there is a third straight line parallel to each of the given lines, then the given lines are parallel to each other.

9.4.2 Skew-lines

Two straight lines in space either *intersect* (they have one point in common) or they are *parallel*, or they *cross*. In the last case they are "skew-lines" and have no point in common, see Figure 9.21.

The angle between two skew-lines is defined as the angle between two intersecting straight lines parallel to the given lines (the angle α in Figure 9.21).

Given two intersecting or two parallel lines, exactly one plane exists which contains both lines. No plane can exist to contain two skew-lines; such lines belong to two different parallel planes, see Figure 9.21.

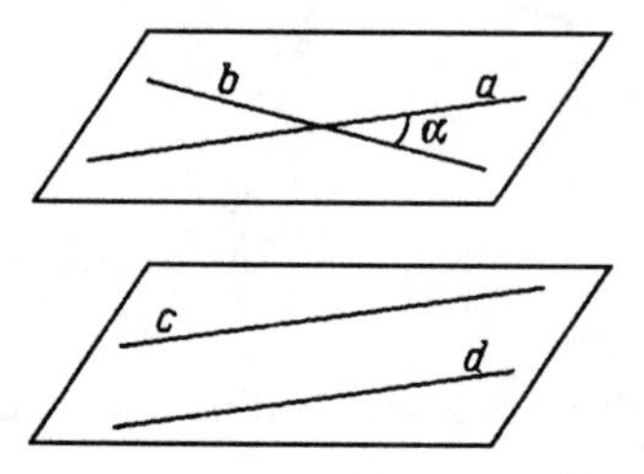

Figure 9.21: Straight lines: a, c, d are parallel, a and b intersect, b and d cross.

There exists exactly one plane to pass through:

- Any three points that do not lie in one straight line.
- A straight line and any point that does not lie in the line.
- Any two parallel straight lines.
- Any two intersecting straight lines.

9.4.3 Perpendicular

A straight line intersecting a plane is said to be *perpendicular* (*orthogonal*, or *normal*) *to the plane* if it is perpendicular to all straight lines lying in the plane (Figure 9.22a).

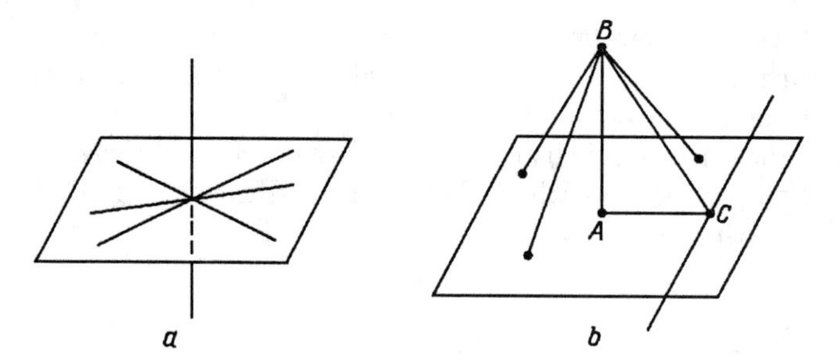

Figure 9.22: a – Perpendicular to the plane, b – perpendicular, oblique lines, and projection of an oblique line on the plane.

A straight line that is perpendicular to any two (nonparallel) straight lines lying in the plane is perpendicular to the plane. Let the straight line intersect the plane at point A and be perpendicular to the plane; the segment AB (Figure 9.22*b*) is called the *perpendicular dropped from the point B to the plane*. The distance from B to the plane is, by definition, the length of AB.

Taking an arbitrary point outside the plane we can drop to the plane one perpendicular and a lot of *oblique lines* (Figure 9.22*b*).

Let AB be the perpendicular and BC be an oblique line, then AC is called the *projection on the plane* of the oblique line, the point A the *base of the perpendicular*, and C the *base of the oblique line*. The angle between a line and a plane is defined as the angle between the line and its projection on the plane.

Theorem about three perpendiculars: if a straight line in a plane is perpendicular to the projection of an oblique line, then it is perpendicular to the oblique line (Figure 9.22*b*). The inverse theorem is also correct.

9.4.4 Angle between two planes

The angle between two intersecting planes may be defined as follows. Take an auxiliary plane that is perpendicular to the line of intersection of the given planes. The angle between the lines of intersection of the auxiliary plane with the given planes is called the angle between the given planes (Figure 9.23*a*). In order to determine this angle, we can take an arbitrary point in the intersection line l of the given planes and, starting from the point, draw the perpendiculars to the line l in both the planes (Figure 9.23*b*).

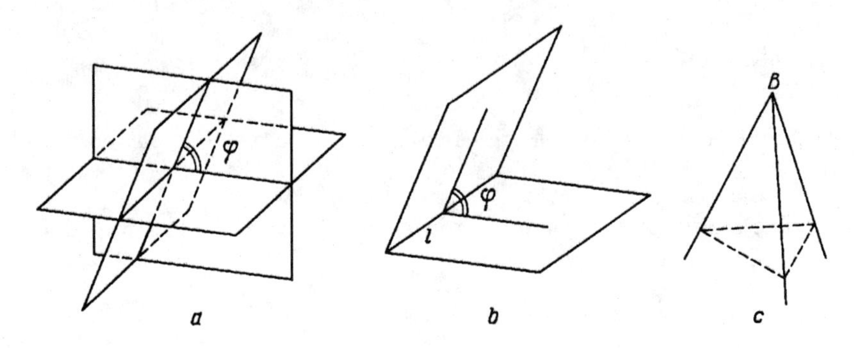

Figure 9.23: *a* – Angle between two planes, *b* – dihedral angle, *c* – trihedral angle.

A criterion for perpendicularity of two planes: the plane containing a straight line perpendicular to another plane is perpendicular to the last plane.

9.4.5 Dihedral angle

A *dihedral* (or two-sided) angle is the union of two half-planes having a common boundary (the common boundary is the *edge* of the dihedral angle, Figure 9.23*b*). Dihedral angles are measured by the corresponding angles between two planes. A *polyhedral* angle is formed by several planar angles having a common vertex (Figure 9.23*c*, the trihedral angle B).

9.4.6 Projection of a figure on a plane

Let us take two nonparallel planes α and α' in space, and a point $A \in \alpha$. The foot A' of a perpendicular dropped from A to the plane α' is called the *projection* of A on α' (Figure 9.24). The set of projections on α' of all points belonging to some line l (or to some figure F) situated in α is called the projection of l (respectively of F) on the plane α'. In Figure 9.24 we can see the segment $A'B'$ as the projection of the segment AB, and the polygon F' as the projection of the polygon F.

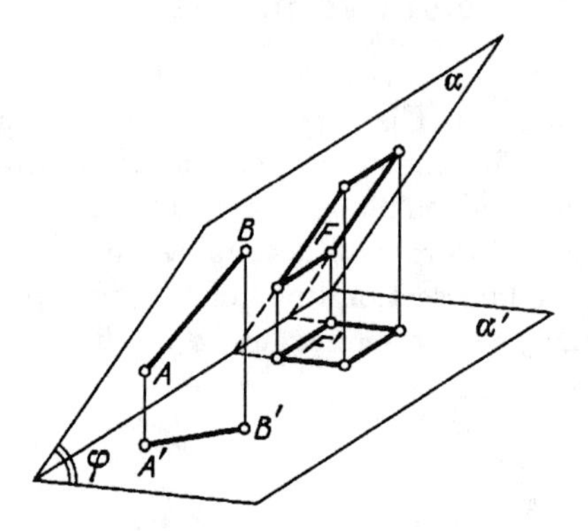

Figure 9.24: Projection of figures situated in the plane α, on the plane α'.

The area S of a polygon lying in α and the area S' of its projection on α' are connected by the formula:

$$S' = S \cos \varphi,$$

where φ is the angle between the planes α and α'. The formula is valid as well for an arbitrary figure and its projection.

Equation of a plane in space

The linear equation in three variables x, y, z:

$$Ax + By + Cz + D = 0$$

is the equation of a plane in the space. Here x, y, z are the Cartesian coordinates (8.3.3) of a variable point of the plane, and A, B, C are constant numbers. The numbers A, B, C may be considered as the coordinates of a vector $\vec{N}(A, B, C)$—the "normal" vector; $\vec{N}$ is perpendicular to the plane.

9.5 Polyhedrons

9.5.1 Polyhedral surface

In this section, the term "polygon" denotes a part of the plane bounded by the sides of a (planar) polygon.

A *polyhedral surface* is a surface constructed of polygons (called the *sides*, or *faces*, of the surface), in the following way. For any polygon, each side of it must coincide with a side of exactly one other polygon (the polygon's sides are the *edges* of the surface); moreover, if we start from any point of an arbitrary chosen face of the surface and go across an edge, to some adjacent face, and so on, we can thus reach any other face.

The vertices of the polygons are called the *vertices of the polyhedral surface.* At any vertex three or more edges meet.

9.5.2 Polyhedron

A *polyhedron* is a spatial body bounded by a closed polyhedral surface. Take a side (or "face") of a polyhedron; a plane containing this side splits the space into two half-spaces. Suppose that the polyhedron, as a whole, is situated in one of the half-spaces. If such a property is valid for every side of the polyhedron, the polyhedron is said to be *convex.*

Theorem of Euler: for every convex polyhedron, the number S of the sides, V of the vertices, and E of the edges are related by the formula:

$$S + V = E + 2 .$$

9.5.3 Prism

A *prism* is a polyhedron whose two opposite sides (the *bases*) are equal polygons lying in parallel planes, while all other sides are parallelograms (the *lateral sides*, Figure 9.25*a*). Those sides of the (lateral) parallelograms that do not lie in the bases are called the *lateral edges* of the prism. Prism may be imagined as a result of a translation (in space) of one base along a lateral edge.

The *height* of a prism is defined as the distance between its bases. The volume V of a prism:

$$V = S_b H,$$

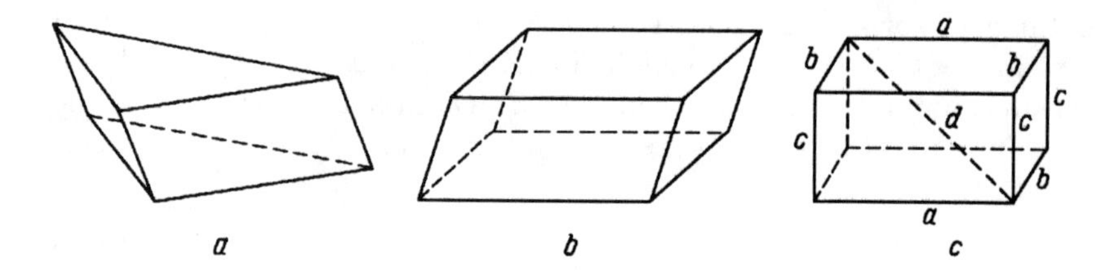

Figure 9.25: a – Prism, b – parallelepiped, c – rectangular parallelepiped.

where S_b is the area of the base, H the height of the prism.

Lateral edges of a *right prism* are perpendicular to the bases. Lateral sides of a right prism are rectangles. The area of the lateral surface S_l of a right prism:

$$S_l = P \cdot H,$$

where P is the perimeter of the base (i.e., the sum of its sides).

A prism is called *regular* if it is a right prism and its bases are regular polygons. The lateral sides of a regular prism are rectangles equal to each other.

9.5.4 Parallelepiped

A *parallelepiped* is a prism whose bases are the parallelograms (Figure 9.25b); hence, the lateral sides are also parallelograms. Any two opposite sides are parallelograms equal to each other. All four internal diagonals of a parallelepiped intersect at one point—the center of symmetry, and each diagonal is split by this point in two.

All sides of a *rectangular*, or *rightangled*, parallelepiped are rectangles, all its edges are perpendicular to corresponding sides (Figure 9.25c), and all four interior diagonals are equal to each other. The *Pythagoras theorem* for a rectangular parallelepiped:

$$d^2 = a^2 + b^2 + c^2$$

(for notations, see Figure 9.25c). In some books the rectangular parallelepiped is called the "rectangular prism" or "rectangular solid".

The volume V and the complete surface S_c of a rectangular parallelepiped:

$$V = abc, \quad S_c = 2(ab + ac + bc).$$

A *cube* is a rectangular parallelepiped whose 12 edges all are equal to each other. All 6 sides of the cube are identical squares. The diagonal of the cube $d = a\sqrt{3}$, a being the edge. The volume V and the complete surface S_c of the cube:

$$V = a^3, \quad S_c = 6a^2.$$

9.5.5 Pyramid

A *pyramid* is a polyhedron whose one side (named the *base*) is a polygon and all other sides (the *slant,* or *lateral, sides*) are triangles having a common vertex (the *apex* of the pyramid); see Figure 9.26*a*. The *height* of the pyramid is the distance from the apex to the basic plane. The volume V of pyramid:

$$V = \frac{1}{3} S_b \cdot H,$$

where S_b is the area of the base and H is the height.

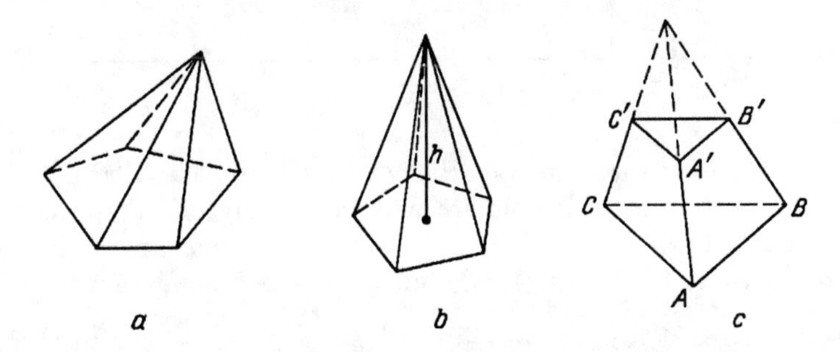

Figure 9.26: Pyramid (a), regular pyramid (b), and a frustum of a pyramid (c).

A pyramid is said to be n-angled if its base is an n-angled polygon. A pyramid is *regular* if its base is a regular polygon and the projection of its apex on the base coincides with the center of the base (Figure 9.26*b*). Slant sides of a regular pyramid are identical isosceles triangles. A regular n-angled pyramid has a symmetry axis of the order n (see Section 8.4.3). The slant (i.e., lateral) surface S_l of a regular pyramid:

$$S_l = \frac{1}{2} P \cdot h,$$

where P is the perimeter of the base and h is the altitude of a lateral side (the *slant height* of the regular pyramid).

9.5.6 Frustum of a pyramid

The *frustum of a pyramid* (or a truncated pyramid) is obtained from a pyramid with the help of truncating from the pyramid a part of it by

Table 9.1: Regular polyhedrons

Name	Number and form of sides	Number of edges	Number of vertices
Tetrahedron	4 triangles	6	4
Cube	6 squares	12	8
Octahedron	8 triangles	12	6
Dodecahedron	12 pentagons	30	20
Icosahedron	20 triangles	30	12

a plane parallel to the base (Figure 9.26*c*, the body $ABCA'B'C'$). The bases of a frustum of a pyramid are two similar polygons, the lateral (or slant) sides are trapezoids. The height of the frustum of a pyramid is the distance between the bases. The volume V of the frustum of a pyramid:

$$V = \frac{1}{3}H\,(S + \sqrt{S \cdot S'} + S) = \frac{1}{3}HS(1 + \gamma + \gamma^2),$$

where $\gamma = |A'B'|/|AB|$ is the similarity coefficient for the bases, S and S' are the areas of the bases $ABCD$ and $A'B'C'D'$, and H is the height.

The bases of a frustum of a regular n-pyramid are similar n-polygons. The area S_l of the slant surface of the frustum of a regular pyramid:

$$S_l = \frac{1}{2}(P + P') \cdot h,$$

where P and P' are the perimeters of the bases and h is the height of a lateral side (the *slant height* of the frustum of a pyramid).

9.5.7 Regular polyhedrons

A *regular polyhedron* is a convex polyhedron, all sides of which are identical regular polygons and all polyhedral angles of which are identical. There exist exactly five regular polyhedrons (see Figure 9.27 and Table 9.1).

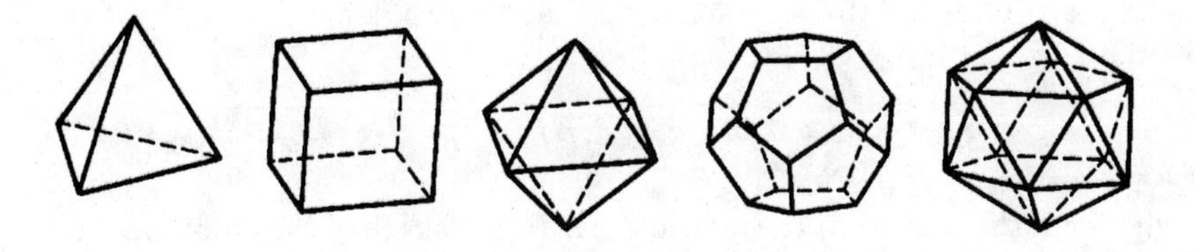

Figure 9.27: Regular polyhedrons

9.6 Bodies of Revolution

9.6.1 Surface of revolution

The *surface of revolution* is the surface obtained by a revolution (in space) of a planar curve about a certain axis lying in the plane.

9.6.2 Cylinder

A *cylinder* (right circular cylinder) is a spatial body bounded by a surface obtained by a revolution of a rectangle about a straight line passing through the midpoints of two opposite sides (Figure 9.28*a*). The surface of the cylinder is a union of two circles (the *bases*) and a *lateral*, or *curved, surface*. The circles are in parallel planes, the distance between them is the *height* of the cylinder. The axis of the cylinder—the straight line passing through the centers of the bases—is the axis of symmetry of infinite order (see Section 8.4.3). The lateral surface is constructed of *elements*, i.e., of identical linear segments parallel to the axis. The lateral surface may be developed onto a plane to obtain a rectangle.

The volume V, the area S_l of the lateral surface, and the area S_c of the complete surface of the cylinder:

$$V = \pi R^2 H, \quad S_l = 2\pi RH, \quad S_c = 2\pi RH + 2\pi R^2,$$

where R is the radius of the base and H is the height of the cylinder.

9.6.3 Cone

A *cone* (right circular cone) is a spatial body bounded by a surface obtained by a revolution of an isosceles triangle about its axis of symmetry (Figure 9.28*b*). The total surface of a cone is a union of a circle (the *base*) and a *lateral* (or *curved*) *surface*. The *vertex* (or *apex*) of a cone may be connected with any point of the basic circumference by an *element*, i.e., by a linear segment lying in the lateral surface. The *height* of the cone is the distance from its vertex to the base.

The volume V, the lateral area S_l, and the complete area S_c of the cone:

$$V = \frac{1}{3} S_b \cdot H = \frac{1}{3} \pi r^2 H, \quad S_l = \frac{1}{2} PL = \pi r L, \quad S_c = \pi r (r + L),$$

where r is the radius, S_l the lateral area, P the length of the circumference of the base, L the length of an element, and H the height of the cone (Figure 9.28*b*).

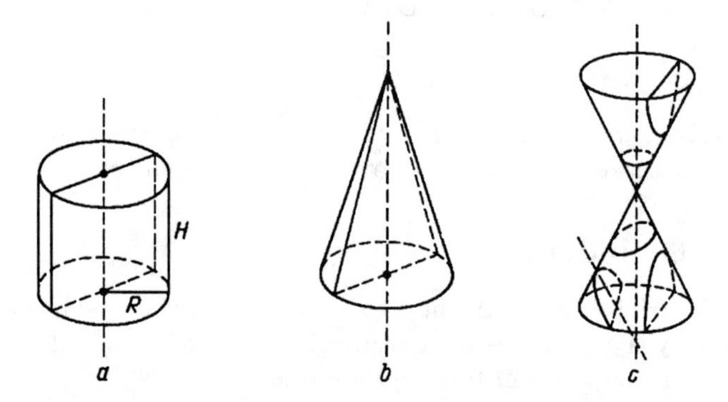

Figure 9.28: Cylinder (*a*), cone (*b*), and conic sections (*c*): circle, ellipse, parabola, hyperbola.

Draw a straight line through the vertex of a cone and a point M of the circumference of the base. Let the point M run along the circumference, then the straight line will describe an infinite "conic surface". Now let us take some plane and consider its intersection with the conic surface—so-called *conics* or *conic sections* (Figure 9.28*c*). There may appear conic sections of three different kinds: an ellipse (in particular, a circle, see Section 9.3.2), a parabola (see Section 9.3.4), or a hyperbola (see Section 9.3.3)—depending on the inclination of the plane to the base of the cone (Figure 9.28*c*).

The *frustum of a cone* is a part of a cone obtained by truncating a part from it with the help of a plane parallel to the base (Figure 9.29*a*). The frustum of a cone may be obtained by rotating an isosceles trapezoid about its axis of symmetry. The bases of the trapezoid, in the process of rotation, describe the bases of the frustum of a cone. The height of the frustum of a cone is the distance between its bases.

The volume V and the lateral area S_l of the frustum of a cone:

$$V = \frac{1}{3} \pi h (r_1^2 + r_2^2 + r_1 r_2), \quad S_l = \pi (r_1 + r_2)\, l,$$

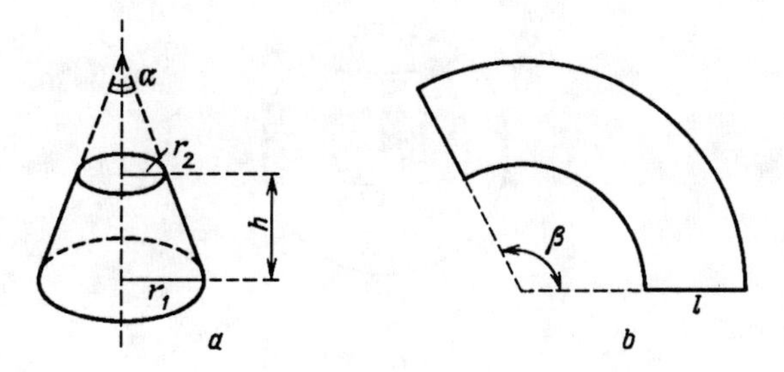

Figure 9.29: Frustum of a cone and its development onto a plane.

where h is the height, r_1 and r_2 the radii of the bases, and l the length of the element of the frustum.

The lateral surface of a cone may be developed onto a plane to obtain a sector whose radius is identical to the element of the cone. Correspondingly, if we develop the lateral surface of the frustum of a cone, we obtain a part of a circular ring (Figure 9.29*b*). Let α be the angle by the vertex of the axial cross-section of the cone and let β be the central angle of the conic development (the angle of the sector); then

$$\beta = 2\pi \sin(\alpha/2).$$

9.6.4 Sphere

A *sphere* is the set of all the points in space that are at a constant distance r from a certain fixed point (the *center* of the sphere). A *ball* is the body bounded by a sphere (Figure 9.30), and it may be obtained as a result of revolution of a circle about its diameter.

The *radius* of the sphere is the distance from its center to a point of the sphere, the *diameter* is the length of a segment that is passing through the center and whose ends are on the sphere. Two radii together are equal to the diameter. Sometimes the terms “radius” and “diameter” are used to denote the corresponding linear segments.

A plane and a sphere in space can have either no points or one point in common (the point of *contact*), or they can intersect along a circumference. Plane cross-sections of a ball are the circles, the maximum of which—the *great circle* —contains the center of a ball. The circumference of the great circle is usually also called the great circle. On the globe, the equator is a great circle, meridians are great half-circles (Figure 9.30). For any two points of a sphere which are not the endpoints of one diameter, there is exactly one great circle passing through the points.

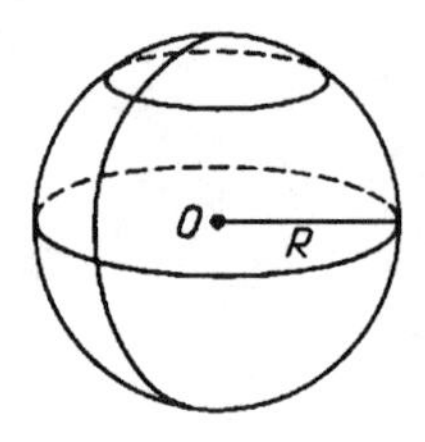

Figure 9.30: Ball.

A tangent plane to the sphere is perpendicular to a radius dropped from the center to the point of a contact.

The volume V of a ball and the area S of a sphere:

$$V = \frac{4}{3}\pi R^3, \quad S = 4\pi R^2.$$

Taking a Cartesian coordinate system in the space, the equation of a sphere is:

$$(x - x_0)^2 + (y - y_0)^2 + (z - z_0)^2 = R^2.$$

Here x_0, y_0, z_0 are the coordinates of the center, x, y, z are the coordinates of a variable point of the sphere, and R is the radius of the sphere.

9.6.5 Spherical segment

A *spherical segment* is a part of a ball cut from it by a plane (Figure 9.31). A *segment spherical surface* is a part of a sphere cut from it by a plane. The volume V of the spherical segment and the area S of the segment spherical surface:

$$V = \frac{1}{3}\pi h^2(3R - h), \quad S = 2\pi Rh,$$

where h is the height of the segment and R is the radius of the ball.

9.6.6 Spherical sector

A *spherical sector* (or solid angle) is a body obtained by a revolution of a planar circular sector (see Section 9.3.1) around its axis of symmetry (Figure 9.32). A spherical sector may be composed of a cone and a spherical segment. The complete surface of a spherical sector consists of the segment spherical surface and the lateral cone surface. The volume V and the area of complete surface S_c of a spherical sector:

$$V = \frac{2}{3}\pi R^2 h, \quad S_c = 2\pi Rh + \pi R\sqrt{2Rh - h^2},$$

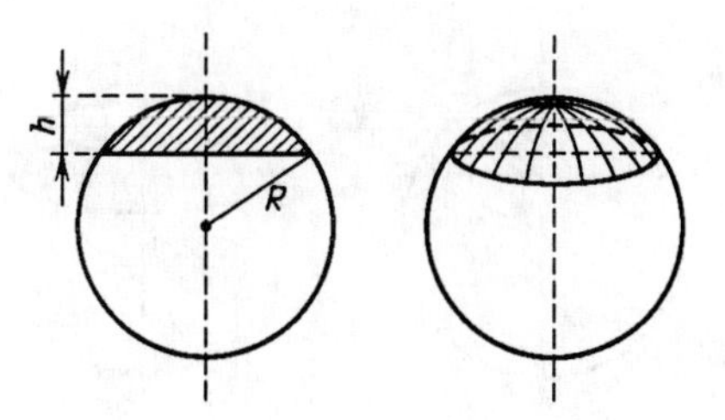

Figure 9.31: Spherical segment.

where R is the radius of the ball and h is the height of the spherical segment.

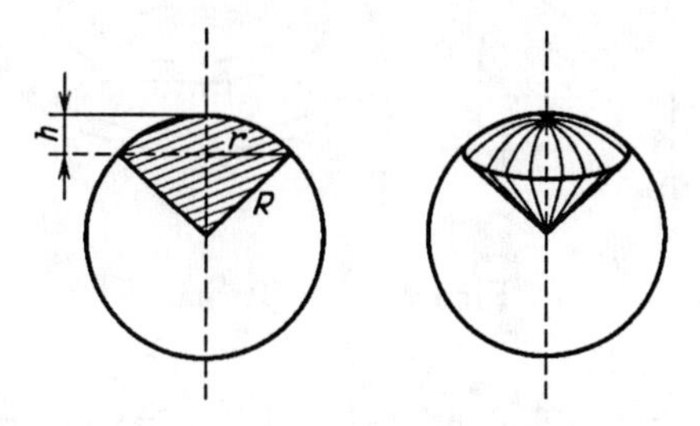

Figure 9.32: Spherical sector.

9.6.7 Spherical layer

A *spherical layer* is a part of a ball cut from it by two parallel planes (Figure 9.33). The volume V and the complete surface area S_t:

$$V = \frac{1}{6}\pi h^3 + \frac{1}{2}\pi h(r_1^2 + r_2^2), \quad S_c = 2\pi Rh + \pi(r_1^2 + r_2^2),$$

where h is the height, and r_1 and r_2 are the radii of the bases of the spherical layer.

9.6.8 Torus

A *torus* is a body obtained by the revolution of a circle about a certain axis situated in the plane of the circle (the axis outside the circle); see Figure 9.34.

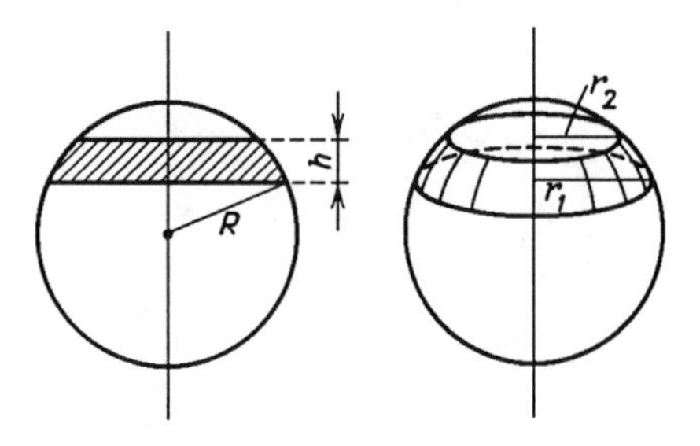

Figure 9.33: Spherical layer.

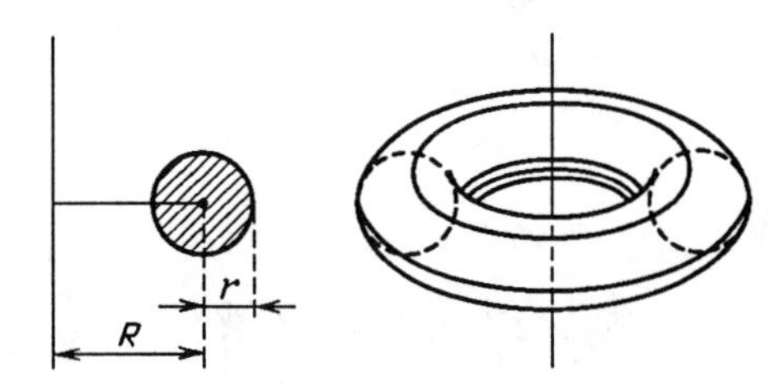

Figure 9.34: Torus.

A tire of a car and a doughnut (bagel) are in the form of a torus. The volume V and the surface area S of a torus:

$$V = 2\pi^2 r^2 R, \quad S = 4\pi^2 rR,$$

where r is the radius of the circle and R is the distance from the center of the circle to the axis of revolution.

9.7 Curvature of a Surface

Consider a smooth surface S in the space (Figure 9.35*a*). Let A_0, A_1, A_2 be some points on the surface which are located near each other and do not lie in one straight line. Draw a plane through the three points; this is the *intersecting* plane. Let A_1 and A_2 both tend to A_0 , then the intersecting plane tends to a limit—to the *tangent plane* to the surface at the point A_0 (the point of *contact*); Figure 9.35*b*.

Note: in the limiting process mentioned above, it is necessary to admit the points A_1 and A_2 to move towards A_0 along two curves that have different tangent lines at the point A_0.

A *convex surface* has exactly one point in common with any tangent plane. Such a surface is situated as a whole from one side of any tangent

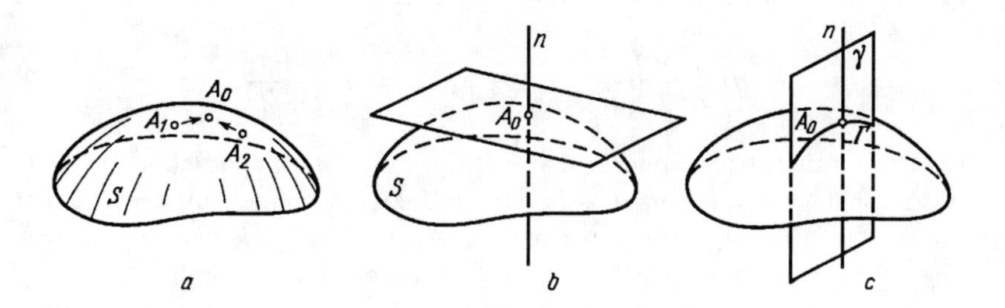

Figure 9.35: (a) – The points A_1 and A_2 tending to A_0, (b) – tangent plane and normal line to the surface, (c) – normal section of the surface.

plane (an analog of a plane convex curve, see Section 5.4.4). Two examples of convex surfaces: sphere and ellipsoid of revolution (the result of the rotation of an ellipse about one of its symmetry axes).

Let A_0 be the point of contact of a plane α and a convex surface S. The *normal* to S at the point A_0 is defined as a straight line which passes through A_0 and is perpendicular to α (Figure 9.35b). Any plane that contains the normal line intersects with the surface along a *normal section* (Figure 9.35c, the normal section Γ is the intersection of S and the plane γ).

Consider the radius R of the curvature (see Section 9.3.5) of a normal section. Rotate the section plane about the normal, then the value of R, generally, will be variable (for the sphere, R is independent of the position of the normal). Denote by R_1 the minimal value of R and by R_2 the maximal value; it is provable that under some conditions the planes of the corresponding normal sections are perpendicular to each other.

The expression

$$H = \frac{1}{2}\left(\frac{1}{R_1} + \frac{1}{R_2}\right) \tag{9.1}$$

is called the *mean curvature* of the surface at the given point, and the expression $\rho = 1/H$ is called the *mean radius*.

Generally, the mean curvature depends on the point of the surface: the curvature changes when we move along the surface. However, there are surfaces of constant mean curvature, for example, a sphere and a cylinder. For a sphere whose radius is R, the values ρ and H are constant: $\rho = R$, $H = 1/R$. For a cylinder, at all points of the surface the minimal radius of curvature equals R, and the maximal one $R_2 = \infty$, hence $H = 1/(2R)$, $\rho = 2R$.

For a torus (see Section 9.6.8, Figure 9.34), the mean curvature depends on the position of the point. For instance, at all points that are farthest from the symmetry axis of the torus,

$$H = \frac{1}{2}\left(\frac{1}{R+r} + \frac{1}{r}\right), \quad \rho = 2r\frac{R+r}{R+2r}. \tag{9.2}$$

For a *non-convex surface*, in a vicinity of a point of contact, some parts of the surface are found in different half-spaces defined by the tangent plane. Calculating the mean curvature in this case, we should consider one of the radii of curvature, say R_1, to be positive, and the other, R_2, negative. For such surfaces, Formula (9.1) may be rewritten as follows:

$$H = \frac{1}{2}\left(\frac{1}{|R_1|} - \frac{1}{|R_2|}\right). \tag{9.3}$$

A *saddle* (Figure 9.36) is an example of a non-convex surface. If we consider the tangent plane at the saddle point (A_0 in Figure 9.36), then some parts of the saddle surface are found above the plane, while other parts are found below the plane. It may happen that the mean curvature

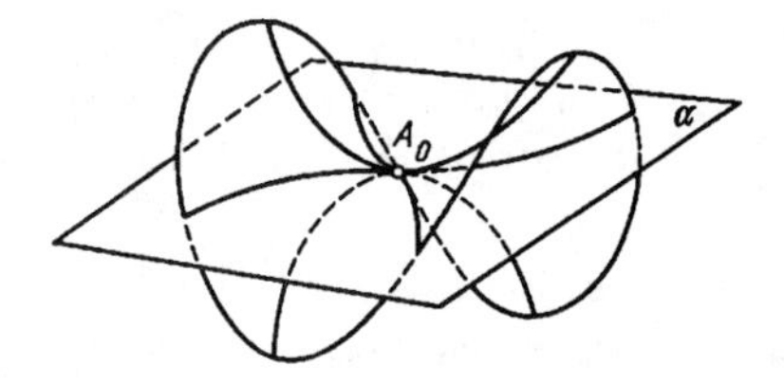

Figure 9.36: Saddle.

is zero. For example, look at a torus, at any of its points nearest to the axis of symmetry. According to (9.2) and (9.3),

$$H = \frac{1}{2}\left(\frac{1}{r} - \frac{1}{R-r}\right).$$

If we take a torus for which $R = 2r$, we obtain $H = 0$.

The concept of the curvature of a surface and the corresponding formulas have various applications in physics (Sections 14.5.2 and 15.1.4).

Chapter 10

Numerical Analysis

The significance of computers for modern society is now very high and it is continuously increasing in parallel to the swift growing of the computer set. To apply the marvelous abilities of computers for solving numerous problems of physics, mathematics, and engineering, a special part of mathematics—the numerical analysis—is used. In this chapter we consider briefly the simplest numerical methods for fitting functions, the approximate calculation of integrals, and the numerical solution to equations, including some differential equations.

10.1 Rounding off and Errors

In solving problems of physics and engineering one has to perform some calculations. For instance, in order to find the mass of a gas contained in a closed cylindrical vessel (see Section 14.1.3), we apply the formula $m = \rho V = \rho \pi R^2 H$ (where ρ is the gas density, R the radius, H the height of the cylinder) and perform, according to it, four multiplications: $\rho \cdot \pi \cdot R \cdot R \cdot H$. Such simple work may be done "by hands", i.e., on a sheet of paper or with the help of a calculator. The rounding-off rules to be applied to such calculations are in Section 1.3.1.

More complicated problems may require many hundreds, thousands, or millions of operations. As an example, imagine the problem of determination of the space stations orbits—here we meet the necessity to solve certain systems of differential equations, which is realizable only with the help of high-capacity computers.

In general, the process of computer solution of a physical problem may be represented as consisting of several stages. First we find a *mathematical model* for the physical problem, i.e., we set a mathematical problem corresponding to the physical one. Then we find a suitable *numerical method* providing us to write down an *algorithm,* i.e., a consequence of formulas and rules that lead to the answer. After that we write a *program* for the computer using one of the algorithmic languages (Basic, Fortran, Pascal, C, Ada, and so on). Finally we input the program (and some additional data) into the computer and obtain an answer in the form of numbers, tables, graphs, pictures on the screen, or some signals to control certain devices.

Consider, for instance, a process which may be described by the differential equation $x'(t) = f(t, x(t))$, together with the initial condition $x(t_0) = A$ (see Section 5.5.3), where t is the time, $x(t)$ an unknown function, and the number A and the function $f(t, x)$ are given by an experiment or theory. Since no formula exists to represent a solution to a differential equation of the general type, we have to use a numerical method, e.g., Euler's method (see Section 10.5.1):

$$x_{k+1} = x_k + hf(t_k, x_k),$$

where $x_0 = A,\ \ t_k = kh,\ \ k = 0, 1, \ldots, n$; the step h is to be chosen small enough. We see that the algorithm of approximate solving of the problem consists of step-by-step calculation of $x(t)$ at the moments $t_1,\ t_2, \ldots, t_n$. It is easy to write a program by using any algorithmic language, and the result of computer calculation will be exact enough for small values of h. A more exact approximation to the solution can be obtained by using the Runge–Kutta method (see Section 10.5.2).

The following three kinds of errors are met in the process of calculations: the input data errors (due to instrumental errors), the numerical method errors (almost all numerical algorithms give non-exact results), and also the round-off errors. The last origin of errors may play a crucial role in spite of the fact that modern computers provide a high accuracy (usually up to 12 or 20 and more decimal digits) and are very fast (many millions of operations per second). In reality, almost any arithmetic operation is performed by the computer using the rounding off procedure, so we obtain, usually, a slightly distorted result. As a rule, the total error is negligible, but at some unfavorable circumstances it can distort the result very strongly. Here are two examples of such cases; a 4-digit computer is taken as a simple model.

Examples

1. $C = (A + B - A) \cdot 10^3$, where $A = 5 \cdot 10^4,\ \ B = 1.0$. The 4-digit computer, when adding $A + B$, at first equalizes the orders of the summands:

$$5.000 \cdot 10^4 + 0.0001 \cdot 10^4 = (5.000 + 0.0001) \cdot 10^4,$$

and then rounds off the numbers up to four digits:

$$(5.000 + 0.000) \cdot 10^4 = 5.000 \cdot 10^4\,.$$

As a result we obtain

$$C = (5.000 \cdot 10^4 - 5.000 \cdot 10^4) \cdot 10^3 = 0.000 \cdot 10^3 = 0,$$

but the exact result should be, of course, $C = 1000$.

2. Suppose that at a certain moment the computer gets the instruction to do: $C = (A - B)/3$, where the variables A and B have the following approximate (up to 4 digits) values: $A = 3.142 \cdot 10^2$, $B = 3.131 \cdot 10^2$. The computer works like this:

$$A - B = 0.011 \cdot 10^2 = 1.100, \quad C = 1.100/3 \to 3.667 \cdot 10^{-1}.$$

May we trust to all the four digits of the result? Evidently not, because the current values of A and B are approximate, and nobody knows their exact 5th, 6th, etc. digits. It is reasonable to write the result as $3.67 \cdot 10^{-1}$ (see Section 1.3.2); however, the computer does not inform us about it!

Such examples illustrate the phenomenon called "loss of precision" in the process of rounding-off calculations. We recommend that one avoid such situations. So, in the first example we could rewrite the formula as $C = (A - A + B) \cdot 10^4$ and obtain as a result: $C = 10^3$, i.e., the exact answer.

Some simplest numerical methods are briefly described in the rest of the chapter. All calculations in the examples below are performed with a 10-digit calculator, the answers are rounded off.

10.2 Approximation of Functions

10.2.1 Approximate formulas

Approximation (or approximate representation) of functions is used in the following cases:

1. When the given functional dependence is too complicated and replacement of it by a simpler one is profitable;
2. When the functional dependence is given by a table.

Many approximate formulas are based on the power *Taylor series* (see Section 5.8) in a neighborhood of a given point:

$$f(x) = f(a) + f'(a)(x-a) + \frac{1}{2!}f''(a)(x-a)^2 + \frac{1}{3!}f'''(a)(x-a)^3 + \ldots.$$

Truncate the series up to some term and then obtain an approximate formula for the function in the form of a polynomial:

$$f(x) \approx f(a) + f'(a)(x-a) + \ldots + \frac{1}{n!}f^{(n)}(a)(x-a)^n. \tag{10.1}$$

Formula (10.1) is local, i.e., it is valid for $a - \varepsilon \le x \le a + \varepsilon$, where $\varepsilon > 0$ is a certain number (Figure 10.1: the dashed line corresponds to a polynomial of a degree $n > 1$; the straight line corresponds to the linear approximation with $n = 1$).

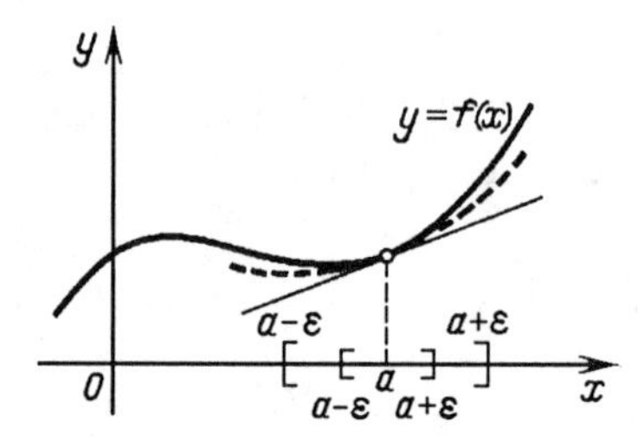

Figure 10.1: Local approximation of a function: by a linear function (straight line) and by a polynomial (dashed line).

The *error* of the approximation formula, i.e., the difference $r_n(x)$ between the exact and the approximate values of the function, can be evaluated in the following way:

$$|r_n(x)| \le \frac{M_{n+1}}{(n+1)!}|x - a|^{n+1},$$

where $M_{n+1} = \max|f^{(n+1)}(x)|$ for $a - \varepsilon \le x \le a + \varepsilon$. In the formulas above, the symbol $n!$ (factorial) denotes the product of all natural numbers from 1 to n (see Section 6.1.2), while $f^{(n)}(x)$ is the n-order derivative of the function $f(x)$ at the point x (see Section 5.4.2).

The most popular is the *linear* approximation:

$$f(x) \approx f(a) + f'(a)(x - a).$$

Its error $r_1(x)$ is evaluated by

$$|r_1(x)| \le \frac{1}{2}(x - a)^2 M_2.$$

The linear approximation geometrically means that the graph of the given function $f(x)$ in a vicinity of the point a is replaced by a straight line $y = f(x)+f'(a)(x-a)$ tangent to the graph of $f(x)$ at $x = a$ (Figure 10.1). Linear approximation is applied in physics, for instance, in simplification of the pendulum equation (Section 16.3.1).

Examples

1. $f(x) = e^x, \quad a = 0, \quad n = 1$. Here we have:

$$e^x \approx 1 + x, \quad |r_1(x)| \le 0.5x^2 e^x.$$

Taking $\varepsilon = 0.1$ we see that the error is not greater than $0.5x^2e^{0.1} \leq 0.6x^2$. Since $|x| \leq 0.1$, then $x_2 \leq 0.01$, hence $|r_1(x)| \leq 0.006$.

2. $f(x) = \ln(1+x)$, $a = 0$, $n = 5$. Here:

$$\ln(1+x) \approx x - x^2/2 + x^3/3 - x^4/4 + x^5/5,$$

$$|r_5(x)| \leq |x|^6 \cdot \frac{5!}{6!} = |x|^6/6.$$

Taking $\varepsilon = 0.5$ we obtain $|r_5(x)| \leq (0.5)^6/6 < 0.03$.

3. $f(x) = \sin x$, $a = 0$, $n = 6$. Here:

$$\sin x \approx x - x^3/6 + x^5/120, \quad |r_6(x)| \leq |x|^7/7!$$

Taking $\varepsilon = 1$ we obtain $|r_6(x)| \leq 0.0002$. Note that the angle is measured here in radians (see Section 9.1.5), so the last approximation formula is true for $|x| \leq 57°$.

Table 10.1 contains approximation formulas for most common functions and also indicates their validity domains and error bounds.

10.2.2 Function given by a table

A function $y = y(x)$ may be given by a table, i.e., by several values $x_1, x_2, \ldots, x_n$ of the argument and the corresponding values $y_1, y_2, \ldots, y_n$ of the function. For instance, look at Table 10.2 which introduces the temperature T of some specimen in dependence on time t.

The numbers x_1, $x_2, \ldots, x_n$ are called the *nodes* of the table, and the points (x_k, y_k) the *node points*. The problem is to find the values of the function $y(x)$ at an intermediate point x located either between the nodes or outside an interval containing all the nodes.

Of course, the problem posed in that way admits infinitely many solutions: an arbitrary number is appropriate as an answer. A reasonable setting of the problem must prescribe a certain class (a set) of functions among which the desired approximation is to be found, then the answer would be definite.

The class of all polynomials up to a given degree is often taken to perform the *interpolation*. A polynomial whose graph passes through all node points is called the *interpolation polynomial*. For any given n-node table, there is exactly one interpolation polynomial of the degree $(n-1)$.

Most usual is the linear interpolation, less usual the quadratic one. Using the linear interpolation, we represent the original function $y(x)$ inside an interval (x_1, x_2) by a linear function whose graph passes through two node points, (x_1, y_1) and (x_2, y_2) (Figure 10.2a):

$$y(x) \approx y_1 + (x - x_1)(y_2 - y_1)/(x_2 - x_1) \qquad (10.2)$$

Table 10.1: Approximation formulas

Approximation formula	Domain of validity	For given value of x, the error is less than	For the whole segment, the error is less than
$\sqrt{1 \pm x} \approx 1 \pm \frac{1}{2}x$	$\lvert x\rvert \leq 0.5$ $\lvert x\rvert \leq 0.1$	$0.2x^2$ $0.15x^2$	$4.3 \cdot 10^{-2}$ $1.5 \cdot 10^{-3}$
$\frac{1}{1 \pm x} \approx 1 \mp x$	$\lvert x\rvert \leq 0.3$ $\lvert x\rvert \leq 0.1$	$1.5x^2$ $1.2x^2$	0.13 $1.2 \cdot 10^{-2}$
$e^x \approx 1 + x$	$\lvert x\rvert \leq 0.5$ $\lvert x\rvert \leq 0.1$	$0.6x^2$ $0.6x^2$	0.15 $5.2 \cdot 10^{-3}$
$\log_e(1 \pm x) \approx x$	$\lvert x\rvert \leq 0.3$ $\lvert x\rvert \leq 0.1$	$0.8x^2$ $0.6x^2$	$5.7 \cdot 10^{-2}$ $5.4 \cdot 10^{-3}$
$\sin x \approx x$	$\lvert x\rvert \leq 1$ $\lvert x\rvert \leq 0.1$	$(1/6)\lvert x\rvert^3$ $(1/6)\lvert x\rvert^3$	0.16 $1.7 \cdot 10^{-4}$
$\cos x \approx 1 - \frac{1}{2}x^2$	$\lvert x\rvert \leq 1$ $\lvert x\rvert \leq 0.1$	$(1/24)x^4$ $(1/24)x^4$	$4.1 \cdot 10^{-2}$ $4.2 \cdot 10^{-6}$
$\sinh x \approx x$	$\lvert x\rvert \leq 1$ $\lvert x\rvert \leq 0.1$	$0.2\,\lvert x\rvert^3$ $0.2\,\lvert x\rvert^3$	0.18 $1.7 \cdot 10^{-4}$
$\cosh x \approx 1 + \frac{1}{2}x^2$	$\lvert x\rvert \leq 1$ $\lvert x\rvert \leq 0.1$	$0.05\,\lvert x\rvert^4$ $0.05\,\lvert x\rvert^4$	$4.4 \cdot 10^{-2}$ $4.2 \cdot 10^{-6}$
$\tan x \approx x$	$\lvert x\rvert \leq 0.5$ $\lvert x\rvert \leq 0.1$	$0.85\,\lvert x\rvert^3$ $0.35\,\lvert x\rvert^3$	$4.7 \cdot 10^{-2}$ $3.4 \cdot 10^{-4}$
$\sin x \approx x - \frac{1}{6}x^3$	$\lvert x\rvert \leq 1$ $\lvert x\rvert \leq 0.1$	$(1/120)\lvert x\rvert^5$ $(1/120)\lvert x\rvert^5$	$8.2 \cdot 10^{-3}$ $8.4 \cdot 10^{-8}$
$\arcsin x \approx x$	$\lvert x\rvert \leq 0.5$ $\lvert x\rvert \leq 0.1$	$0.25\lvert x\rvert^3$ $0.2\lvert x\rvert^3$	$2.4 \cdot 10^{-2}$ $1.7 \cdot 10^{-4}$
$\arctan x \approx x$	$\lvert x\rvert \leq 0.5$ $\lvert x\rvert \leq 0.1$	$\frac{1}{3}\lvert x\rvert^3$ $\frac{1}{3}\lvert x\rvert^3$	$4.2 \cdot 10^{-2}$ $3.4 \cdot 10^{-4}$

Table 10.2: Result of an experiment

t, s	0	10	20	30	40	50	60
T, °C	54.15	54.22	54.21	54.18	54.19	54.21	54.31

or, which is the same,

$$y(x) \approx y_2 - (x_2 - x)(y_2 - y_1)/(x_2 - x_1). \qquad (10.2a)$$

Formula (10.2)—or (10.2a)—is used also for *extrapolation*, i.e., for $x < x_1$ or $x > x_n$.

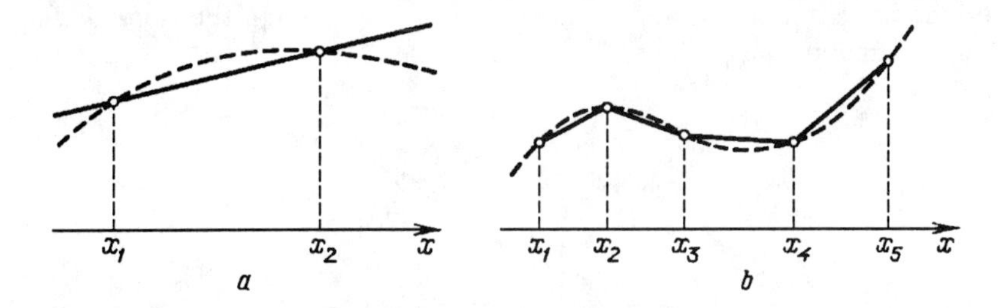

Figure 10.2: *a* – Linear interpolation on a segment, *b* – linear spline interpolation.

Remark. A formula for theoretical evaluation of interpolation error is known, which, for the interval (x_1, x_2), can be written as

$$|r_1(x)| \leq \frac{1}{8}(x_2 - x_1)^2 \cdot \max_{x_1 \leq x \leq x_2} |y''(x)|.$$

Unfortunately, this formula is of little practical meaning as soon as the function $y(x)$ is, in reality, unknown between the nodes x_1 and x_2.

If the table contains more than two points we can use the linear interpolation for each interval bounded by two adjacent nodes. So we obtain a linear polygonal line passing through all node points (Figure 10.2*b*). Linear interpolation is often used in working with mathematical tables.

An interesting way to approximate functions is related to the so-called *splines*, that is, with smooth piecewise polynomial functions. In order to introduce the concept of a spline, split the interval $[a, b]$ into $(n-1)$ parts $A_k = [x_k,\, x_{k+1}]$ by the points $x_2,\, x_3, \ldots, x_{n-1}$:

$$[a,\, b] = \Delta_1 \cup \Delta_2 \cup \ldots \cup \Delta_{n-1},$$

and choose, for each part, some polynomial of the degree not greater than m. A function $S_m(x)$ which is equal to the corresponding polynomial on every partial interval Δ_k and which possesses $(m-1)$ continuous derivatives at each division point $x_2, \ldots, x_{n-1}$, is called the *spline of order* m. In particular, the linear spline $S_1(x)$ is a continuous piecewise linear function, and its graph is a linear polygonal line (Figure 10.2*b*).

The quadratic spline $S_2(x)$ is a continuous function that coincides, in each partial interval, with some quadratic trinomial; at every interior node point, the graph of $S_2(x)$ is smooth and it admits a tangent line because the derivative $S_2'(x)$ is continuous at such points (Figure 10.3). It is worthy to note that the quadratic spline is not completely determined by the requirement to pass through all node points; the spline becomes unique if we impose an additional condition, e.g., assign the slope to the tangent line at one of the node points.

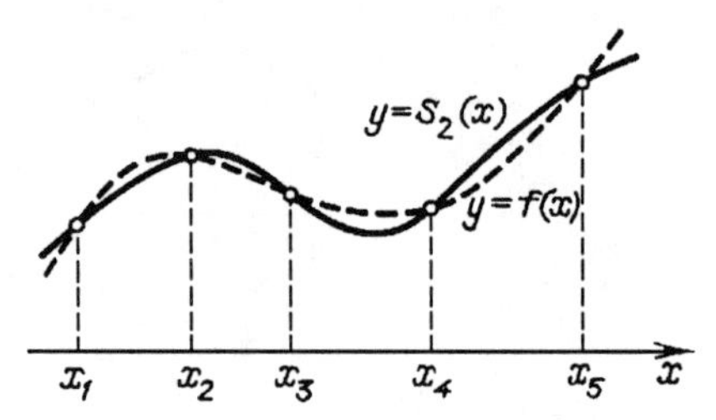

Figure 10.3: Approximation by a quadratic spline.

The cubic spline $S_3(x)$ is defined by analogy with the quadratic one: in each partial interval it coincides with a cubic polynomial, and at every interior node point the spline $S_3(x)$ itself and its derivatives $S_3'(x)$ and $S_3''(x)$ are continuous. To determine completely the cubic spline whose graph passes through all node points, we can assign, for instance, the slope to the tangent line at the extreme node points.

10.2.3 Method of least squares

The *method of least squares* (MLS) is applied often in approximation of functions, especially in the processing of measurement data. Let a table of a function $y = y(x)$ be given: (x_1, y_1),$(x_2, y_2), \ldots, (x_n, y_n)$. We approximate $y(x)$ by a polynomial of a low degree, usually by the first one: $y(x) \approx ax + b$, more seldom by 2nd or 3rd degree. The parameters a

and b in MLS are to be determined to minimize the "quadratic deviation":

$$\varphi(a,b) = \sum_{k=1}^{n} (y_k - (ax_k + b))^2.$$

This condition leads to the linear *normal system*:

$$\begin{cases} \alpha_1 a + \beta_1 b = \gamma_1, \\ \alpha_2 a + \beta_2 b = \gamma_2, \end{cases}$$

where

$$\alpha_1 = \sum_{k=1}^{n} x_k, \quad \beta_1 = n, \quad \gamma_1 = \sum_{k=1}^{n} y_k,$$

$$\alpha_2 = \sum_{k=1}^{n} x_k^2, \quad \beta_2 = \alpha_1, \quad \gamma_2 = \sum_{k=1}^{n} x_k y_k.$$

The normal system always has a unique solution (a, b) which may be found by using the Cramer formulas or the Gauss method (see Section 3.6.1). As a result, we obtain a linear approximation $y(x) \approx ax + b$, most close (in the sense of quadratic deviation) to the given table function. The graph of the approximating function $(ax + b)$ passes near the node points (Figure 10.4). If a polynomial of the degree m is used in MLS, then the normal system is of m equations.

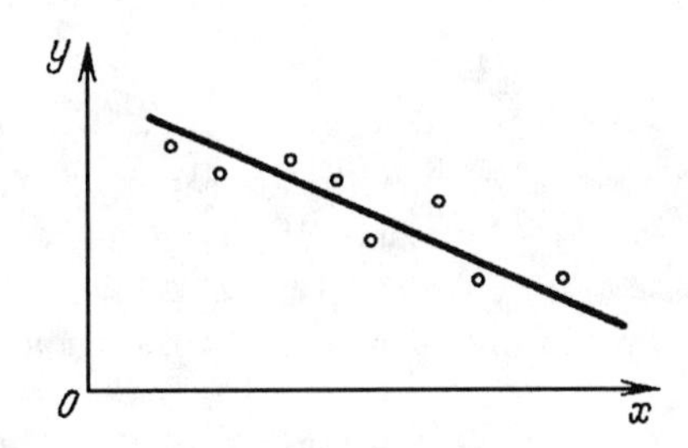

Figure 10.4: Least squares approximation: the polynomial passes near the node points.

10.3 Numerical Integration

We can find definite integrals $\int_a^b f(x)\,dx$ by using the Newton–Leibnitz formula (5.7.2) in those cases only, where we know explicitly the indefinite integral of $f(x)$; see Section 5.6. In other cases we have to calculate the integrals approximately. For numerical integration, we can use so-called *quadrature formulas,* in particular, the trapezoid rule, the rectangular formula, and Simpson's formula. The general concept to deriving such formulas is in replacing the given integrand by a simpler one that is easily integrable.

10.3.1 Trapezoid rule

In order to obtain the trapezoid rule let us split the interval of integration $[a, b]$ into n identical parts with the help of points $x_k = a + kh$, where $k = 0, 1, \ldots, n$, and $h = (b - a)/n$ is the *integration step*. The points $x_0 = a,\ x_1 = a + h, \ldots, x_n = b$ are called the *nodes*.

Replace, on each partial interval, the integrand $f(x)$ by a linear function whose graph passes through the points $(x_k, f(x_k))$, $(x_{k+1}, f(x_{k+1}))$, as well as the graph of $f(x)$. Thus the exact value of the integral (which is equal to the area of a curvilinear trapezoid, see Section 5.7.1) would be replaced by a sum of small rectangular trapezoids (Figure 10.5a, $n = 4$).

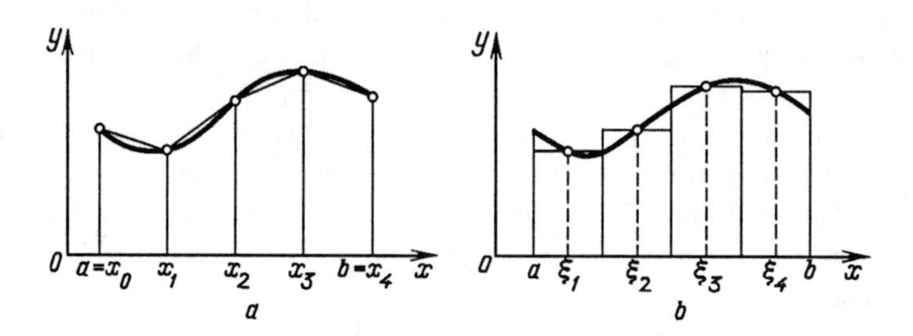

Figure 10.5: Approximation of curvilinear trapezoid: (a) – by a sum of rectangular trapezoids, (b) – by a sum of rectangles.

The calculation is performed according to the *trapezoid rule*:

$$\int_a^b f(x)\,dx \approx h\left[\frac{f(a)+f(b)}{2} + \sum_{k=1}^{n-1} f(a+kh)\right], \qquad (10.3)$$

or, in an expanded form,

$$\int_a^b f(x)\,dx \approx h\,[(f(a)+f(b))/2 + f(a+h) + \\ f(a+2h) + \ldots + f(b-2h) + f(b-h)]\,.$$

The error R_n of this formula (i.e., the difference between the exact and the approximate values of the integral) is evaluated by the inequality

$$|R_n| \le M_2/(12n^2), \quad \text{where} \quad M_2 = \max_{a \le x \le b} |f''(x)|.$$

It is seen that the increasing of n results in the decreasing of the error like $1/n^2$. The trapezoid rule usually provides a good accuracy for values of n already of order 10 – 200.

Example

$I = \int_0^{\pi/2} x \sin x \, dx$. Exactly: $I = -[x \cos x]_0^{\pi/2} + \int_0^{\pi/2} \cos x \, dx = 1$.

The approximation due to (10.3):

$$I \approx 1.013 \text{ for } n = 4, \quad I \approx 1.003 \text{ for } n = 8.$$

10.3.2 Formula of rectangles

Split the interval $[a, b]$ in the same way as in Section 10.3.1 and take the values of the integrand at the midpoints $\xi_k = a - h/2 + kh$ of partial intervals, where $k = 1, 2, \ldots, n$. Thus we obtain the *rectangles formula*:

$$\int_a^b f(x) \, dx \approx h \sum_{k=1}^{n} f(a - h/2 + kh), \tag{10.4}$$

or, in an expanded form,

$$\int_a^b f(x) \, dx \approx h \left[f(a + \frac{1}{2}h) + f(a + \frac{3}{2}h) + \ldots + f(b - \frac{1}{2}h) \right].$$

The rectangles formula means geometrically that the area of the curvilinear trapezoid is approximated by the sum of areas of small rectangles (Figure 10.5*b*, $n = 4$).

The error R_n of this formula is evaluated as follows:

$$|R_n| \le M_2/(24n^2), \quad \text{where } M_2 = \max_{a \le x \le b} |f''(x)|.$$

The rectangles formula usually provides a good accuracy for values of n already of order 10 – 200.

Example. Putting $n = 4$, we obtain:

$$\int_0^{\pi/2} x \sin x \, dx \approx \frac{\pi}{8} \left[f\left(\frac{\pi}{16}\right) + f\left(\frac{3\pi}{16}\right) + f\left(\frac{5\pi}{16}\right) + f\left(\frac{7\pi}{16}\right) \right] \approx 0.9935 \,.$$

10.3.3 Simpson's formula

Simpson's formula (or "parabolas formula") is based upon the quadratic interpolation of the integrand: on each partial interval we replace $f(x)$ by a quadratic function whose graph (a part of a parabola) passes through the following three points:

$$(x_k, f(x_k)), \ (\xi_k, f(\xi_k)), \ (x_{k+1}, f(x_{k+1})),$$

where $\xi_k = a - h/2 + kh = x_k + h/2$ is the midpoint of the partial interval. As a result, the area of the curvilinear trapezoid will be replaced approximately by a sum of small curvilinear trapezoids (see Figure 10.6, $n = 2$).

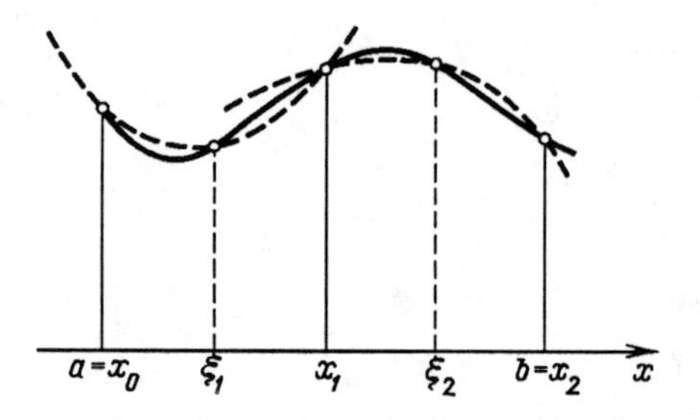

Figure 10.6: Simpson's formula.

The calculation is performed according to Simpson's formula:

$$\int_a^b f(x)\,dx \approx \frac{h}{6}[f(a) + f(b) + 2\,(f(a+h) + f(a+2h) + \ldots + f(b-h)) + 4\,(f(a+h/2) + f(a+3h/2) + \ldots + f(b-h/2))] \quad (10.5)$$

or, briefly,

$$\int_a^b f(x)\,dx \approx \frac{1}{3}(S_1 + S_2) + \frac{2}{3}S_4, \quad (10.6)$$

where
$S_1 = \frac{h}{2}\,[f(a) + f(b)]$, $S_2 = h\sum_{k=1}^{n-1} f(a + kh)$,
$S_4 = h\sum_{k=1}^{n} f(a - h/2 + kh)$,
and $h = (b-a)/n$ is the step of splitting.

Note that $S_1 + S_2$ is the quadrature sum for the trapezoid rule (10.3), and S_4 the quadrature sum for the rectangles formula (10.4).

Error R_n is evaluated as follows:

$$|R_n| \le M_4/(2880n^4), \quad \text{where } M_4 = \max_{a \le x \le b} |f^{(4)}(x)|.$$

We see that the increasing of n results in diminishing of the error like $1/n^4$, i.e., much more rapidly than that of the trapezoid rule or of the rectangles formula. That is the reason why Simpson's formula is mostly used in the approximate calculation of integrals.

Example

$I = \int_0^{\pi/2} x \sin x\, dx = 1.$

By using Formula (10.5) for $n = 4$ we obtain: $I \approx 0.999975$.

Remark. Sometimes Simpson's formula is written in the alternative form:

$$\int_a^b f(x)\,dx \approx \frac{h_1}{3}\left[f(a) + f(b) + 2\sum_{i=1}^{n-1} f(\gamma_{2i}) + 4\sum_{i=1}^{n} f(\gamma_{2i-1})\right],$$

where $h_1 = h/2 = (b-a)/(2n)$ is the *integration step*, and $N = 2n-1$ the total number of all *interior nodes* $\gamma_i \in (a,b)$, $\gamma_i = a + ih_1$, $i = 1, 2, \ldots, N$.

10.3.4 Taylor approximation

Another method of numerical integration is based on the use of the Taylor formula (see Section 10.2.1): expand the integrand $f(x)$ in powers of $(x-c)$, where c is some point of $[a,b]$, then the integral $\int_a^b f(x)\,dx$ is approximately equal to the sum

$$S_n = f(c)(b-a) + \frac{1}{2!}f'(c)[(b-c)^2 - (a-c)^2] + \ldots$$
$$\frac{1}{(n+1)!}f^{(n)}(c)[(b-c)^{n+1} - (a-c)^{n+1}].$$

Example

$I = \int_0^1 \sin x\,dx \approx \int_0^1 (x - x^3/3! + x^5/5! - x^7/7!)\,dx = \frac{1}{2} - \frac{1}{4!} + \frac{1}{6!} - \frac{1}{8!} = 0.45969742\ldots$ The exact value: $I = 1 - \cos 1 = 0.4596769\ldots$.

The Taylor expansion method provides good accuracy when the interval of integration is not large. Find the basic expansions in Section 5.8.4.

10.3.5 Improper integrals

Improper integral (see Section 5.7.3) of the form $I = \int_a^\infty f(x)\,dx$, with a continuous function $f(x)$, may be approximately calculated with the help of the "truncation" method. Split the interval $[a,\infty)$ into two subintervals, $[a,b]$ and $[b,\infty)$, choosing the number b large enough in order to make $\int_b^\infty f(x)\,dx$ be a small number (e.g., less than 10^{-4}). Then $I \approx \int_a^b f(x)\,dx$, and now we can use one of the quadrature formulas given above.

Example

$I = \int_0^\infty \frac{e^{-x}}{1+x^2}\,dx$. We write: $I = \int_0^{10} f(x)\,dx + \int_{10}^\infty f(x)\,dx$, where $f(x) = e^{-x}/(1+x^2)$.

Since $\int_{10}^{\infty} f(x)\,dx < \int_{10}^{\infty} e^{-x}\,dx = e^{-10} \approx 4.54 \cdot 10^{-5} < 5 \cdot 10^{-5}$, the error is small. By Simpson's formula, for $n = 2$, we obtain $I \approx 0.87154$.

10.4 Approximate Solution of Equations

10.4.1 Bisection method

The *bisection method* is applied to equations of the general form:

$$f(x) = 0. \tag{10.7}$$

Here we assume the function $f(x)$ to be continuous on the segment $[a, b]$ and to be of different sign at the end-points:

$$f(a) \cdot f(b) < 0.$$

By Cauchy's theorem (see Section 5.3.1), there is at least one root of the equation inside the interval: $\xi \in (a, b)$. According to the bisection method, we take the midpoint x_0 of the interval as an *initial approximation*: $x_0 = (a + b)/2$. Then we calculate $f(x_0)$ and check the sign of the product $f(a) \cdot f(x_0)$: if $f(a) \cdot f(x_0) < 0$, then $\xi \in (a, x_0)$, and the *first approximation* x_1 is defined as $x_1 = (a + x_0)/2$; if $f(x_0) = 0$, then the exact root is found: $\xi = x_0$; ultimately, if $f(a) \cdot f(x_0) > 0$, then $\xi \in (x_0, b)$ and $x_1 = (x_0 + b)/2$. So, the interval containing the root is either (a, x_0) or (x_0, b), and x_1 may be regarded as a new initial approximation.

Now we repeat the procedure described above: calculate $f(x_1)$ and check the sign of $f(x_1) \cdot f(x_0)$, and so on. The error of the n-th approximation is evaluated in the following way: $|x_n - \xi| \leq (b - a)/2^{n+1}$.

The bisection method is very simple and also reliable; however, the convergence $x_n \to \xi$ is rather slow.

10.4.2 Iteration method

The *iteration method* may be applied to equations of the special form

$$x = f(x). \tag{10.8}$$

Let ξ be a root of (10.8). The iterative process consists of the following operations: we choose a number x_0 as the *initial approximation*, then calculate the *first approximation* (or first iteration) by the formula $x_1 = f(x_0)$, then the *second approximation* (second iteration) by $x_2 = f(x_1)$, and so on. The general formula for the iterations:

$$x_{n+1} = f(x_n). \tag{10.9}$$

Conditions sufficient for convergence of the iterations:

Let ξ be a root of Equation (10.8), the function $f(x)$ and its derivative $f'(x)$ be continuous in a neighborhood $D_\varepsilon = [\xi-\varepsilon, \xi+\varepsilon]$ of the root, and the "contraction condition" be fulfilled in D_ε : $|f'(x)| < 1$. Then, for arbitrary initial approximation $x_0 \in D_\varepsilon$, it is true that $\lim_{n\to\infty} x_n = \xi$; moreover, the root is unique in D_ε.

Example

Kepler's equation is of the form $x = a\sin x + b$, where a and b are some given constants. If we plot the graphs of the left and right sides of the equation, it would be easily seen that there exists at least one root. Here we have:

$$f(x) = a\sin x + b, \quad f'(x) = a\cos x, \quad |f'(x)| \leq a.$$

Assume that $|a| < 1$, then we may assert that the root ξ is unique and that, for arbitrary choice of x_0, the sequence of iterations $x_1, x_2, \ldots$ converges to the root. For instance, let $a = 0.3$, $b = 1$, and $x_0 = 0$, then the iterations are

1, 1.2524, 1.2849, 1.2878, 1.288069, 1.2880894, 1.2880911, 1.2880913, 1.288091312, 1.2880911313, 1.2880911313.

The iteration method is applicable as well in the calculation of complex roots (see Section 7.5). Practical application of this method is not free of a specific difficulty: starting from an equation that is given in the general form $F(x) = 0$, one has to rewrite it to the special form (10.8) and ensure that the convergence conditions are satisfied.

10.4.3 Newton's method

Newton's method (or the *method of tangents*) is applicable to equations of the general form $f(x) = 0$, where $f(x)$ and its derivative $f'(x)$ are continuous functions. The algorithm is the following one.

At first we choose an *initial approximation* x_0 and calculate, at this point, the values $f(x_0)$ and $f'(x_0)$, then the *first approximation* x_1 is defined by the formula $x_1 = x_0 - f(x_0)/f'(x_0)$. Then we calculate $f(x_1)$ and $f'(x_1)$, and define the *second approximation*, $x_2 = x_1 - f(x_1)/f'(x_1)$, and so on. The general formula of Newton's method:

$$x_{n+1} = x_n - \frac{f(x_n)}{f'(x_n)}. \tag{10.10}$$

It is known that if ξ is a simple root (i.e., $f(\xi) = 0$, $f'(\xi) \neq 0$) and, moreover, x_0 is close enough to ξ, then the sequence of approximations $x_0, x_1, x_2, \ldots$ does converge to the root ξ.

The proper choice of x_0 is very important in Newton's method: for x_0 taken insufficiently near to ξ, the sequence of x_n may converge to another root or not converge at all.

Fortunately, if we have some additional information concerning the behavior of the function $f(x)$, we can confidently choose x_0. Namely, let the following sufficient conditions be satisfied:

$$\begin{aligned} &f(a) \cdot f(b) < 0, \\ &f'(x) \neq 0 \text{ for } a \leq x \leq b, \\ &f''(x) \neq 0 \text{ for } a \leq x \leq b. \end{aligned}$$

Let us choose x_0 in such a way that the following inequality holds:

$$f(x_0) \cdot f''(x_0) > 0.$$

Then it is true that $\lim_{n\to\infty} x_n = \xi$.

In practice, it is convenient to take $x_0 = a$ in the case where $f(a) \cdot f''(a) > 0$ and $x_0 = b$ in the case where $f(a) \cdot f''(a) < 0$.

Newton's method is based on the concept of the replacement, at every step, of the equation $f(x) = 0$ by a linear one. Geometrically speaking, such a procedure means the drawing of a straight line tangent to the graph of $f(x)$ at point x_0 and seeking a point x_1 of its intersection with the x-axis, then drawing a new tangent line at point x_1 and seeking a new point x_2 of its intersection with the x-axis, and so on (Figure 10.7).

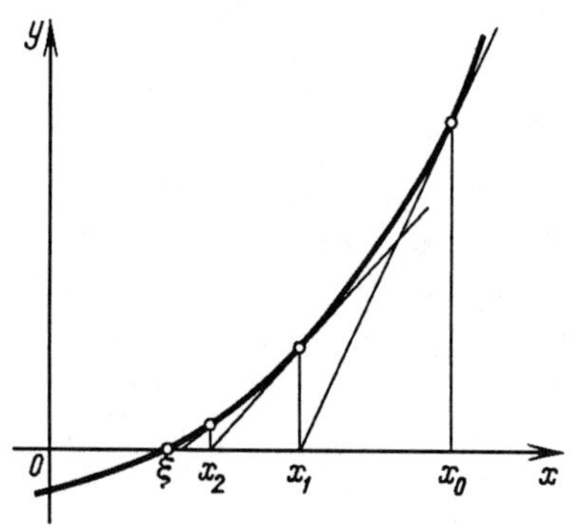

Figure 10.7: Newton's method.

Convergence of approximations $x_0, x_1, x_2, \ldots$ calculated according to this method is very rapid and is accelerating as x_n is coming closer to the root. Moreover, Newton's method is applicable to the calculation of complex roots. Unfortunately, the fact of convergence of x_n depends upon a proper (or not) choice of the initial approximation x_0. Another difficulty: it may be not profitable to calculate the derivative of the given function at every step of the procedure.

Example

The equation $x^3-3x+1=0$ can be solved by using Newton's method. First, look at the graph of the left side (Figure 5.10). We see that there are three real roots ξ_1, ξ_2, ξ_3, and that

$$-2<\xi_1<-1,\quad 0<\xi_2<1,\quad 1<\xi_3<2.$$

Let us choose the initial approximation $x_0=-2$ for the root ξ_1, $x_0=0.1$ for ξ_2, and $x_0=2$ for ξ_3. Then find the first, the second, the third, etc. approximations according to Formula (10.10). Having performed 5 steps of Newton's procedure, we obtain (approximately):

$$\xi_1=-1.879385241,\quad \xi_2=0.3472963553,\quad \xi_3=1.532088886.$$

In all cases the error does not exceed 10^{-9}.

10.5 Approximate Solution of Differential Equations

10.5.1 Euler's method

Euler's method is the simplest one to solve approximately the Cauchy problem (see Section 5.5.3) for the differential equation $y'=f'(x,y(x))$ with an initial condition $y(x_0)=y_0$.

The algorithm is the following one. Choose a small number h (the integration step) and calculate the value $y_1=y_0+hf(x_0,y_0)$ as an approximation for the solution $y(x)$ at point $x_1=x_0+h$: $y_1\approx y(x_1)$. Then calculate $y_2=y_1+hf(x_1,y_1)$ to obtain $y_2\approx y(x_1+h)$, and so on. In this way we get the sequence of numbers y_1, $y_2,\dots,y_n$ that approximate the solution $y(x)$ at points x_1, $x_2,\dots,x_n$:

$$\begin{aligned} x_k &= x_0+kh,\quad k=1,2,\dots,n,\\ y_k &= y_{k-1}+hf(x_{k-1},y_{k-1}). \end{aligned}$$

As a result we have, instead of the exact solution $y(x)$, a sequence of points (x_k,y_k) in the plane (x,y); these points are located near the graph of the exact solution $y(x)$, provided h is small enough (Figure 10.8).

The concept of Euler's method is to use, on each step, the linear approximation (see Section 10.2.1) for the solution: $y(x+h)\approx y(x)+y'(x)h$. The calculation scheme of the method is very simple, but the accuracy is not so high: at every step, the error is of order h^2 as $h\to 0$. To make the error small, one should choose a small enough h.

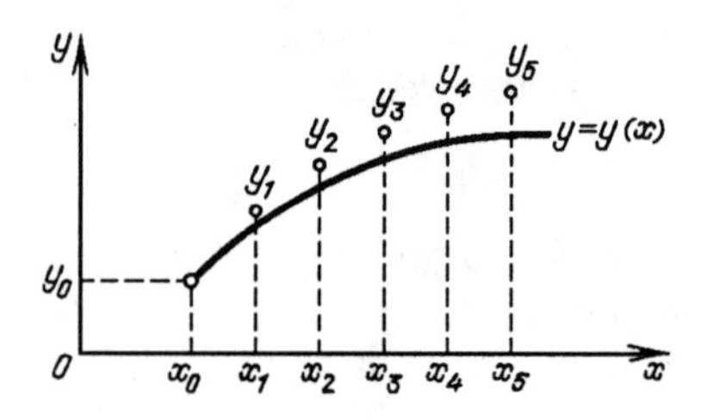

Figure 10.8: Euler's method of finding of an approximate solution.

10.5.2 Runge–Kutta method

The *Runge–Kutta method* is more accurate than Euler's method. There are several modifications of the algorithm, the most popular is the following one. Choose a small number h and calculate the following four values:

$$\begin{aligned} p_1 &= hf(x_0, y_0), \\ p_2 &= hf(x_0 + 0.5h, y_0 + 0.5p_1), \\ p_3 &= hf(x_0 + 0.5h, y_0 + 0.5p_2), \\ p_4 &= hf(x_0 + h, y_0 + p_3), \end{aligned}$$

then the approximate value y_1 of the solution at the point $x_1 = x_0 + h$ is found as

$$y_1 = y_0 + (p_1 + 2p_2 + 2p_3 + p_4)/6.$$

Approximate values y_2, $y_3, \ldots$ at the corresponding points x_1, $x_2, \ldots$ are to be calculated by the same procedure. The error for each step is of order h^5 as $h \to 0$. The Runge–Kutta method is commonly used in practice.

Chapter 11

Probability Theory

Probability theory is widely applied in statistical physics, quantum mechanics, mathematical physics, in mathematical processing of measurement data, and also in many problems of radio communication, information transmitting, economics, statistics, insurance, and theory of queues.

11.1 Random Events and Probabilities

11.1.1 Random event

An event is said to be *random* if it may happen or not at given circumstances. Some examples of random events: the rain on 4th of July in New York; appearance of the front or the back side in coin-tossing; appearance of an even side in die-tossing.

How to consider a given event—as a random one or not—may depend upon the information that we possess. Thus, the arrival of a plane to the airport in the time interval from 10:15 until 10:25 is regarded as a random event by John, who is not informed of the schedule, and as a non-random event by Mary, who is informed. Consider the experiment of coin-tossing: if we knew the initial position and velocity of the coin and its mass, we could (in principle) compute the trajectory of the coin and predict the side which must appear after the coin falls down.

A so-called *certain* event happens for sure, as the appearance at least of one score in die-tossing. An *impossible* event cannot happen, for example, the appearance of seven scores in die-tossing (with a normal die whose sides are numerated from one to six). Two events are said to be *incompatible* (i.e., mutually exclusive) if they cannot both appear in one experiment. For example, let us toss once a die: the appearance of a score greater than two and, at the same time, the appearance of an even side are two compatible events, while the appearance of score equal to two and, in this case, appearance of an odd side are two incompatible events.

11.1.2 Probability

The concept of probability of a random event is commonly met in our everyday life when we evaluate the chance for such an event to occur. The *probability of an event* A is a number $P(A)$ that characterizes the possibility of this event to occur. By definition, $0 \leq P(A) \leq 1$, the probability of a certain event is 1, and the probability of an impossible event is 0. Probability may be measured in percentages as well. In some evident situations we can easily assign probability to a given random event. For instance, in tossing a (symmetric) coin, it is natural to consider both possible results—the front or the back side—as having an identical probability equal to 0.5, or 50%. In die-tossing, any result—from 1 to 6 scores—is an equiprobable event whose probability equals $1/6$.

In general, when a given experiment can bring n different results and there is no reason to consider one of the results preferable comparatively with an other one, it is ordinarily said that each result is of probability $1/n$. If the event A appears here as a consequence of one of m equiprobable results, then we set $P(A) = m/n$. So, in die-tossing, let the event A be the appearance of an even side (score number: 2, 4, or 6), here $m = 3$, $n = 6$, and we obtain: $P(A) = 3/6 = 0.5$.

In situations like the above mentioned, the probability is referred to as an "apriori" probability. In more complicated situations (say, in statistical physics, see Section 14.3.1), the probabilities of random events are to be calculated in different ways based on some additional consideration of the process in detail.

We define the *sum of events* A *and* B, and denote it by $A + B$, as the event "appearance at least one of these events". The sum of three or more events is defined by analogy. For example, the appearance of an even side of a die is the sum of three events: the appearance of 2, or 4, or 6.

We define the *product of events* A *and* B, and denote it by $A \cdot B$, as the event of the appearance both A and B. For instance, in the tossing of two coins, let A be the appearance of the front side of the 1st coin and B be the same for the 2nd coin, then the product $A \cdot B$ means the appearance of the front sides of both coins. Obviously, the product of two incompatible events is an impossible event. The sum and the product of events look like the union and the intersection of sets (see Section 2.1.1).

The probability of the sum of incompatible events is equal to the sum of their probabilities:

$$P(A + B) = P(A) + P(B).$$

In the general case we have:

$$P(A + B) = P(A) + P(B) - P(A \cdot B). \tag{11.1}$$

Example

George and Clive are shooting at a target using bows. Let George alone be able to hit with the probability of 0.8, while Clive alone be able hit with the probability of 0.5. Now let the two boys shoot together simultaneously at the same target (once). What is the probability that at least one of them hits?

Solution. Let the event A be a hit due to George, and the event B be the same to Clive. Then $A + B$ is the hit due to at least one of the boys, and $A \cdot B$ is the double hit (due to both the boys). By Formula (11.1) we obtain:

$$P(A + B) = 0.8 + 0.5 - P(A \cdot B).$$

Here the events A and B may be regarded as independent (see below!), hence:

$$P(A \cdot B) = P(A) \cdot P(B) = 0.8 \cdot 0.5 = 0.4\,.$$

The answer: $P(A + B) = 0.9$.

Define the ***conditional probability*** as the probability of the event A provided that the event B has already appeared; such a probability is denoted by $P(A|B)$. Now we can find, in terms of conditional probability, the probability of the product of events by using the following formula:

$$P(A \cdot B) = P(A) \cdot P(B|A) = P(B) \cdot P(A|B). \tag{11.2}$$

Example

There are 7 red and 5 blue balls in an urn. Mr. Longman takes at random one ball from the urn and doesn't put it back. Then he takes at random the second ball. The question is: what is the probability of twice taking the blue balls?

Solution. Let the event A be the appearance of a blue ball on the first time, and let the event B be the appearance of a blue ball on the second time, then $A \cdot B$ will be the event of taking blue balls twice. Evidently, $P(A) = 5/12$. Furthermore, if the first ball taken is blue, then the conditional probability of B equals $P(B|A) = 4/11$, since just before taking the second ball, there were 11 balls in the urn (7 red and 4 blue). Now we can calculate the probability of twice taking blue balls by using Formula (11.2):

$$P(A \cdot B) = P(A) \cdot P(B|A) = \frac{5}{12} \cdot \frac{4}{11} = \frac{5}{33}.$$

We say that events A and B are *independent* in the case when the conditional probability $P(A|B)$ equals the probability $P(A)$. In other words, for independent events the appearance of one event does not change the probability of the other event. Thus, in the example above, if Mr. Longman had put the first taken ball back into the urn before taking the second one, the events A and B would be independent, i.e., $P(B)$ would be the same—regardless the color of the 1st ball.

For independent events, the probability of their product is equal to the product of their probabilities:

$$P(A \cdot B) = P(A) \cdot P(B).$$

In practice we watch independent events rather often because the causal connection between phenomena in many cases is absent or unessential.

Example

When somebody is tossing a coin n times in a sequence, the probability of appearance, in each trial, of the front side is evidently independent on the results of other trials, therefore the results of these n trials may be considered as independent events.

11.1.3 Bernoulli formula

Let us imagine an experiment that consists of n independent trials, each trial having two possible results, A and non-A, and assume that for each trial, the possibility $P(A)$ is the same and equals p. There is the *Bernoulli formula* (or *binomial distribution*), which finds the probability $P_{m,n}$ of appearance of A exactly m times:

$$P_{m,n} = C_n^m p^m (1-p)^{n-m},$$

where $C_n^m = \binom{n}{m}$ is the binomial coefficient, or the total amount of combinations of m elements from given n (see Section 6.1.6). The Bernoulli formula is used in physics (Section 14.3.6).

Example

Santa likes coin-tossing. One day she had tossed a coin $n = 2m$ times. Find the probability that the "tails" appeared exactly m times; take $m = 1$, then $m = 2$, $m = 5$.

Solution. By the Bernoulli formula,

$$P_{m,2m} = C_{2m}^m \left(\frac{1}{2}\right)^{2m} = \frac{(2m)!}{2^{2m}(m!)^2}.$$

For $m = 1$, $P_{1,2} = 0.5$; for $m = 2$, $P_{2,4} = 0.375$; for $m = 5$, $P_{5,10} \approx 0.25$.

For large values of m and n the Bernoulli formula may be transformed (by using Stirling formula for $n!$, see Section 6.1.3) to the special form:

$$P_{m,n} \approx \frac{1}{\sigma\sqrt{2\pi}} e^{-x^2/2},$$

where $\sigma = \sqrt{np(1-p)}$, $x = (m - np)/\sigma$.

This formula is called the *Gaussian distribution,* or the *normal distribution.*

For p being a small number, Poisson's formula

$$P_{m,n} \approx \frac{y^m}{m!} e^{-y}, \quad \text{where } \; y = np,$$

provides a more accurate approximation than the normal distribution formula.

Example

An automatic machine in a factory is making nails. Let the probability of a defective nail appearing be $p = 0.1\%$. Find the probability of the appearance of not more than two defective nails in a series of 1000 nails.

Solution. Consider the following types of series: with no defective nails, with 1 defective nail, and with 2 such nails; the desired probability:

$$Q = P_{0,1000} + P_{1,1000} + P_{2,1000}.$$

According to Poisson's formula (with $y = 1$):

$$Q \approx e^{-1}(1 + 1 + 0.5) = 2.5e^{-1} \approx 92\%.$$

11.1.4 Large numbers' law

Let the event A appear m times in the series of n trials, then the quantity $\omega = m/n$ is called the *frequency* (or the frequency ratio) of the event A. While n is increasing, the frequency of A is approximating, in a certain sense, to the probability p of A. For this case, the *theorem of Bernoulli* (or the *large numbers' law*) is applicable: let the probability p of A in each trial be constant, and the trials be independent, then for an arbitrarily chosen number $\varepsilon > 0$, the probability that the difference between ω and p is less than ε, tends to 1 as $n \to \infty$:

$$\lim_{n\to\infty} P(|\omega - p| < \varepsilon) = 1. \tag{11.3}$$

The relation (11.3) means that provided the number of trials is large enough, the frequency of the event is close to its probability (more exactly speaking, the probability of a large difference between ω and p tends to zero as $n \to \infty$). Thus, when tossing a coin many times, we "almost certainly" get an approximate equal appearance of the front and back sides of the coin.

11.2 Random Variables and Distributions

11.2.1 Random variable

A *random variable* (or stochastic variable) is such a variable that its values depend on the occasion. Some examples of random variables: atmospheric pressure at 10 p.m. on July 4th in Boston, score number to appear in die-tossing, speed of a gas molecule at a given moment, etc.

To characterize a random variable, we have to know the set of its possible values and also the corresponding probabilities of the values. For instance, the distribution of score number in die-tossing may be described by identical probabilities 1/6 for each score number from 1 to 6.

We distinguish the *discrete* and the *continuous* random variables. Values of a discrete random variable form a finite (or a countable infinite) set, while the values of a continuous random variable cover all of the x-axis (or some its subintervals).

A discrete variable may be specified by a table containing its values and their probabilities. A continuous random variable A must be specified in the other way: let $\rho(x)$ be a given continuous (or piece-wise continuous) function, and let the probability of the fact that the values of A belong to the interval $[x, x + \delta x]$, be

$$P = \int_{x}^{x+\delta x} \rho(\xi)\, d\xi .$$

This formula is the description of the random variable A. The function $\rho(x)$ is called the *probability density* of the variable A. In statistical physics $\rho(x)$ is called usually the *distribution function of A* (see Section 14.3.3). The formula above denotes, in particular, that the probability for A to be found in a small interval $[x, x + \delta x]$, is approximately equal to $\rho(x)\delta x$.

It is a standard to assume that:

1. $\rho(x) \geq 0$,

2. $\int_{-\infty}^{+\infty} \rho(x)\, dx = 1$.

The property 2 is usually referred to as the "norming condition".

In physics we meet often the *normal*, or *Gaussian*, distribution (see Section 14.3.3), where:

$$\rho(x) = \frac{1}{\sigma\sqrt{2\pi}} \exp[-\frac{(x-a)^2}{2\sigma^2}]. \qquad (11.4)$$

Graphs of $\rho(x)$ for different values of the parameter σ are shown in Figure 11.1. As an important example of normally distributed values note the random errors in measurements.

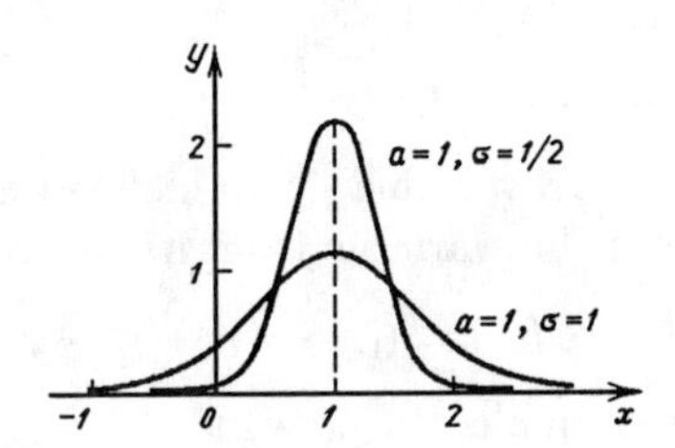

Figure 11.1: Gaussian curves.

11.2.2 Mean value and dispersion

The most essential characteristics of a random variable are its *mean value* (or *mathematical expectation*) and its *dispersion*.

The mean value of a discrete random variable A whose possible values are $x_1, x_2, \ldots, x_n$ with identical probabilities $p = 1/n$, is defined by the expression

$$\langle A \rangle = \frac{1}{n}\sum_{k=1}^{n} x_k.$$

In the case of a general discrete random variable A whose values are $x_1, x_2, \ldots, x_n$ and probabilities $p_1, p_2, \ldots, p_n$, the mean value is defined as

$$\langle A \rangle = \sum_{k=1}^{n} p_k x_k, \quad \text{where} \quad \sum_{k=1}^{n} p_k = 1. \qquad (11.5)$$

For a continuous random variable A having the distribution function $\rho(x)$ we define:

$$\langle A \rangle = \int_{-\infty}^{+\infty} \rho(x) x \, dx.$$

Example

Bill is going to buy a lottery ticket for \$20. He reads an information text that there will be prizes of three kinds: \$100, \$1,000, and \$10,000, with probabilities to win, respectively, 3%, 0.5%, and 0.01% (per one ticket). Bill desires to evaluate his possible winnings.

Solution. Bill is recommended to find the mean value of the winnings by using Formula (11.5):

$$\$100 \cdot 0.03 + \$1,000 \cdot 0.005 + \$10,000 \cdot 0.00001 = \$9\,.$$

The answer, of course, does not mean that after the results of the lottery published, Bill gets exactly $9. His real winnings may be from $0 to $10,000, but now he understands that his intention to participate in the lottery is risky enough for his purse.

A physical example of the calculation of the average speed of a gas molecule is given in Section 14.3.3.

Dispersion of a random variable A is defined as the mean value of the squared deviation of this variable from its mean value:

$$D(A) = \langle (A - \langle A \rangle)^2 \rangle.$$

For a discrete random variable, it follows that

$$D(A) = \sum_{k=1}^{n} p_k (x_k - \langle A \rangle)^2,$$

while for a continuous one,

$$D(A) = \int_{-\infty}^{+\infty} \rho(x)(x - \langle A \rangle)^2 \, dx.$$

Here [and also in (11.6)] we assume the improper integral to be convergent.

We mention a special property of the dispersion:

$$D(A) = \langle A^2 \rangle - \langle A \rangle^2.$$

The dispersion of a random variable A estimates the scattering of A near its mean value. A determinate variable, i.e., a variable whose values all are equal to the same number x_0, has the mean value equal to x_0, and the dispersion is equal to 0.

Note the physical meaning of the dispersion: the parameter $\sigma = \sqrt{D(A)}$ is the measure of fluctuation of the physical value A (see Section 14.3.6). As an illustration, consider a vessel containing some amount of a gas which is in thermodynamic balance. Choose mentally a fixed volume inside the vessel, then the number n of molecules in the volume is a random variable. For a given mean value $n_0 = \langle n \rangle$, the fluctuation of n is equal to $\delta n = \sqrt{\langle (n - n_0)^2 \rangle}$, i.e., the square root of the dispersion. It may be proved that if n_0 is increasing, the fluctuation δn is increasing like $\sqrt{n_0}$, while the relative fluctuation, or $\delta n / n_0$, is decreasing like $1/\sqrt{n_0}$.

Part II

PHYSICS

Chapter 12

Physical Quantities and Systems of Units

Laws studied by physics lie in the basis of the whole natural science. Physics is an *exact science*, i.e., the laws of nature established in physics can be expressed quantitatively. The laws of physics rely on facts obtained experimentally as a result of *measuring*.

Any measurement implies the usage of some units of measurement. *Units of physical quantities* are divided into system units (i.e., those included in a certain system of units) and nonsystem units (e.g., units of energy electron-volt or kilowatt-hour). System units are subdivided into *base* or *fundamental* units (chosen arbitrarily by a convention) and *derived* units. Derived units are expressed through the fundamental units. This chapter deals with two systems of units widely used in physics: the Gaussian system (CGS) and the International system (SI).

Physical quantities are characterized by their *dimensions*. Analysis of dimensions is sometimes a good theoretical method of investigation of physical phenomena.

12.1 Basic Concepts. Laws of Physics

12.1.1 Physical quantities and measurements

We give names to various characteristics of objects as well as to natural phenomena. In this way *concepts* are formed, their definitions giving answers to the question what is what. Sometimes a concept can be associated with *a physical quantity*. The corresponding characteristic of an object, a phenomenon, or a process must permit a *quantitative expression*, i.e., for this quantity a *procedure of measurement* can be *defined* (introduced by definition) and a *unit of measurement* can be determined.

For example, the concept of spatial extent is associated with a physical quantity called the distance. For the measurement of distances a certain procedure is conventionally assumed by definition and a certain unit (a meter) is chosen.

To *measure* a physical quantity means to find out how many times a quantity of the same physical nature as the latter and considered as a *unit quantity* contains in the measured quantity. The corresponding number is called the *numerical value* of the physical quantity.

12.1.2 Equations in physics

Different physical quantities are connected by objectively existing relations, or *physical laws.* Sometimes these laws can be conveyed in the form of mathematical equations. Physics is called an *exact science* to emphasize the fact that correlations established in physics usually can be expressed in the *quantitative form.* However, it is worth mentioning that, as a rule, these correlations (physical laws) are approximate and valid in limited ranges.

Opposite to "pure" mathematics, where quantities by definition have the characteristics arbitrarily attributed to them, in physics it is impossible to attribute characteristics to physical quantities; it is necessary to discover objectively existing characteristics of physical quantities.

Values of physical quantities included into equations of physical laws, as well as the symbols (letters) designating them, should always be considered as *products of a numerical value and the corresponding unit of measurement.* If, in an equation for physical quantities, mathematical functions of type log, sin, or exp are encountered, their argument can only be a dimensionless number (e.g., a ratio of quantities of equal dimensions, see Section 12.3).

12.1.3 Physical models

For a theoretical investigation of physical phenomena a reasonable well-grounded choice of *a physical model* is necessary. Any physical model implies a certain *idealization* in which the most important traits of the phenomenon should be preserved. Examples of physical models are a material point, an absolutely rigid body, an ideal gas, etc.

Validity of a certain physical model depends mainly on the questions to be answered rather than on the characteristics of the real system in question. It is very important to build up the feeling of a necessary degree of mathematical rigorism while using a physical model: it makes little sense to search for an exact solution within the scope of a rather rough physical model.

12.2 Systems of Units

12.2.1 Base and derived units

Any measurement implies a comparison of the measured quantity with another one that has the same physical nature as the latter and is assumed to be the unit of the corresponding quantity. Actually, units for all quantities can be chosen independently from one another. However, this is inconvenient, since in this case numerical coefficients will occur in all equations expressing relations between different physical quantities. Besides, one would have to introduce a separate standard for every physical quantity.

The main peculiarity of modern systems of units is the existence of certain relations between units of different quantities. These relations are determined by the physical laws or definitions connecting the measured quantities. For example, the unit of velocity is chosen so that it is expressed through the units of length and time. In this choice of the unit of velocity the definition of velocity is used. The unit of force is defined with the help of Newton's second law and is expressed through the units of acceleration and mass. This means that for several arbitrarily chosen physical quantities, units are defined independently. Such units are called *base* or *fundamental units.* Units for the other quantities are expressed through fundamental ones and called *derived units.*

The number of base units and their choice itself may differ in different systems of units. Thus, in Gaussian units (CGS), three units are chosen as fundamental ones: units of length—centimeter (L), time—second (T) and mass—gram (M). The international system of units (SI) has seven base units and two supplementary units, all other units being derived from these units. Units chosen as base ones: units of length—meter (L), mass—kilogram (M), time—second (T), thermodynamic temperature—kelvin (θ), amount of substance—mole (N), electric current—ampere (I), and luminous intensity—candela (J). Dimensionless supplementary units are the units of plane angle—radian, and of solid angle—steradian. Definitions of fundamental, supplementary, and derived units in SI are given in Supplement II.

12.2.2 Dimensions of physical quantities

Arbitrarily chosen quantities are not only the physical quantities whose units are assumed as fundamental, and *the scale* (the size) of these units, but also the *coefficients of proportionality* in the formulas by which derived units are introduced. Let us consider an example of the unit of area. If the meter is chosen as a unit of length, it is possible to choose either the square meter (the area of a square with its side equal to one meter) or the "round" meter (the area of a circle with its diameter equal to one meter). In the first case, the area of a square with the side l is expressed by the formula $S = l^2$, and the area of a circle of the diameter l is expressed by the formula $S = \pi l^2/4$. In the second case, a more simple formula is for the area of a circle: $S = l^2$, while the formula for the area of a square contains π: $S = 4l^2/\pi$.

The above-considered possibilities of introducing units of area which differ by a numerical coefficient are based on the same geometric relationship between areas of similar figures and their linear dimensions. This relationship states that the area S of a figure is proportional to the square of the corresponding length l: $S \propto l^2$. However, when a derived unit of some quantity is introduced, not only is the numerical coefficient in its definition arbitrarily chosen, but also is the physical law that defines the relation of a derived unit with fundamental ones. For example, the unit of force is usually defined with the help of Newton's second law $F = ma$. In this case, the expression for the unit of force through fundamental units, i.e., the *dimensions* of force, has the form

$$\dim F = MLT^{-2}. \tag{12.1}$$

However, with the same fundamental units (L, M, T) it is possible to define the unit of force in another way. Instead of Newton's second law, one can use the law of gravitation considering the proportionality coefficient to be dimensionless and equal, say, to unity: $F = m_1 m_2/r^2$. In this case, for the unit of force, such a force is taken with which two material points whose masses each equal a unit of mass, being separated by a unitary distance, are mutually attracted. The dimensions of this force have the form

$$\dim F = M^2 L^{-2}. \tag{12.2}$$

Of course, when the unit of force is chosen in such a way, in Newton's second law a dimensional coefficient will appear, like it appears in the law of gravitation (*the gravitational constant*) when the choice of the unit of force is based on second Newton's law.

The considered example shows that *the dimensions of a physical quantity depends on how the system of units is constructed.*

Thus, there is a large arbitrariness in the method of constructing a system of units. However, in practice one has to reckon with a whole number of requirements limiting the choice. It would be inconvenient to have too many base units because dimensional coefficients would appear in many physical formulas, and a large number of standards would have to be established. On the other hand, too small numbers of fundamental units would make it difficult to use in practice derived units made on their basis. Usually we use in practice systems with the number of base units from three to seven.

12.2.3 Standards of base units

In the choice of base units it is very important to take into account the possibilities of creating the standards in order to ensure the constancy of the unit and to provide a reproduction of a standard against the risk of its loss. The safest way to solve the problem is to entrust the nature itself to "preserve" the standards. Thus, the contemporary standard of time is based on the period of the specific oscillations occurring in an atom of a cesium-133 isotope. By definition, the unit of time *second* equals the duration of 9,192,631,770 periods of these oscillations. Atoms of the same isotope are identical, therefore nature gives us practically an unlimited number of absolutely identical "clocks."

To define the base unit of length, today the same standard as that for the unit of time is used: by definition, a *meter* is the length of the path traveled by light in vacuum during a time interval of 1/299,792,458 seconds.

For the mass standard suitable for macroscopic bodies it is still impossible to use the mass of some atomic particle, because the accuracy of the determination of the number of particles in a macroscopic body yields to that of the determination of mass by weighting. The standard of mass *kilogram* is the international prototype—a weight made of platinum and iridium, kept in the International Bureau of Weights and Measures at Sevres, near Paris.

Today both in educational and scientific literature, mainly two systems of units are widely used: *the International system of units* SI and the *Gaussian system* (or *symmetrical system* CGS—the first letters of the names of fundamental units Centimeter, Gram, Second). Therefore, it is necessary to understand the principles of their construction. Units of mechanical quantities in these two systems differ only in scale, because their fundamental units are chosen on the basis of the same physical quantities—length, time, and mass. Therefore, in mechanics all formulas and equations expressing physical laws and definitions are the same in both systems of units.

The case is somewhat different in electrodynamics. The unit of electric charge is derived and expressed through the fundamental units in the Gaussian system with the help of Coulomb's law, the proportionality coefficient in the latter being chosen as dimensionless and equal to unity: $F = q_1 q_2 / r^2$. This gives the following dimensions for electric charge:

$$\dim q = M^{1/2} L^{3/2} T^{-1}. \tag{12.3}$$

The unit of electric charge in the Gaussian system has no special name. The other electric quantities in this system have their units expressed through the unit of electric charge and thus through the base units. For example, the dimensions of electric current

$$\dim I = M^{1/2}L^{3/2}T^{-2}. \qquad (12.4)$$

The derived units for the strength of the electric field, for potential, capacity, and derived units for other electric quantities are introduced in an analogous way.

12.2.4 Units of magnetic quantities in Gaussian system of units

In the Gaussian system, the units of magnetic quantities are introduced as follows. Let us consider a magnetic field created by a straight infinite conductor in which current I is flowing. According to the law of Bio–Savar–Laplace, in the observation point located at a distance r from an element Δl of this conductor, this element creates the magnetic field induction ΔB (magnetic flux density) equal to

$$\Delta B = k\frac{I\Delta l \sin\alpha}{r^2},$$

where coefficient k depends on the choice of units. Addition of the fields created by all elements of the conductor gives, for the resulting magnetic field induction in the observation point, the following expression:

$$B = k\frac{2I}{r}. \qquad (12.5)$$

It is possible to detect the magnetic field by its influence on another conductor with current. If this conductor is located parallel to the conductor creating the magnetic field, then the force influencing the former is proportional to the induction of the magnetic field B, current I' in the conductor, and its length l:

$$F = k'I'Bl. \qquad (12.6)$$

The coefficient k in Equation (12.5) can be chosen arbitrarily, because the unit of the magnetic field induction B is not yet established. But after this coefficient k in Equation (12.5) is chosen (and thus the induction unit B is chosen), the coefficient k' in Equation (12.6) cannot be chosen arbitrarily, but should be determined experimentally.

Of course, it is possible to choose another way, namely by using Equation (12.6) to introduce the unit of the magnetic field induction, assuming $k' = 1$; then the coefficient k in Equation (12.5) should be determined experimentally. In the Gaussian system (in the symmetrical system CGS) they do as follows: the coefficient k in Equation (12.5) is chosen to be equal to the coefficient k' in Equation (12.6).

If the magnetic field induction B is substituted into Equation (12.6) from Equation (12.5), then we shall get the following expression for the force of interaction of two parallel conductors with currents I and I', separated by distance r:

$$F = kk'\frac{2II'l}{r}. \tag{12.7}$$

In the Gaussian system, $kk' = k^2$. Since the units for all the quantities encountered in this formula are already chosen, the coefficient k^2 has the dimensions $L^{-2}T^2$, inverse to those of the velocity squared. This coefficient must be determined experimentally by measuring the force of interaction of two parallel conductors separated by a certain distance, when certain currents flow in them. The experiment has proved that the numeric value of k^2 is equal to $1/c^2$, where c is the velocity of light in a vacuum:

$$c = 2.997,924,58 \cdot 10^{10}\ \text{cm/s} \approx 3 \cdot 10^{10}\ \text{cm/s}.$$

Thus, in the Gaussian system of units the law of Bio–Savar–Laplace and Ampere's law are written in the form

$$\Delta B = \frac{1}{c}\frac{I\Delta l \sin\alpha}{r^2}, \qquad F = \frac{1}{c}I'Bl. \tag{12.8}$$

The unit of the magnetic field induction (magnetic flux density) is defined on the grounds of the last formula. This unit is called a *gauss*. One gauss is the induction of such a magnetic field that acts with the force of $1/c$ dyne upon one centimeter of a conductor in which an electric current of one unit of current (1 ampere) flows, if the conductor is perpendicular to the induction lines of the magnetic field.

12.2.5 Units of magnetic quantities in SI

Unlike the Gaussian system of units, where the unit of electric current is derived and expressed through the base units by means of Coulomb's law, in SI the unit of electric current is fundamental. This unit is chosen on the basis of the law of interaction of two parallel currents, expressed by Equation (12.7). The base unit of electric current is called an *ampere* (symbol A). An ampere is defined as a constant current that, being maintained in two straight parallel conductors of an infinite length and negligible cross-section, would produce a force of interaction between the conductors of $2 \cdot 10^{-7}$ newton per meter of the conductor's length, the conductors being located in a vacuum at a distance of 1 meter apart. Like a *meter*, *second*, and *kilogram*, an *ampere* is one of the base units of SI (see Supplement II). The unit is named after A.M. Ampere.

After the unit of electric current is chosen, the dimensions of the coefficient kk' in Equation (12.7) can be determined. It is equal to N/A^2, i.e., $MLT^{-2}I^{-2}$. The coefficient kk' in SI is represented by $\mu_0/4\pi$, and the quantity μ_0 is called the *magnetic constant.*

The numerical value of the magnetic constant can be calculated with the help of Equation (12.7) and the definition of the ampere. Substituting $F = 2 \cdot 10^{-7}$ N, $I = I' = 1$ A, $l = r = 1$ m, and $kk' = \mu_0/4\pi$ into Equation (12.7), we find

$$\mu_0 = 4\pi \cdot 10^{-7}\ \text{N/A}^2 = 12.566,370,614 \cdot 10^{-7}\ \text{N/A}^2. \tag{12.9}$$

The numerical factor 4π is introduced in Equation (12.9) for so-called rationalization of the system of units. By virtue of this rationalization, the coefficient 4π disappears in many formulas that are frequently used (though at the same time it appears in some other formulas). Introduction of the factor 4π here is absolutely analogous to the above example of transition from square meters to "round" meters in the measurement of area.

It should be emphasized that the numerical value of the constant μ_0, Equation (12.9), in SI is a direct consequence of the definition of the ampere, unlike the experimentally determined coefficient $1/c^2$ in the Gaussian system. It is so because in SI the unit of electric current is fundamental (arbitrarily chosen), whereas in the Gaussian system it is a derived unit.

Introduction of the unit of electric current *ampere* defines uniquely only the product of coefficients k and k' entering Equations (12.5) and (12.6): $kk' = \mu_0/4\pi$. After the product kk' is determined, there remains a freedom in the choice of the co-factors. In SI $k' = 1$ is assumed, therefore k becomes equal to $\mu_0/4\pi$. As a result, Ampere's law describing the action of a magnetic field on a conductor with current I' is written in SI in the form

$$F = I'Bl. \tag{12.10}$$

In SI this law is used for the introduction of the unit of magnetic field induction B (or the magnetic flux density). This unit is named *tesla* (symbol T). It is named after Nikola Tesla, an american electrical engineer. One tesla is the induction of such a magnetic field that acts on 1 meter of a conductor carrying the constant current 1 ampere, with the force 1 newton, if the conductor is perpendicular to the induction lines of the magnetic field. It follows from Equation (12.10) that B has the dimensions $\text{N}/(\text{A} \cdot \text{m})$, i.e., $MT^{-2}I^{-1}$.

The law of Bio–Savar–Laplace, which is the basis for the calculation of the induction of the magnetic field created by a conductor with current, in SI contains the coefficient $k = \mu_0/4\pi$:

$$\Delta B = \frac{\mu_0}{4\pi}\frac{I\Delta l \sin\alpha}{r^2}.$$

In SI, Coulomb's law that describes the interaction of point charges contains a dimensional coefficient because the unit of charge, the *coulomb*, is defined in SI independently of Coulomb's law on the basis of the unit of electric current ampere: $1\text{C} = 1\text{A} \cdot \text{s}$. The dimensional coefficient in Coulomb's law is written in the form $1/(4\pi\varepsilon_0)$, where the quantity ε_0 is called the *electric constant*, and the numerical factor 4π is introduced for the rationalization:

$$F = \frac{1}{4\pi\varepsilon_0} \frac{q_1 q_2}{r^2}. \qquad (12.11)$$

It follows from Equation (12.11) that the electric constant ε_0 has the dimensions $\text{C}^2/(\text{N} \cdot \text{m}^2)$, i.e., $M^{-1}L^{-3}T^4I^2$. Its unit is named *farad per meter* (F/m). The numerical value of the electric constant ε_0 is determined experimentally. For example, the value of ε_0 can be found by measurement of the force of interaction of certain point charges located at a certain distance from each other. The measurements give the following value for ε_0:

$$\varepsilon_0 = 8.854,187,817 \cdot 10^{-12}\ \text{C}^2/(\text{N} \cdot \text{m}^2). \qquad (12.12)$$

12.2.6 Relationship between SI and Gaussian units

To determine the relationship between units of charge (or current) in the Gaussian system and in SI, the following method is possible. Let electric currents of 1 ampere flow in parallel conductors separated by the distance of 1 meter. Then the force acting upon 1 meter of the conductor's length is equal to $2 \cdot 10^{-7}$ N, i.e., $2 \cdot 10^{-2}$ dyne. The formula for the calculation of this force (Equation (12.7)) in the Gaussian system of units has the form

$$F = \frac{1}{c^2} \frac{2I^2 l}{r}.$$

Let us substitute $F = 2 \cdot 10^{-2}$ dyne into the left part of this formula, and $l = r = 100$ cm, $c = 2.997,924,58 \cdot 10^{10}$ cm/s—into its right part. Then we shall find that such a force of magnetic interaction is ensured by a current that numerically equals $0.1\,c$ of the electric current unit in the Gaussian system. Thus,

$$1\text{A}/(1 \text{ unit of Gaussian system}) = 2.997,924,58 \cdot 10^9 \approx 3 \cdot 10^9.$$

The relationship between units of current in the Gaussian system and in SI is expressed through a constant c determined experimentally.

The relationship between electric charge units in these systems is the same:

$$1\ \text{C} \approx 3 \cdot 10^9 \text{ units of charge of the Gaussian system.}$$

Knowing this relationship, it is possible to calculate the value of the electric constant ε_0. Two point charges of 1 coulomb each, located at the distance of 1 meter apart, interact with the force equal to $1/(4\pi\varepsilon_0)$ newton. In the Gaussian system of units this force is

$$F = \frac{q^2}{r^2} = \frac{(3 \cdot 10^9)^2}{(100)^2} \text{ dyne } = 9 \cdot 10^9 \text{ N}.$$

Thus, the numerical value of the coefficient $1/(4\pi\varepsilon_0)$ is equal to $9 \cdot 10^9$. Therefore,

$$\varepsilon_0 = \frac{1}{4\pi \cdot 9 \cdot 10^9} \text{C}^2/(\text{N} \cdot \text{m}^2) = 8.854 \cdot 10^{-12} \text{C}^2/(\text{N} \cdot \text{m}^2).$$

Note that in fact we have not "calculated" the electric constant ε_0 theoretically, but rather only expressed it through another constant c determined experimentally. It means after all that the electric constant is determined experimentally. In this respect the electric constant ε_0 differs essentially from the magnetic constant μ_0, whose value is not measured, but rather prescribed by the definition of ampere. This difference between ε_0 and μ_0 is due to the fact that in SI the introduction of the electric charge unit is based on the magnetic interaction of currents rather than on the electrostatic interaction of charges.

The relationship between experimentally determined constants ε_0 and c can be revealed in a somewhat different way. Let us compare expressions for forces of an electrostatic interaction of charges and a magnetic interaction of currents, written in the Gaussian system and in SI:

$$F_e = \frac{q^2}{r^2}, \qquad F_e = \frac{1}{4\pi\varepsilon_0}\frac{q^2}{r^2}, \tag{12.13}$$

$$F_m = \frac{l}{c^2}\frac{2I^2 l}{r}, \qquad F_m = \frac{\mu_0}{4\pi}\frac{2I^2 l}{r}. \tag{12.14}$$

The dimensionless relation F_e/F_m must be the same in the both systems of units. Making the ratio of the right sides of Equations (12.13) and (12.14) and equating the values of this ratio in both systems of units, one can ensure that

$$1/c^2 = \varepsilon_0\mu_0. \tag{12.15}$$

As it was already mentioned, the appearance of some formulas in the Gaussian system and SI differs not only in dimensional coefficients, but also in numerical factors, which appear by virtue of the rationalization accepted in SI. Notations of the most important formulas of electrodynamics in the two systems of units are compared in Supplement V.

12.3 The Method of Dimensional Analysis

12.3.1 Dimensionless and dimensional units

Physical quantities whose numerical values are independent from the chosen scale (the size) of units are called *dimensionless*. Examples of dimensionless quantities: angle (ratio of the length of an arc to the radius), refractive index of light (the ratio of the velocity of light in a vacuum to the velocity of light in a substance). Physical quantities whose numerical values change as the scale of units is changed are called *dimensional*. Examples of dimensional quantities: length, velocity, and energy. Expressions for a derived unit of a physical quantity through the fundamental ones are called the *dimensions* of the unit (or the dimensional formula). For example, the dimensions of momentum

$$\dim p = LMT^{-1}.$$

Another notation for the dimensions of a physical quantity—square brackets—is also used instead of dim. For example,

$$[p] = LMT^{-1}.$$

Dimensions of physical quantities in SI are given in Supplement II.

Knowledge of the dimensions of the physical quantities is useful for checking the correctness of results obtained while solving physical problems: right and left parts of the obtained expressions, and individual terms as well, must have equal dimensions. The method of dimensional analysis can also be used for derivation of formulas and equations as soon as the physical parameters, on which this unknown quantity may depend, are known. It is easier to understand the essence of the method by considering concrete examples.

12.3.2 Example: velocity in free fall

Let us determine the velocity v, with which a body of the mass m freely falling from the height h falls on the ground. Since the unknown quantity may depend on the acceleration of free fall g, the height h, and the mass m, we can search for the expression for v in the form

$$v = Ch^x g^y m^z, \tag{12.16}$$

where C is some dimensionless constant, and x, y, and z are the numbers to be determined. Let us equate the dimensions of the left and right parts of Equation (12.16):

$$LT^{-1} = L^x (LT^{-2})^y M^z.$$

The indices of powers must be equal for L, M, and T in the left and the right parts, therefore

$$\begin{aligned} L: &\quad 1 = x + y, \\ T: &\quad -1 = -2y, \\ M: &\quad 0 = z. \end{aligned}$$

Thus, $z = 0,\ y = 1/2,\ x = 1/2$, and Equation (12.16) takes the form

$$v = C\sqrt{gh}. \tag{12.17}$$

The proper value of the velocity is $\sqrt{2gh}$, i.e., the dimensional analysis has given us the opportunity to determine the character of dependence of v on g, h, and m, but has left the value of numerical factor C indefinite.

12.3.3 Example: flight range

Let us determine the flight range of a bullet shot horizontally at the initial velocity v at a height h above the ground. We search the expression for s in the form

$$s = Cv^x g^y h^z m^u. \tag{12.18}$$

The equality of dimensions:

$$L = (LT^{-1})^x (LT^{-2})^y L^z M^u. \tag{12.19}$$

Equating indices of power, we obtain

$$\begin{aligned} L: &\quad 1 = x + y + z, \\ T: &\quad 0 = -x - 2y, \\ M: &\quad 0 = u. \end{aligned}$$

Thus, $x = -2y,\ z = 1 + y$, and Equation (12.18) will take the form

$$s = Cv^{-2y} g^y h^{1+y} = Ch(gh/v^2)^y. \tag{12.20}$$

The dimensional analysis has allowed us to determine that the flight range s does not depend on the mass of the bullet, but depends on the height h and some unknown power y of the dimensionless combinations of parameters gh/v^2. If we know (e.g., from experiment) the dependence of s on even one of the parameters, we can immediately determine y. Assume that we know that $s \propto v$, then $y = -1/2$, and for s from Equation (12.20) we get

$$s = Cv\sqrt{h/g}.$$

This expression coincides with the proper value of $s = v\sqrt{2h/g}$ accurate to the constant factor C. The dimensional analysis did not allow us to determine completely the character of dependence because the number of parameters on which the flight range s could depend was more than the number of fundamental units of the used system of units.

In this example, the complete determination of the character of dependence is possible, if so-called vectorial units of length are used. We shall measure length in horizontal and vertical directions in different units, and label their dimensions by L_h and L_v. Then, taking into account that

$$\dim v = L_h T^{-1}, \qquad \dim g = L_v T^{-2}, \qquad \dim h = L_v,$$

instead of Equation (12.19) we get

$$L_h = (L_h T^{-1})^x (L_v T^{-2})^y L_v^z M^u.$$

Equating the power indices, we get

$$\begin{aligned} L_h: &\quad 1 = x, \\ L_v: &\quad 0 = y + z, \\ T: &\quad 0 = -x - 2y, \\ M: &\quad 0 = u. \end{aligned}$$

Therefore, $y = -1/2$, $z = 1/2$, and for s we obtain

$$s = Cvg^{-1/2}h^{1/2} = Cv\sqrt{h/g}.$$

An increase in the number of fundamental units enhances the possibilities of the method of dimensional analysis.

12.3.4 Example: viscous flow

Let us consider a laminar flow of a viscous liquid in a tube (Figure 12.1). The volume of liquid V flowing through the tube cross-section during the time t is proportional to the time t and depends on the difference of pressure Δp on the ends of the tube, on viscosity η of the liquid, and on the radius R of the tube:

$$V = C(\Delta p)^x \eta^y l^x R^u t.$$

It is also convenient here to measure the length along and across the tube in different units with dimensions $L_\parallel$ and $L_\perp$, respectively. Then

$$\dim l = L_\parallel, \qquad \dim R = L_\perp, \qquad \dim V = L_\parallel L_\perp^2,$$

$$\dim \Delta p = M L_\parallel L_\perp^{-2} T^{-2}, \qquad \dim \eta = M L_\parallel^{-1} T^{-1},$$

and the equality of dimensions takes the form

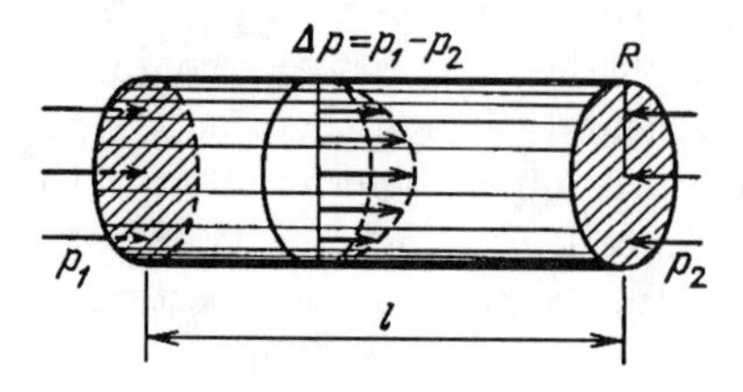

Figure 12.1: Laminar flow of a viscous liquid in a tube.

$$L_{\parallel} L_{\perp}^{2} = (M L_{\parallel} L_{\perp}^{-2} T^{-2})^{x} (M L_{\parallel}^{-1} T^{-1})^{y} L_{\parallel}^{z} L_{\perp}^{u} T.$$

Equating the indices of power,

$$\begin{aligned} M: &\quad 0 = x + y, \\ L: &\quad 1 = x - y + z, \\ L: &\quad 2 = -2x + u, \\ T: &\quad 0 = -2x - y + 1, \end{aligned}$$

we find $x = 1$, $y = -1$, $u = 4$, and for V we obtain

$$V = C \frac{\Delta p R^4}{\eta l} t.$$

Thus, the volume of liquid V flowing through the cross-section of a tube during the time t is proportional to the difference of pressures per unit of the tube length $\Delta p / l$, and inversely proportional to viscosity, which is obvious enough even without the cited calculation. However, the conclusion that the volume of liquid is proportional to the fourth power of the tube radius (i.e., to the squared area of its cross-section) is by no means obvious. The deduced law is valid for a tube with a constant cross-section of an arbitrary shape. In the case of a round cross-section, the dynamic calculation gives $C = 8/\pi$ (see Poiselle's formula, Eq. (2.92)).

12.3.5 Example: speed of sound

Let us determine the dependence of the speed of sound (i.e., longitudinal elastic waves, see Section 16.7.4) on the characteristics of the medium. It can be supposed that this velocity depends on elastic properties of the media, defined by the Young modulus of elasticity E (see Section 14.6.2), on inertia properties described by the density ρ, and on the wavelength λ. Dimensions of these quantities:

$$\dim E = L^{-1}MT^{-2}, \qquad \dim \rho = ML^{-3}, \qquad \dim \lambda = L.$$

Writing the expression for the unknown velocity of sound u in the form $u = CE^x \rho^y \lambda^z$, we get the following equality of dimensions:

$$LT^{-1} = (ML^{-1}T^{-2})^x (ML^{-3})^y L^z,$$

whence

$$\begin{aligned} L: \quad & 1 = -x - 3y + z, \\ M: \quad & 0 = x + y, \\ T: \quad & -1 = -2x, \end{aligned}$$

i.e., $x = 1/2$, $y = -1/2$, and $z = 0$. Thus we get

$$u = C\sqrt{E/\rho}. \tag{12.21}$$

The velocity of sound does not depend on the wavelength λ. Dynamic calculation gives the same result with $C = 1$.

12.3.6 Example: velocity of waves

Let us determine the velocity of waves on the water surface (see Section 16.7.13). The velocity of capillary waves depends on the surface tension σ (see Section 14.5.1), water density ρ, and wavelength λ. Writing the expression for u_c in the form $u_c = C\sigma^x \rho^y \lambda^z$ and taking into account that $\dim \sigma = MT^{-2}$, we get

$$LT^{-1} = (MT^{-2})^x (ML^{-3})^y L^z,$$

whence $x = 1/2$, $y = z = -1/2$. Therefore,

$$u_c = C\sqrt{\sigma/(\rho\lambda)}.$$

The dynamic consideration gives $C = \sqrt{2\pi}$.

The velocity of heavy waves in deep water ($h \gg \lambda$) may depend only on g and λ. Making the equality of dimensions for the formula $u_h = Cg^x \lambda^y$, we find $x = y = 1/2$, i.e.,

$$u_h = C\sqrt{g\lambda}.$$

The dynamic consideration gives $C = 1/\sqrt{2\pi}$.

The velocity of ultimately long waves ($\lambda \gg h$) and of waves in shallow water must be independent of the wavelength λ, but it may depend on the depth h of the pool. The equality of dimensions for the formula $u_h = Cg^x \lambda^y$ gives

$$u_h = C\sqrt{gh}. \tag{12.23}$$

In the given case, $C = 1$.

12.3.7 Example: microscopic model of a real gas

For a real gas (see Section 14.4.1) it is necessary to take into account the finite size of molecules. The simplest model of such a system is a set of solid spheres of a radius r_0, interacting only at contiguity with each other. Besides the radius r_0, the system is characterized by the mass m of a molecule (a sphere), the concentration n of molecules, and the average heat energy kT of a molecule. The proximity of properties of such a system to those of the ideal gas depends on the ratio of the energy of interaction of molecules to the energy of their thermal motion kT, i.e., it is defined by the dimensionless parameter composed of r_0, m, n, and kT. Since the time is included only into kT, the product kT cannot enter the dimensionless parameter γ. But the mass m cannot enter it, either. Only one independent parameter can be composed of the remaining quantities r_0 and n, namely $\gamma = nr_0^3$. Its physical sense is the ratio of the volume of the sphere to the average volume per one molecule in the gas. Strong rarefaction corresponds to the limiting case of the ideal gas, i.e., $\gamma \ll 1$. The ratio of the energy of interaction of molecules to the energy of their thermal motion kT in such a model appeared to be independent from kT. This means that the energy of molecules interaction in the model of solid spheres, like the kinetic energy, is proportional to the temperature.

However, experiments have shown that in real gases the degree of ideality depends on temperature, in contradiction with the prediction of the model of solid spheres. It means that in the microscopic model of the real gas it is necessary to take into account interaction of molecules both at a distance and at contiguity.

12.3.8 Example: time of relaxation in a gas

Relaxation time is the time of establishment of the equilibrium state (see Section 14.1.1). The system is characterized by four parameters r_0, m, n, and kT (Section 12.3.7). Dimensional analysis shows that the molecule's mass m and the energy kT of thermal motion enter the quantity τ (which has the dimensions of time) in certain powers in the combination $\sqrt{kT/m}$, while the dependence of τ on r_0 and n is determined up to an arbitrary function f of the dimensionless parameter $\gamma = r_0^3 n$:

$$\tau = \frac{n^{-1/3}}{\sqrt{kT/m}} f(r_0^3 n). \tag{12.24}$$

As $\sqrt{kT/m}$ has the physical sense of the velocity of thermal motion of molecules, the obtained results means that relaxation in a gas consisting of neutral molecules can occur only due to their thermal motion, and the characteristic time τ is always inversely proportional to the square root of the temperature.

12.3.9 Example: time of relaxation in plasma

Instead of the parameter r_0 characterizing the size of neutral particles, in plasma the square of the charge e^2 is present because Coulomb's interaction of charged particles covers an infinitely long distance. Here it is more convenient to use the Gaussian system of units, where electric charge dimensions can be expressed in terms of L, M, and T. Dimensional analysis shows that of e^2, n, m, and kT, only one independent dimensionless parameter may be composed, namely $\gamma = e^2 n^{1/3}/(kT)$. It has the meaning of the ratio of the potential energy of interaction of particles located at an average distance $n^{-1/3}$ from each other to the energy of thermal motion kT. Therefore, properties of classic (not quantum) plasma are more similar to those of the ideal gas, the less concentration n of charged particles and the more the temperature T. By analogy with Equation (12.24) it is easy to write the parameter τ with the dimensions of time:

$$\tau = \frac{n^{-1/3}}{\sqrt{kT/m}} f\left(\frac{e^2 n^{1/3}}{kT}\right). \tag{12.25}$$

Like in the neutral gas, here relaxation processes exist that are determined by the thermal motion of plasma particles: $f \sim 1$ and $\tau \propto n^{-1/3}/\sqrt{kT/m}$. However, in plasma some processes are also possible whose characteristic time is independent of temperature: for $f(x) = C/\sqrt{x}$ we get, with the help of Equation (12.25),

$$\tau = C\sqrt{m/(ne^2)}. \tag{12.26}$$

The relaxation time τ does not depend on temperature, i.e., on the velocity of particles. The only mechanical motion of a particle whose characteristic time is independent of the velocity is a harmonic oscillation described by Equation (16.2). Oscillations in plasma with the period defined by Equation (12.26) occur with a local disturbance of its electric neutrality (when in a certain place there are unequal concentrations of particles with charges of opposite sign). The frequency ω of such *plasma oscillations* depends on electron concentration n and, as a dynamic consideration shows, is equal to $\sqrt{4\pi ne^2/m}$ in the Gaussian system of units ($\omega = \sqrt{ne^2/(m\varepsilon)}$ in SI).

12.3.10 Example: temperature dependence of black-body radiation

Thermal radiation can be considered as the gas of photons. The volume density of the radiation energy w is equal to the product of concentration n of photons by the average energy of a single photon $\langle E \rangle$, which is equal to kT by the order of magnitude in thermal equilibrium. Since $\dim n = L^{-3}$, it is necessary to make a combination of quantities kT, c, and $\hbar$, which has the dimensions of length. There is only one such combination, namely, $c\hbar/(kT)$. Therefore, concentration n of photons is proportional to $(kT/c\hbar)^2$, and we obtain for the density of energy

$$w = C(kT)^4/(\hbar c)^3,$$

where C is a dimensionless coefficient. The total energy emitted from the black body is proportional to the fourth power of the absolute temperature.

Chapter 13

Mechanics

Mechanics deals with the simplest form of motion—the so-called *mechanical motion*. The mechanical motion of a body consists of changing its position relative to other bodies.

For a mathematical description of the mechanical motion of an object (or objects), a definite frame of reference is required. A *frame of reference* is formed by a material body (or by a set of bodies that are at rest with respect to each other), together with instruments to measure space distances and time intervals.

The *relativity* of mechanical motion is revealed in its dependence on the frame of reference—the same motion has different forms in various frames of reference. The concepts of rest and motion of a body are essentially relative: a body being at rest in some reference frame is at the same time in motion with respect to other frames of reference.

On the basis of the study of mechanical motion, the concepts of the *physical space* and *time* are formed. These concepts are fundamental, i.e., it is impossible to reduce them to more simple notions.

The following properties of space and time are discovered experimentally: the *physical space is three-dimensional, homogeneous, and isotropic; time is one-dimensional and homogeneous*.

The homogeneity of time reveals itself in invariability of physical laws in time: an experiment performed in identical conditions at different times—long ago, now, or in the future—gives the same results. With the homogeneity of time, conservation of energy (see Section 13.4.7) is associated. The homogeneity and isotropy of the physical space are revealed in the independence of physical phenomena in an isolated physical system of the spatial position and orientation of the system as a whole. With the homogeneity of space, conservation of momentum is associated, and with isotropy of space, conservation of angular momentum is associated (see Section 13.4.3).

13.1 Kinematics

The aim of *kinematics* is to describe mechanical motions in their relationship with space and time by means of quantitative mathematical methods without going into the physical causes of the motion. The physical models used in kinematics are particle, solid body, and continuous medium.

13.1.1 Kinematics of a particle

Particle or *material point* —any physical body whose shape and size are insignificant in the motion under consideration. In particular, any rigid body in a translation motion (see Section 13.1.6) can be considered as a material point.

The spatial position of a particle in some fixed reference frame is determined by its *radius-vector* (or *position-vector*) $\mathbf{r}$ drawn from the origin of a coordinate system (see Section 8.1.1) to the location of the particle. To specify a radius-vector $\mathbf{r}$ is the same as to indicate three corresponding scalar quantities, e.g., three *cartesian coordinates* (see Section 8.3.3) x, y, z of the point (Figure 13.1).

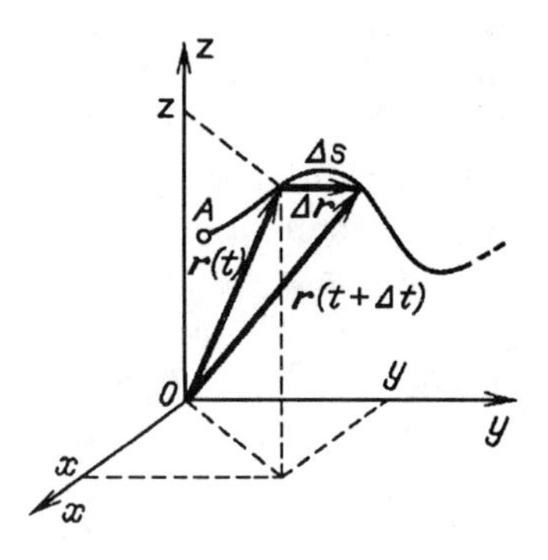

Figure 13.1: Radius-vector $\mathbf{r}$, coordinates x, y, z, trajectory, displacement $\Delta\mathbf{r}$, path Δs.

The number of independent quantities that are necessary to indicate the location in space of a physical system is called the *number of the degrees of freedom* of the system. A free particle has three degrees of freedom.

While a particle moves, its radius-vector and coordinates vary in time. If we can indicate a vector function of time $\mathbf{r} = \mathbf{r}(t)$, or three scalar functions $x = x(t)$, $y = y(t)$, and $z = z(t)$, which are equivalent to $\mathbf{r} = \mathbf{r}(t)$, we say that the motion of the particle is specified. The curve traced out in space by a moving particle is called the *trajectory* of the particle. Motions are classified as *rectilinear* and *curvilinear*, depending on the form of the trajectory.

13.1.2 Example: motion along an ellipse

Let coordinates of a particle be given as functions of time corresponding to harmonic oscillations along coordinate axes x and y, with the same frequency ω, and amplitudes b and d, respectively:

$$x(t) = b\cos\omega t, \qquad y(t) = d\sin\omega t, \qquad z(t) = 0. \tag{13.1}$$

These relations can be treated as an implicit equation of the trajectory. To obtain an equation of the trajectory in an explicit form it is necessary to exclude time t from Equation (13.1). Using the identity $\cos^2\omega t + \sin^2\omega t = 1$ (see Section 9.3.2), we get

$$\frac{x^2}{b^2} + \frac{y^2}{d^2} = 1. \tag{13.2}$$

It is an equation of ellipse, with semiaxes b and d (Figure 13.2). Equations (13.1) describe the counter-clockwise motion of a particle in xy-plane along this ellipse. Information about the direction of motion is lost in transition from Equations (13.1) to (13.2). We can determine this direction from Equation (13.1) in the following way: at $t = 0$ the particle is in the position A, and then at $t > 0$ its coordinate x decreases and y increases, so the particle moves towards the position B.

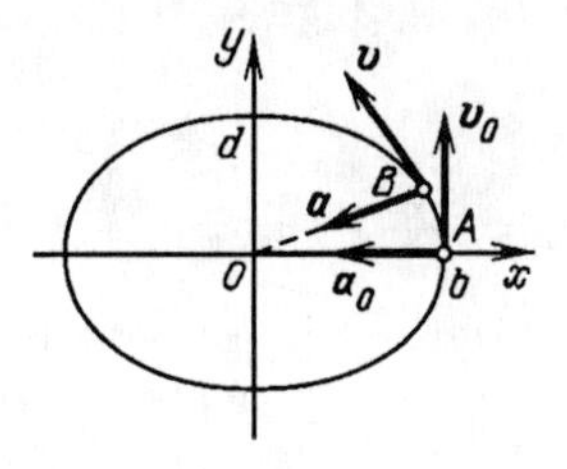

Figure 13.2: Motion along an ellipse.

13.1.3 Velocity and acceleration

Displacement of a particle during a time interval Δt is the vector $\Delta\mathbf{r}$, joining the particle positions at the time instants t and $t + \Delta t$. It is clear from Figure 13.1 that $\mathbf{r}(t + \Delta t) = \mathbf{r}(t) + \Delta\mathbf{r}$.

The *path* Δs travelled by the particle during the same time interval Δt is the length of the corresponding part of its trajectory. For rectilinear motion in one direction $\Delta s = |\Delta\mathbf{r}|$, for curvilinear motion, $\Delta s > |\Delta\mathbf{r}|$.

The whole path $s(t)$ travelled by a particle during the time interval between $t = 0$ and t is the length of its trajectory measured from the origin A where the particle was located at $t = 0$, up to its position at the time instant t. If the particle changed its direction of motion along the same trajectory, its path is defined as the whole length of the way travelled along the trajectory.

Average velocity $\mathbf{v}_{av}$ is defined as the vector quantity $\mathbf{v}_{av} = \Delta\mathbf{r}/\Delta t$. *Average speed* $v_s = \Delta s/\Delta t$ is by definition a scalar quantity. For a rectilinear motion in one direction these quantities are equal: $|\mathbf{v}_{av}| = v_s$, for a curvilinear motion, $|\mathbf{v}_{av}| < v_s$.

Velocity $\mathbf{v}$ of a moving particle at time instant t (or *instantaneous velocity*) is the limit to which the average velocity $\mathbf{v}_{av}$ tends as the time interval Δt tends to zero. In other words, velocity $\mathbf{v}$ is the derivative (see Section 5.4.1) of the particle's radius-vector with respect to time:

$$\mathbf{v} = \lim_{\Delta t \to 0} \frac{\Delta\mathbf{r}}{\Delta t} = \frac{d\mathbf{r}}{dt}. \tag{13.3}$$

Velocity characterizes the rate of displacement. The vector of velocity in any point of a trajectory is directed along the tangent to the trajectory at this point. The components of velocity vector are time derivatives of coordinates x, y, z of the moving particle:

$$v_x = \frac{dx}{dt}, \quad v_y = \frac{dy}{dt}, \quad v_z = \frac{dz}{dt}. \tag{13.4}$$

Acceleration characterizes the rate of velocity alteration. It is a vector quantity defined as the first derivative of the vector of velocity $\mathbf{v}$ with respect to time t:

$$\mathbf{a} = \frac{d\mathbf{v}}{dt} \quad \text{or} \quad a_x = \frac{dv_x}{dt}, \quad a_y = \frac{dv_y}{dt}, \quad a_z = \frac{dv_z}{dt}. \tag{13.5}$$

The vector of acceleration is the second derivative of radius-vector with respect to time, and its components are the second derivatives of coordinates

$$\mathbf{a} = \frac{d^2\mathbf{r}}{dt^2} \quad \text{or} \quad a_x = \frac{d^2x}{dt^2}, \quad a_y = \frac{d^2y}{dt^2}, \quad a_z = \frac{d^2z}{dt^2}. \tag{13.6}$$

Example: motion along an ellipse. For the motion described by Equations (13.1) we can write:

$$\begin{aligned} v_x &= -\omega b \sin\omega t, & a_x &= -\omega^2 b \cos\omega t, \\ v_y &= \omega d \cos\omega t, & a_y &= -\omega^2 d \sin\omega t, \\ v_z &= 0, & a_z &= 0. \end{aligned}$$

At the point A (Figure 13.2) $v_x = 0$, $v_y = \omega d$, i.e., velocity $\mathbf{v}$ is directed along the y-axis; and $a_x = -\omega^2 b$, $a_y = 0$, i.e., acceleration $\mathbf{a}$ is directed in the negative direction of the x-axis.

13.1.4 Tangential and radial acceleration

In the case of an arbitrary *curvilinear motion* (Figure 13.3) the vector of acceleration can be decomposed into two orthogonal components: *tangential acceleration* $\mathbf{a}_\tau$, directed along the tangent to the trajectory, and *radial* or *normal acceleration* $\mathbf{a}_n$, directed to the center of curvature (see Section 9.3.5) of the trajectory (along the normal to the trajectory):

$$\mathbf{a} = \mathbf{a}_\tau + \mathbf{a}_n, \qquad a_\tau = \frac{dv}{dt}, \qquad a_n = \frac{v^2}{R}, \tag{13.7}$$

where R is the *radius of curvature* of the trajectory at the point. Tangential acceleration $\mathbf{a}_\tau$ characterizes the rate of variation of the velocity magnitude, normal acceleration $\mathbf{a}_n$ characterizes the rate of variation of the velocity direction.

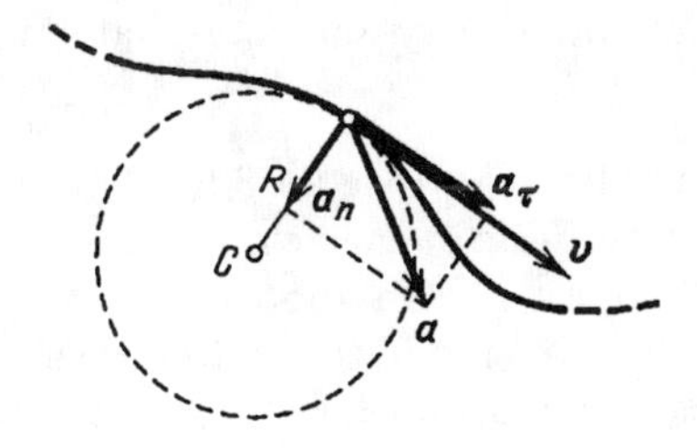

Figure 13.3: Acceleration in a curvilinear motion.

13.1.5 Rectilinear motion

For a rectilinear motion it is possible to choose the direction of one of the coordinate axes, e.g., of x-axis, along the trajectory. Thus y- and z-components of the displacement, velocity, and acceleration vectors receive zero values. In the following formulas x-components of the corresponding vectors are denoted simply by x, v, and a.

In a *uniform motion* velocity is constant, therefore

$$a = 0, \qquad v = v_0, \qquad x = x_0 + v_0 t. \tag{13.8}$$

On the plot of the velocity (Figure 13.4*a*) the shaded area equals numerically the path travelled during the time interval from $t = 0$ to t. The tangent of the slope angle α of the coordinate graph (Figure 13.4*b*) equals numerically the value of velocity v_0.

In *uniformly varied motion* (motion with a constant acceleration)

$$a = a_0, \quad v = v_0 + a_0 t, \qquad x = x_0 + v_0 t + \frac{a_0 t^2}{2}. \tag{13.9}$$

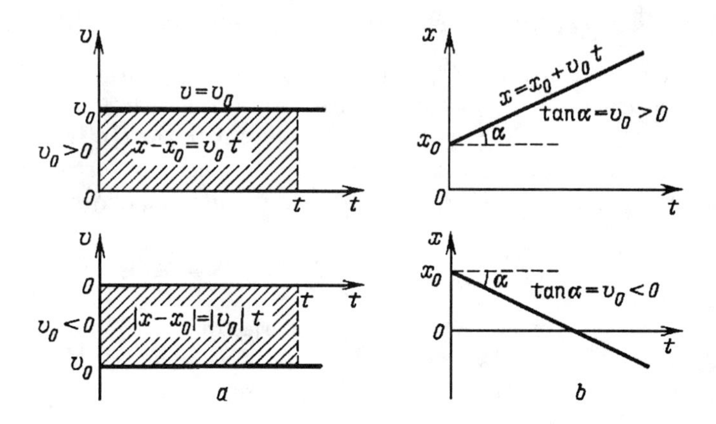

Figure 13.4: The plots of velocity and coordinate in a uniform motion.

On the plot of acceleration (Figure 13.5a) the shaded area equals numerically the absolute value of the velocity increment during the time interval from $t = 0$ up to t. On the plot of velocity (Figure 13.5b) the shaded area equals numerically the path travelled during this time interval from $t = 0$ to t. The time-dependent plot of the coordinate is a parabola (Figure 13.5c).

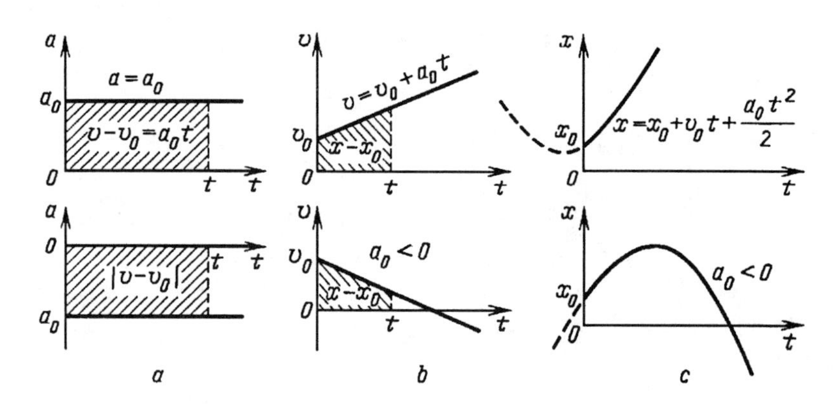

Figure 13.5: The plots of acceleration, velocity, and coordinate in a uniformly varied motion.

In a special case of motion with the initial velocity of zero ($v_0 = 0$) from the origin ($x_0 = 0$):

$$v = a_0 t, \qquad x = \frac{a_0 t^2}{2}. \tag{13.10}$$

Eliminating time t from Equations (13.10) we get the following relation between velocity v and coordinate x:

$$v^2 = 2a_0x. \tag{13.11}$$

In a special case of $v_0 \neq 0$, $x_0 = 0$ the above formulas take the following form:

$$v = v_0 + a_0t, \qquad x = v_0t + \frac{a_0t^2}{2}, \qquad v^2 - v_0^2 = 2a_0x. \tag{13.12}$$

13.1.6 Circular uniform motion

In a uniform circular motion, i.e., a motion with constant speed $v = \text{const}$ along a circumference, the vector of velocity has a constant magnitude but its direction is continuously changing. Therefore, in this case the tangential acceleration (see Section 13.1.3) $a_\tau = dv/dt = 0$, and the vector of acceleration is directed towards the center of the circular trajectory, its magnitude being equal to normal, or *centripetal* acceleration:

$$a_n = \frac{v^2}{R}. \tag{13.13}$$

Angular velocity ω is expressed in terms of (linear) velocity v and radius R of the circumference by the following relation:

$$\omega = \frac{v}{R}. \tag{13.14}$$

A uniform circular motion can be characterized also by the *frequency* ν (or f) of rotation $\nu = \omega/2\pi$ (ν equals the number of cycles per unit time) and by *period* of rotation $T = 1/\nu = 2\pi/\omega$. For centripetal acceleration a_n (2.13) the following formulas are valid:

$$a_n = \omega^2R = 4\pi^2\nu^2R = \frac{4\pi^2}{T^2}R. \tag{13.15}$$

13.1.7 Kinematics of a rigid body

A *rigid body* (or a *solid*) is a system of material points forming a definite configuration in which all distances between the points do not change when the system moves. A solid is characterized by six degrees of freedom: to specify its position it is sufficient to indicate three coordinates of some point of the body, and three angles which determine the orientation of the solid in space.

A motion of rigid body during which its orientation does not change is called *translational motion*. In translational motion all points of a solid move along identical trajectories with equal velocities (Figure 13.6), so the whole solid can be regarded as one material point, and its translational motion is characterized by three degrees of freedom (about translation in mathematics see Section 8.4.1).

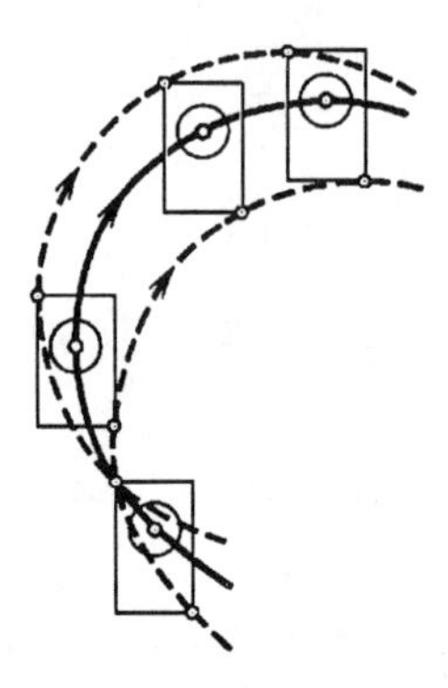

Figure 13.6: Translational motion.

Other important special cases of a rigid body motion are rotation around a fixed axis, plane motion, and rotation around a fixed point.

13.1.8 Rotation about a fixed axis

In *rotation around a fixed axis* (Figure 13.7) all points of a solid perform circular motion. Velocity $\mathbf{v}$ of any of them equals the vector product (see Section 8.2) of the angular velocity vector ω directed along the rotation axis and of radius-vector $\mathbf{r}$ of the point:

$$\mathbf{v} = \omega \times \mathbf{r}. \tag{13.16}$$

It is assumed here that the origin O (see Figure 13.7) is chosen somewhere on the axis of rotation. Magnitude of the velocity $v = \omega R$, where $R = r \sin\theta$ is the distance of the point from the axis of rotation (Figure 13.7).

Rotation around a fixed axis is characterized by only one degree of freedom.

13.1.9 Plane motion of a solid

A familiar special case of *plane motion* is the rolling of a cylinder (Figure 13.8). This motion can be represented as a sum of the rotation around some axis parallel to the cylinder's axis and of translational motion, with the velocity equal to the velocity of the cylinder's points lying on this axis. At any choice of the axis the given roll is characterized by the same angular velocity vector ω. As an axis of rotation it is convenient to choose either the axis O of the cylinder, or the line O' along which the cylinder

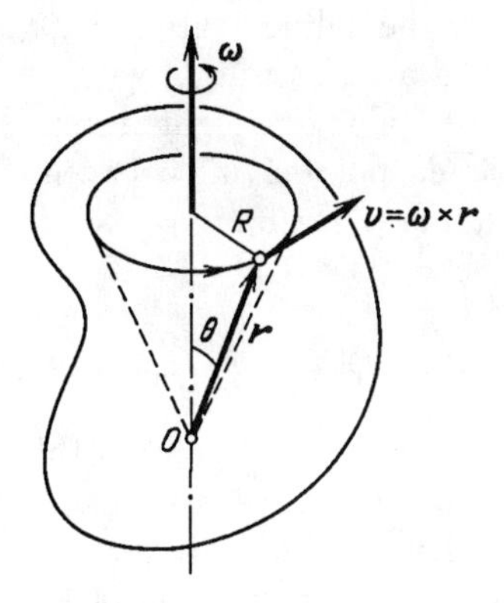

Figure 13.7: Rotation about a fixed axis.

touches the ground surface. If the cylinder is rolling *without sliding*, then velocities of its points lying on the line O' of contact with the ground are zero. The velocity of any other point is the same, as in a pure rotation with the same angular velocity ω around the fixed axis coinciding with the line of contact O' (the *instantaneous axis of rotation*).

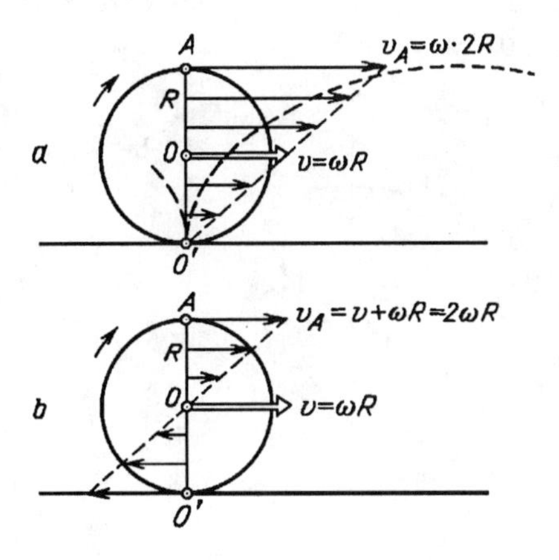

Figure 13.8: Rolling of a cylinder.

Velocities of different points lying on the vertical diameter are shown in Figure 13.8*a*. The velocity of the point A is equal in magnitude to $\omega \cdot 2R$ because this point is located at the distance $2R$ from the instantaneous axis of rotation O'. The same velocity can be represented as the sum of velocity v of the cylinder's axis O and velocity ωR, which this point A

is due to the rotation of the cylinder around its axis O (Figure 13.8*b*): $v_A = v + \omega R = 2\omega R$ (because at rolling without sliding $v = \omega R$). All points of the axis O move along straight lines; points of the cylinder's surface move along cycloids; points lying between the axis and the surface move along trochoids (curves resembling a cycloid but with smoothed angles).

13.1.10 Rotation about a fixed point

An example of *rotation about a fixed point* is the rolling without sliding of a cone (cone B in Figure 13.9) over the surface of an immovable cone (cone A in Figure 13.9) having the same vertex (point O). This motion can be represented either as pure rotation of the cone B, with angular velocity ω around the instantaneous axis passing through the vertex O along the line of contact, or as the sum of two rotations: one with angular velocity ω_0 around its own axis and the other with angular velocity $\mathbf{\Omega}$ around the axis of immovable cone A:

$$\omega = \omega_0 + \mathbf{\Omega}. \tag{13.17}$$

The velocity $\mathbf{v}$ of any point of a rolling cone can be calculated with the help of Equation (13.16) in which we can substitute ω with the expression given by Equation (13.17). All points of the axis of a rolling cone are moving along circumferences, and other points are tracing rather complex wavy sinuous trajectories.

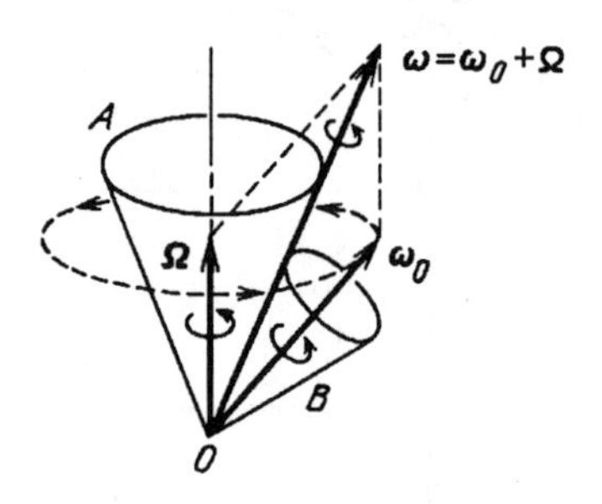

Figure 13.9: Rotation about a fixed point.

13.2 Dynamics

13.2.1 Basic concepts of classical dynamics

Dynamics is a branch of mechanics revealing the relationship between mechanical motion and the cause which produces it. The study of motion of material bodies in dynamics is based on the concept of their *interaction.* Interaction is the physical cause that modify the *velocity* of motion, i.e., it is the cause of *acceleration.* The acceleration of a body in contrast with its velocity can not be given arbitrarily: its value at a given time instant does not depend on the preceding motion of the body and is determined by position and motion of surrounding bodies. Classical dynamics relies on Newton's three laws of motion.

Newton's *first law* (the *law of inertia*) permits us to choose the frames of reference in which the laws of motion have the most simple form: according to this law the frames of reference (called *inertial frames*) exist in which the *motion of a free body is uniform and rectilinear.*

In inertial frames of reference the state of rest, as well as a uniform rectilinear motion ("inertial motion"), are equally natural states of a body—natural in the sense that they need not have any cause. The aim of dynamics is to explain the alteration of this state, i.e., to explain acceleration of a body. Strictly speaking, a free body is an idealization: there are no bodies that are not subjected to the influence of other bodies. Nevertheless, due to the decreasing of all known interactions with distance a free body can be realized with any desired accuracy. Experimental facts indicate that the *heliocentric reference frame* (associated with the sun and "immovable" stars) is an example of an inertial frame within the limits of presently attainable precision.

Newton's *second law* relates the acceleration of a body with the forces that cause it and the mass of the body: in an inertial frame of reference, the acceleration of a body is proportional to the vector sum of the forces acting on the body and inversely proportional to its mass:

$$\mathbf{a} = \frac{\sum_i \mathbf{F}_i}{m}. \tag{13.18}$$

Force is a physical quantity characterizing quantitatively the interaction of bodies. *Measurement* of forces independently of their physical nature can be based on the property of a force to cause *deformation of elastic bodies.* The measuring instrument using this principle is called *dynamometer.*

Experiment indicates that force is a vector quantity, i.e., the rules of operation with vectors (see Section 8.1.1) are applicable to forces. In particular, under the action of several forces (even of different physical nature) the motion of a body occurs just in the same way as if there was only one force (*resultant force*) equal to the vector sum of all separate forces.

One and the same force imparts various accelerations to different bodies. The less the acceleration, the more the *inertia* of the body. Inertia is the property of matter that causes it to resist any change in its motion. The physical quantity characterizing the property of inertia is called *mass.* Measurement of the mass of a body can be based on comparison of accelerations of this body and of some standard body produced by the same force: the ratio of masses by definition equals the inverse ratio of magnitudes of the accelerations. Such dynamical measurement of masses of atoms and molecules is performed by means of *mass-spectrometers*, whose principle of operation uses deflection of the beams of ions by static electric and magnetic fields. In these measurements *atomic mass unit* (a.m.u.) is used which is equal to 1/12 of the mass of an atom of isotope carbon-12:

$$1 \text{ a.m.u.} = 1.660,540 \cdot 10^{-27} \text{ kg}.$$

Masses of macroscopic bodies, in practice, are measured by *weighting*, i.e., by comparison of the gravity force acting on the body with the gravity force acting in the same gravity field on a standard body (on a weight). Measurement of mass by weighting is based on the law of proportionality of inertial and gravitational masses (see Section 13.3.1).

Experiment testifies that the mass of a body is an additive scalar quantity independent of its position and velocity (providing the speed of the body is much less than the speed of light, see Section 18.3.2). The ratio of mass m of a body to its volume V is called *density* ρ:

$$\rho = \frac{m}{V}.$$

For an inhomogeneous body this formula defines the average density. Densities of some substances are indicated in Section 13.8.5.

Newton's *third law* characterizes quantitatively an important property of any interaction: the forces which two interacting bodies exert on one another are equal by magnitudes and opposite by directions:

$$\mathbf{F}_{12} = -\mathbf{F}_{21}. \tag{13.19}$$

In particular, the forces of gravitational interaction (Newton's law of gravitation, see Section 13.3.1) and the forces of electrostatic interaction (Coulomb's law, see Section 15.1.1) satisfy Newton's third law. The forces of interaction of particles (of material points) are *central forces*, i.e., these forces act along the straight line passing through the positions of the particles. Newton's third law implies instantaneous propagation of interaction, and so for distant interacting bodies in motion it is valid only for relatively slow motions (compared with the speed of light, see Section 18.1).

It is worthwhile to note that the logical scheme of classical dynamics presented above is not the only one possible. As a matter of fact, it is impossible to introduce such important concepts of dynamics as force and mass without appealing to the laws of dynamics, i.e., beyond the scope of dynamics. As a consequence, there is some arbitrariness in the choice of statements which ought to be treated as definitions of corresponding quantities, and statements which ought to be regarded as physical laws that can be proved experimentally. For example, the procedure for measuring mass can be introduced by definition on the basis of Newton's third law: we can assume the ratio of masses to be equal to the inverse ratio of magnitudes of accelerations of two bodies during their interaction:

$$\frac{m_1}{m_2} = \frac{a_2}{a_1}.$$

For such a measurement of the mass of a body, the concept of force acting on the body is not utilized. Now the statement of Newton's second law concerning the inverse proportionality of acceleration to the mass of the body becomes a statement which we can verify by an experiment, and not the definition of mass. But the physical sense of Newton's third law in such a logical scheme reduces to the following statement: the ratio of magnitudes of the accelerations of two interacting bodies are always the same independently of the character of their interaction, and the vectors of accelerations are directed oppositely.

One more possible consistent logical scheme of dynamics arises when the method of measurement of forces is defined not on the property of forces to cause elastic deformation, but on the proportionality of acceleration to the force producing it.

13.2.2 Momentum

The *momentum* of a particle is a vector quantity defined as the product of its mass and velocity:

$$\mathbf{p} = m\mathbf{v}.$$

Since $\mathbf{a} = d\mathbf{v}/dt$, the equation of Newton's second law in the cases when mass m remains constant can be written in the following form:

$$\frac{d\mathbf{p}}{dt} = \sum_i \mathbf{F}_i \tag{13.20}$$

—the rate of change of momentum equals the applied resultant force. Newton's second law, in the form expressed by Equation (13.20), is valid for particles moving with arbitrary velocities including relativistic velocities, approaching to the limiting speed c (speed of light). But for relativistic motions, momentum of a particle is not simply proportional to its velocity—momentum depends on the velocity in a more complex way (see Section 18.3.1).

13.2.3 Determination of force on the basis of given motion

There are two different kinds of problems in classical dynamics which can be solved with the help of Newton's second law. One kind of problem consists in the determination of forces from the known motion of a body. A typical example of such a problem is the determination of the dependence of the gravitational force on the distance between the bodies on the basis of knowledge (from astronomical observations) the laws of planets' motions. In particular, Kepler's third law (see Section 13.5.1) states that for circular orbits the square of each planet's period of revolution is proportional to the cube of the orbit radius: $T^2 \propto r^3$. In uniform circular motion $v = 2\pi r/T$, acceleration $\mathbf{a}$ is directed to the center and its magnitude equals v^2/r (see Section 13.1.3). Applying Newton's second law to this motion, we get the following expression for the force:

$$F = ma = m\frac{v^2}{r} = m\frac{(2\pi r/T)^2}{r}. \tag{13.21}$$

Taking into account that $T^2 \propto r^3$ we find from Equation (13.21) that $F \propto m/r^2$—the gravitational force is inversely proportional to the square of the planet's distance from the sun and directly proportional to the mass m of the planet. Since the roles of the sun and the planet in their gravitational interaction are equivalent, this force must be proportional also to the mass M of the sun:

$$F = G\frac{mM}{r^2}$$

(see also Section 13.3.1). The value of the gravitational constant G cannot be determined from astronomical observations: to do this, a laboratory experiment is necessary (Cavendish's experiment).

13.2.4 Motion caused by given forces

Problems of the second kind consist in the determination of characteristics of motion of a body in conditions when exerted forces and an *initial mechanical state* (the position and velocity at the initial moment) are specified. The simplest example of such a problem—motion of a particle (a mass point) in a homogeneous constant field. In particular, for the motion of a particle in the field of gravity near the surface of the earth in the absence of air drag, Newton's second law reduces to the form

$$\frac{d\mathbf{v}}{dt} = \mathbf{g}.$$

Integration (see Section 5.6) of this equation with the initial conditions

$$\mathbf{v}(0) = \mathbf{v}_0, \qquad \mathbf{r}(0) = \mathbf{r}_0$$

allows us to determine the mechanical state of a moving particle at any subsequent time instant t:

$$\mathbf{v}(t) = \mathbf{v}_0 + \mathbf{g}t, \qquad \mathbf{r}(t) = \mathbf{r}_0 + \mathbf{v}_0 t + \frac{\mathbf{g}t^2}{2}. \tag{13.22}$$

Trajectory of the particle lies in the plane determined by vectors $\mathbf{v}_0$ and $\mathbf{g}$ (Figure 13.10).

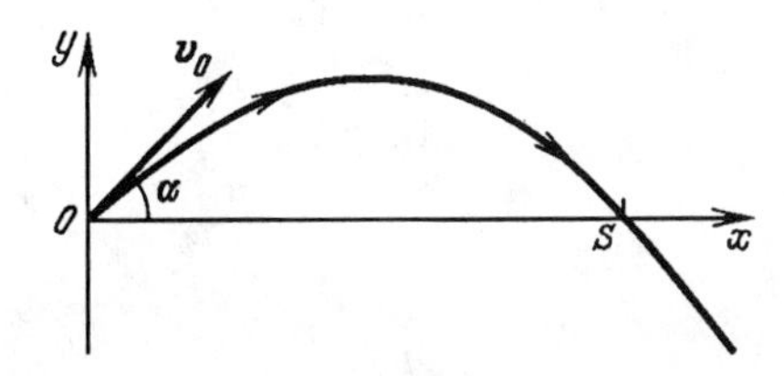

Figure 13.10: A trajectory in a homogeneous field of gravity.

Introducing in this plane horizontal axis x and vertical axis y, we can get from Equation (13.22) corresponding components of the velocity vector and coordinates of the particle as functions of time:

$$v_x(t) = v_0 \cos\alpha, \qquad v_y(t) = v_0 \sin\alpha - gt, \tag{13.23}$$

$$x(t) = (v_0 \cos\alpha)t, \qquad y(t) = (v_0 \sin\alpha)t - \frac{gt^2}{2}, \tag{13.24}$$

where α is the angle between the vector of initial velocity and the x-axis. Eliminating time t from Equation (13.24) we get the equation of the trajectory (see Figure 13.10) in the explicit form:

$$y = x\tan\alpha - \frac{gx^2}{2v_0^2}(1 + \tan^2\alpha). \tag{13.25}$$

Equation (13.25) describes the family of parabolic trajectories depending on two parameters: α and v_0. Various problems concerning free fall reduce to the analysis of this equation.

Examples

1. The range s of flight in the horizontal direction is found from Equation (13.25) if we assume there $y(s) = 0$ (see Figure 13.10):

$$s = \frac{v_0^2}{g}\sin 2\alpha. \tag{13.26}$$

Maximal range at a given value v_0 of the initial velocity is achieved at $\alpha = 45° : \quad s_{\max} = v_0^2/g$.

2. The boundary of attainable targets for a given magnitude of the initial velocity is determined from the equation of the trajectory, Equation (13.22), if for a fixed value of x we find the highest point that can be reached by the launched body. To find the point, we should determine the maximum of y as a function of the angle α. This problem reduces to the analysis of a quadratic trinomial (see Section 2.6.2) with respect to the tangent of α. The trinomial reaches maximum at $\tan\alpha = v_0^2/(gx)$. Substitution of this value into Equation (13.25) gives the equation of the desired boundary (Figure 13.11):

$$y = \frac{v_0^2}{2g} - \frac{gx^2}{2v_0^2}. \tag{13.27}$$

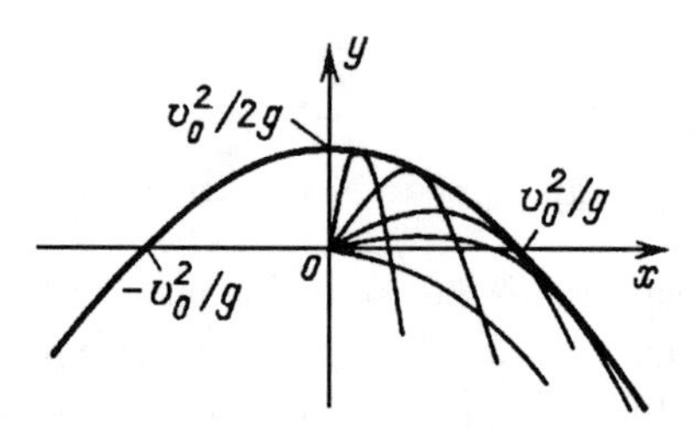

Figure 13.11: The boundary of attainable targets.

The boundary, Equation (13.27), is the envelope of the family of parabolic trajectories given by Equation (13.25) that correspond to a fixed value of v_0 and various values of the initial angle α. This envelope curve is also a parabola with vertex located at $x = 0$, $y = v_0^2/(2g)$. The boundary intersects the x-axis at the points $\pm v_0^2/g$ in agreement with the maximum range value $s_{\text{max}} = v_0^2/g$ determined in the previous example.

13.2.5 Restricted motion

There are problems in dynamics that do not refer to any of the two kinds considered above: some of the exerted forces are given but the other—usually the *forces of reaction* exerted by the bindings that restrict the motion—are unknown and ought to be found in the process of solving. To solve problems of this kind, it is necessary, along with Newton's second law, to take into account kinematic restrictions imposed by the bindings of the motion under consideration.

Example. A conic pendulum of length l rotates with the angular velocity ω (Figure 13.12). The force of gravity mg is known, but as for the force $\mathbf{F}$ of the string tension, we can indicate only its direction (along the string), and the magnitude of this force is unknown. From the character of motion of the pendulum it follows that the acceleration **a** is directed horizontally to the center of the circular path of the bob. Its magnitude equals $\omega^2 R$. From Newton's second law it follows that $m\mathbf{a} = \mathbf{T} + m\mathbf{g}$. Using the similarity of triangles in Figure 13.12 we can write the proportion $T/(ma) = l/R$. Substituting here $a = \omega^2 R$ we get the magnitude of the tension force: $T = m\omega^2 l$. Taking into account that T is the hypotenuse and mg is a leg of a right triangle, we can state that $T > mg$, whence $\omega^2 > g/l$—only when this inequality is fulfilled, the conic motion of the pendulum is possible. Otherwise (at $\omega^2 \leq g/l$) the string of the pendulum must be in the vertical position. The angle α between the string and the vertical line (see Figure 13.12) is determined by the relation $\cos\alpha = mg/T = g/(\omega^2 l)$.

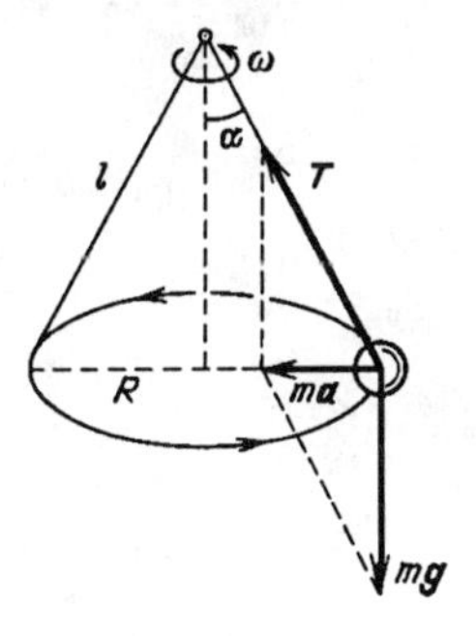

Figure 13.12: A conic pendulum.

13.3 Forces of Gravitation, Friction, and Elasticity

The laws of Newtonian mechanics determine the motion of a body independently of the nature of exerted forces that cause its acceleration. All the variety of interactions in nature can be reduced to four different types called fundamental interactions: *gravitational, electromagnetic, weak*, and *strong*. The scope of strong and weak interactions is restricted to various processes in atomic nuclei and to transmutations of elementary particles (see Section 18.6.2 and Section 18.7). Unification of these four types of interaction into one theoretical model is the aim of the *unified field theory*. This aim has not yet been achieved, although progress has been made in the unification of weak and electromagnetic interaction in the so-called electroweak theory, treating them as different manifestations of a unified *electroweak* interaction.

All macroscopic phenomena in the surrounding world are determined by long-range gravitational and electromagnetic interactions.

13.3.1 The law of gravitation

Gravitational interaction obeys Newton's *law of gravitation*: any two massive particles (point masses) attract each other with the force proportional to the product of their masses and inversely proportional to the square of distance between them:

$$F = G\frac{m_1 m_2}{r^2}. \tag{13.28}$$

The proportionality factor G is called the *gravitational constant.* Its value $G = 6.672\,59 \cdot 10^{-11}\text{m}^3/(\text{kg} \cdot \text{s}^2)$. Newton's law of gravitation can be expressed in the vector form (Figure 13.13):

$$\mathbf{F}_{12} = Gm_1 m_2 \frac{\mathbf{r}_2 - \mathbf{r}_1}{|\mathbf{r}_2 - \mathbf{r}_1|^3}. \tag{13.29}$$

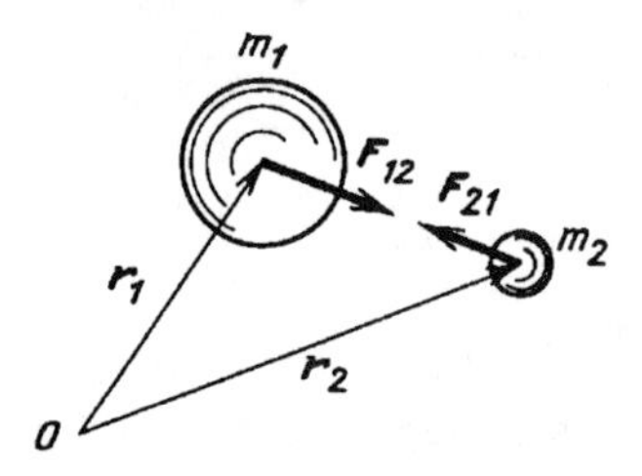

Figure 13.13: The law of gravitation.

Equations (13.28) and (2.29) are also valid for the interaction of a point mass with a body having spherical symmetry of mass distribution, as well as for the interaction of two such bodies. In this case r in Equation (13.28) or Equation (13.29) is the distance between their centers. In other words, real bodies with spherical symmetry act as point masses positioned at their centers of mass. Motion of planets and satellites governed by gravitational forces is discussed in Section 13.5.2.

The source of the gravitational field is the *gravitational mass* (or gravitational charge) of a body m_{gr}, similarly to the electric charge q, which is the source of the electric field. This is clearly seen from comparison of Newton's law of gravitation, Equation (13.28), and Coulomb's law of interaction of electric charges (see Section 15.1.1). Experiment testifies that the gravitational mass m_{gr} of a body is proportional to its inertial mass m_{in}, which characterizes its resistance to acceleration in Newton's second law. This proportionality permits us to measure gravitational mass m_{gr} and inertial mass m_{in} in the same units. The reason for the appearance of a dimensional gravitational constant G in Equation (13.28) is associated with the fact that the units of all quantities in Newton's law of gravitation are already chosen independently of this law.

Proportionality of inertial and gravitational masses is verified experimentally with relative error smaller than 10^{-12}. The intrinsic relation of physical quantities that characterize inertial and gravitational properties is demonstrated by the fact that all bodies in the same gravitational field have the same acceleration of free fall.

Acceleration g, which the gravitational field of the earth imparts to a body (acceleration of free fall, the strength of earth's gravitational field), can be expressed in terms of mass M of the earth and the distance r of the body from the center of the earth, if we substitute Equation (13.28) for the force into the right side of Newton's second law, Equation (13.18):

$$g(r) = G\frac{M}{r^2}. \tag{13.30}$$

The distance r from the center of the earth equals the sum $R + h$, where R is the earth's radius, and h is the height of a body over the earth's surface. Substituting $r = R + h$ into Equation (13.30) we obtain the dependence of free fall acceleration g on the height h over the surface:

$$g(h) = \frac{g_0}{(1 + h/R)^2} \approx g_0(1 - 2h/R), \tag{13.31}$$

where $g_0 = GM/R^2 = 9.81\ \text{m/s}^2$ is the acceleration of free fall near the earth's surface. In reality due to the oblateness of the earth near the poles and due to its rotation around the axis the acceleration of free fall depends on the latitude of the place: the first reason decreases g on equator by 0.18% in comparison with its value on a pole, and the second reason decreases g yet by 0.34% more.

13.3.2 Friction and elasticity

The forces of friction and elasticity are different exhibitions of electromagnetic interaction.

There are three kinds of friction between bodies being in contact: kinetic friction, static friction, and rolling friction. The *force of kinetic friction* resists the motion of one surface relative to another with which it is in contact. It is directed along the surface of contact opposite to the direction of relative velocity of touching bodies (Figure 13.14). The force of kinetic friction $\mathbf{F}_{\text{fr}}$ and a normal contact force $\mathbf{N}$ (normal force of elastic reaction) can be considered as components of one force $\mathbf{Q}$ exerted by the surface of one body on the surface of the other. Magnitudes of these forces are related by the experimentally established approximate *Coulomb's law of friction*:

$$F_{\text{fr}} = \mu N. \tag{13.32}$$

The proportionality factor μ (the ratio of F_{fr} to N) is called the *coefficient of kinetic friction.* Its value depends on the nature of the surfaces and is independent of the surface area of the body and is almost independent of the relative velocity of the body, and the surface. This dependence is usually neglected. Kinetic friction is inevitably accompanied by dissipation of mechanical energy (its conversion into internal heat energy, see Section 14.1.4).

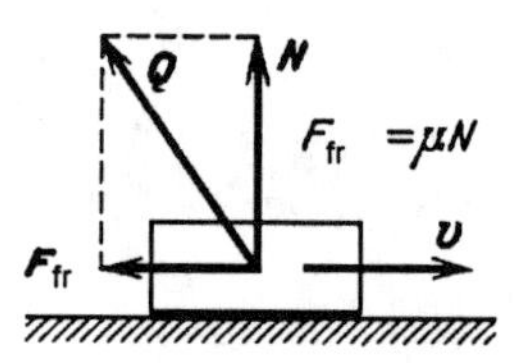

Figure 13.14: The force of kinetic friction.

The *force of static friction* is equal and opposite to the external tangent force applied with the intention to move the body along the surface of contact. Its value is determined by the conditions of equilibrium (see Section 13.6.1). If the external tangent force is increased until the body just moves, the value of the force of static friction will also increase until it reaches the *limiting frictional force.* This maximum value of the force of static friction is proportional to the normal contact force N pressing the body to the surface, i.e., it also obeys Coulomb's law, Equation (13.32). The *coefficient of static friction* is usually accepted to be equal to the coefficient of kinetic friction for the same surfaces (though in reality it is usually greater).

The cause of friction is that surfaces, however smooth they may look to the eye, on the microscopic scale have many humps and crests. Therefore, the actual area of contact is very small. This causes very high local pressure, which leads to welding of the surfaces by forming molecular bonds. During motion these bonds are broken and restored continually, causing resistance to the motion.

The *rolling friction* is caused by distortion of the plane surface in front of the rolling body and by destruction of temporarily formed molecular bonds in the place of contact. The *force of rolling friction* is proportional to the normal force pressing the rolling body to the surface and inversely proportional to the radius of the body. Usually the force opposing rolling is appreciably smaller than the force of kinetic friction opposing sliding, but at a very high speed of rolling approaching to the speed of propagation of distortions (the speed of sound) this force sharply increases.

Example: motion with friction. A heavy box lying on horizontal rough surface is submitted to a force **F** directed at an angle α to the horizon. In Figure 13.15 all the forces applied to the box are shown.

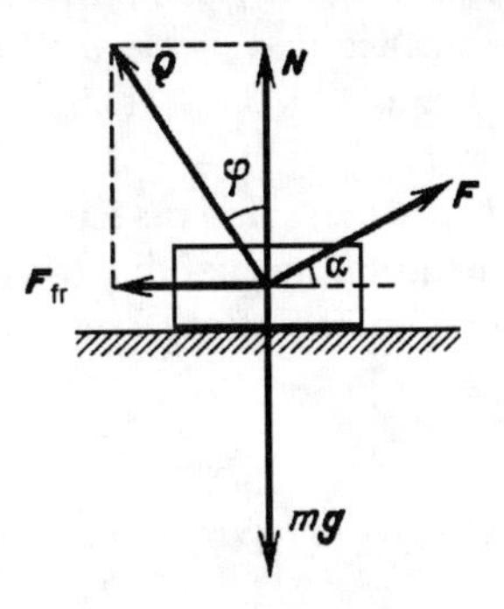

Figure 13.15: Forces applied to the box.

The equation of Newton's second law in this case gives

$$\mathbf{F} + m\mathbf{g} + \mathbf{N} + \mathbf{F}_{\text{fr}} = m\mathbf{a}. \tag{13.33}$$

If the box remains at rest or moves along the surface, the vertical component of its acceleration is zero and projection of Equation (13.23) to the y-axis permits us to determine the normal contact force N:

$$F \sin\alpha - mg + N = 0, \qquad N = mg - F \sin\alpha. \tag{13.34}$$

The box remains on the surface until $N > 0$, e.g., if $F \sin\alpha < mg$.

Projection of Equation (13.33) to the x-axis permits us to find acceleration of the box if it is moving, or the force of static friction if it is at rest:

$$F \cos\alpha - F_{\text{fr}} = ma. \tag{13.35}$$

If the box is moving, the force of kinetic friction can be expressed in the following way:

$$F_{\text{fr}} = \mu N = \mu(mg - F \sin\alpha).$$

Substituting this value into Equation (13.35), we find:

$$a = \frac{F}{m}(\cos\alpha + \mu \sin\alpha) - \mu g.$$

The force F can produce motion of the box if $a > 0$, e.g., if

$$F > \frac{\mu}{cos\alpha + \mu \sin\alpha} mg.$$

In the case of the opposite sense of the inequality, the box remains at rest (a = 0), and Equation (13.35) determines the force of static friction: $F_{\rm fr} = F\cos\alpha$. Note that the forces $\mathbf{N}$ and $F_{\rm fr}$ can be regarded as components of one force $\mathbf{Q}$ applied to the box by the surface (Figure 13.15). When the box slides along the surface and $F_{\rm fr} = \mu N$, th tangent of the angle φ formed by the force $\mathbf{Q}$ and the vertical line equals the coefficient of kinetic friction μ: $\tan\varphi = F_{\rm fr}/N = \mu$.

The *forces of elasticity* arise in some materials when a stress is applied to them. These forces depend on mutual disposition of interacting bodies and are determined by their deformation. For more detailed information, see Section 14.6.2.

13.4 Conservation Laws

13.4.1 Conservation of momentum

The *momentum* $\mathbf{P}$ of a system of particles (mass points) is the vector sum of momenta $\mathbf{p}_i = m_i\mathbf{v}_i$ of individual particles:

$$\mathbf{P} = \sum_i m_i\mathbf{v}_i, \qquad i = 1,\ 2,\ \ldots,\ n. \tag{13.36}$$

The rate of change of momentum of an individual particle is determined by Newton's second law:

$$m_i\frac{d\mathbf{v}_i}{dt} = \mathbf{F}_i + \sum_{k\neq i}\mathbf{F}_{ik}, \qquad i,\ k = 1,\ 2,\ \ldots,\ n, \tag{13.37}$$

where $\mathbf{F}_{ik}$ is the force exerted by the particle of mass m_k on the particle of mass m_i, and $\mathbf{F}_i$ is the resultant force exerted on the particle with mass m_i by all bodies which are not included in the system (*external force*).

According to Newton's third law, $\mathbf{F}_{ik} = -\mathbf{F}_{ki}$. Summing the Equations (13.37) for all particles and taking into account the cancellation of all inner forces we get the law that determines the rate of change of momentum of the system:

$$\frac{d\mathbf{P}}{dt} = \sum_i \mathbf{F}_i, \tag{13.38}$$

i.e., the rate of change of total momentum of a system of particles is determined by the sum of only *external forces* exerted on the particles of the system.

If the external forces remain constant during a time interval Δt, then Equation (13.38) can be expressed in the following way:

$$\Delta \mathbf{P} = \sum_i \mathbf{F}_i \Delta t.$$

The product $\mathbf{F}_i \Delta t$ is called the *impulse of the force* $\mathbf{F}$. We get that the change of total momentum of a system of particles during some time interval equals the resultant impulse of all external forces corresponding to the same time interval.

In an *isolated* or *closed* physical system there are no external forces, and total momentum $\mathbf{P}$ of the system remains the same ($\mathbf{P} = \text{const}$) in spite of the fact that momenta of individual particles can change due to interaction of the particles (the *law of momentum conservation*). If in a non-isolated physical system the projection of external forces on some direction is zero, the corresponding component of momentum remains constant.

13.4.2 The center of mass

The *center of mass* (center of inertia) of a system is the point whose location is determined by the following formula:

$$\mathbf{R} = \frac{\sum_i m_i \mathbf{r}_i}{\sum_i m_i}, \tag{13.39}$$

where $\mathbf{r}_i$ is the radius-vector of the particle with mass m_i. The center of mass of two identical particles is located at the middle of the rectilinear segment connecting the particles; in the case of different particles the center of mass divides the segment into two parts that are inversely proportional to their masses.

If a homogeneous body has the center of symmetry, an axis, or a plane of symmetry, its center of mass is located at the center of symmetry, on the axis, or in a plane of symmetry respectively (see Section 8.4).

In some cases the center of mass is the point at which the whole mass of a body may be considered to be concentrated. In particular, if a body is situated in a uniform gravitational field, the center of mass coincides with the *center of gravity*—the point at which the whole weight of a body may be considered to be applied (the point where the resultant of the gravitational forces acting on separate parts of the body is applied).

There is a simple practical method that permits us easily determine the location of the center of mass of a flat figure: let us suspend the body in gravity field allowing it to turn freely about the point of suspension O_1 (Figure 13.16). In equilibrium the center of mass C is located below the point of suspension on the same vertical line with it because the torque (see Section 13.6.1) of gravity force (which can be considered to act at the center of mass) is zero. Changing the point of suspension, we find by the same method one more straight line O_2C passing through the center of mass. The location of the center of mass is determined by the point of intersection of these lines.

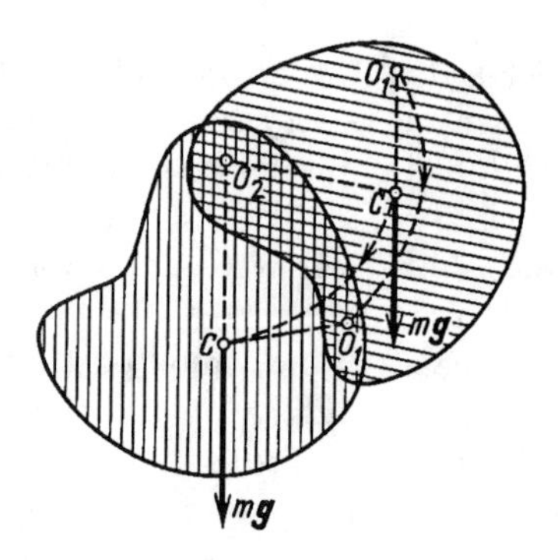

Figure 13.16: Location of the center of mass of a flat figure.

Velocity of the center of mass:

$$\mathbf{V} = \frac{d\mathbf{R}}{dt} = \frac{\sum_i m_i \mathbf{v}_i}{\sum_i m_i}. \tag{13.40}$$

The momentum of the system of particles, Equation (2.36), equals the product of mass $M = \sum_i m_i$ of the whole system and the velocity V of the center of mass of the system:

$$\mathbf{P} = \sum_i m_i \mathbf{v}_i = M\mathbf{V}. \tag{13.41}$$

The latter expression shows that the center of mass characterizes the motion of a system of particles as a whole.

13.4.3 The law of motion of the center of mass

The law that determines the change of momentum of a system of particles, Equation (13.38), is essentially the *law of motion of the center of mass* (the center of inertia). The rate of change of momentum equals the product of the whole mass M of the system and the acceleration of the center of mass:

$$\frac{d\mathbf{P}}{dt} = M\frac{d\mathbf{V}}{dt}.$$

From Equation (13.38) it follows that the center of mass of a system moves in just the same way as a particle of mass M would move under the action of a force that equals the vector sum of all external forces acting on the particles of the system:

$$M\frac{d\mathbf{V}}{dt} = \sum_i \mathbf{F}_i. \tag{13.42}$$

The center of mass of an *isolated system* (in the absence of external forces) in an inertial frame of reference moves with a constant velocity **V** or rests, though the velocities of individual particles of the system can change due to their interaction.

The law of conservation of momentum is a *universal* law, i.e., it is fulfilled in all isolated physical systems at *any kinds of interactions*. This universality is ensured by the fact that the conservation of momentum is related not with some particular kinds of interactions but with the general property of homogeneity of the physical space.

13.4.4 Jet propulsion

The law of momentum conservation explains the principle of *jet propulsion* or *reaction propulsion*—the propulsion of a body produced by discharging a fluid in the form of a jet. The main use of jet propulsion is in aircraft and spacecraft. It is the only known method of propulsion in space. When the engine of a rocket discharges the products of combustion with a velocity $\mathbf{v}_{\mathrm{rel}}$ with respect to the rocket, the motion of the rocket is described by the following equation:

$$m\frac{d\mathbf{v}}{dt} = \mathbf{v}_{\mathrm{rel}}\frac{dm}{dt} + \mathbf{F}, \tag{13.43}$$

where m is the changing mass of the rocket, dm/dt is the rate of its change ($dm/dt < 0$ because the mass of the rocket decreases), and $\mathbf{F}$ is the resultant of all external forces exerted to the rocket (drag of the atmosphere, the force of gravity). When external forces are absent and the relative velocity of discharge $\mathbf{v}_{\mathrm{rel}}$ is constant, Equation (13.43) reduces to $dm/dv = -(1/v_{\mathrm{rel}})m$ (see Section 5.5.1), and the velocity gained by a rocket is related to its initial (m_0) and final (m) masses by the Tziolkovsky formula:

$$m = m_0 \exp(-v/v_{\mathrm{rel}}). \tag{13.44}$$

13.4.5 Work and kinetic energy

The *work* done by a force **F** acting on a body is the scalar product of the force and displacement vectors:

$$\Delta A = \mathbf{F}\Delta\mathbf{r} = F\Delta r \cos\alpha. \tag{13.45}$$

The work is positive if the angle α between the force **F** and the displacement $\Delta\mathbf{r}$ of the body is acute; it is negative if the angle α is obtuse;

if the force $\mathbf{F}$ acts perpendicularly to the displacement $\Delta\mathbf{r}$ the work is zero.

The definition of Equation (13.45) is reasonable if the force $\mathbf{F}$ remains constant in magnitude and in direction while the body is subjected to the displacement $\Delta\mathbf{r}$. If this requirement is not fulfilled, we can use the definition to sufficiently small parts of a trajectory and define the work of the force $\mathbf{F}$ on the whole path as the sum of the works on individual small parts of it.

If several forces $\mathbf{F}_1$, $\mathbf{F}_2$, ..., $\mathbf{F}_n$, are exerted on a body, the work of the resultant force $\mathbf{F}$ (i.e., of the vector sum $\mathbf{F}_1 + \mathbf{F}_2 + \ldots + \mathbf{F}_n$) equals the sum of the works of separate forces: $\Delta A = \Delta A_1 + \Delta A_2 + \ldots + \Delta A_n$.

The *Kinetic energy* E_k of a particle is the physical quantity defined by the following relation:

$$E_\mathrm{k} = \frac{mv^2}{2} = \frac{p^2}{2m}. \tag{13.46}$$

Kinetic energy of a system is defined as an additive quantity: it equals the sum of kinetic energies of individual particles:

$$E_\mathrm{k} = \frac{m_1 v_1^2}{2} + \frac{m_2 v_2^2}{2} + \ldots + \frac{m_n v_n^2}{2}. \tag{13.47}$$

The *work-energy theorem* states that the change ΔE_k of kinetic energy of a particle equals the work of the resultant force:

$$\frac{mv^2}{2} - \frac{mv_0^2}{2} = A. \tag{13.48}$$

Equation (13.48) means that kinetic energy $mv^2/2$ of a body equals the amount of work that ought to be done by an exerted force in order to bring a free body from rest to motion with given speed v.

The change of kinetic energy of a system of particles equals the algebraic sum of works done by *all forces* acting in the system, both external forces and internal forces of interaction. In the case of a *rigid system* (a *solid*) the sum of works done by internal forces is equal to zero. So for a solid the change of kinetic energy equals the algebraic sum of works done only by external forces.

These statements are direct consequences of Newton's second law. Kinetic energy and work change their values at transition from one inertial frame of reference to another, but Equation (13.48) is valid in all frames of reference.

13.4.6 Potential energy

All forces exerted on the particles of a system can be divided into potential and non-potential. Forces are called *potential* if their work at any change of positions of the particles does not depend on trajectories of the particles but is determined only by the initial and final configurations of the system. Examples of such forces are the forces of gravity, Coulomb's forces of electrostatic interaction of charged particles, and forces of elasticity. The work of non-potential forces depends on the form of trajectory. The force of friction is an example of a non-potential force.

The *potential energy* U of a particle located at some point is a physical quantity equal to the work of potential forces which is done at transfer of the particle from the given point to a definite point, where potential energy is assumed to be zero. Since this work does not depend on the form of trajectory, potential energy depends only on the position of the particle. The choice of the point of zero potential energy is arbitrary, so potential energy is defined to within an additive constant depending on this choice. This ambiguity does not manifest itself in applications because only the differences of potential energy at various points have real physical sense. These differences are independent of this choice.

The work A_{12} of potential forces at transfer of a particle from point 1 to point 2 equals the difference of initial (U_1) and final (U_2) values of potential energy:

$$A_{12} = U_1 - U_2, \qquad \text{or} \quad A = -\Delta U, \tag{13.49}$$

where $\Delta U = U_2 - U_1$ is the change of potential energy.

The potential energy of a system of interacting particles equals the work done by all potential forces (both external and internal) during transition of the system from the given configuration to some configuration in which the potential energy is assumed to be zero.

The potential energy of a system of particles consists of the potential energy of their interaction and of their potential energy in an external field. In contrast to additive potential energy of particles in an external potential field (this energy equals the sum of potential energies of individual particles in the external field), th potential energy of the interaction of particles cannot be represented as a sum of energies of individual particles. This part of potential energy depends on the configuration of the system (on distances between particles).

The potential energy of gravitational attraction of two point masses or bodies with spherical symmetry is given by the expression:

$$U = -G\frac{m_1 m_2}{r}. \tag{13.50}$$

Here the potential energy is assumed to be zero at infinitely large distance r between the bodies. The energy increases as the distance is increased.

For a body in the gravitational field of the earth we can rewrite this formula in a more convenient form:

$$U = -mg\frac{R^2}{r}, \tag{13.51}$$

where $g = GM/R^2$ is the acceleration of free fall (the strength of the gravitational field) near the surface of the earth, R is the radius, and M is the mass of the earth.

In the homogeneous field of gravity ($\mathbf{g} = \text{const}$) the potential energy of a body of mass m linearly depends on the height h over the surface:

$$U = mgh. \tag{13.52}$$

Here h is the height over the level of zero potential energy. For the earth's gravitational field, Equation (13.52) is approximately valid for $h \ll R$. It can be considered as linearization of exact formula expressed by Equation (13.51), in which zero level of potential energy is transferred from infinity to the earth's surface.

The potential energy of Coulomb's interaction (see Section 15.1.7) of two point electric charges:

$$U = \frac{1}{4\pi\varepsilon_0}\frac{q_1 q_2}{r}. \tag{13.53}$$

In the case of repulsion of the charges ($q_1 q_2 > 0$), potential energy is positive; in the case of attraction ($q_1 q_2 < 0$) it is negative.

The potential energy of a body subjected to an elastic deformation that conforms to Hook's law (see Section 14.6.2) is proportional to the square of the deformation. For example, a stretched (or stressed) spring or elastic rod has the potential energy

$$U = \frac{1}{2}k(\Delta l)^2. \tag{13.54}$$

13.4.7 Conservation of mechanical energy

The *Mechanical energy* E of a system is by definition the sum of kinetic energy E_k and potential energy U:

$$E = E_\text{k} + U. \tag{13.55}$$

Mechanical energy depends on positions and velocities of all the particles of the system, i.e., it is a function of the mechanical state of the system. From this definition, Equation (13.55), and the work-energy theorem (see Section 13.4.5) it follows that the change of mechanical energy is equal to the algebraic sum of the works done by all non-potential forces acting in the system (both external and internal):

$$E_2 - E_1 = A_\text{non-potential}. \tag{13.56}$$

Equation (13.56) follows from Equation (13.48) if the work of potential forces in the right side of Equation (13.48) is expressed in terms of the change of potential energy according to Equation (13.49). Note that it is also possible to keep the work of external forces in the right side of Equation (13.56). However, in this case we must treat the mechanical energy of the system as the sum of its kinetic energy and potential energy only of the interaction of particles of the system with one another—the potential energy of the system should not include in this case the potential energy of the particles in the external field.

The *law of energy conservation* is valid for mechanical systems where there is no non-potential forces, or the work of these forces equals zero for all possible motions. Physical systems in which mechanical energy does not change are called *conservative systems*. It is the property of systems where non-potential forces are absent and external forces are constant.

13.4.8 Collisions

Various processes of short-lived impulsive interaction between physical bodies are called *collisions*, provided that, before their mutual approach (and after it if the bodies go apart), at large distances between them the bodies can be regarded as free. As a result of collision the bodies can join in one body (*perfectly inelastic collision*), new bodies can be formed (decay of particles or reactions, see Section 18.6.3), and colliding bodies can scatter (change abruptly their velocities) and go apart with or without a change of their intrinsic state. In the latter case a collision is called *perfectly elastic scattering*.

The variations of velocities of colliding bodies are usually large and occur during a very short time. Hence the accelerations and consequently the forces involved are considerable. So only these forces of interaction are taken into account in the analysis of instantaneous collisions (all other external forces can be neglected) and the system is regarded as isolated.

In an absolutely *inelastic collision* the velocity of the newly formed body of mass $m_1 + m_2$ is determined by the law of momentum conservation: this final velocity equals the velocity $\mathbf{V}$ of the center of mass of the system that does not change in collision:

$$\mathbf{V} = \frac{m_1\mathbf{v}_1 + m_2\mathbf{v}_2}{m_1 + m_2}. \tag{13.57}$$

In the analysis of an *elastic collision* it is convenient to consider one of the particles being at rest before collision: $\mathbf{v}_2 = 0$. In this case the laws of momentum and energy conservation take the form:

$$m_1\mathbf{v}_1 = m_1\mathbf{v}_1' + m_2\mathbf{v}_2', \qquad \frac{m_1v_1^2}{2} = \frac{m_1v_1'^2}{2} + \frac{m_2v_2'^2}{2}, \tag{13.58}$$

where the velocities after the collision are marked with a prime. For *head-on collision*, the final velocities are directed along the same line as the velocity of the projectile particle before collision, and from Equation (13.58) we obtain

$$v_1' = \frac{m_1 - m_2}{m_1 + m_2} v_1, \qquad v_2' = \frac{2m_1}{m_1 + m_2} v_1. \tag{13.59}$$

For $m_1 > m_2$, the projectile particle moves after the collision in the former direction with a reduced velocity; for $m_1 < m_2$ it bounces backwards; for $m_1 = m_2$ it comes to rest while the target particle moves off with the velocity equal to that of the projectile before collision.

For a *non-central elastic collision* of particles of equal masses ($m_1 = m_2$), the velocity vectors of particles after the collision form a right angle, and the *angle of scattering* (the change in the direction of the projectile velocity) can assume any value between 0 and 90°, depending on the *aiming parameter* (the distance between the center of the target and the straight line continuing the original velocity of the projectile). If the projectile particle is lighter than the target ($m_1 < m_2$), the projectile can be scattered through any angle (including 180°), and the angle between the velocities after the collision is obtuse. A heavy particle in a collision with a light one ($m_1 > m_2$) cannot be scattered at an angle greater than $\varphi_{\max} = \arcsin(m_2/m_1)$, and the angle between the velocities after a collision is acute. For example, in the case of elastic scattering of deuterons on resting protons ($m_2/m_1 = 1/2$) the angle of scattering can not exceed 30°.

A moving particle in an elastic collision with a particle of an equal mass being at rest can transmit to the latter a significant part of its energy (at head-on collision the whole energy is transmitted: it follows from Equation (13.59) that the projectile particle comes to rest and the target particle gets velocity equal to that of the projectile before collision).

However, in elastic collisions of particles whose masses differ considerably (more precisely, under conditions of large or small ratio of masses, as in the case of an electron and ion) the exchange of energy is hampered: the light particle bounces away from a heavy particle at rest as if it is reflected from a wall. Therefore, the light particle transmits to the heavy one a negligible part (of the order of $m_1 \ll m_2$) of its energy. That is why in a mixture of such particles (e.g., in plasma), thermodynamic equilibrium is established in a short time in every subsystem (light and heavy), but equalization of mean kinetic energies (see Section 14.2.4) of light and heavy particles takes a long time because a large number of collisions is required. Mean kinetic energy of a particle is related to absolute temperature T (see Section 14.2.4) of the system. Therefore, plasma is often characterized by two different temperatures simultaneously: one is the temperature of electrons T_e and the other of ions T_i.

13.5 Motion in a Central Gravitational Field

13.5.1 Kepler's laws of planetary motion

Many problems concerning motion of planets, comets, and satellites in a central gravitational field can be solved on the basis of conservation laws, without solving the differential equations of motion. Using Equation (13.50) for the potential energy of a body in the central gravitational field produced by a heavy body of mass M having spherical symmetry (sun, earth, or any other star or planet), we can write the law of energy conservation for a body of mass $m \ll M$ moving in this field with a velocity v:

$$\frac{mv^2}{2} - G\frac{mM}{r} = E_0. \tag{13.60}$$

Another physical quantity remaining constant for a particle moving in any central field is the ***angular momentum*** that is equal to the vector product (see Section 8.2.2) of the radius-vector and momentum (***linear momentum***) of the particle:

$$\mathbf{r} \times m\mathbf{v} = \mathbf{L}_0. \tag{13.61}$$

An example of motion in a central gravitational field is planetary motion around the sun. For this motion Kepler's three laws are valid (if perturbations caused by mutual gravitational attraction of planets are neglected). These laws were formulated by Johannes Kepler in about 1610 on the basis of astronomical observations made by Tycho Brahe. Kepler's laws state that (Figure 13.17):

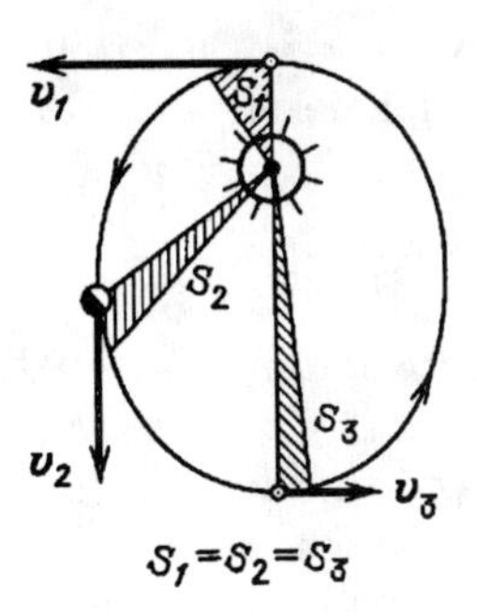

Figure 13.17: Elliptical orbit of a planet.

1. The orbits of planets are elliptical with the sun at one focus of the ellipse.

2. Each planet revolves around the sun so that its radius-vector (with the origin in the sun) sweeps out equal areas in equal time intervals.
3. The ratio of the square of each planet's period to the cube of the major axis of its orbit is the same for all planets:

$$T_1^2/T_2^2 = a_1^3/a_2^3. \tag{13.62}$$

Motion of a planet along a closed elliptical (in particular circular) orbit takes place at negative values of the total mechanical energy, Equation (13.60): $E_0 < 0$. At $E_0 = 0$ the trajectory of a body in a central gravitational field is a parabola; at $E_0 > 0$ it is a hyperbola whose near focus is located at the center of the force (in the sun for orbits of comets). The statements of Kepler's first law (that orbits are *conics*, i.e., curves formed by the intersection of a plane and a cone, see Section 9.6.3), and of the third law are specific for the inverse square central fields (where the force depends on the distance r as $\sim 1/r^2$). On the other hand, the second law is a direct consequence of the law of conservation of the angular momentum (Equation (13.61)) and is valid for all central fields (with arbitrary dependence of the force on r).

13.5.2 Cosmic velocities

The *first cosmic velocity* or the *circular velocity* v_1 is the initial velocity imparted to a projectile in horizontal direction at which the projectile becomes an earth's satellite moving around in a circular orbit:

$$v_1 = \sqrt{\frac{GM}{r}}, \tag{13.63}$$

where M is the earth's mass, r is the radius of the satellite's circular orbit, and G is the gravitational constant. For a low circular orbit, when $r \approx R$ (R is the earth's radius), the value of circular orbital velocity is:

$$v_1 = \sqrt{GM/R} = \sqrt{gR} \approx 7.9\,\text{km/s}. \tag{13.64}$$

Here $g = GM/R^2$ is the acceleration of free fall near the earth's surface.

The *second cosmic velocity* v_2, or the *escape velocity*, or the *parabolic velocity* is the minimal speed that must be imparted to a projectile near the earth's surface in order to send it away to an infinitely long distance from the earth (without taking into account the gravitation of the sun). The escape velocity is determined by the law of energy conservation:

$$v_2 = \sqrt{2}\, v_1 = \sqrt{2gR} \approx 11.2\,\text{km/s}. \tag{13.65}$$

The projectile with such initial velocity will withdraw to infinity independently of the orientation of its velocity. Direction of the initial velocity influences only the form of its parabolic trajectory.

Velocities in the range $v_1 < v < v_2$ are called *elliptic* (they produce elliptic orbits); velocities exceeding the second cosmic velocity ($v > v_2$) are called *hyperbolic*.

The *third cosmic velocity* corresponds to liberation of a projectile from the gravitational field of the sun. If the projectile is located at the distance r from the sun that is equal to the radius of the earth's orbit ($r \approx 1.5 \cdot 10^{11}$ m), the minimal necessary velocity is $\sqrt{2}$ times greater than the orbital velocity of the earth $v_{\rm orb} \approx 29.8$ km/s:

$$v = \sqrt{2} v_{\rm orb} \approx 42.1\,\text{km/s}.$$

We can use the orbital motion of the earth around the sun in order to facilitate gathering this velocity for the projectile. Then the minimal additional velocity with which the projectile must escape from the gravitational field of the earth ought to be $(\sqrt{2}-1)v_{\rm orb} \approx 12.3$ km/s. To obtain it, we must provide on the surface of the earth an initial velocity v_3, which can be calculated on the basis of the law of energy conservation:

$$v_3 = [(\sqrt{2}-1)^2 v_{\rm orb}^2 + v_2^2]^{1/2} \approx 16.6\,\text{km/s}. \tag{13.66}$$

13.5.3 Example: elliptic orbit of a satellite

A satellite is launched to an elliptic orbit with an initial velocity v_0 ($v_1 < v_0 < v_2$) in horizontal direction near the surface of the earth. Let us find the distance r from the center of the earth to the apogee of its elliptic orbit (Figure 13.18).

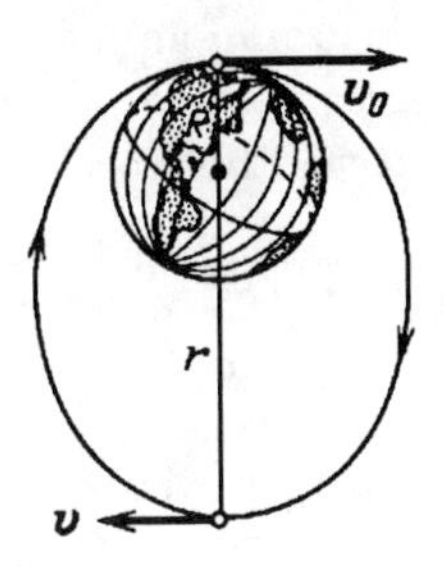

Figure 13.18: Location of the apogee of a satellite's orbit.

The law of energy conservation, Equation (13.60),

$$\frac{mv^2}{2} - mgR = \frac{mv^2}{2} - mg\frac{R^2}{r}, \tag{13.67}$$

gives the equation with two unknown quantities: the desired distance r between the earth's center and the apogee, and velocity v at the apogee. The second equation we get from the law of conservation of the angular momentum (from Kepler's second law): taking into account in Equation (13.61) that at the perigee and at the apogee of the elliptic orbit the vector of velocity is perpendicular to the radius-vector of the satellite, we can write

$$v_0 R = vr. \tag{13.68}$$

(The perigee of the orbit is located at the initial point and its distance from the center of the earth is approximately equal to the earth's radius R). Substituting v from Equation (13.68) into Equation (13.67) we get a quadratic equation relative to r, the two solutions of which are R and $v_0^2 R/(2gR - v_0^2)$. The first root corresponds to the perigee, the second to the apogee of the elliptic orbit. When $v_0 \to v_2 = \sqrt{2GR}$ the apogee recedes to infinity: an elliptic orbit transforms into the parabolic one.

13.6 Mechanical Equilibrium

13.6.1 Conditions of equilibrium

Statics is the branch of mechanics concerned with bodies that are acted upon by balanced forces so that they remain at rest or in unaccelerated motion (remain in *equilibrium*). In many practical cases it is possible to use the model of a perfectly *rigid body* to solve the problems of static, i.e., we can consider the shape and the sizes of the body uninfluenced by the applied forces.

A body is in static equilibrium in an inertial frame of reference if the vector sum of all external forces applied to the body and the vector sum of their moments (torques) are both zero:

$$\sum_i \mathbf{F}_i = 0, \tag{13.69}$$

$$\sum_i \mathbf{r}_i \times \mathbf{F}_i = 0. \tag{13.70}$$

In the definition of the moment of a force as the vector product $\mathbf{r}_i \times \mathbf{F}_i$ (see Section 8.2) the choice of the origin of radius-vectors $\mathbf{r}_i$ of the points of application is arbitrary but must be common for all the forces. By virtue of this arbitrariness this origin can be chosen at any convenient point: Equation (13.70) will be simpler if as many forces as possible will have zero moments.

When the condition of Equation (13.69) is fulfilled, the acceleration of the center of mass of the body is zero. When the condition of Equation (13.70) is fulfilled, the angular acceleration of rotation is zero.

13.6.2 Plane system of forces

For a *plane system of forces* (when all external forces acting on a body lie in one plane), vectorial Equation (13.69) reduces to only two scalar equations:

$$\sum_i F_{ix} = 0, \tag{13.71}$$

$$\sum_i F_{iy} = 0, \tag{13.72}$$

if we choose the axes x, y in the plane where the forces act. The vectors of moments of these forces are directed perpendicularly to this plane (if the common origin of the radius-vectors of their points of application lies in this plane). Therefore vectorial Equation (13.70) reduces to the single scalar equation, which states that the algebraic sum of torques of all the forces is zero. In this sum the torques that tend to turn the body clockwise around the chosen origin enter with one sign (say, plus), while those that tend to turn it counter-clockwise enter with the opposite sign.

If the body is acted upon only by three forces, then in equilibrium all three lines of their action intersect at the same point: otherwise the moment of the third force relative to the point of intersection of the lines of action of the first two forces would be non-zero.

In the problems of statics usually only several of external acting forces are given, and the other forces are to be determined. These unknown forces are usually the *forces of reaction* of the support or of bonds preventing a motion of the body. In the absence of friction the forces of reaction are perpendicular to the surface of contact.

13.6.3 Example: determination of forces of reaction

A very light portable step-ladder (Figure 13.19) consists of a pair of similar steps with a hinge joint at the upper ends connected by a rope at the lower ends. At the middle of one side a person of weight P stays. The problem is to determine the forces of reactions N_1 and N_2 acting from the floor, the force of tension of the rope T, and the forces of interaction in the joint.

The system consists of two rigid bodies—left and right steps. We can apply the conditions of equilibrium both for the whole system and separately to each of the steps. From the conditions of equilibrium of the whole system we can find the external reaction forces N_1 and N_2 (see Figure 13.19). In the absence of friction Equation (13.72) gives

$$N_1 + N_2 = P.$$

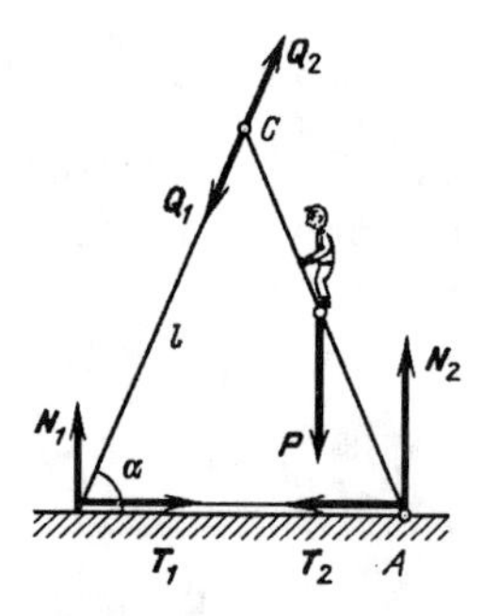

Figure 13.19: Determination of the forces of reaction.

The second required equation follows from the condition of balance of the torques. Considering point A as the origin, we obtain

$$N_1 \cdot 2l \cos\alpha = P\frac{l}{2}\cos\alpha .$$

From these equations we get $N_1 = P/4$, $N_2 = 3P/4$.

To find the internal reaction forces (the tension T of the rope and the force Q of interaction in the joint), we should consider the conditions of equilibrium for separate interacting parts of the system, e.g., for the left steps. The equation of balance of the torques relative the point C gives:

$$N_1 l \cos\alpha = Tl\sin\alpha,$$

and thus we find

$$T = N_1 \cot\alpha = \frac{P}{4}\cot\alpha .$$

The force $\mathbf{Q}_1$ is directed along the left steps downward, because in equilibrium it must be aimed at the point of intersection of the two other forces $\mathbf{N}_1$ and $\mathbf{T}_1$ acting on the left steps. Since the vector sum of $\mathbf{Q}_1$, $\mathbf{N}_1$, and $\mathbf{T}_1$ equals zero, we obtain

$$Q = \frac{P/4}{\sin\alpha}.$$

The force $\mathbf{Q}_2$ acting at the joint on the right steps according to Newton's third law is equal to $-\mathbf{Q}_1$.

Sometimes in the problems of statics the model of absolutely rigid body is inapplicable. For example, it is impossible to determine the forces of reaction acting on a heavy beam that rests on three supports (Figure 13.20) without taking into account its deformation.

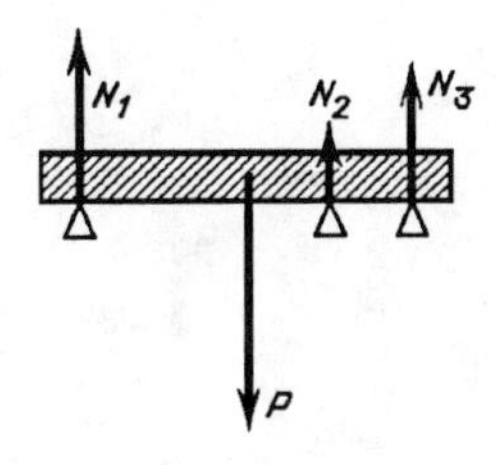

Figure 13.20: A heavy beam on three supports.

13.6.4 Statics and energy conservation

Certain problems of statics can be solved with convenience by applying the law of energy conservation to idealized simple machines (such as a pulley, block and tackle, lever, wheel and axle, inclined plane, etc.) operating without friction. All these devices are used to transform forces. They are characterized by *mechanical advantage*—the ratio of the output force of the machine to the force applied to the input of the machine. No such machine can produce work exceeding the work done at its input (the *golden rule of mechanics*). If we neglect frictional losses in the machine, the ratio of distances through which output and input forces act in doing work is reciprocal to its mechanical advantage.

Example. A heavy weight P is suspended by means of a light hinge with three links (Figure 13.21). Let us calculate the tension of the thread joining the points A and B. If we unfasten the thread at point A and pull it up to raise point B to some height, the weight will uplift to the height three times greater. Equating the work done by the tension of the thread to the increase of potential energy of the weight, we get $T = 3P$.

13.6.5 The stability of equilibrium

Mechanical equilibrium of a system can be stable, unstable, or indifferent (Figure 13.22). The equilibrium of a body (or a system) is *stable*, if after a slight displacement from the equilibrium the forces arise that tend to return the body (or the system) to the original position; the equilibrium is *unstable* if the forces tend to move it farther from original position to some new position. At *indifferent equilibrium* the adjacent positions are also equilibrium positions.

A minimum of potential energy of the system with respect to its values in adjacent positions corresponds to a stable equilibrium position; a maximum of potential energy corresponds to an unstable position.

Example: equilibrium of a column. A vertical freely standing column whose weight is the only downward force is in stable equilibrium because at small inclinations (Figure 13.23) its center of mass upraises.

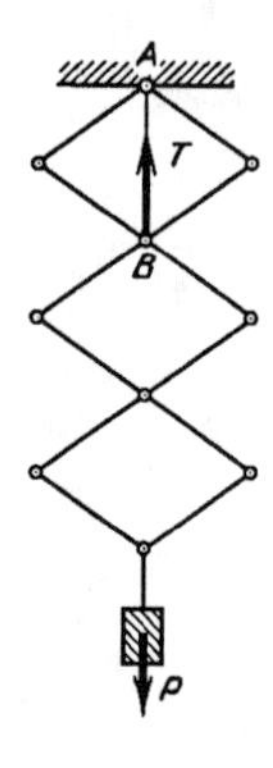

Figure 13.21: A hinge with three links.

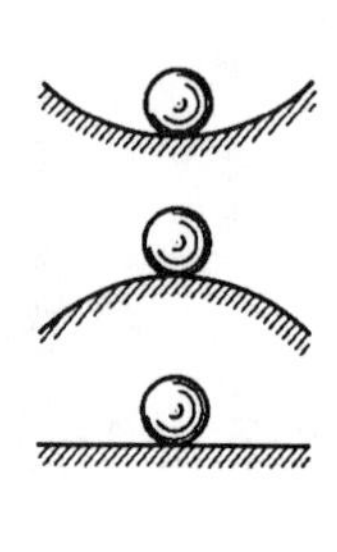

Figure 13.22: Stable, unstable, and indifferent equilibrium.

This is the case in which the vertical line through the center of gravity passes through the base of the column, i.e., while the angle of deflection of the column from the vertical line does not exceed some definite maximal value. In other words, the interval of stability extends from the minimum of potential energy (in vertical position) up to the nearest maximum (see Figure 13.23). If the center of mass is located exactly over the base edge, the column is also in equilibrium, but this equilibrium is unstable. A much more wide interval of stability corresponds to the column lying horizontally.

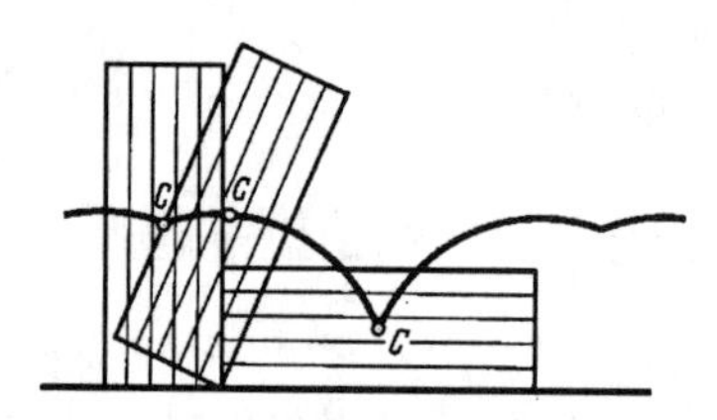

Figure 13.23: Position of the center of mass.

13.7 Dynamics of a Solid

13.7.1 The principal laws

Dynamics of a rigid body is based on the *law of momentum variation* (see Section 13.4.3) and on the *law of angular momentum variation* of a body considered as a system of material points:

$$\frac{d\mathbf{P}}{dt} = \sum_i \mathbf{F}_i, \qquad (13.73)$$

$$\frac{d\mathbf{L}}{dt} = \sum_i \mathbf{r}_i \times \mathbf{F}_i. \qquad (13.74)$$

Momentum **P** and angular momentum **L** of a body are formed by addition of the momenta and angular momenta of discrete mass points that may be considered as small parts of the body. The mutual configuration of the mass points forming a rigid body remains invariable when the body moves. The six scalar equations, Equations (13.73) and (13.74), correspond to the six degrees of freedom of a rigid body (see Section 13.1.6).

When all external forces $\mathbf{F}_i$ acting on a body are given, with the help of Equation (13.73) we can find the law of motion of the center of mass of the body (see Section 13.4.3), and Equation (13.74) permits us to find the law of rotation of the body around the center of mass.

13.7.2 Moment of inertia

Rotation of a rigid body about a fixed axis (see Section 13.1.7) is characterized by a single degree of freedom. In this case, in the component of Equation (13.74), which is parallel to the axis of rotation, the torques of unknown forces of reaction acting on the body in bearings turn to zero, and so this component of Equation (13.74) permits us to find the angular acceleration of the body. The remaining five equations of Equations (13.73) and (13.74) serve to find the forces of reaction of bearings.

The component of angular momentum along the axis of rotation can be represented in the form $I\omega$, i.e., as a product of the moment of inertia I and angular velocity ω. The *moment of inertia* I of a body with respect to an axis is the sum of moments of inertia of all its separate mass points. The moment of inertia of a discrete mass point equals the product of its mass Δm and the square of its distance r from the axis: $\Delta I = \Delta m r^2$. Therefore, for a body,

$$I = \sum_i \Delta m_i r_i^2. \qquad (13.75)$$

The moment of inertia I characterizes inertial properties of a body that manifest themselves in rotation. It measures the *inertia of rotation*, whereas mass M measures the *inertia in translation*. The moment of inertia depends not only on the mass of a body but also on its distribution in the body.

For a ring (a hoop) or of a circular tube with thin walls, the moment of inertia with respect to its axis equals mR^2 because all points of the ring (or a tube) are located at the same distance R from the axis. The moment of inertia of a solid homogeneous cylinder or a disk: $I = \frac{1}{2}mR^2$; of a solid spherical ball: $I = \frac{2}{5}mR^2$; of a homogeneous rod of length L with respect to the axis perpendicular to the rod passing through its center: $I = \frac{1}{12}mL^2$ (Figure 13.24*a*). The moment of inertia I_0 with respect to the axis passing through the center of mass is related to the moment of inertia I with respect to another parallel axis being at a distance l apart (Figure 13.24*b*) by Huygens' theorem:

$$I = I_0 + ml^2. \tag{13.76}$$

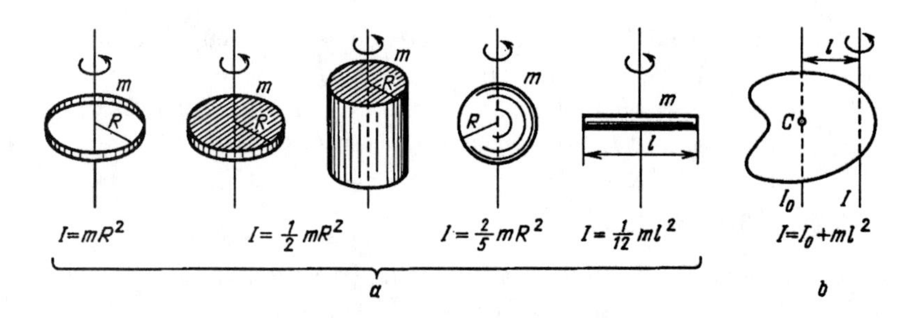

Figure 13.24: Moments of inertia of different homogeneous bodies (*a*) and relation of the moments of inertia with respect to parallel axes (*b*).

Example. A long light thread is winded around a massive cylindrical pulley of the radius R. The pulley can rotate freely about the horizontal axis O (Figure 13.25). A downward force F is applied to the thread. The angular momentum of the pulley with respect to the axis O equals $I\omega$, where ω is its angular velocity. The moment of the force F with respect to the axis O equals RF. The component of Equation (13.74) parallel to the axis takes the form:

$$I\frac{d\omega}{dt} = RF. \tag{13.77}$$

The angular acceleration $d\omega/dt$ of the pulley is proportional to the torque of the force F and inversely proportional to the moment of inertia I of the pulley.

The force Q of reaction of the bearings exerted on the axis of the pulley can be found from Equation (13.73) if we take into account that the acceleration of the center of mass is zero: $Q = mg + F$.

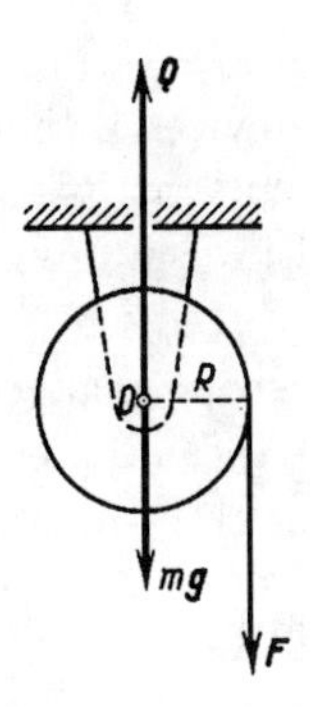

Figure 13.25: Forces applied to rotating cylinder.

13.7.3 Energy of rotation

The *kinetic energy* of a rigid body rotating about a fixed axis is given by the formula

$$E_{\mathrm{k}} = \frac{1}{2} I \omega^2. \tag{13.78}$$

Kinetic energy of a rigid body in a plane motion equals the sum of kinetic energy of rotation about the axis passing through the center of mass and kinetic energy of translational motion of the body with the velocity V which equals the velocity of the center of mass in this plane motion:

$$E_{\mathrm{k}} = \frac{1}{2} I_0 \omega^2 + \frac{1}{2} m V^2, \tag{13.79}$$

where I_0 is the moment of inertia with respect to the axis passing through the center of mass.

When a hoop or a cylindrical tube with thin walls is rolling without sliding, its kinetic energy is equally divided between the energy of rotation and the energy of translation. When a solid homogeneous cylinder is rolling, its kinetic energy of translation is twice the kinetic energy of rotation.

13.7.4 A gyroscope

A heavy rigid body whose shape has the axial symmetry is called a *gyroscope* if it is spinning rapidly about its axis of symmetry. The angular momentum $\mathbf{L}$ of such a body is directed along the axis of rotation. In the absence of external torques the axis of a spinning gyroscope retains its orientation with respect to an inertial frame of reference. The operation of instruments used in inertial navigation systems is based on this property of a gyroscope. In such devices a spinning disc with a heavy rim is mounted in a double gimbal so that its axis can adopt any orientation. When the gimbals are turned the spinning disc maintains its orientation in space.

The axis of a spinning gyroscope changes its direction under the action of external torques. When the torques are moderate the axis turns rather slowly, and with good accuracy we can consider the angular momentum to be directed along the gyroscope's axis in spite of (slow) rotation of the gyroscope around a perpendicular axis. This means that behavior of a gyroscope's axis is approximately described by Equation (13.74), which governs the behavior of the vector **L**. When a gyroscope is subjected to a torque that tends to alter the direction of its axis, the gyroscope turns about an axis that is perpendicular both to the axis about which the torque was applied and to its main axis of spin. The resulting motion is called *precession*.

Example: Precession of a spinning top. The axis of a heavy spinning top is tilted (Figure 13.26) through some angle with the vertical line. The moment of the gravity force $\mathbf{r} \times m\mathbf{g}$ is directed perpendicularly to the vertical plane in which the top's axis lies. According to Equation (13.74), the increment vector $\Delta\mathbf{L}$ of the vector **L** of the angular momentum has the same direction:

$$\Delta\mathbf{L} = (\mathbf{r} \times m\mathbf{g})\,\Delta t.$$

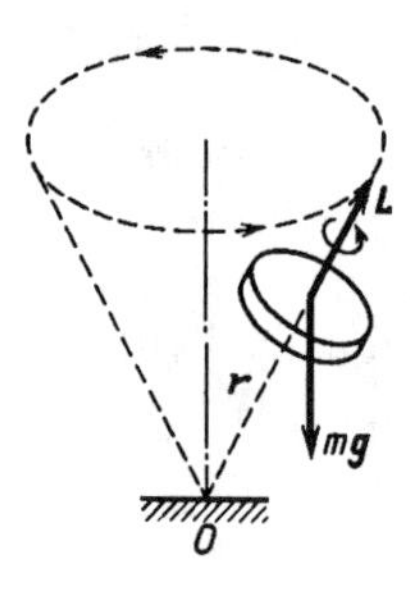

Figure 13.26: Precession of a gyroscope.

As a result, the vector **L** (and the main axis of spin) executes precession, i.e., the vector **L** describes a cone in space as it is shown in Figure 13.26. The torque of the gravitational force tilting the top makes its axis to turn about a perpendicular axis.

13.8 Hydrostatics

Hydrostatics is concerned with the study of liquids at rest (in *equilibrium*) and with the mechanical action of a liquid on submerged bodies.

Slow alteration of the shape of a liquid without the change of its volume can be produced by the action of arbitrarily small forces. In the field of gravity a liquid has no shape of its own, but acquires the shape of the vessel. The surface of a liquid at rest is perpendicular to the force of gravity (that is, the surface is horizontal) independently of the shape of its container. In connected vessels a liquid of constant density has the free surface at the same level (Figure 13.27).

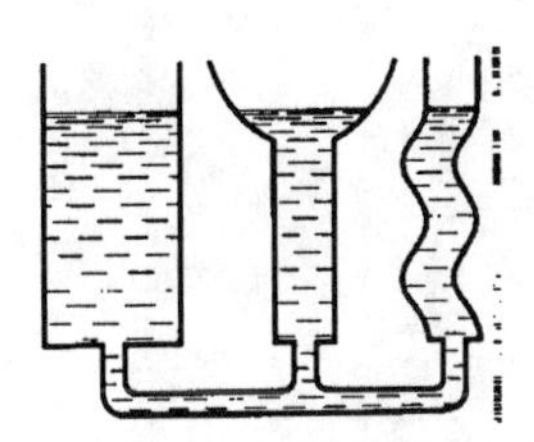

Figure 13.27: Connected vessels.

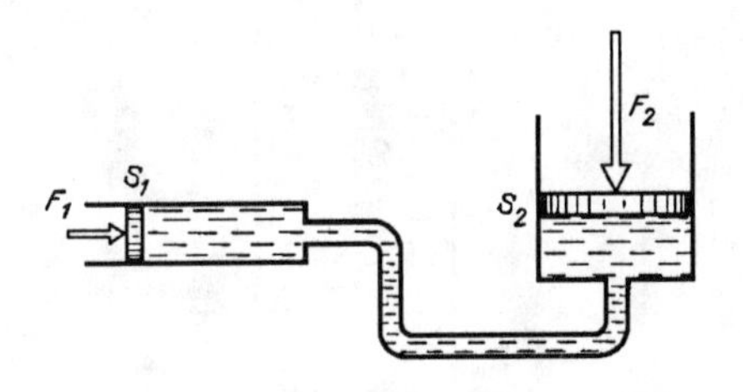

Figure 13.28: A hydraulic press.

13.8.1 Pressure in a liquid

The *pressure* p in a liquid is the ratio of magnitude F of the force exerted by the liquid normally on an area element to its area S:

$$p = \frac{F}{S}. \tag{13.80}$$

Pressure is a scalar quantity. Its value is independent of the orientation of the area element. According to *Pascal's principle* a pressure exerted anywhere on an enclosed fluid by external forces is transmitted equally in all directions. The operation of many hydraulic devices widely used in engineering depends on Pascal's principle. Examples of such devices are hydraulic presses, jacks, drives of vehicle brakes, earth-moving machinery, etc., usually with oil as the working fluid.

In a *hydraulic press* (Figure 13.28) a moderate force F_1 applied to a small piston of area S_1 is transmitted through a fluid to a larger piston of area S_2, where it gives rise to a larger force F_2:

$$F_2 = F_1 \frac{S_2}{S_1}, \tag{13.81}$$

because $p = F_1/S_1 = F_2/S_2$.

13.8.2 Hydrostatic pressure

Under a field of gravity the pressure in a liquid increases with depth due to the weight of the liquid itself: as you go down in sea depth, the weight of the water above you exerts additional pressure. For an incompressible liquid with one and the same density ρ at any depth (Figure 13.29*a*), the pressure linearly depends on the depth h:

$$p_2 = p_1 + \rho g h. \tag{13.82}$$

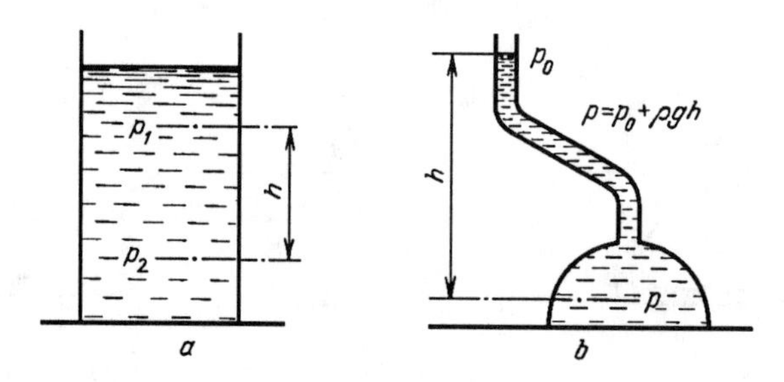

Figure 13.29: Pressure in a liquid at different depths.

For a compressible liquid or a gas the dependence of pressure on depth becomes more complex (see Section 14.3.2).

The resulting pressure p produced by external forces acting on the surface of a liquid (p_0) and produced by the weight of the liquid column ($\rho g h$) is called the (absolute) ***hydrostatic pressure*** (Figure 13.29*b*):

$$p = p_0 + \rho g h. \tag{13.83}$$

The ***hydrostatic paradox*** exhibits itself in the fact that the force of pressure produced by the weight of liquid acting on the bottom of a vessel ($\rho g h S$) may differ from the weight of the liquid in the vessel. In widening vessels the force of pressure is less than the weight of the liquid, in narrowing vessels it is greater. If a liquid is poured up to the same level in vessels of different shape, but with equal bottom area (Figure 13.30), then in spite of different weights of the liquid the force of weight pressure is the same for all vessels and it coincides with the weight of the liquid in the cylindrical vessel.

The explanation of the paradox relies on the existence of a vertical component of the pressure force acting on inclined walls. Such a component is directed downward in a widening vessel, and it is directed upward in a narrowing vessel.

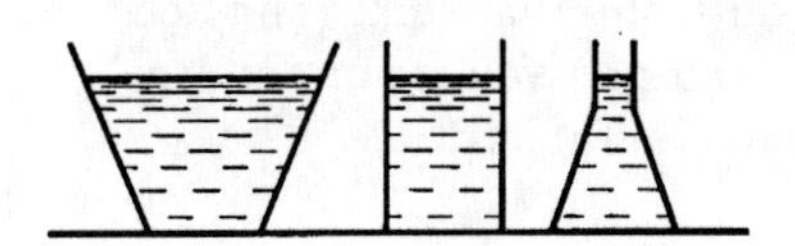

Figure 13.30: The hydrostatic paradox.

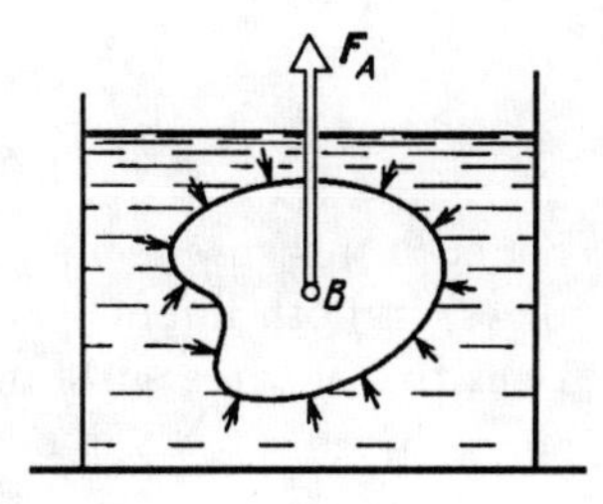

Figure 13.31: The buoyant force (Archimedes' force).

13.8.3 Archimedes' principle and buoyant force

The *Archimedes' principle* states that the upward thrust force exerted by a liquid on an body immersed (partially or totally) in the liquid (the *buoyant force*) is directed upright and its magnitude is equal to that of the weight of the liquid displaced (i.e., of the liquid occupying the same volume as the submerged part of the body). The center of mass B of this homogeneous volume (Figure 13.31) can be considered as the point of application of the buoyant force. In the general case (when the floating body is inhomogeneous) this point does not coincide with the center of mass of the submerged part of the body.

The buoyant force is the resultant (the vector sum) of pressure forces exerted by the liquid on all area elements of the surface of the body. A non-zero net buoyant force is produced because the pressure of liquid on the lower surface of the submerged body is greater than the pressure on the upper surface.

13.8.4 Measurement of density

The density ρ of a solid can be found by measuring the weight P of this body in the air and its weight $P_1 = P - F_A$ in a liquid with the known density ρ_1 by means of a *hydrostatic balance*. The desired density is calculated by the formula:

$$\rho = \rho_1 \frac{P}{P - P_1}. \tag{13.85}$$

The body must be submerged totally when it is weighed in the liquid.

In order to measure the density ρ_2 of a liquid by this method it is necessary to weight a body three times to measure its weight P in the air, the weight P_1 of the body in a liquid with known density, and its weight P_2 in the liquid whose density we are going to determine. The desired density is calculated by the following formula:

$$\rho_2 = \rho_1 \frac{P - P_2}{P - P_1}. \tag{13.86}$$

For such measurements we can use any solid that sinks (and does not dissolves) in both liquids. We need not know its volume or density.

Densities of some substances are listed in Table 13.1.

Table 13.1: Densities of some substances (in 10^3 kg/m^3)

Solids			
Aluminum	2.70	Lead	11.34
Asbestos	2.0–2.8	Lithium	0.534
Beryllium	1.84	Magnesium	1.74
Bronze	8.7–8.9	Mercury (-39 °C)	14.9
Bismuth	9.78	Nickel	8.8
Chromium	6.92	Platinum	21.45
Copper	8.89	Potassium	0.86
Cork	0.22–0.26	Silicon	2.35
Ebonite	1.15	Silver	10.5
Gallium	5.93	Sodium	0.97
Germanium	5.46	Steel	7.8
Glass	2.4–2.8	Tin	7.3
Gold	19.3	Titanium	4.5
Ice (0 °C)	0.917	Tungsten	19.3
Invar	8.0	Vanadium	5.6
Iridium	22.42	Wood	0.5–0.9
Iron	7.7–7.88	Zinc	7.1
Liquids			
Acetone	0.792	Mercury (20 °C)	13.55
Ether	0.736	Petrol	0.899
Ethyl alcohol	0.791	Water (4 °C)	1.000
Glycerin	1.26	Water (20 °C)	0.998
Kerosene	0.8	Water (seawater)	1.025

13.8.5 Floating on the surface

If at complete submergence of a body the buoyant force exceeds the weight of the body ($F_A > P$), the body rises to the surface and ***floats***, being only partially submerged in the liquid. In mechanical equilibrium ($F_A = p$) the center of mass of a floating body and the point of application of the buoyant force lie on the same vertical line because the moments of acting forces in equilibrium are also compensated.

The *stability* of floating depends on disposition of the point called the metacenter of a boat relative to its center of mass (Figure 13.32). The *metacenter* is the point of intersection of the medial plane of symmetry

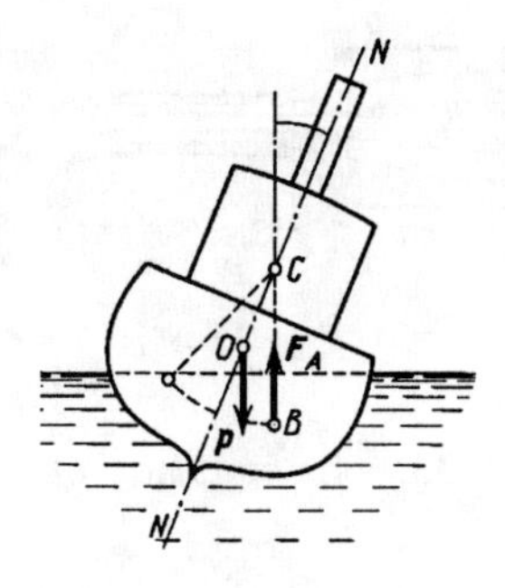

Figure 13.32: Metacenter C.

of a boat with the vertical line passing through the point of application of the buoyant force at the lateral heel of the boat. If the metacenter C lies above the center of mass of a boat, its floating is stable because the torques arising at a heel of the boat tend to return it to vertical position.

13.9 Hydrodynamics

13.9.1 Equation of continuity

A fluid (liquid or gas) in motion is characterized by ***streamlines***, i.e., the lines whose tangents at any point indicate the direction of motion of the fluid particles. In the case of a ***stationary*** flow of a fluid the streamlines retain their configuration in time and coincide with the trajectories of fluid particles.

In the case of a stationary flow in a tube, equal ***masses*** of fluid pass through any cross-section during equal time intervals. This condition implies the relation combining the density ρ of the fluid, velocity v of its motion, and the area S of the cross-section (Figure 13.33) called the *equation of continuity*:

$$\rho_1 S_1 v_1 = \rho_2 S_2 v_2. \tag{13.86}$$

In the case of ***incompressible fluid*** (ρ =const), equal ***volumes*** of the fluid flow through any cross-section in equal time intervals:

$$S_1 v_1 = S_2 v_2. \tag{13.87}$$

It follows from this equation that through a narrow part of the tube with small cross-sectional area the fluid flows faster than through a wide part, and vice versa.

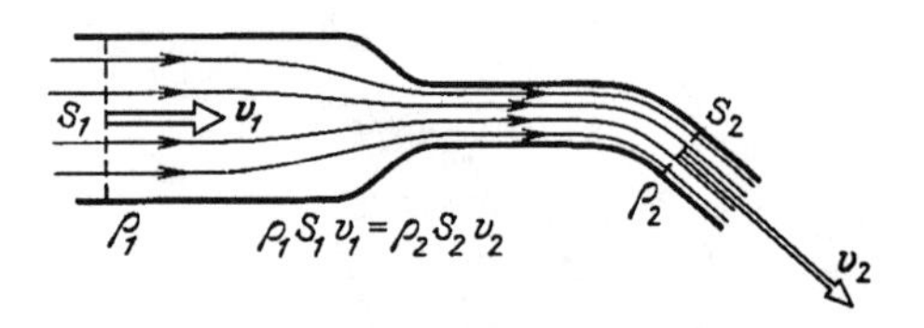

Figure 13.33: The equation of continuity.

13.9.2 Bernoulli's principle

In conditions when the forces of internal friction (viscosity) are insignificant, the model of an *ideal fluid* can be used. In the opposite case we use the model of a *viscous fluid*. In the flow of an ideal fluid there is no dissipation of mechanical energy. The conservation of mechanical energy of an ideal fluid is expressed by *Bernoulli's principle*, stating that at any point in a pipe through which a fluid is flowing the sum of the following three terms has the same value:

$$p_1 + \rho g h_1 + \frac{\rho v_1^2}{2} = p_2 + \rho g h_2 + \frac{\rho v_2^2}{2}. \tag{13.88}$$

Here p_1 and p_2 are the values of the pressure that a pressure gauge moving with the fluid would read in corresponding points, h_1 and h_2 are the heights of these points measured relative to some common level, v_1 and v_2 are the values of velocities of the fluid at these points.

The following examples illustrate the use of Bernoulli's principle.

1. To find the speed of the stream emerging from a very narrow muzzle of the needle of a syringe (Figure 13.34) we can consider Bernoulli's equation with $h_1 = h_2$ and assume the velocity v_1 near the plunger to be zero. The difference of pressures $p_1 - p_2$ in this case equals F/S_1, where F is the force applied to the plunger. Substituting these values into Equation (13.88) we get:

   $$v_2 = \sqrt{\frac{2F}{\rho S_1}}. \tag{13.89}$$

   If it is impossible to neglect the cross-sectional area S_2 of the muzzle with respect to the area S_1 of the plunger, i.e., if the condition $S_2/S_1 \ll 1$ is not fulfilled, the speed of the emerging stream is given by the following expression:

   $$v_2 = \sqrt{\frac{2FS_1}{\rho(S_1^2 - S_2^2)}}. \tag{13.90}$$

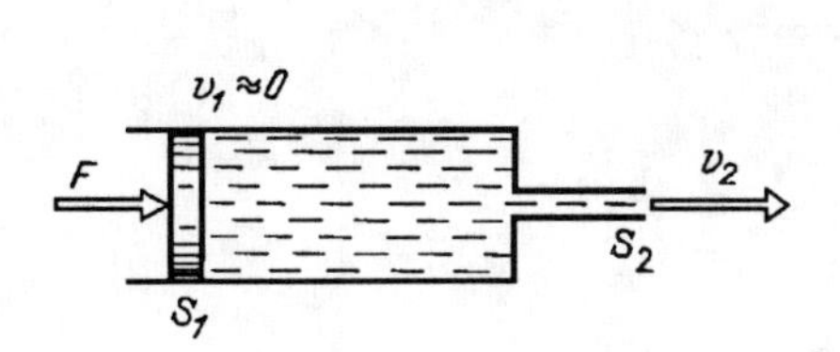

Figure 13.34: A syringe.

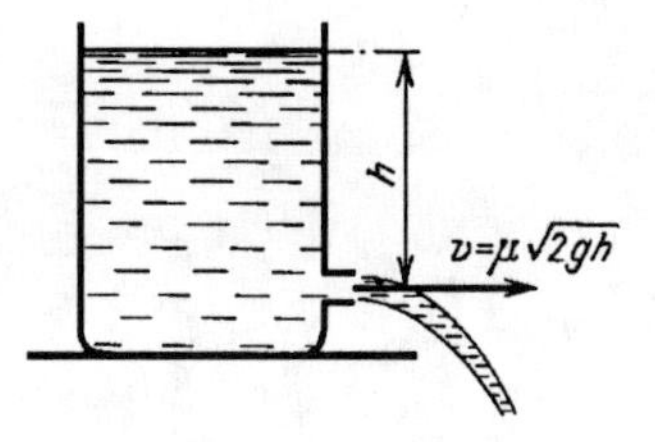

Figure 13.35: Torricelli's formula.

2. For the speed of the stream emerging from a tiny hole in the wall of a broad open vessel (Figure 13.35) Bernoulli's equation, Equation (13.88), leads to *Torricelli's formula*:

$$v = \sqrt{2gh}, \tag{13.91}$$

where h is the height of the free level of liquid over the hole.

The speed of the emerging stream does not depend on the density of the liquid and coincides with the speed of a free fall from the height h. When the shape of the muzzle edges is non-streamlined, in reality the velocities of fluid particles in the hole are not parallel to each other and have a component directed toward the axis of the stream. As a result of this, the emerging stream shrinks, and the effective speed of the stream is always smaller than the speed of a free fall. The effective speed can be written in the form

$$v = \mu\sqrt{2gh}, \tag{13.92}$$

where $\mu < 1$ is the outflow factor depending on the shape of the hole's edges (see Figure 13.36).

In a moving fluid the pressure acting on a resting area element depends on its orientation in the stream. Relation of the measured pressure with the velocity in each case is determined by Bernoulli's principle. If the opening of the pressure gauge is located on the side surface of the gauge pipe (tangent to the stream, see Figure 13.37*a*), its reading is the same as that of the pressure gauge moving along with the fluid (the *static pressure*). If the pipe with the open front edge is turned towards the stream (*Pitot tube*, see Figure 13.37*b*), the reading of the gauge (*dynamic pressure*) exceeds the static pressure p by $\rho v^2/2$. A combination of the two tubes (called *Prandtl's tube*, Figure 13.38*a*) permits one to measure the difference of dynamic and static pressures Δp and to determine the speed v of the fluid:

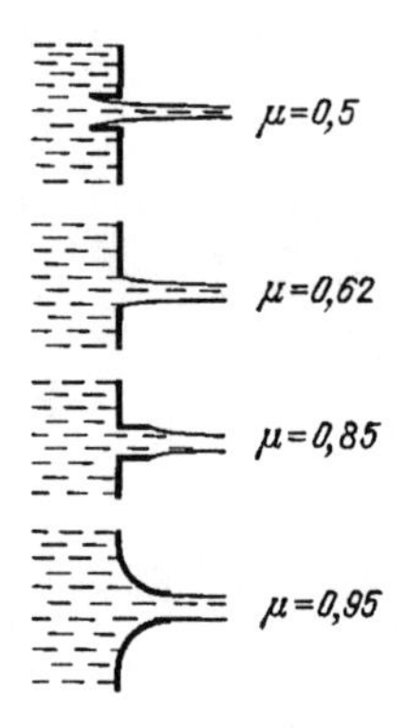

Figure 13.36: The outflow factor for holes of different shapes.

$$v = \sqrt{\frac{2\Delta p}{\rho}}.$$

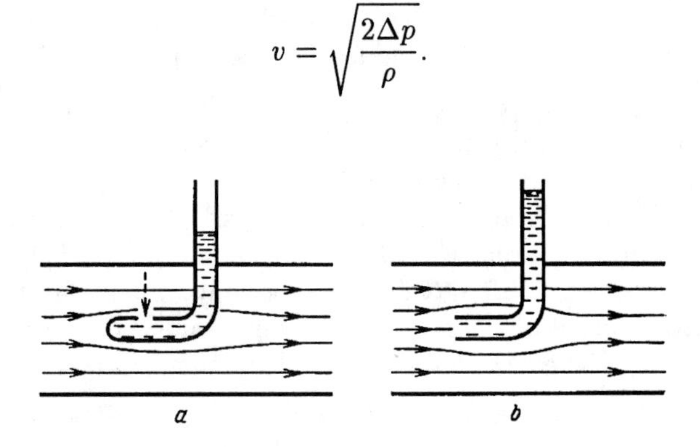

Figure 13.37: The gauge tube in a fluid flow (*a*) and Pitot tube (*b*).

The *Venturi tube* (Figure 13.38*b*) is a device that measures the pressure drop between two different cross-sections of a tube, which is related to the speed of the fluid in one of the cross-sections by the following formula:

$$p_1 - p_2 = \frac{\rho v_1^2}{2} \left[\left(\frac{S_1}{S_2} \right)^2 - 1 \right]. \tag{13.93}$$

13.9.3 Motion of a viscous fluid

A flow of a viscous fluid is called *laminar* or *streamline* if the fluid moves in layers without fluctuations so that successive particles passing the same

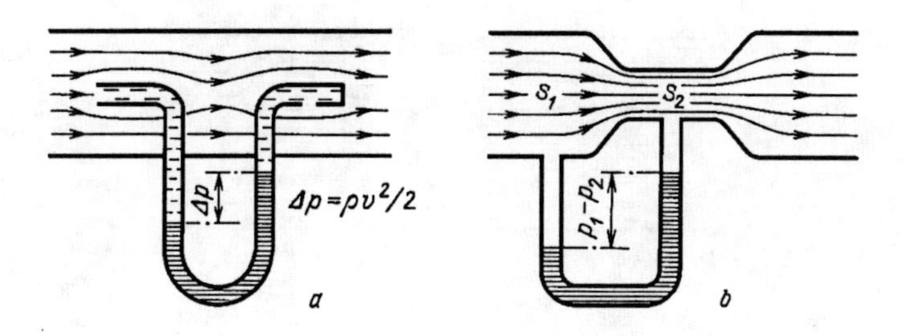

Figure 13.38: Prandtl's tube (a) and a Venturi tube (b).

point have the same velocity. Along the streamlines, velocity of the fluid alters in a predictable and regular way. A flow of a viscous fluid is called *turbulent* if the particles of the fluid move in a disordered manner in irregular paths resulting in mixing of layers, in forming of random whirls, and in a transfer of momentum from one portion of the fluid to another.

An example of laminar flow is the motion of a rather thin layer of liquid confined between two parallel plates, one of which is moving in a parallel way along the other with a moderate velocity v (Figure 13.39). This example gives an idealized scheme of the flow of lubricating oil in bearings. To maintain the uniform motion of the plate, a force F is necessary which is proportional to the area S of the plate, its velocity v, and inversely proportional to the separation d between the plates:

$$F = \eta \frac{Sv}{d}. \tag{13.94}$$

The factor η is a constant called the *coefficient of viscosity* of the fluid. The coefficient of viscosity strongly depends on temperature. For water it reduces to nearly one half when the temperature rises from 0 to 20 °C.

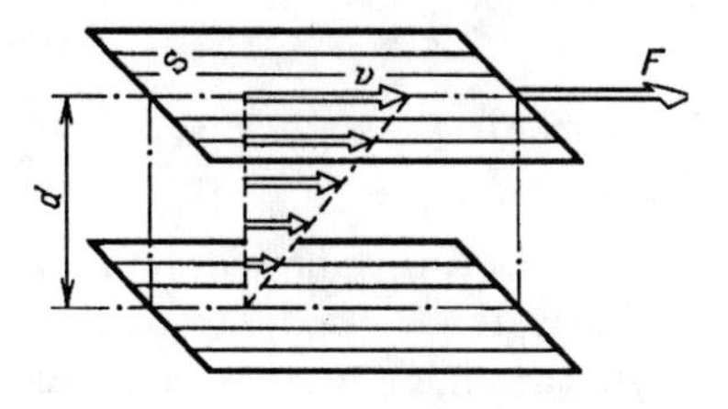

Figure 13.39: Viscous (internal) friction.

For a laminar flow of a viscous fluid in a pipe of circular cross-section the profile of velocities has the form of a paraboloid (Figure 13.40a).

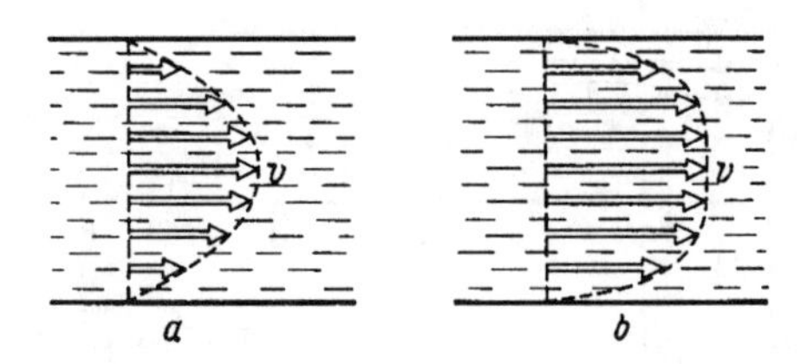

Figure 13.40: Profiles of velocities at laminar flow (a) and at turbulent flow (b).

In this case the volume V of a fluid passing through the tube in time t is given by *Poiseuille's law*:

$$V = \frac{\pi}{8} \frac{\Delta p R^4}{\eta l} t, \tag{13.95}$$

where Δp is the pressure drop between the ends of the tube, R is the radius of the tube, and l is its length. Note that the flow V is proportional to the fourth power of the tube's radius, that is, to the *square* of the cross-sectional area. In a tube of constant cross-section the pressure gradient is constant and the pressure reduces linearly along the tube (Figure 13.41).

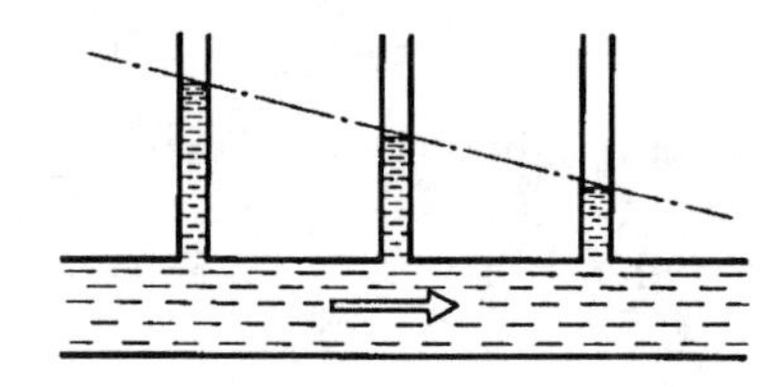

Figure 13.41: Pressure drop at a flow of viscous fluid in a tube.

The force exerted on a resting spherical ball placed in a laminar flow of viscous fluid or, similarly, the frictional force on a spherical ball moving with the same velocity v through a viscous medium being at rest, is given by *Stokes' law*:

$$F = 6\pi\eta R v, \tag{13.96}$$

where R is the radius of the ball.

13.9.4 Turbulent flow of viscous fluid

When the speed of viscous fluid in a stream reaches some critical value, laminar flow becomes unstable and the flow of fluid becomes *turbulent*. At a stationary turbulent flow the particles of the fluid move in a disordered manner in irregular paths, and their velocities change randomly in magnitude and direction. But the average velocity at a given place of the tube (see Figure 13.40*b*) remains constant in magnitude and is directed along the tube. The profile of the average velocity differs considerably from the parabolic one, which is characteristic of a laminar flow (see Figure 13.40*a*). It is characterized by a more rapid increment of the speed near the walls (in the *boundary layer*) and by smaller curvature near the axis of the tube. A laminar flow is typical for the motion of viscous liquid at low speed in narrow tubes (e.g., for the flow of blood in the capillaries). A turbulent flow is characteristic for the motion of fluid with low viscosity at large speed in wide tubes.

At large speeds v the force that resists to the motion of a solid through a fluid medium is determined mainly not by the coefficient of viscosity η of the fluid, but by its density ρ. In this case the drag F experienced by a moving body is proportional to the square of the speed:

$$F = CS\rho v^2, \tag{13.97}$$

where S is the cross-sectional area of the body. The coefficient C depends on the form of the body. For example, for a thin flat disc oriented perpendicularly to the flow $C \approx 0.55$; for a spherical ball the value of C lies in the limits $0.05 - 0.2$, depending on the character of the surface (rough or smooth); for a passenger car $C \approx 0.2$.

Behavior of a moving fluid (either laminar or turbulent) depends on the relative role of dynamic resistance and viscosity. This relative role is characterized by the dimensionless *Reynolds number*:

$$\mathbf{Re} = \frac{l^2 \rho v^2}{\eta l v} = \frac{l \rho v}{\eta}, \tag{13.98}$$

where l is a linear size that characterizes the flow. For a body placed in a stream and flowed around, the size l is its length or its transversal size; for a fluid in a tube, l is the diameter of the tube. The *law of similarity* states that the values of coefficients C in Equation (13.97) for the drag are equal for geometrically similar bodies if the corresponding Reynolds numbers are equal. If this condition is fulfilled, the character of the streamlines is the same at different densities, viscosities, velocities, and sizes of the bodies. The law of similarity permits us to use small-scale models of bodies flowed around in order to make physical measurements and experimental investigations in engineering and design of prototypes.

At small speeds of flow, while the Reynolds number is essentially smaller than some critical value, the flow of any fluid will be laminar.

When the critical value of the Reynolds number is achieved, the laminar flow will be replaced by a turbulent flow. Critical values of Reynolds numbers for a flow in tubes of circular cross-section depend essentially on the state of inner surface of the tube and on the way the fluid flows into the tube. Typical values lie in the interval 1200 – 10,000.

Chapter 14

Molecular Physics and Thermodynamics

Therc are two different approaches to the investigation of the properties of macroscopic systems (i.e., systems, consisting of large number of particles)—statistical mechanics and thermodynamics. The aim of *statistical mechanics* is to establish the laws of behavior of macroscopic systems on the basis of certain dynamic laws of motion of individual particles that form the system, i.e., of atoms and molecules. Statistical mechanics establishes the relations between experimentally measured *macroscopic parameters*, that characterize the system as a whole (e.g., pressure, volume, temperature, electric field strength), and the *microscopic parameters* of the system (masses and charges of individual particles of the system, their coordinates and momentums, etc.).

On the contrary, the *thermodynamic* approach is based not on the model concepts that assume atomic-molecular structure of matter, but rather on general laws such as the law of energy conservation, reliably established in experiment. Basic concepts of thermodynamics are introduced on the basis of experimental data and hence thermodynamics deals only with macroscopic parameters.

The most comprehensive knowledge about properties of macroscopic systems is obtained by a combined use of thermodynamics and statistical mechanics. For example, for one mole of an ideal gas, statistical mechanics permits us to establish the following relation between the product of two macroscopic parameters—pressure p and molar volume V_μ—and the mean value of the microscopic parameter $\langle\varepsilon\rangle$—mean kinetic energy of chaotic translational motion of one molecule (see Section 14.2.4):

$$pV_\mu = \frac{2}{3}\langle\varepsilon\rangle N_A, \tag{14.1}$$

where $N_A = 6.0221367 \cdot 10^{23}\ \mathrm{mol}^{-1}$ is the number of molecules per mole (Avogadro constant, see Section 14.2.3).

In thermodynamics the following relation among three macroscopic parameters of a gas—p, V_μ, and thermodynamical temperature T—was established experimentally for one mole of a gas:

$$pV_\mu = RT, \tag{14.2}$$

where $R = 8.314510\ \mathrm{J/(mol \cdot K)}$ is the universal gas constant (see Section 14.1.3). Comparison of Equation (14.1) and Equation (14.2) allows us to clarify the physical meaning of thermodynamical temperature, i.e., to connect it with the mean value of kinetic energy of the particle's thermal motion:

$$\langle \varepsilon \rangle = \frac{3}{2} kT, \tag{14.3}$$

where $k = R/N_a = 1.380658 \cdot 10^{-23}\ \mathrm{J/K}$ is the Boltzmann's constant.

14.1 Principles of Thermodynamics

14.1.1 Thermal equilibrium

Thermodynamics is a phenomenological theory of heat, based on experimentally established main laws ("principles" of thermodynamics). The intrinsic atomic-molecular structure of real systems is not considered in thermodynamics. A *thermodynamic system* is an arbitrary macroscopic body or a set of bodies. A system is called *closed* (or *isolated*) if it does not interact in any way with surrounding bodies. The physical state of a thermodynamic system is determined by several *macroscopic parameters*. The number of necessary parameters depends on the system under consideration and is determined experimentally. The quantities that do not depend on the history of a system and are determined only by its physical state at a given instant are called *functions of the state.*

The state of an isolated system is called the *equilibrium* state, if the values of macroscopic parameters of the system do not vary with time (fluctuations are not taken into account here; see Section 14.3.6). Macroscopic parameters also do not change in time in a *stationary* (or *steady*) state of an open system, but it always turns out to be so by virtue of some external (with respect to the system) process. For example, the state with some constant distribution of temperature along a rod whose edges are kept at two different temperatures is *stationary* (but not equilibrium).

The *zeroth law of thermodynamics* states that any isolated thermodynamic system finally reaches an equilibrium state characterized by equal values of temperature of all its macroscopic parts. Time needed for transition of a system to the equilibrium state is called *relaxation time* (see Section 12.3.8).

14.1.2 Parameters of the equilibrium state

Thermodynamic (or heat) equilibrium has the property of transitivity: if a system A is in thermodynamic equilibrium with a system B, and the system B is in equilibrium with a system C, then the systems A and C are also in thermodynamic equilibrium. This property gives an opportunity of introducing the concept of temperature, which describes thermodynamic equilibrium between bodies being in thermal contact. Bodies in thermodynamic equilibrium have equal temperatures. The concept of thermodynamic temperature that does not depend on thermometric substance is based on *the second law of thermodynamics* (see Section 14.1.13). This idea of an absolute temperature scale creation was proposed by William Thomson (lord Kelvin) and, therefore, the thermodynamic scale of temperature is called also Kelvin's scale.

Empirical temperature scales are usually based on the phenomenon of thermal expansion of some substances (mercury, alcohol, etc., see Section 14.6.3). The Celsius scale is obtained by division of the temperature region between the fixed point of ice melting, assumed as 0 °C, and the point of water boiling at normal atmospheric pressure, into one hundred equal parts (degrees). Readings of two thermometers with different thermometric substances coincide, generally speaking, only at 0 °C and at 100 °C.

Pressure p (scalar quantity) in thermodynamics is defined as the magnitude of the average force, acting normally on a surface of unit area.

There are two kinds of macroscopic parameters of any system: *external* (whose values are determined by experimental conditions) and *internal* (determined by properties of the system with fixed external parameters). For example, the volume of a gas, contained in a vessel of fixed volume, is an external parameter. The pressure of the gas is an internal parameter in this case. If a gas is contained in a vertical vessel under a freely moving piston loaded by a weight, then the pressure becomes an external parameter and the gas volume should be considered as an internal one.

Internal parameters that are proportional to the number of particles in a system are called *extensive* or *additive* parameters (such as the internal energy of the whole system, its magnetic moment, etc.). Internal parameters determined by a number of particles per unit volume (i.e., parameters independent of the total number of particles in the system) are called *intensive* parameters (pressure, temperature, etc.). Extensive parameters describe a system as a whole, intensive parameters characterize every macroscopic part of it.

Internal parameters appear to be some functions of temperature and of the external parameters of a physical system that is in the state of thermodynamic equilibrium. This situation is described by an *equation of state*. An equation of state for a given system either is obtained by experiment, or it is deduced theoretically within the limits of statistical approach on the basis of some model of the system.

14.1.3 Equation of state for the ideal gas

The *Clapeyron–Mendeleev equation*, which is valid within some intervals of thermodynamic parameters of any gas, is an example of an equation of state, which is established in experiments on real gases. For one mole of gas (see Section 14.2.3) the equation has the form

$$pV_\mu = RT. \tag{14.4}$$

Since the volume is an extensive parameter, $V = V_\mu\, m/M$, then from Equation (14.4) we obtain the equation of state for an arbitrary amount of the gas (for an arbitrary number of moles $\nu = m/M$):

$$pV = \frac{m}{M}RT, \quad \text{or} \quad pV = \nu RT. \tag{14.5}$$

Here m is the mass of the gas, M is the molar mass of the gas (see Section 14.2.3), and $R = 8.314510$ J/(mol·K) is the universal gas constant.

The Clapeyron–Mendeleev equation envelopes all *gas laws* that were first established experimentally for real gases:

- *Boyle's–Mariotte's law:* for a given mass of a certain gas the product of gas pressure p and its volume V is a constant quantity if the temperature remains the same:

$$pV = \text{const} \qquad (T,\ m,\ M\ -\ \text{const}). \tag{14.6}$$

  Curves, corresponding to Boyle's law in the pV-diagram, are called *isotherms* (Figure 14.1). They are hyperbolas (see Section 9.3.3).

- *Charles' law:* the volume V of a given mass of definite gas is proportional to the thermodynamic temperature T when pressure is kept constant, i.e.,

$$V = V'\frac{T}{T'} \qquad (p,\ m,\ M\ -\ \text{const}), \tag{14.7}$$

  where V' is the volume of the gas at a temperature T'. The plots of Charles' law at pV-diagram are called *isobares* (Figure 14.2).

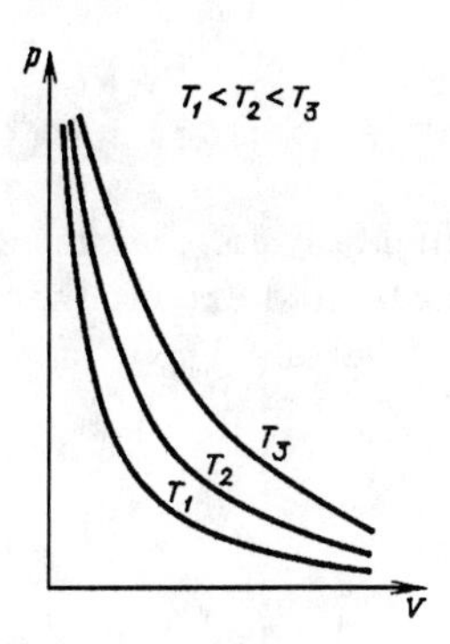

Figure 14.1: Isotherms of the ideal gas.

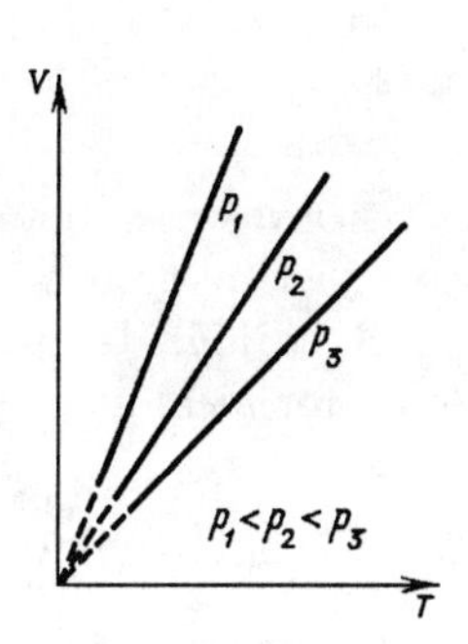

Figure 14.2: Isobars of the ideal gas.

- *Gay-Lussac's law:* for a given mass of a gas that is kept at a constant volume the pressure p is proportional to the thermodynamic temperature T:

$$p = p'\frac{T}{T'} \qquad (V,\ m,\ M\ \text{– const}), \tag{14.8}$$

where p' is the pressure of the gas at a temperature T'. The plots of Gay-Lussac's law on pV-diagram are called *isohores* (Figure 14.3).

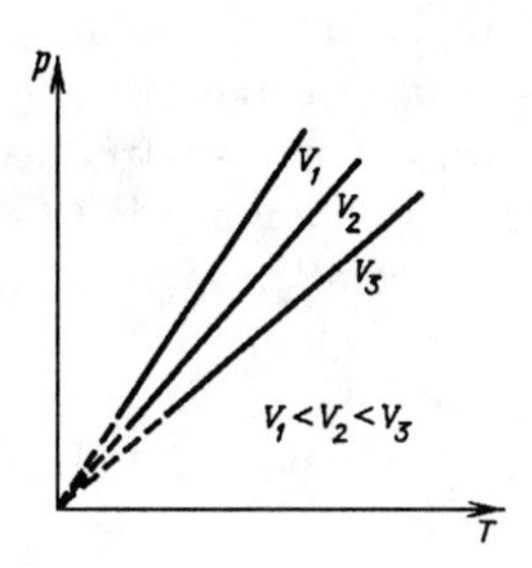

Figure 14.3: Isohores of the ideal gas.

Clapeyron's equation can be derived from any two of the above gas laws:

$$\frac{pV}{T} = \frac{p'V'}{T'} \qquad (m,\ M\ \text{– const}). \tag{14.9}$$

Gas density ρ is determined by the following equation, which follows from Equation (14.5):

$$\rho = \frac{m}{V} = \frac{pM}{RT}. \tag{14.10}$$

The gas laws were first established experimentally for real gases although they are obeyed by real gases to only a limited extent; they are obeyed best at high temperatures and low pressures. These laws are precise only for ideal gases (see Section 14.2.4).

14.1.4 Gas thermometer

When the Celsius scale of temperature is used, Gay-Lussac's and Charles' laws are written in the following way:

$$V = V_0(1 + \alpha t), \qquad p = p_0(1 + \alpha t). \tag{14.11}$$

Here V_0 and p_0 are the volume and pressure of a gas at 0°C and α is the coefficient of thermal expansion (thermal pressure coefficient), which has the same value for all gases:

$$\alpha = (1/273.15)\mathrm{K}^{-1}.$$

A gas as a thermometric substance is convenient for creating a temperature scale (*gas thermometer*) because the expansion of different gases at heating is practically the same, and the thermal expansion of the material of the container where thermometric gaseous substance is heated, is negligible. In order to have proportionality to temperature instead of linear dependence on temperature t measured in the Celsius scale in Gay-Lussac's and Charles' laws (see Equations (14.7)–(14.8)), the International Practical Temperature Scale is introduced. This scale conforms as closely as possible to the thermodynamic (Kelvin) temperature scale (see Section 14.1.2). The minimal possible temperature in the Celsius scale is −273.15° C.

A gas thermometer provides the most precise measuring of thermodynamic temperature in a wide region of temperatures. The relation of temperature T in the Kelvin scale and temperature t in the Celsius scale is

$$\frac{T}{K} = \frac{t}{^\circ\mathrm{C}} + 273.15. \tag{14.12}$$

Thermodynamic temperature T enters practically in all thermodynamic relations.

14.1.5 Components and phases

Thermodynamic systems are divided into homogeneous and heterogeneous ones. Properties of ***homogeneous*** systems do not vary, or vary continuously from one spatial point to another. *Heterogeneous* systems consist of some homogeneous subsystems so that the system has internal boundaries with abrupt breaks of continuity in somc properties.

A homogeneous part of a heterogeneous system, which is separated from other parts by partition surface, is called a *phase*. Chemically individual substances that can be extracted from the system are called *components*. For example, the mixture of different gases is a single-phase, but many-component system. Ice–water mixture is a two-phase but single-component system. A system with two components is called binary and a system with three components is called a ternary one.

14.1.6 Reversible and irreversible processes

A *thermodynamic process* is a transition of a system from one state to another. When external conditions vary rather slowly (i.e., the rate of the process occurring in the system is considerably less than the rate of relaxation), the process can be represented as a chain of nearby equilibrium states. Such a process is described by the same macroscopic parameters as an equilibrium state. These slow processes are called *equilibrium* (or *quasistatic*) processes. They are called also *reversible* processes, because the series of equilibrium states can be passed in direct and reverse successions. Real processes are *irreversible* and can be considered approximately as quasistatic with more or less precision.

Here are some examples of irreversible processes: heat transfer under a finite temperature difference between the bodies, gas expansion into a vacuum, diffusion of one substance into another (see Section 14.2.1). In these processes a system is so far from the equilibrium state that it cannot be described as a whole by definite values of temperature, pressure, or concentration of particles.

Circular processes or cycles take a particular place in thermodynamics. *Cycle* is a process after executing which the system returns to the initial state. Cycles play a dominant role in performance of heat engines.

14.1.7 Internal energy

One of the main characteristic of any thermodynamic system is its *energy*. Energy does not appear and does not disappear: it is only transmitted from one system to some other or transforms from one form to another in natural phenomena. This is the main point of the *energy conservation law*—one of the basic laws of nature.

Total energy E_{tot} of a system consists of mechanical energy E and internal energy U (see Section 14.2.4). Thermodynamics deals usually with bodies at rest, whose mechanical energy does not change. There are three different ways to change the internal energy of a chosen physical system: performing *work* by external forces, transferring *heat* to the system, and *exchanging substance* with the surrounding medium.

Neither macroscopic work of external forces nor heat are the forms of energy, but rather the modes of energy exchange and transference of energy between two systems. Work and heat are not functions of state of the system. Heat and work describe a change of a state, i.e., processes occurring in a system.

The *first law of thermodynamics*: the change of internal energy of a system ΔU in transition from state 1 to state 2 is equal to the sum of the work A performed over the system by external forces, and of the amount of heat Q transferred to the system:

$$\Delta U = U_2 - U_1 = A + Q. \tag{14.13}$$

It is supposed that there are no changes of mechanical energy and of number of particles in the system, i.e., that a substance exchange with the surrounding medium is absent. A and Q are algebraic quantities (i.e., they can be positive or negative). Thus, $Q > 0$ indicates an amount of heat transferred to the system, while $Q < 0$ indicates a heat that is withdrawn from the system. Sometimes the first law of thermodynamics is written as follows:

$$Q = \Delta U + A', \tag{14.14}$$

where A' is the amount of work, done *by* the system, i.e., by forces acting upon the surroundings: $A' = -A$.

The first law of thermodynamics states the equivalence of different forms of energy: energy is converted from one form to another in strictly equivalent amounts.

14.1.8 Heat capacity

The work $\Delta A'$ done by any system at pressure p while its volume V changes by ΔV is given by the formula

$$\Delta A' = p\Delta V. \tag{14.15}$$

The work is equal to the area of shaded rectangle with p as the height and ΔV as its width on the pV-diagram (Figure 14.4).

The work done by a system, expanding from a state 1 to a state 2 by a path 1 – 2, is equal to the area of the whole curvilinear trapezoid (Figure 14.4).

Heat capacity is the ratio of Q (amount of heat transferred to the system) to the temperature change ΔT during the heat transfer. Heat capacity characterizes a process which a physical system undergoes. For example, the heat capacity of one mole of an ideal monoatomic gas at $V =$ const (see Equation (14.35) for U) is given by the following expression:

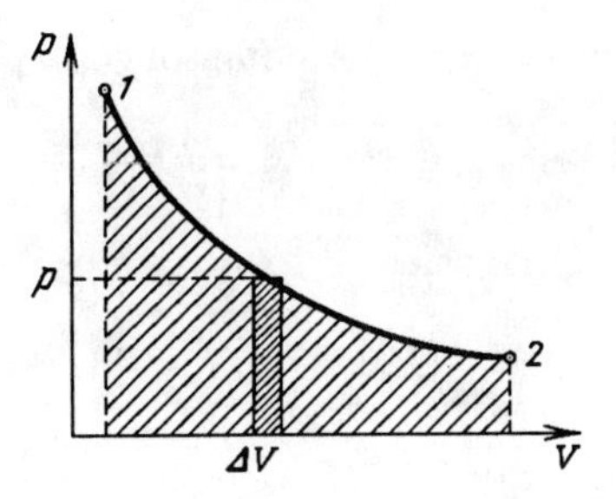

Figure 14.4: Work on the pV-diagram.

$$C_V = \frac{3}{2}R. \tag{14.16}$$

For diatomic gas (V = const):

$$C_V = \frac{5}{2}R. \tag{14.17}$$

For an isobaric process (p = const) in an ideal gas:

$$C_p = C_V + R. \tag{14.18}$$

If gas pressure in the process is proportional to the volume ($p = kV$),

$$C = C_V + \frac{R}{2}. \tag{14.19}$$

Heat capacities of solids and liquids, in contrast to gases, weakly depend on external conditions (see Section 14.7.1), i.e., $C_p \approx C_V$.

14.1.9 Isoprocesses in the ideal Gas

- *Isohoric process:* V = const. During an isohoric process no work is performed:

$$A' = 0,$$

because $\Delta V = 0$. For an arbitrary gas quantity ν the heat Q transferred to the gas in an isohoric process

$$Q = \nu C_V (T_2 - T_1).$$

For an isohoric process

$$Q = \Delta U,$$

as it follows from the first law of thermodynamics, Equation (14.14).

- *Isobaric process:* $p = \text{const}$. Amount of work performed by the gas:

$$A' = p(V_2 - V_1) = \nu R(T_2 - T_1).$$

Amount of heat Q transferred to the gas:

$$Q = \nu C_p(T_2 - T_1).$$

- *Isotermal process:* $T = \text{const}$. The work done by the gas (see Section 5.7):

$$A' = \int_{V_1}^{V_2} p\,dV = \nu T \ln \frac{V_2}{V_1}.$$

The internal energy U of an ideal gas depends only on its temperature and does not change during an isothermal process ($\Delta U = 0$). Thus, as it follows from the first law of thermodynamics, the work A' performed by the gas in an isothermal process is equal to the heat Q transferred to the gas:

$$A' = Q.$$

- *Adiabatic process:* $Q = 0$, $C = 0$. From the first law of thermodynamics it follows that

$$A' = -\Delta U.$$

Pressure and volume of an ideal gas undergoing an adiabatic process are related by the Poisson equation:

$$pV^{\gamma} = \text{const}, \tag{14.20}$$

where $\gamma = C_p/C_V$. An adiabatic curve on the pV-diagram is steeper than an isothermal curve (Figure 14.5).

14.1.10 Efficiency of a heat engine

A schematic diagram of a *heat engine* is represented in Figure 14.6. Some heat quantity (Q_1) is taken from a heater (a heat reservoir at a high temperature T_1) and is partially transformed by a working substance into mechanical work. The remaining heat O_2 is rejected to a cooler (a heat reservoir with a lower temperature $T_2 < T_1$). The ratio η of the work done A to the heat input Q_1,

$$\eta = \frac{A}{Q_1} = \frac{Q_1 - Q_2}{Q_1}, \tag{14.21}$$

is called the *efficiency* of a heat engine.

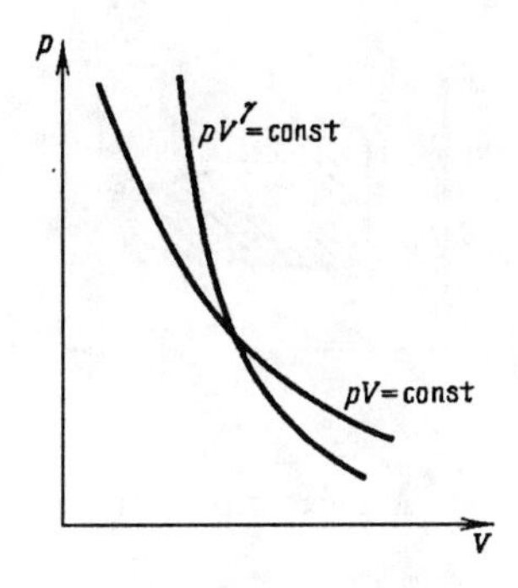

Figure 14.5: Diagrams of adiabatic and isothermal processes.

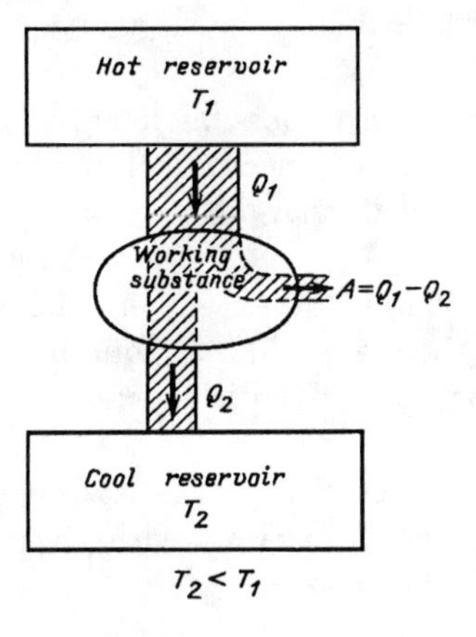

Figure 14.6: A heat engine.

Cycles (see Section 14.1.6) play a dominant role in heat engine operation. The work done by a heat engine during one cycle is represented by the corresponding area in a pV-diagram enclosed by the plots of the processes that constitute the cycle (Figure 14.7). If a system undergoes a cyclic process in the *clockwise* manner, thermal (internal) energy is converted into mechanical work. If a system undergoes a cycle in the *counter-clockwise* manner, mechanical work is converted into internal energy.

14.1.11 The second law of thermodynamics

The *second law of thermodynamics* states the qualitative non-equivalence of different forms of energy concerning their ability of conversion into other forms. There are several different (but equivalent) formulations of this law.

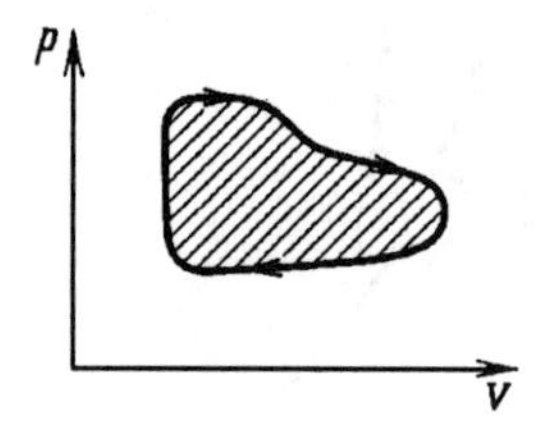

Figure 14.7: The work performed during a cyclic process.

1. *Clausius:* a process of which the only result is a transfer of heat from any body with low temperature to another body of higher temperature is impossible.

2. *Thomson* (*Kelvin*): it is impossible to create a periodically working heat engine that performs mechanical work consuming heat from a hot reservoir without discharging some heat to a cold reservoir.

3. *Caratheodori:* in the vicinity of any equilibrium state of a thermodynamic system there are adjacent equilibrium states that can not be reached through an adiabatic process.

4. *Carnot theorem:* the efficiency of a heat engine operating periodically between two heat reservoirs with given temperatures T_1 and T_2 cannot exceed the value

$$\eta_{\max} = \frac{T_1 - T_2}{T_1}. \tag{14.22}$$

The maximal value $\eta_{\max}$ defined by Equation (14.22) is reached by a heat engine executing a reversible process. This value is independent of the nature of a working substance. A cycle is called *reversible* if it consists of reversible processes that can be performed in both directions through the same succession of equilibrium states.

The *Carnot cycle* is the only reversible cycle that can be executed between two heat reservoirs with fixed temperatures; it consists of two isotherms and two adiabates. Graphic representation of such a cycle for an ideal gas is shown in Figure 14.8. Process 1 – 2: isothermal expansion of an ideal gas at T_1 (heater temperature). Process 2 – 3: the gas expands adiabatically, its temperature is lowered and becomes equal to T_2 (the temperature of cool reservoir). Process 3 – 4: isothermal compression of the gas at T_2. Process 4 – 1: adiabatic compression of the gas until its temperature reaches the value T_1.

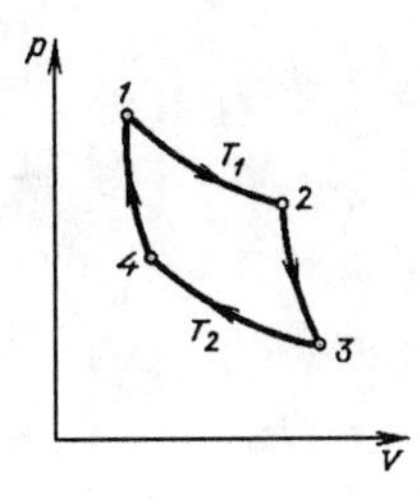

Figure 14.8: The Carnot cycle.

14.1.12 A refrigerator machine

A *refrigerator* serves to cool a substance and to maintain it at a temperature lower than that of its surroundings. Therefore, a refrigerator must transfer heat from a cold reservoir (the cooled substance) to the surrounding medium that can be considered as a hot reservoir. The operation of a refrigerator is based on a cycle performed in a counter-clockwise manner (Figure 14.9). The efficiency of a refrigerator can be defined by using the analogy with the efficiency of a heat engine:

$$\varepsilon = \left| \frac{Q_{\text{taken}}}{A} \right|. \tag{14.23}$$

If the cycle is reversible, then

$$\varepsilon = \frac{T_2}{T_1 - T_2}. \tag{14.24}$$

In contrast to refrigerator a ***heat pump*** (Figure 14.10) must reject as much heat as possible to a hot substance (for example, to the water in a heating system). The transformation coefficient that characterizes the efficiency of a heat pump is defined by the following expression:

$$\varepsilon = \frac{Q_{\text{add}}}{A}. \tag{14.25}$$

If the cycle is reversible, then

$$\varepsilon = \frac{T_1}{T_1 - T_2} > 1. \tag{14.26}$$

14.1.13 Thermodynamic temperature

The second law of thermodynamics for reversible and irreversible cycles can be expressed with the use of Equations (14.21) and (14.22) in the following form:

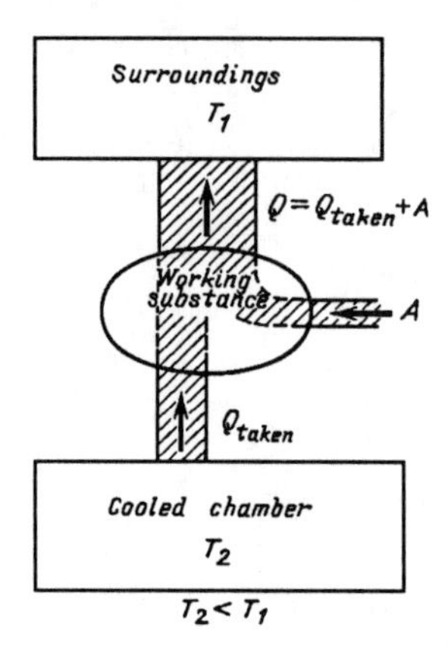

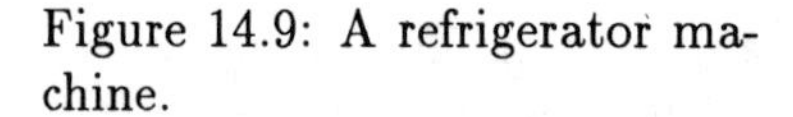

Figure 14.9: A refrigerator machine.

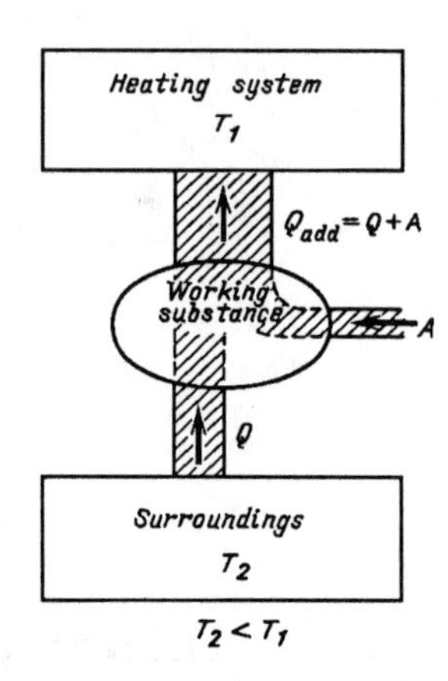

Figure 14.10: A heat pump.

$$\frac{Q_1 - Q_2}{Q_1} \leq \frac{T_1 - T_2}{T_1}.$$

This inequality can be written in another form:

$$\frac{Q_1}{T_1} - \frac{Q_2}{T_2} \leq 0. \tag{14.27}$$

Here the sign of equality corresponds to reversible processes. Q is an amount of heat taken from the heater at T_1, and Q_2 is an amount of heat discharged to the cold reservoir at T_2.

Equation (14.27) does not depend on the nature of the working substance (see Section 14.1.2). That is why it is possible to use it in order to introduce a *thermodynamic temperature* that does not depend on properties of a thermometric substance. For unambiguous definition of the *thermodynamic temperature scale* it is necessary to define the value of temperature at some fixed reference point. The triple point of water is assumed to be such a point (see Section 14.4.5).

14.1.14 Enthropy

The second law of thermodynamics gives a possibility of introducing some definite function of state of the system S called the *enthropy*. The value of S does not change during equilibrium adiabatic processes. The differential dS of this function for a reversible process is defined as follows:

$$dS = \frac{\delta Q}{T}. \tag{14.28}$$

Here δQ is an elementary amount of heat received by the system at temperature T as a result of a reversible process. Irreversible processes in an isolated system lead to growth of the enthropy, while in reversible processes the enthropy remains constant:

$$\Delta S \geq 0$$

(here the sign of equality refers to reversible processes).

Example

A thermoinsulated vessel is subdivided into two equal parts. One part contains an ideal gas, the other is empty. Let us find the enthropy change of the gas if the partition is suddenly removed and the gas fills the whole vessel.

Since the gas does not receive heat and does not do any work (it expands in a vacuum), its internal energy remains constant. Internal energy of an ideal gas is proportional to its temperature and does not depend on its volume. Therefore, gas acquires the same temperature when it reaches the finite equilibrium state. The gas undergoes here an irreversible process. But enthropy of the finite state of the gas is a function of the state. It does not depend on the path leading to the state. Therefore, instead of the irreversible process in question, we can consider a reversible isothermal process which gas undergoes, expanding and moving the partition until it fills all the volume.

In this isothermal process the gas receives the amount of heat Q which is equal to the work A' done by it (see Section 14.1.9):

$$Q = A' = \nu RT \ln \frac{V_2}{V_1} = \nu RT \ln 2,$$

because $V_2/V_1 = 2$. Since in the process $T = \text{const}$, for the change of the entropy ΔS we obtain

$$\Delta S = \frac{Q}{T} = \nu R \ln 2 = \nu k N_A \ln 2 = kN \ln 2,$$

where $N_A = 6.022 \cdot 10^{23}\ \text{mol}^{-1}$ is Avogadro's number. For solution of this problem on the basis of statistical approach, see Section 14.3.8.

14.2 The Principles of Statistical Mechanics

14.2.1 Thermal motion

All macroscopic systems consist of atoms and molecules that are in the state of incessant chaotic heat motion.

Diffusion is a process by which different substances mix as a result of the random heat motions of their component atoms, molecules, and ions.

Atoms or molecules of one substance penetrate into the spaces between molecules and atoms of another. *Self-diffusion* is a confusion of molecules of one sort.

Brownian movement is a continuous random heat motion of small solid particles suspended in liquids and gases. Their sizes are much greater than atomic and molecular sizes. The mean displacement of a particle from the origin (the point where the particle was localized at the initial time) during time interval t is proportional to $\sqrt{t}$. This displacement increases with the thermodynamic temperature T.

Phenomena of diffusion and Brownian motion give reliable evidence in support of thermal random motion of atoms and molecules.

14.2.2 Molecular interaction

Linear sizes of typical molecules are of the order of 10^{-8} cm. The size of molecules of high-molecular compounds is much larger (for example, the characteristic length of a rubber molecule is about 0.02 mm).

Due to the electric neutrality of atoms and molecules the forces of interaction between molecules are *short-range*. These forces do not reveal themselves on distances that are several times larger than the size of molecules. The *forces of intermolecular interaction* are resultant of the forces of attraction and repulsion that compensate each other at some equilibrium distance between molecules. The forces of repulsion practically vanish when the separation of molecules is two or three times larger than the molecular diameter. These forces increase sharply at small distances when electron shells of atoms (see Section 7.5.2) begin to overlap. The forces of attraction decrease with the separation much slower than the forces of repulsion. Dependence of the forces of attraction and repulsion on the separation between the centers of molecules is shown approximately by curves in Figure 14.11.

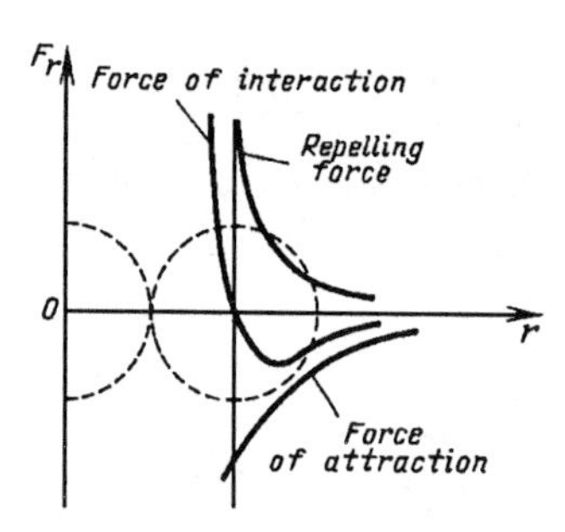

Figure 14.11: Forces of interaction between molecules.

The *relative molecular* (or *atomic*) *mass* M_r (or A_r) of a substance is a ratio of the mass m_0 of a molecule (or an atom) of the substance to

1/12 of the mass m_C of an atom of carbon-12, i.e., the ratio of m_0 to the atomic mass unit:

$$M_r = \frac{m_0}{\frac{1}{12} m_C}. \tag{14.29}$$

It is clear from this definition that M_r is a dimensionless quantity. *The atomic mass unit* (a.m.u.) is 1/12 of the mass of an atom of the isotope carbon-12:

$$1 \text{ a.m.u.} = \frac{1}{12} m_C = 1.6605402 \cdot 10^{-27} \text{ kg}.$$

The mass of one molecule of any substance in kilograms:

$$m_0 = 1.6605402 \cdot 10^{-27} M_r \text{ kg}.$$

14.2.3 Amount of substance

An *amount of substance* is a physical quantity defined as a measure of the number of specified entities (structure elements—atoms, molecules, electrons, etc.) in a substance. The SI unit of an amount of substance is the mole.

A *mole* is the amount of a substance that contains as many specified entities (structure elements) as there are atoms in 0.012 kg of carbon-12. The amount of substance ν of any physical system is determined by a number of moles in the system.

The *molar mass* M of a substance is defined as the mass of one mole of this substance. The *Avogadro number* N_A is the number of molecules (or atoms) in one mole of a substance:

$$N_A = 6.0221367 \cdot 10^{23} \text{ mol}^{-1}.$$

The molar mass M of a substance is equal to the product of the mass m_0 of one molecule and the Avogadro number:

$$M = m_0 N_A.$$

This quantity is expressed in kilograms per mole (kg/mol). The mass m of a system is proportional to the number of moles in it:

$$m = \nu M.$$

14.2.4 Kinetic theory of an ideal gas

An *ideal gas* is a simplest physical model of a real gas. In this model it is possible to neglect the linear size of a molecule and to assume that molecules obey the laws of classical mechanics and interact with each other only during collisions that are considered to be elastic. The pressure of a gas on a wall of the vessel in this model is considered as an averaged result of strokes that molecules produce on the walls. In these strokes some momentum is transferred to the walls by the molecules.

The basic equation of kinetic theory of an ideal gas expresses the pressure p in terms of the mean value of the gas molecules' velocity squared $\langle v^2 \rangle$:

$$p = \frac{1}{3} n m_0 \langle v^2 \rangle, \tag{14.30}$$

where $n = N/V$ is the mean number of molecules per unit volume (*concentration* of molecules). The mean value of the squared velocity $\langle v^2 \rangle$ is determined by the relation

$$\langle v^2 \rangle = \frac{1}{N}({v_1}^2 + {v_2}^2 + \ldots + {v_N}^2). \tag{14.31}$$

If we introduce the mean kinetic energy (see Section 2.4.5) of chaotic heat motion of molecules

$$\langle \varepsilon \rangle = \frac{1}{2} m_0 \langle v^2 \rangle,$$

it is possible to rewrite Equation (14.30) in the following way:

$$p = \frac{2}{3} n \langle \varepsilon \rangle. \tag{14.32}$$

Multiplying both sides of this equality by the volume V of a gas we get

$$pV = \frac{2}{3} N \langle \varepsilon \rangle.$$

If we compare this expression with the Clapeyron–Mendeleev equation, Equation (14.5), and take into account that $N = \nu N_A$, we obtain Equation (14.3):

$$\langle \varepsilon \rangle = \frac{2}{3} kT, \tag{14.33}$$

where $k = R/N_A = 1.380658 \cdot 10^{-23}$ J/K is the Boltzmann's constant.

Equation (14.33) gives the mean value of the energy of chaotic heat motion of molecules for a monoatomic gas. For polyatomic gases, Equation (14.33) gives the mean value of kinetic energy associated only with translational chaotic heat motion of molecules. Equation (14.32) can be rewritten with the help of Equation (14.33) in the following way:

$$p = nkT. \tag{14.34}$$

Internal energy of a mole of a monoatomic ideal gas does not depend on the volume of the gas and is equal to the sum of kinetic energies of all its molecules, or to the product of the number of molecules N_A and the mean energy of one molecule $\langle\varepsilon\rangle$:

$$U = N_A\langle\varepsilon\rangle = \frac{2}{3}kN_AT = \frac{2}{3}RT. \qquad (14.35)$$

The kinetic theory of an ideal gas provides the theoretical foundation for the following experimentally established laws, which describe the properties of real gases:

- *Avogadro's law:* equal volumes of any gas at equal pressure and temperature contain the same number of molecules. In particular, a mole of any gas under normal conditions (T= 273.15 K, p= 101325 Pa) occupies a volume equal to $22.414 \cdot 10^{-3}$ m^3/mol.

- *Dalton's law:* the pressure of a mixture of gases is equal to the sum of their partial pressures (see Sections 14.4.3 and 14.4.4)

14.3 Statistical Distributions

14.3.1 Distributions of different quantities

In the state of thermodynamic equilibrium the macroscopic parameters of a system do not change with time. But the coordinates and momentums of molecules vary continuously due to chaotic heat movement. Nevertheless the complete chaos that characterizes the heat motion of molecules obeys its own laws: in the state of thermodynamic equilibrium any physical system is characterized by definite mean values of different physical quantities and by a definite law of distribution of the values of these quantities of individual molecules. In particular, certain distributions of molecules with respect to the values of their coordinates and velocities exist and these distributions do not vary with time. Knowledge of these distributions permits us to calculate the mean values of corresponding microscopic parameters of a system, i.e., the mean value of squared velocity of a molecule, determined by Equation (14.31).

A similar problem appears in the *mathematical statistics*, where mean values of *random quantities* are to be calculated. Random quantities can assume arbitrary values from a certain region (see Section 11.2). For example, the simultaneous tossing of two dice can give from 1 to 12 pips. In mathematical statistics the law of distribution of random quantities is derived either on the basis of experimental data or on the basis of heuristic considerations (for example, the hypothesis of an equal probability of turning up any side of the dice). In statistical mechanics the laws of distribution of physical quantities are derived on the basis of dynamic equations of motion for the particles of the system under consideration and some statistical hypothesis (for example, the one of equal probabilities for different microscopic states).

For any macroscopic system in thermodynamic equilibrium *the statistical distributions* of physical quantities have the universal form, established by J. Gibbs. In particular, the law of distribution of molecules with respect to any quantity that describes their state (a coordinate, a velocity, etc.) is of an exponential form. The argument of the exponent is the ratio of a typical energy of a molecule to the product kT, i.e., to the quantity, proportional to the mean value of kinetic energy of chaotic heat motion. The ratio must be taken with the sign "minus".

Particular cases of *Gibbs distributions* are the Maxwell distribution of velocities of the ideal gas molecules and the *Boltzmann distribution* of positions of the molecules in space in the presence of any potential field.

14.3.2 Maxwell distribution

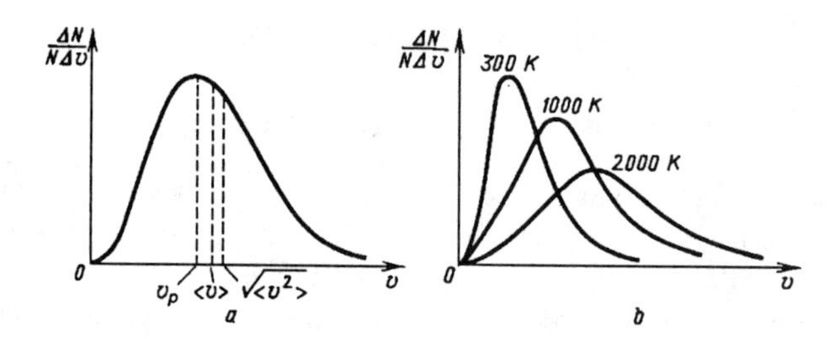

Figure 14.12: Maxwell distribution of speeds among the molecules of a gas (a) and its variation with temperature (b).

Consider the total macroscopic number ΔN of ideal gas molecules for which components v_x, v_y, and v_z of the velocity are in the intervals $(v_x, \ v_x + \Delta v_x)$, $(v_y, \ v_y + \Delta v_y)$, and $(v_z, \ v_z + \Delta v_z)$. The fraction $\Delta N/N$ is given by the following expression:

$$\frac{\Delta N}{N} = \left(\frac{m_0}{2\pi kT}\right)^{3/2} \exp\left(-\frac{m_0({v_x}^2 + {v_y}^2 + {v_z}^2)}{2kT}\right) \Delta v_x \Delta v_y \Delta v_z. \quad (14.36)$$

The fraction $\Delta N/N$ for the ideal gas molecules having the magnitude of velocity in the interval $(v, \ v + \Delta v)$ is given by the expression

$$\frac{\Delta N}{N} = 4\pi\left(\frac{m_0}{2\pi kT}\right)^{3/2} v^2 \exp\left(-\frac{m_0 v^2}{2kT}\right) \Delta v. \quad (14.37)$$

14.3.3 Probabilities

It is possible to treat Equation (14.36) as a probability for an arbitrarily chosen gas molecule to have the components of the velocity in the indicated interval. Similarly Equation (14.37) can be treated as a probability for an arbitrarily chosen molecule to have the magnitude of its velocity in the indicated interval (see Section 11.1).

In statistical mechanics the ***probability*** to find a system in a definite state l is determined as the quantity

$$P(l) = \lim_{q \to \infty} \frac{n(l)}{q}, \tag{14.38}$$

where q is the total number of measurements performed upon the system, and $n(l)$ is the number of cases when the system was found in a state l. In practice, the quantity q should be chosen large enough for the ratio $n(l)/q$ not to vary essentially when q is increased by several times (see Section 11.1.4). It is obvious that $P(l) > 0$ and the sum of probabilities to find the system in one of its possible states is equal to unity (the *normalizing condition*):

$$\sum_l P(l) = 1. \tag{14.39}$$

14.3.4 Calculation of mean values

Let the physical quantity A (for example, an energy or a magnetic moment) of the system in the state l have the value $A(l)$. The mean value (the *mathematical expectation*) $\langle A \rangle$ of the quantity A is determined by the expression (see Section 11.2.2)

$$\langle A \rangle = \sum_l P(l) A(l). \tag{14.40}$$

It is clear that Equation (14.36) corresponds to the definition of probability, Equation (14.38), and satisfies the normalizing condition, Equation (14.39), if the sum $\sum_l$ in it is treated as an integral over all possible values of the velocity components:

$$\left(\frac{m_0}{2\pi kT}\right)^{3/2} \int_{-\infty}^{\infty} \exp\left(-\frac{m_0({v_x}^2 + {v_y}^2 + {v_z}^2)}{2kT}\right) dv_x dv_y dv_z = 1. \tag{14.41}$$

The analogous statement is valid for the expression (14.37):

$$4\pi\left(\frac{m_0}{2\pi kT}\right)^{3/2} \int_0^{\infty} v^2 \exp\left(-\frac{m_0 v^2}{2kT}\right) dv = 1. \tag{14.42}$$

The mean value of the velocity squared $\langle v^2 \rangle$ can be calculated in correspondence with Equations (14.40) and (14.37):

$$\langle v^2 \rangle = 4\pi \left(\frac{m_0}{2\pi kT} \right)^{3/2} \int_0^\infty v^4 \exp\left(-\frac{m_0 v^2}{2kT} \right) dv = \frac{3kT}{m_0}. \tag{14.43}$$

This calculation is in agreement with the definition of Equation (14.31) for $\langle v^2 \rangle$.

The Maxwell distribution makes possible the calculation of mean values of different quantities, depending on the velocity of a molecule. For example, the mean value $\langle v^2 \rangle$ of the magnitude of the molecule's velocity is determined by the expression:

$$\langle v \rangle = \frac{1}{N}(v_1 + v_2 + \ldots + v_N). \tag{14.44}$$

It is equal to

$$\langle v \rangle = 4\pi \left(\frac{m_0}{2\pi kT} \right)^{3/2} \int_0^\infty v^3 \exp\left(-\frac{m_0 v^2}{2kT} \right) dv = \left(\frac{8kT}{\pi m_0} \right)^{1/2}. \tag{14.45}$$

The most probable value v_p of the magnitude of a molecule's velocity corresponds to the maximum (see Section 5.4.3) of the distribution function, Equation (14.37) and Figure 14.12, and is given by the expression:

$$v_p = \left(\frac{2kT}{m_0} \right)^{1/2}. \tag{14.46}$$

The graph of the Maxwell distribution function for molecules' velocities changes its shape with increasing temperature as it is shown in Figure 14.12. The maximum of the curve shifts with temperature to the region of greater values of v. At the same time the maximum value $\Delta N/N$ decreases so that the area under the curve does not change and equals unity in agreement with Equation (14.42).

14.3.5 Boltzmann distribution

The *Boltzmann distribution* is the equilibrium distribution of molecules in space in a potential field:

$$n(\mathbf{r}) = n_0 \exp\left(-\frac{\varepsilon_p(\mathbf{r})}{kT} \right), \tag{14.47}$$

where $\varepsilon_p(\mathbf{r})$ is the potential energy of a molecule at the point with radius-vector $\mathbf{r}$, $n(\mathbf{r})$ is the concentration of molecules at this point, and n_0 is the concentration at the point with the zero value of potential energy. In particular, for a homogeneous field of gravity, where $\varepsilon_p = m_0 g z$ (m_0—the mass of a molecule), we have

$$n(z) = n_0 \exp\left(-\frac{m_0 g z}{kT}\right). \tag{14.48}$$

14.3.6 Fluctuations

Thermodynamic equilibrium is always a dynamic equilibrium because thermal motion of atoms and molecules never terminates. In such a state macroscopic parameters of a system are not strictly constant, but perform small chaotic oscillations near some mean values. *Fluctuations* are the chaotic deviations of physical quantities from their mean values, caused by the thermal motion of the particles of a system.

Main laws of fluctuations can be illustrated by the following example of spatial distribution of an ideal gas molecules inside a vessel at the equilibrium state. Gas fills the whole volume uniformly only on the average, but at any instant this assumption is correct to some extent, because the molecules inevitably wander from one part of the vessel to another.

The probability $P(n)$ for n molecules of the total number N to be located in a definite half of a vessel can be easily found assuming that the probability for one molecule to be in this part does not depend on the location of other molecules at the same instant. The probability $P(n)$ is the ratio of the number of possible distributions for which any n molecules are located in the chosen part of the vessel, to the total number of possible distributions of all N molecules in the two parts of a vessel. The latter equals 2^N. The number of such distributions is equal to the number of ways to choose n molecules from their total number N, i.e., to the number C_N^n of possible combinations of n elements from N ones (or to the binomial coefficient C_N^n, see Section 6.1.6):

$$C_N^n = \frac{N!}{n!(N-n)!}. \tag{14.49}$$

Therefore, the probability $P(n)$ for any n molecules to be in a half-part of a vessel is given by the expression

$$P(n) = \frac{N!}{2^N n!(N-n)!}. \tag{14.50}$$

This distribution is called *binomial distribution* (see Section 11.1.3). The distribution of probabilities for a gas of $N=$ 20 molecules is shown in Figure 14.13*a*. This distribution can be characterized by a smooth curve (the envelope). The curve becomes more and more sharp with the increasing of the total number of molecules and turns into the Gaussian distribution (see Section 11.1.3). This can be seen in Figure 14.13*b* for $N=$ 100. Relative fluctuation the smaller, the greater the size of a system (relative fluctuation is of the order of $1/\sqrt{N}$, see Section 11.2.2).

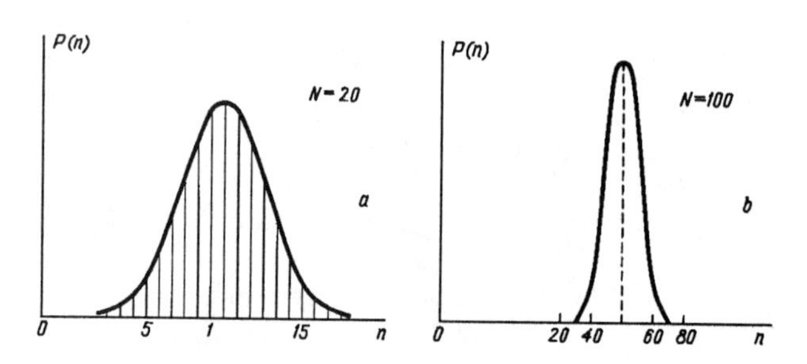

Figure 14.13: The probability distribution for gases with $N = 20$ (*a*) and $N = 100$ (*b*).

14.3.7 Physical reasons for irreversibility

The statistical approach discovers the reasons of irreversibility of real processes and the reasons for a definite direction of energy transformations in nature. In an isolated system all processes develop in the direction corresponding to the change of less probable states of the system with more probable states. An *irreversible process* is a process with very small probability of the corresponding reverse process.

14.3.8 Statistical meaning of enthropy

The *thermodynamic probability* or the *statistical weight* of certain macroscopic state is determined to be the number of different microscopic states that realize the macroscopic state under consideration. Let the macroscopic state be determined by a number n of molecules that are in a certain half of a vessel. Its statistical weight $W(n)$ is the number of modes of distribution of N molecules, Equation (14.49), that realize the given macroscopic state.

In statistical mechanics the *entropy* S of a system is defined as the following quantity proportional to the statistical weight W of its state:

$$S = k \ln W, \tag{14.51}$$

where $k = 1.380658 \cdot 10^{-23}$ J/K is the Boltzmann's constant (see Section 14.2.4). Determined in such a way, the entropy of a system coincides with the entropy introduced in thermodynamics (see Section 14.1.14). In isolated systems the entropy never decreases: $\Delta S = 0$ for reversible processes and $\Delta S > 0$ for irreversible processes.

Example

An ideal gas occupies one half of a thermoinsulated vessel, which is divided into two equal parts by a partition. Let us determine the change of the entropy of the system when the partition is suddenly removed and the gas fills the whole vessel (see Section 14.1.14).

The statistical weight of the initial state is equal to unity: $W_1 = 1$ because there is only one microscopic state when all molecules are in one half of the vessel. In the final state all molecules are distributed in the vessel almost homogeneously: the numbers of molecules in both half-parts of the vessel are equal if we neglect fluctuations. Due to the extremely sharp shape of the curve of the probability of various molecular distributions among both parts of the vessel, it is admissible to consider that all other $(2^N - 1)$ microscopic states of the system correspond to the final macroscopic state:

$$W_2 \approx 2^N - 1 \approx 2^N .$$

Hence, according to Equation (14.51),

$$\Delta S = k(\ln W_2 - \ln W_1) = kN \ln 2.$$

Exact solution needs the use of Equation (14.49) for $n = N/2$:

$$W_2 = C_N^{N/2} = \frac{N!}{[(N/2)!]^2},$$

and the Stirling formula (see Section 6.1.3):

$$n! \approx (2\pi n)^{1/2}(n/e)^n .$$

Thus we get the following result:

$$\ln W_2 \approx N \ln 2 - \frac{1}{2} \ln N.$$

Since $N \approx 10^{23}$, we get $\ln N \approx 50$, and the former value of the entropy is valid (see Section 14.1.14):

$$\Delta S = kN \ln 2.$$

14.4 Real Gases

14.4.1 Van der Vaals equation

Corrections connected with finite sizes of molecules and with interactions between them give rise to approximate *equations of state* for real gases. *Van der Vaals equation* is one of such equations. For one mole of a substance it is written in the following form:

$$\left(p + \frac{a}{V_\mu^2}\right)(V_\mu - b) = RT. \tag{14.52}$$

Phenomenological constants a and b in Equation (14.52) are determined from the requirement to fit experimental data as best as possible, that is, to provide the best descriptions of the gas behavior by the equation.

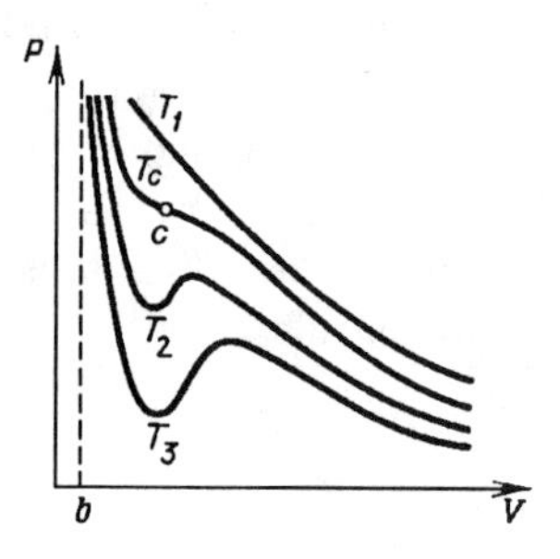

Figure 14.14: Van der Vaals isotherms.

Van der Vaals isotherms corresponding to different values of temperature ($T_1 > T_C > T_2 > T_3$) are shown in Figure 14.14. A certain temperature T_c exists which corresponds to a curve that separates the monotone isotherms lying above the curve from the "humpbacked" ones lying below. Such temperature and a corresponding curve are called *critical* (see Section 14.4.3). The critical isotherm has one point of inflection C (see Section 5.4.4).

14.4.2 Experimental isotherms and phase transitions

An isotherm for carbonic acid gas CO_2, obtained experimentally for a temperature smaller than the critical one is shown in Figure 14.15; the second curve in Figure 14.15 is the Van der Vaals isotherm for the same temperature. On the coinciding part AB of both isotherms corresponding to large values of volume and low values of pressure, behavior of the carbonic acid gas is very similar to that of an ideal gas. If we compress the gas at the same constant temperature further, on the next part BC of the compression region liquid carbonic acid appears, whose amount increases as the volume is decreased further until all the gas condenses at the point C. The part BC corresponds to the equilibrium between liquid carbonic acid and its saturated vapor. At a given temperature T some definite pressure p corresponds to this equilibrium. The dependence $p(T)$ on the pT-diagram determines *the equilibrium curve* for liquid and gaseous phases of the substance. The further compression is accompanied by rapid growth of pressure at a negligible decrease of the volume. It corresponds to small *compressibility* of liquids.

The rectilinear part BC of experimental isotherms becomes shorter with the increase of temperature, and at critical temperature T_c shrinks

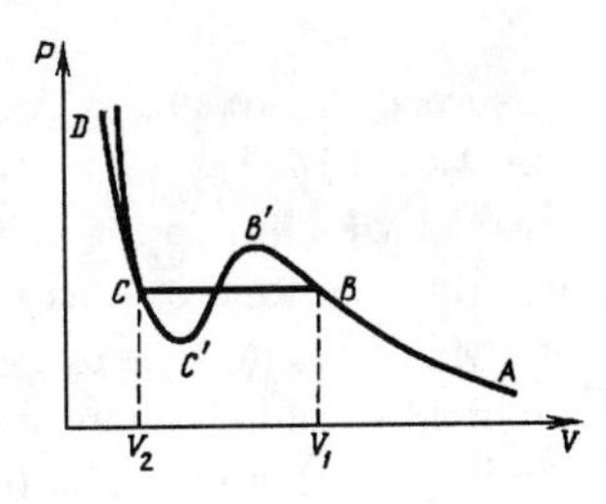

Figure 14.15: The carbonic acid gas isotherms.

into one point C (Figure 14.14). This means that the existence of the liquid phase at $T > T_c$ is impossible even at very high values of pressure. The pressure p_c and the volume V_c corresponding to the point C are called *critical*. The state of substance at point C is called the *critical state*.

The parts BB' and CC' of the van der Vaals isotherm can be realized in experiment. The first of them corresponds to oversaturated vapor and can be realized at a gradual compression of a gas in a volume free from dust and ions. At the appearance of *condensation centers* (dust particles or ions) the oversaturated vapor instantly turns into a mist. Oversaturated vapor can be obtained not only by isothermal compression, but also by cooling of a saturated vapor. It is sometimes called a supercooled vapor. The part CC' corresponds to a *superheated* liquid. It can be realized by heating a liquid at constant pressure in conditions that exclude the rise of boiling—a liquid itself and the walls of a vessel should not contain dissolved gases. The part $B'C'$ corresponds to absolutely unstable states of the substance that cannot be realized experimentally.

Thus, the empirical van der Vaals equation obtained with the aim of introducing small corrections to the equation of state of an ideal gas turned to be effective in a more wide region. It predicts the existence of the critical temperature and of the superheated liquid and explains qualitatively the small compressibility of liquids.

A gas can be condensed into the liquid state by compression without cooling only at a temperature smaller than the critical temperature T_c. A substance in a gaseous state at a temperature lower than the critical one is called a *vapor*.

14.4.3 Phase transitions

Vaporization. The presence of other gases or vapors over the liquid surface does not influence the process of vaporization of the substance. The pressure of its own vapor over the surface is called *partial*; its maximal value is equal to the pressure of the saturated vapor of the corresponding liquid.

Dalton's law: the pressure in a mixture of gases is equal to the sum of partial pressures (see Section 14.2.4).

Evaporation is a process in which a vapor is created by the escape of molecules with sufficient values of kinetic energy from the surface of a liquid or a rigid body. The value of kinetic energy should be sufficient for overcoming the attraction of surrounding molecules. The vapor that is in dynamic equilibrium with its liquid is called *saturated*. Partial pressure of the vapor that is in the state of dynamic equilibrium with the liquid is called the *pressure of saturated vapor*. This pressure does not depend on the volume occupied by the vapor. The pressure of saturated vapor increases with temperature.

Boiling is a process of vaporization taking place inside a liquid into growing vapor bubbles when the pressure of the saturated vapor in the bubbles becomes equal to the pressure in the liquid. The temperature at which boiling occurs is called the *boiling point*. Its value depends on the external atmospheric pressure.

Critical temperature is the temperature corresponding to the disappearance of a difference in physical properties between a liquid and its saturated vapor. Typical dependence of densities of a liquid and its saturated vapor on temperature is shown in Figure 14.16. The curves merge at the critical temperature $T = T_c$.

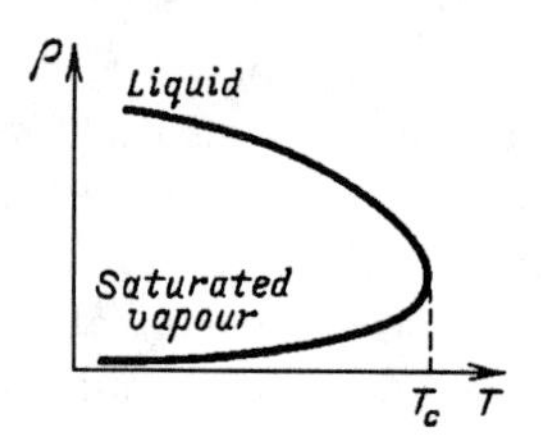

Figure 14.16: Temperature dependence of the density.

14.4.4 Humidity of air

Absolute humidity is the partial pressure of the water vapor, i.e., the pressure of the water vapor if all other gases are absent. *Relative humidity* is the ratio (expressed as a percentage) of partial pressure of the water vapor, contained in the air at a given temperature, to the pressure p_0 of the saturated vapor at the same temperature:

$$\varphi = \frac{p}{p_0} \cdot 100\%. \tag{14.53}$$

The *condensation point* is the temperature that corresponds to saturation of vapor contained in the air. The condensation point gives an opportunity to determine the partial pressure of the water vapor and relative humidity by using the tables of the saturated vapor pressure at different temperatures (Table 14.1).

Table 14.1: Pressure and density of saturated water vapor at different temperatures

t, °C	p		ρ_{max} 10^{-3} kg/m^3	t, °C	p		ρ_{max} 10^{-3} kg/m^3
	kPa	mm of mercury			kPa	mm of mercury	
-5	0.401	3.01	3.25	12	1.401	10.51	10.67
-4	0.437	3.28	3.53	13	1.497	11.23	11.36
-3	0.463	3.47	3.83	14	1.597	11.98	12.08
-2	0.517	3.88	4.14	15	1.704	12.78	12.84
-1	0.563	4.22	4.49	16	1.817	13.63	13.65
0	0.611	4.58	4.85	17	1.937	14.53	14.50
1	0.656	4.92	5.20	18	2.062	15.47	15.39
2	0.705	5.59	5.57	19	2.196	16.47	16.32
3	0.757	5.68	5.95	20	2.377	17.53	17.32
4	0.813	6.10	6.37	21	2.486	18.65	18.35
5	0.872	6.54	6.80	22	2.642	19.82	19.44
6	0.935	7.01	7.27	23	2.809	21.07	20.60
7	1.005	7.54	7.79	24	2.984	22.38	21.81
8	1.072	8.04	8.28	25	3.168	23.76	23.07
9	1.148	8.61	8.83	26	3.361	25.21	24.40
10	1.227	9.20	9.41	27	3.565	26.74	25.79
11	1.312	9.84	10.02	28	3.780	28.35	27.26

14.4.5 Equilibrium of phases

A *triple point* is a point where the three curves of phase equilibrium meet on a pT-diagram:

1. the plot of boiling temperature as a function of pressure (the curve of liquid – vapor equilibrium);

2. the plot of fusion temperature as a function of pressure (the curve of liquid – solid equilibrium);

3. the plot of vapor pressure upon the solid phase as a function of temperature (the curve of equilibrium of gaseous and solid phases).

At the triple point the three phases of substance (vapor, liquid, and solid) are in equilibrium.

Figure 14.17 demonstrates these curves for H_2O. For water the triple point occurs at T = 273.16 K (0.01 °C), p = 610.6 Pa (4.58 mm of mercury). The triple point of water is used for the determination of the basic unit of thermodynamic temperature in SI *kelvin:* kelvin is equal to 1/273.16 part of the temperature of the triple point.

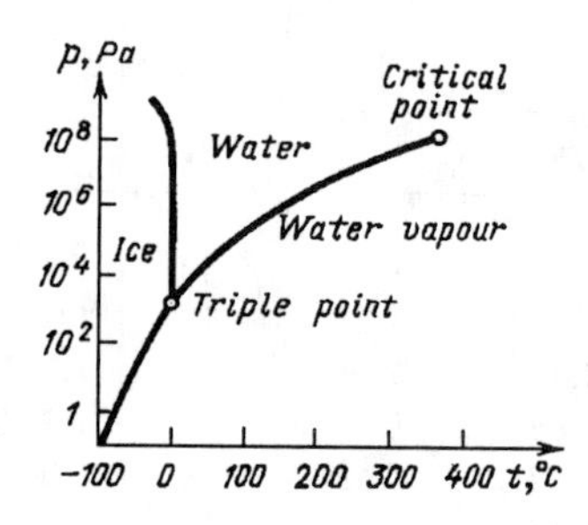

Figure 14.17: Phase equilibrium.

14.5 Liquids

14.5.1 Surface tension

Surface energy is a surplus of potential energy of molecules in a surface layer compared with the energy of molecules inside the phases (far from the boundaries), caused by the difference between intermolecular interactions in these phases. For instance, a molecule in the interior of a liquid experiences interactions with other molecules equally from all sides, whereas a molecule at the surface is only affected by molecules below in the liquid. The force of surface tension acts along the surface of liquid perpendicularly to the line that restricts the surface. The *surface tension* σ is defined as the force that acts on a unit length of the line, i.e., it is the ratio of the magnitude F of the force of surface tension acting on a boundary of the surface layer to the length l of the boundary:

$$\sigma = \frac{F}{l}. \tag{14.54}$$

The surface tension is measured in N/m. The surface tension σ can equally be defined as the energy required to increase the surface area isothermally by the unit of area (1 m^2), or, similarly, as the ratio of the surface energy to the area of the surface, and therefore it can be measured in J/m^2 (which is equivalent to N/m).

Due to the surface tension, a liquid accepts the spherical form in the absence of external forces. The sphere has the minimal surface area at a given volume, and hence it provides the minimal value of the surface energy. The property of surface tension is responsible also for the formation of liquid drops, bubbles, and meniscuses, as well as for the rise of liquids in capillary tubes, and the absorption of liquids by porous substances.

Values of surface tension for some liquids are presented in Table 14.2.

Table 14.2: Surface tension for some liquids

(in 10^{-2} N/m at 20° C)

Acetone	2.37	Gasoline	2.89	Turpentine	2.7
Alcohol	2.23	Glycerin	6.4	Water	7.27
Chloroform	2.72	Kerosene	2.6		

14.5.2 Capillary phenomena

Wetting is a phenomenon that arises as a result of the interaction of the molecules of a liquid with one another (the forces of *cohesion*), and by virtue of their interaction with molecules of an adjoining solid (the forces of *adhesion*). This phenomenon is responsible for the bend of the surface of a liquid near the surface of a solid. The liquid surface lines itself up perpendicularly to the resultant of these two forces. Three different cases are possible. In the first, the forces of adhesion are strong enough to produce concave upward curvature of the liquid on the solid surface. To describe this situation, we say that the liquid is "wetting" the surface. In the second case the force of adhesion is small and the resultant force on the liquid at the junction remains directed back into the liquid. The resulting curvature of the liquid surface is convex, away from the solid container. In the third case the resultant of the forces of adhesion and cohesion is parallel to the wall and hence the liquid surface is perpendicular to the wall.

In the first case the angle θ between the plane tangential to the surface of the liquid (see Section 9.7) and the surface of adjacent solid (wetting angle) is acute (Figure 14.18*a*), in the second case this angle is obtuse (Figure 14.18*b*), in the third case $\theta = \pi/2$. The value $\theta = 0$ corresponds to a total wetting, $\theta = \pi/2$ to a total non-wetting.

The curved surface of a liquid in a cylindrical tube is called a *meniscus*. In order to establish a meniscus with a curvature so that the surface is perpendicular to the resultant of adhesion and cohesion forces, the

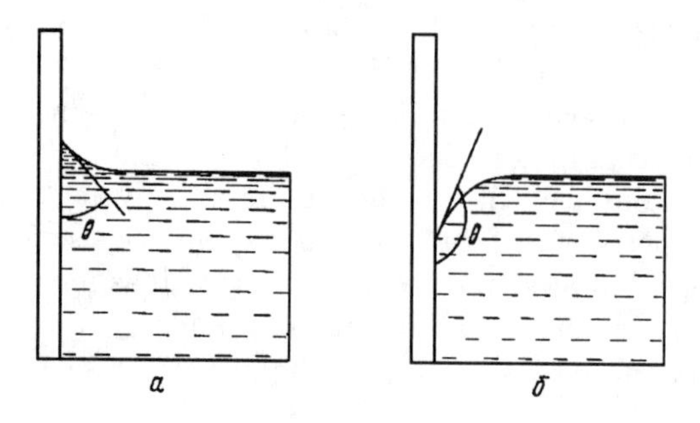

Figure 14.18: Wetting angles.

edge of the liquid must rise or fall some distance along the solid surface (Figure 14.19). Hence an additional *capillary pressure* Δp appears that is connected with the mean radius of curvature r of the surface (see Section 9.7) by the Laplace equation:

$$\Delta p = p_1 - p_2 = \frac{2\sigma}{r}, \tag{14.55}$$

where p_1 is a pressure inside a liquid and p_2 is a pressure in the phase which is in contact with the liquid.

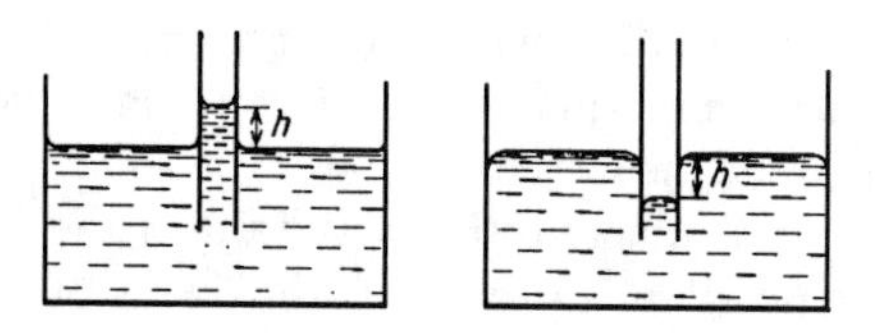

Figure 14.19: Liquids in capillars.

For a spherical surface of the radius R, the mean radius of curvature $r = R$, for a cylindrical surface $r = 2R$. In the case of a concave surface $r < 0$, $p_1 < p_2$, and $\Delta p < 0$; in the case of a convex surface $r > 0$, $p_1 > p_2$, and $\Delta p > 0$. The height of a rise (or a fall) of a liquid in a cylindrical tube is determined by the condition of a balance between the capillary pressure and the hydrostatic pressure of a liquid column due to the difference of levels in the vessel and in the tube:

$$\Delta p = \rho g h, \tag{14.56}$$

where $h < 0$ (the level of the liquid is higher in the tube) for a wetting liquid ($r < 0$) and $h > 0$ (the level of the liquid is higher in the vessel) for a non-wetting liquid ($r > 0$).

At full wetting the height of the column of liquid in the capillary tube of the radius R is determined by the condition $r = R$:

$$h = \frac{2\sigma}{\rho g R}.$$

This height can also be determined if we write down the condition of balance between the downward force of the weight of the column and the upward capillary force, acting along the boundary of the surface layer.

14.6 Solids

14.6.1 Crystals and amorphous bodies

There are two kinds of solids—crystalline and amorphous bodies. *Crystals* are solids with a certain ordered distribution of their atoms and molecules in space. In such solids, atoms and molecules form a *crystalline lattice* (see Section 8.4.3). Crystals are *anisotropic* bodies, their physical properties being dependent on a direction inside a crystal.

Two types of crystals are distinguished—monocrystals and polycrystals. *Monocrystals* are large single crystals, *polycrystals* consist of a lot of separate irregularly oriented small (but macroscopic) crystalline granules. Monocrystals occur in nature to be of different size—from large (up to hundreds of kilograms) quartz crystals (rock crystal), fieldspar, and other minerals, to tiny diamond crystals. Technical metals, alloys, and many natural rocks are polycrystals.

The violations of strict regularity of atomic (or molecular) positions in a crystalline lattice are called *defects in crystals*. Imperfections can be of two types—point defects and line defects. Line defects are called *dislocations*. Imperfections in crystals influence their physical properties and determine their firmness and strength.

Amorphous substances are solids without a strict order in disposition of atoms or molecules. An analogous situation is typical for liquids. Only nearest neighbors are arranged in an order which is called the *short-range order,* in contrast to the *long-range order* in crystals. Amorphous substances are macroscopically *isotropic*—their physical properties do not depend on a direction in space.

14.6.2 Elastic deformations

Deformations that disappear after removal of applied external forces are called *elastic deformations*. Deformations that do not disappear entirely when the applied stress is removed are called *plastic deformations*.

The deformation of *longitudinal extension* (or *contraction*) of a rod subjected to *tensile stress* (or *compressive stress*) is characterized by the *linear strain* (or *tensile strain*), which is the ratio of the change in length to the original length of the rod: $\varepsilon = \Delta l / l_0$. Here l_0 is the original length and l is the final length of the rod. *Hooke's law* is valid in the case of an elastic deformation. It states that linear strain is proportional to the stress:

$$\frac{|\Delta l|}{l_0} = \frac{1}{E}\frac{F}{S}, \quad or \quad \sigma = E|\varepsilon|, \tag{14.57}$$

where $\sigma = F/S$ is the *stress* in the rod, equal to the ratio of the applied force to the area S of the cross-section. The *Young's modulus of elasticity* E (the modulus of longitudinal elasticity) characterizes elastic properties of material.

Hooke's law can be written in a form

$$F = k|\Delta l|,$$

where k is the *rigidity* of a rod connected with the Young's modulus and the sizes of a rod by the relation

$$k = SE/l_0.$$

Mechanical tension in the longitudinal direction causes not only the lengthening, but also a *lateral contraction* of a rod; the following relation is valid:

$$(\Delta d)/d = -\mu(\Delta l)/l, \quad \text{or} \quad \varepsilon_{tr} = -\mu\varepsilon, \tag{14.58}$$

where μ is the *Poisson's ratio*, d is the transverse size of the rod (diameter, thickness of a rod, etc.), Δd is the change of the transverse size, and ε_{tr} is the relative change of the transverse size of the rod.

The Young's modulus and the Poisson's ratio characterize entirely the elastic properties of an isotropic material; they do not depend on the size and shape of a sample. All other elastic constants can be expressed in terms of E and μ.

The relative change of the volume $\Delta V/V$ at *hydrostatic compression* of a sample is proportional to the pressure p:

$$\frac{\Delta V}{V} = -\frac{1}{K}p, \tag{14.59}$$

where K is the *bulk modulus*. It can be expressed in terms of E and μ by the following relation:

$$K = \frac{E}{3(1-2\mu)}. \tag{14.60}$$

Equation (14.59) is valid for samples of arbitrary shape. The values of elastic constants for some materials are represented in Table 14.3.

Table 14.3: The values of elastic constants for some substances

Substance	Young's modulus (10^{10} Pa)	Bulk modulus (10^{10} Pa)	Poisson's ratio
Aluminum	7.0	7.5	0.34
Bronze	10.5	9.0	0.36
Copper	11.8	13.5	0.34
Cronglass	6.0	5.0	0.25
Gold	8.0	16.5	0.42
Iron	21.0	16.0	0.28
Silver	7.7	10.5	0.37
Steel	22.0	16.0	0.28
Zinc	8.0	3.5	0.3
Cast iron	11.0	9.5	0.27
Brass	9.0	6.0	0.36
Tin	5.3	5.3	0.33

Shear deformation is shown in Figure 14.20. A shear deformation occurs when tangential forces of opposite directions act on parallel surfaces of a sample. Shear deformation is characterized by a shear angle γ measured in radians. In the case of elastic deformation the angle γ is proportional to the shear stress F/S:

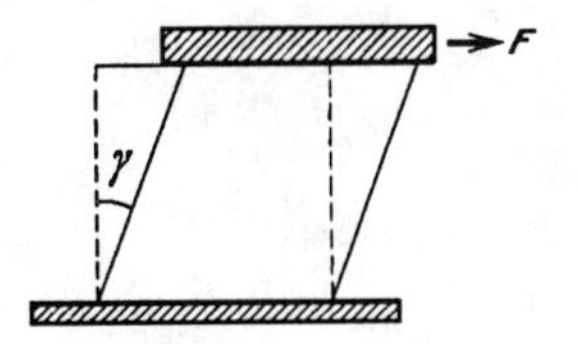

Figure 14.20: Shear deformation.

$$\gamma = \frac{1}{G}\frac{F}{S}, \quad \text{or} \quad \tau = G\gamma, \tag{14.61}$$

where F is a force parallel to the plane S, S is the area of the surface, τ is the tangential stress, and G is the *shear modulus*. The shear modulus is expressed in terms of the Young's modulus E and the Poisson's ratio

μ by the relation

$$G = \frac{E}{2(1+\mu)}. \tag{14.62}$$

Deformation of *bending* can be reduced to an inhomogeneous linear strain or compression of different values in various parts of the sample (Figure 14.21).

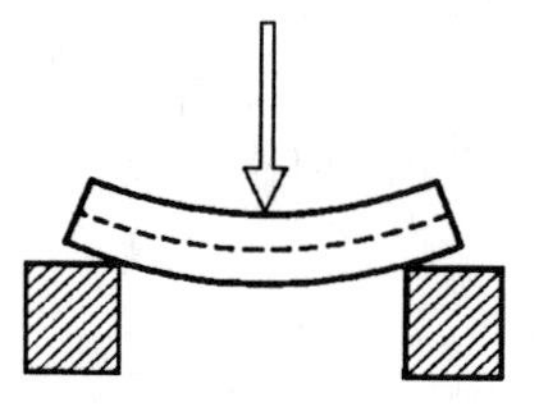

Figure 14.21: Deformation of bending.

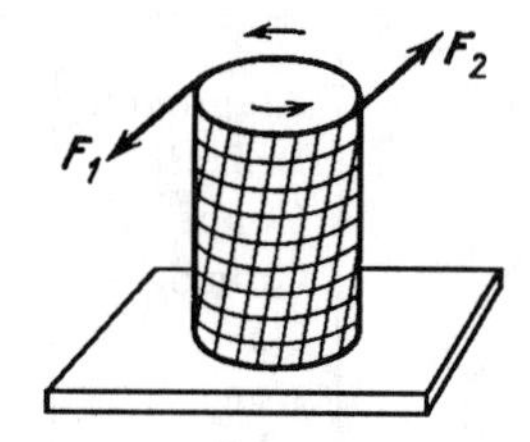

Figure 14.22: Deformation of twisting.

The deformation of *twisting* is an inhomogeneous shift: all the layers of a sample stay parallel to each other, but they rotate through some angle relatively to each other (Figure 14.22). A torque M that is required for twisting a cylindrical rod through some angle φ is proportional to the angle:

$$M = D\varphi,$$

where D is the *torsion modulus* that can be expressed in terms of the shift modulus G, the length l, and the radius R of the cross-section of a rod:

$$D = \frac{\pi G}{2l} R^4.$$

14.6.3 Thermal expansion

At heating solids *expand* in all directions. Rods and wires mainly increase their length, but relative change of the length $\Delta l/l$ does not depend on direction (the geometric similarity survives during heating). *Linear expansion* (see Section 2.6.1) is described by the following formula:

$$l_2 = l_1(1 + \alpha\Delta t), \tag{14.63}$$

where l_1 is the length of the rod at a temperature t_1, l_2 is the length at a temperature t_2, and $\Delta t = t_2 - t_1$, α is the *linear expansivity*. *Volume expansion* can be considered as a superposition of linear expansions in three orthogonal directions:

$$V_2 = V_1(1 + \beta \Delta t), \tag{14.64}$$

where V_1 is the volume of a specimen at a temperature t_1, V_2 is the volume at a temperature t_2, and β is the *volume expansivity:*

$$\beta = 3\alpha.$$

Linear and volume expansivities of solids weakly depend on temperature (Table 14.4).

Table 14.4: Volume expansivities of some substances

Substance	Temperature, K	Volume expansivity 10^{-6} K−1
Aluminum	25	1.5
	293	69
	600	85
Carbon (Graphite)	100	0.15
	293	3.0
	600	9.0
Iron	25	0,6
	293	3.6
	600	45
Copper	25	1.8
	293	50
	600	57
Lead	25	43
	293	86
	500	96
Mercury (solid)	100	111
Mercury (liquid)	273–573	180
Melted quartz	23	-2
(non-crystallyne SiO_2)	293	1.2
	600	1.8
Water (solid)	73	1.8
	173	93
	273	167
Water (liquid)	277–373	430

Liquids expand at heating much larger than solids do. The following expression is valid:

$$V_2 = V_1(1 + \beta \Delta t). \tag{14.65}$$

Bulk expansivity of liquids β weakly depends on temperature. Usual water is an exclusion from this rule. Volume expansivity of water strongly depends on temperature and has a negative value in the interval from 0° C up to 4° C.

Volume and density of a liquid are inversely proportional:

$$\rho_2/\rho_1 = V_1/V_2,$$

hence it is valid for an expanding liquid:

$$\rho_2 = \frac{\rho_1}{1+\beta\Delta t} \approx \rho_1(1-\beta\Delta t). \quad (14.66)$$

14.7 Heat Exchange. Phase Transitions

14.7.1 Thermal capacity

The amount of heat required for heating the sample is proportional to its mass m and to the temperature change Δt:

$$Q = cm\Delta t. \quad (14.67)$$

Here $c = C/m$ is the specific heat capacity (C is the heat capacity of the whole specimen of the substance; see Section 14.1.8). Due to relatively low values of volume expansivities of liquids and solids (in contrast to gases) their specific heat capacities at constant volume c_V and at constant pressure c_p are nearly equal.

Values of specific heat capacity for some substances are presented in Table 14.5.

Table 14.5: Specific heat capacity of some substances

Solids				Liquids	
Aluminum	0.896	Copper	0.385	Acetone	2.160
Wolfram	0.134	Nickel	0.448	Water	4.190
Iron	0.465	Tin	0.218	Glycerin	2.390
Gold	0.130	Lead	0.13	Mercury	0.138
Ice (0° C)	2.090	Zinc	0.389	Alcohol	2.390

During the process of establishing of thermal equilibrium between two substances that have different temperatures, an exchange of heat between these substances takes place. The *equation of heat balance* valid for equalizing their temperatures is written in the form:

$$c_1 m_1(t_1 - t) = c_2 m_2(t - t_2), \quad t_1 > t_2, \quad (14.68)$$

where t is an established common temperature for both substances.

14.7.2 Heat of combustion

The energy liberated in the burning (or oxidization) of a fuel and transferred as a heat to surroundings is proportional to the mass m of completely burned fuel:

$$Q = Hm, \tag{14.69}$$

where H is called the *specific heat of combustion*.

14.7.3 Latent heat of phase transitions

Under standard conditions of constant pressure the phase transition of a pure substance from solid state into a liquid (fusion or melting of a crystal) and backwards (crystallization) occurs at a definite temperature t_m. This temperature is called the *melting point* or the *fusion point* of the substance. The melting point depends on pressure. During phase transition from solid state to liquid at a definite temperature a certain quantity of heat Q is absorbed and the same quantity of heat is released at the backward phase transition:

$$Q = qm, \tag{14.70}$$

where q is the *specific latent heat of fusion*, and m is the mass of melted (or crystallizated) substance.

Under a given external pressure a pure liquid always boils at the same temperature t_b, which remains constant until all the liquid is evaporated. This temperature is called the *boiling point* of the liquid under the given pressure. The boiling point of a liquid depends strongly on the external atmospheric pressure. For standard atmospheric pressure (76 cm of mercury) the boiling point is called the *normal* boiling point. This temperature depends on the liquid. A definite quantity of heat Q is absorbed during the phase transition from liquid into gas (at evaporation) and an equal quantity of heat is released during the backward transition (condensation) under the same external pressure:

$$Q = \lambda m, \tag{14.71}$$

where λ is the *latent heat of vaporization*, and m is the mass of the vaporized liquid.

Characteristics of phase transition for some substances are presented in Table 14.6.

Example

Some amount of water vapor at the temperature $t_v = 100^\circ$ C is let into an ideal calorimeter, containing an ice sample at the temperature 0° C. The whole ice sample melts and the temperature t is established.

Table 14.6: Characteristics of phase transitions for some substances

Substance	Melting point, °C	Latent heat of fusion, kJ/kg	Boiling point, °C	Latent heat of vapori-zation, kJ/kg
Alcohol	-115	108	78	840
Aluminum	660	1060	2500	12000
Copper	1083	205	2590	4790
Gold	1063	66	2700	1650
Hydrogen	-259	59	-253	454
Iron	1535	277	2735	6340
Lead	327	23	1750	8600
Magnesium	650	1090	1095	6000
Mercury	-39	12	357	275
Nickel	1453	303	2800	6480
Plutonium	640	125	3350	1430
Silicon	1415	2760	3250	16000
Titan	1670	1500	3300	10000
Uranium	1130	200	1200	2250
Water	0	334	100	2256
Zinc	420	111	907	1755
Tin	232	60	2430	2450

Determine the mass m_v of the vapor if the mass of the ice sample is equal to m.

In this case the equation of heat balance is written in the form

$$qm + cmt = \lambda m_n + cm_n(t_v - t),$$

hence

$$m_v = \frac{q + ct}{\lambda + c(t_v - t)} m,$$

where c is the specific heat capacity of water.

Chapter 15

Electricity and Magnetism

Electromagnetic interaction is the most prevalent in the surrounding world and manifests itself in many various natural phenomena. The reason is that all bodies are built of electrically charged particles, and electromagnetic interaction among them is much stronger than the gravitational one. Electromagnetic interaction is long-range in contrast to the short-range nuclear (strong) interaction. Electromagnetic interaction determines the structure and the properties of matter—atoms, molecules, crystals, liquids, etc.

15.1 Electrostatics

15.1.1 Interaction of electric charges

There exist two kinds of *electric charges*, positive and negative ones. When glass is rubbed with silk, the glass becomes *positively* charged while the silk is negative. The name *negative* is assigned to the charge that appears on a hard rubber when it is rubbed with wool or hair.

Charge is a fundamental property of some elementary particles. Electric charge gives rise to an electromagnetic interaction between particles and to macroscopic material phenomena described as *electrical.*

The carrier of a negative charge is an *electron*. The carrier of a positive charge is a *proton*.

The magnitudes of electron and proton charges are equal to the *elementary charge* $e = 1.602,177,33 \cdot 10^{-19}$ C. It is the natural unit of electric charge.

The main property of an electric charge is expressed by the law of *charge conservation*: the algebraic sum of all the charges in any isolated system is constant. It is one of the fundamental laws of nature.

Electric charges interact one with another. Separated charges of the same kind repel, opposite charges attract one another.

The interaction of stationary electric charges is governed by *Coulomb's law*. This law gives the dependence of the influence between two electric point charges in vacuum on the magnitudes $|q_1|$, $|q_2|$ and their separation r:

$$F = \frac{1}{4\pi\varepsilon_0}\frac{|q_1||q_2|}{r^2}, \quad \text{(in SI)} \tag{15.1}$$

where $\varepsilon_0 = 8.854187817 \cdot 10^{-12}$ F/m is the *electric constant*, also called the absolute permittivity of free space (see Section 12.2.6).

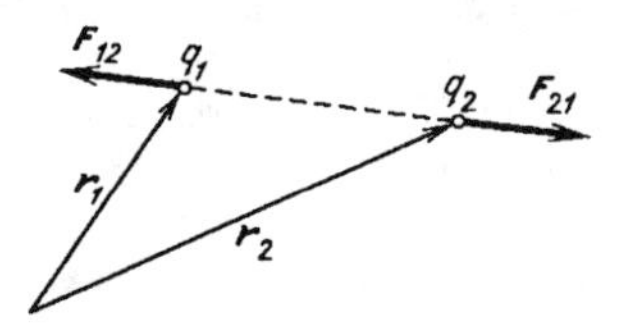

Figure 15.1: Interaction of charges of the same kind.

Submerged in a medium, electric charges act one on another with the force

$$F = \frac{1}{4\pi\varepsilon}\frac{|q_1||q_2|}{r^2} = \frac{1}{4\pi\varepsilon_0\varepsilon_r}\frac{|q_1||q_2|}{r^2}, \tag{15.2}$$

where ε is the absolute permittivity of the intervening medium: $\varepsilon = \varepsilon_r\varepsilon_0$. The factor ε_r is the relative permittivity or the *dielectric constant* of the material.

The law of charge interaction can be expressed in a vector form:

$$\mathbf{F}_{12} = -\mathbf{F}_{21} = \frac{1}{4\pi\varepsilon}\frac{q_1 q_2}{|\mathbf{r}_1 - \mathbf{r}_2|^3}(\mathbf{r}_1 - \mathbf{r}_2), \tag{15.3}$$

where $\mathbf{r}_1$ and $\mathbf{r}_2$ are the radius-vectors of the charges q_1 and q_2. Interaction between charges of the same kind is shown in Figure 15.1.

15.1.2 Electrostatic field

Electric charges provide the surrounding space with peculiar properties—they produce an *electric field*. The field of stationary charges that do not vary in time is called *electrostatic field*.

Field strength or *electric intensity* at any point in an electric field is the vector quantity defined to be the force per a unit positive test charge at that point:

$$\mathbf{E} = \frac{\mathbf{F}}{q}. \tag{15.4}$$

The *electric potential* at a point is the scalar quantity defined to be the work per unit positive test charge, which is done by the electric field while the charge is moved from the point where the potential is being specified, to another point where potential is ascribed to be zero:

$$\varphi = \frac{A}{q}. \tag{15.5}$$

The difference of the electric potential between two points is called the *voltage*:

$$U = \varphi_1 - \varphi_2 = -\Delta\varphi.$$

The amount of *work* done by the field force while the electric charge is moved depends on the positions of the initial and final points, but not on the form of the trajectory:

$$A_{1\to 2} = q(\varphi_1 - \varphi_2) = qU. \tag{15.6}$$

An electric field satisfies the *principle of superposition*. The individual contributions of different charges must be added as vectors to produce the net field strength at the point:

$$\mathbf{E} = \mathbf{E}_1 + \mathbf{E}_2 + \ldots + \mathbf{E}_n. \tag{15.7}$$

Resultant potential at any point in the electric field produced by charges $q_1, q_2, \ldots, q_n$ is the algebraic sum of potentials that the charges would separately produce at the point:

$$\varphi = \varphi_1 + \varphi_2 + \ldots + \varphi_n. \tag{15.8}$$

The point of zero potential is common for all the charges. Usually it is a point removed to infinity.

The graphic sketch of an electric field can be given either by field lines or by equipotential surfaces. Field lines arise from positive charges, never intersect each other, and end on negative charges or go to infinity. The tangent to a field line shows the direction of the force experienced by a positive test charge at the point. An equipotential surface is the locus of points with the same value of potential.

Field lines are always perpendicular to equipotential surfaces.

The electric field surrounding a single isolated point charge, called the *Coulomb field*, is spherically symmetric (Figure 15.2). The magnitude of its field strength and the electric potential are given by the expressions

$$E = \frac{1}{4\pi\varepsilon_0\varepsilon}\frac{|q|}{r^2}, \quad \varphi = \frac{1}{4\pi\varepsilon_0\varepsilon}\frac{q}{r}. \tag{15.9}$$

The field lines are directed radially and the equipotential surfaces are the concentric spheres centered at the point charge.

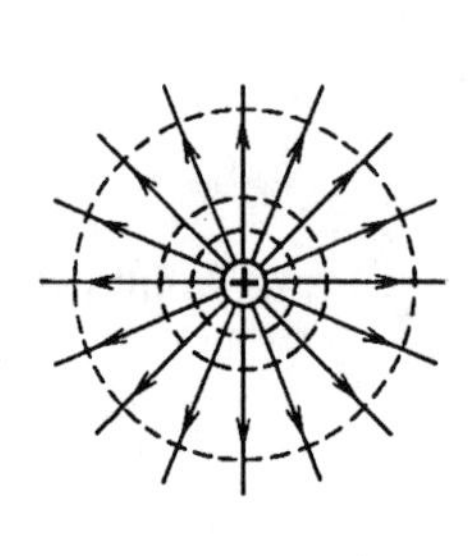

Figure 15.2: Electric field of a positive point charge.

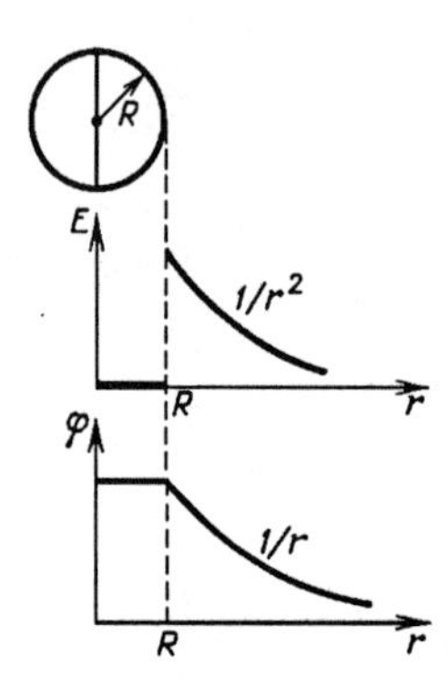

Figure 15.3: Electric field of a charged sphere.

The electric field of a charged metal sphere in the surrounding space coincides with the Coulomb field of a point charge of the same magnitude located at the center of the sphere. Inside the sphere the field strength is equal to zero and the electric potential is the same at all points (Figure 15.3).

Pictures of the electric fields of two equal point charges of opposite polarities and of the same polarity are shown in Figure 15.4. A system of two near-by charges with equal magnitudes of opposite sign is called a *dipole*. The electric *dipole moment* p is the product of one (positive) charge and the distance between them:

$$p = |q|\, l.$$

The potential φ of the electric field created by a dipole in a vacuum (Figure 15.5) in the case $l \ll r$ is given by the formula

$$\varphi = \frac{1}{4\pi\varepsilon_0}\frac{p}{r^2}\cos\theta. \tag{15.10}$$

The strength $\mathbf{E}$ of the dipole's field is the vector sum of two mutually perpendicular components $\mathbf{E}_r$ and $\mathbf{E}_\theta$ with the magnitudes

$$E_r = \frac{1}{4\pi\varepsilon_0}\frac{2p}{r^3}\cos\theta, \quad E_\theta = \frac{1}{4\pi\varepsilon_0}\frac{p}{r^3}\sin\theta. \tag{15.11}$$

The magnitude of the vector $\mathbf{E}$ is

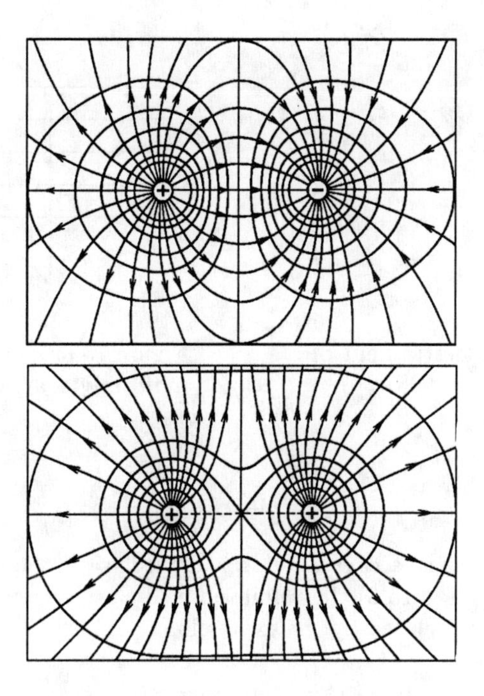

Figure 15.4: Field lines and equipotential surfaces for two equal point charges of opposite polarities and of the same polarity.

$$E = \frac{1}{4\pi\varepsilon_0}\frac{p}{r^3}\sqrt{1+3\cos^2\theta}. \tag{15.12}$$

Equations (15.10) and (15.11) in the case of large distances ($r \gg l$) are valid not only for a dipole but also for any electrically neutral system of charges with noncoinciding centers of positive and negative charges (e.g., for a polar molecule).

At sufficiently large distances any system with a complicated distribution of charge seems to be simply a point charge with the value Q, equal to the total charge of the system. This model can be often applied to an ion of an atom or a molecule.

At large distances the electric field of a neutral body ($Q = 0$) with some complicated distribution of charges can be represented with sufficient accuracy as a field of a suitable dipole. This is the case of an electric field produced by a neutral molecule.

In a uniform electric field the field strength $\mathbf{E}$ is the same at all points and field lines are parallel to each other. In particular, such a field is produced by an infinite uniformly charged plane:

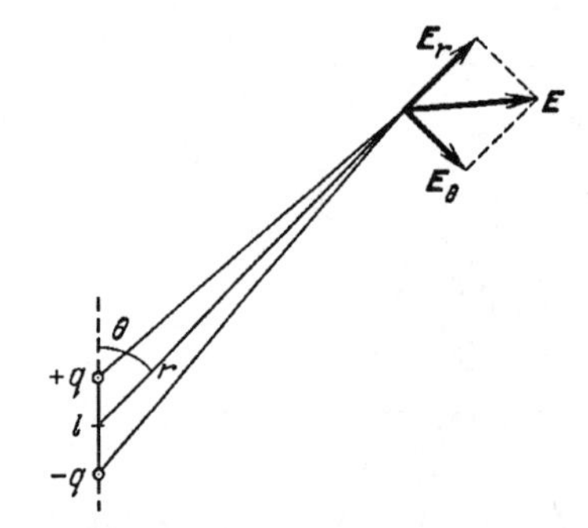

Figure 15.5: Electric field of a dipole.

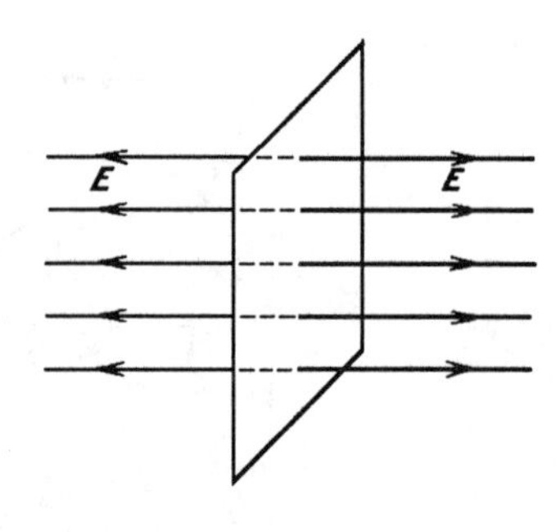

Figure 15.6: The field lines of a uniformly charged plane.

$$E = \frac{\sigma}{2\varepsilon_0}, \tag{15.13}$$

where σ is the *surface charge density* defined to be the charge per unit surface area (Figure 15.6). In a uniform field the electric voltage and the field strength are related by the expression

$$U = \varphi_1 - \varphi_2 = El, \tag{15.14}$$

where l is the distance along the field line between two points with potentials φ_1 and φ_2.

15.1.3 Electrostatic field in dielectrics

The field strength in a *dielectric* is weaker than in a vacuum due to the polarization of the dielectric. The dielectric constant, or the *relative permittivity* ε_r of a medium, is defined as the ratio of the field strength in a vacuum, E_0, and the field strength, E, in a uniform dielectric placed in that field provided that the surface of the dielectric is perpendicular to the field lines:

$$\varepsilon = E_0/E.$$

Values of dielectric constants of some materials are given in Table 15.1.

15.1.4 Electric field near conductors

Inside a *conductor* the field strength is equal to zero and electric potential is the same at all points. Outside a conductor the field lines near the surface are perpendicular to the surface. Non-compensated charges are located only on the surface of a conductor. The field strength near the surface is proportional to the surface charge density σ:

Table 15.1: Dielectric constants of some materials

Material	Constant	Material	Constant
Air	1.0006	Polyvinyl	4–8.5
Ebonite	2.7	Rubber	3.3
Alcohol	2.6	Teflon	2.0
Glass	4–7	Distilled water	81
Cardboard	5	Wood	2.5–7
Paraffin	2.1		

$$E = \frac{\sigma}{\varepsilon_0}. \tag{15.15}$$

In the case of a solitary conductor, the surface charge density is the same at the points where the surface curvature is the same. For a sphere or a cylinder the density is the same everywhere. The surface charge density and consequently the field strength are greater at points where the surface curvature (see Section 9.7) is greater.

15.1.5 Capacitors

The potential of a conductor measured with respect to some point (e.g., to the earth) increases while the conductor is being charged positively. The potential increases proportionally to the charge already stored on the conductor: $\varphi \sim q$. Therefore, the charge of a conductor is proportional to its potential: $q = C\varphi$. The proportionality factor is called the *electric capacitance* of the conductor:

$$C = \frac{q}{\varphi}. \tag{15.16}$$

The *capacitance* is the property of a conductor or a system of conductors that describes its ability to store electric charge.

A *capacitor* (or *condenser*) is an arrangement of conductors separated by a thin layer of insulator (vacuum or dielectric) used to store electric charge. In its simplest form it consists of two parallel conducting plates (armatures) with equal and opposite charges. The *charge of a capacitor* is defined as the charge stored on the inner surface of one of its plates.

The capacitance of a capacitor is defined as the ratio of its charge and the voltage between its plates:

$$C = \frac{q}{U}. \tag{15.17}$$

The capacitance is determined by the size and shape of a capacitor and does not depend on its charge. Capacitors are classified as plane, cylindrical, and spherical according to the shape of their plates, and as an air, vacuum, paper, ceramic, and electrolytic according to the material used as a dielectric.

The capacitance of a *plane* capacitor is defined by the formula:

$$C = \frac{\varepsilon_0 \varepsilon S}{d}, \tag{15.18}$$

where S is the useful surface area of one plate (the smaller one if they are not equal), d is the gap between the plates, and ε_r is the relative permittivity of the dielectric placed between the plates (see Section 15.1.3).

The capacitance of a *cylindric* capacitor and a coaxial cable is as follows:

$$C = \frac{2\pi\varepsilon_0 \varepsilon l}{\ln (b/a)}, \tag{15.19}$$

where b and a are the radii of the outer and inner cylinders and l is the length of the capacitor.

The capacitance of a *spherical* capacitor is as follows:

$$C = \frac{4\pi\varepsilon_0 \varepsilon}{1/a - 1/b}, \tag{15.20}$$

where a and b are the radii of the inner and outer spheres.

The above formulas for cylindrical and spherical capacitors are valid when the outer cylinder and the outer sphere are grounded. When the gap between the armatures is small in comparison with their sizes, the expressions (15.19) and (15.20) turn out to be equivalent to the expression (15.18) for a plane capacitor.

The capacitance of a solitary sphere equals $4\pi\varepsilon_0\varepsilon R$, where R is its radius. The capacitance of the planet earth is approximately equal to 700 μF.

15.1.6 Connection of capacitors

When several capacitors are arranged *in parallel* (Figure 15.7), the potential difference across each capacitor is the same and the total charge stored is the sum of the individual charges on the separate capacitors. Therefore, for capacitors in parallel, the total capacitance C is the sum of the capacitances:

$$C = C_1 + C_2 + \ldots + C_n. \tag{15.21}$$

When several capacitors are arranged *in series* (Figure 15.8), the charges on all capacitors are the same and the total potential difference is the sum of voltages across the individual capacitors. This is true for the circuit of capacitors, none of which had been charged before they

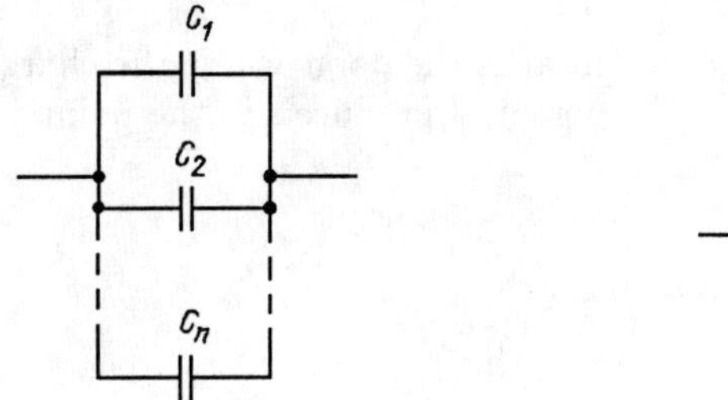

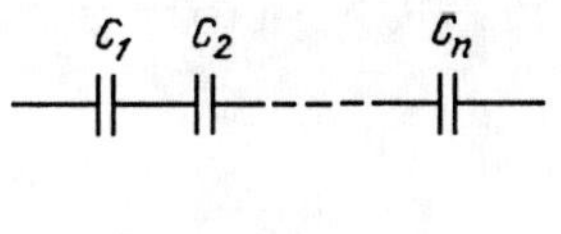

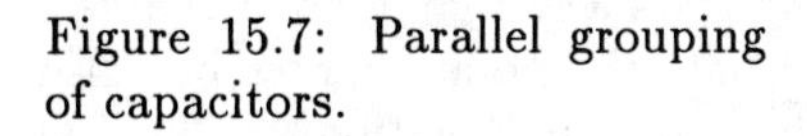

Figure 15.7: Parallel grouping of capacitors.

Figure 15.8: Several capacitors arranged in series.

were connected. For capacitors in series the total capacitance C is given by the formula

$$\frac{1}{C} = \frac{1}{C_1} + \frac{1}{C_2} + \ldots + \frac{1}{C_n}. \tag{15.22}$$

For capacitors in series the total capacitance is less than the smallest individual capacitance in the circuit.

15.1.7 Energy of electric field

The *energy* stored in a capacitor may be treated either as the energy of the electric field between the armatures or as the energy of interaction of the charges on its plates. This energy can be calculated as below:

$$W = \frac{1}{2}qU = \frac{1}{2}CU^2 = \frac{1}{2}\frac{q^2}{C}. \tag{15.23}$$

In a plane (parallel plate) capacitor the electric field is uniform and the energy is given by the formula:

$$W = \frac{1}{2}\varepsilon_0\varepsilon E^2 V, \tag{15.24}$$

where $V = Sd$ is the volume occupied by the electric field between the plates. The quantity

$$w = \frac{1}{2}\varepsilon_0\varepsilon E^2$$

can be interpreted as the *energy density* of electric field.

The energy of interaction of N stationary point charges in a vacuum is given by the formula

$$W_N = \frac{1}{2}\sum_{i=1}^{N} q_i\varphi_i, \tag{15.25}$$

where φ_i is the electric field potential at the point where the charge q_i is located. The potential φ_i is the potential produced at the point i by all the charges except the charge q_i:

$$\varphi_i = \frac{1}{4\pi\varepsilon_0} \sum_{k=1,\, k\neq i}^{N} \frac{q_k}{r_{ik}}, \tag{15.26}$$

where r_{ik} is the distance between the charges q_i and q_k.

The expression (15.25), giving the energy of a system of point charges, is valid also for systems containing not only the point charges but also charged or neutral conductors. For this case, in Equation (15.25) the charge q_i in the terms corresponding to conductors has the sense of the total charge of the i-th conductor, and φ_i has the sense of its potential produced by all the charges of the system including q_i.

Earnshaw theorem: equilibrium of a system of charges interacting only by electrostatic forces is always unstable.

15.2 Electric Current

15.2.1 Ohm's law

An *electric current* is an ordered motion (a flow) of electric charges. The *direction of current* is considered to be the direction of motion of positively charged particles independently of the kind of particles that carry the charge in reality. For instance, in metals an electric current is produced by a flow of negatively charged electrons, and the current direction is opposite to the direction of the electrons' flow.

A *direct current* is the current that does not change in time, i.e., an equal amount of electric charge passes through any cross-section per equal time intervals. The rate of flow of charge at a particular cross-section is characterized by the physical quantity called the *current*. Let Δq be the charge passing through a given cross-section during a time interval Δt. The limit

$$I = \lim_{\Delta t \to 0} \frac{\Delta q}{\Delta t} = \frac{dq}{dt} \tag{15.27}$$

is called the *current strength* or, simply, the *current*. The current strength is a scalar quantity.

The *current density* $\mathbf{j}$ is connected with the current I, passing through a given cross-section S, by the relation

$$I = (\mathbf{j} \cdot \mathbf{S}), \tag{15.28}$$

where $\mathbf{S}$ is the vector having the length S and directing along the normal to the cross-section. Of course, there are two possibilities to choose the direction of this normal. In Equation (15.28) we assume that the angle between **S** and the direction of the current is acute.

Let the current be produced by the motion of charged particles of the same kind. Then the current density is given by the expression:

$$\mathbf{j} = ne\langle \mathbf{v} \rangle, \tag{15.29}$$

where e is the charge of one of the moving charged particles, n is the number of these particles (charge carriers) in unit volume (concentration), and $\langle \mathbf{v} \rangle$ is the mean speed of the particles' motion. The value of this speed is proportional to the field strength. The proportionality coefficient, u, is called the *mobility*. In terms of the mobility one can write

$$\mathbf{j} = neu\mathbf{E} = \sigma\mathbf{E}. \tag{15.30}$$

The value $\sigma = neu$ is called the *specific conductance.* Equation (15.30) gives Ohm's law written in the *differential form.*

The *voltage* between two points of an electric circuit is equal to the difference of the potential values between these points.

For a segment of a homogeneous circuit Ohm's law can be written in the following form:

$$I = \frac{U}{R}, \tag{15.31}$$

where R is *resistance* of the segment. This formula is valid when no external (non-electric) forces are acting on the charge carriers inside the above segment.

If charge carriers in a conductor are electrons, we say that it is a *conductor of the first kind.* Most of metals are conductors of this type.

Consider a conductor having a constant cross-section area S and being made of a homogeneous material. Let its length be equal to l. Resistance of this conductor can be represented in the form:

$$R = \rho\frac{l}{S}, \tag{15.32}$$

where ρ is the *specific resistance* equal to resistance of the conductor made of the same material and having the unit cross-section and the unit length. The values ρ and σ are connected by the relation $\rho = 1/\sigma$. Both the specific conductance σ and the specific resistance ρ characterize the properties of material and do not depend on form and size of a conductor (if the conductor's size is macroscopic; see Chapter 14).

For many metals the growth of temperature leads to an increase of specific resistance. There exists an approximate relation

$$\rho = \rho_0(1 + \alpha t), \tag{15.33}$$

Table 15.2: Specific resistance and temperature resistance coefficient for some metals

Metal	ρ_0, 10^{-6} Om·m	α, 10^{-3} K^{-1}	Metal	ρ_0, 10^{-6} Om·m	α, 10^{-3} K^{-1}
Aluminum	0.027	4.3	Wolfram	0.055	4.1
Gold	0.022	3.9	Copper	0.0172	3.8
Nickel	0.087	6.5	Platinum	0.107	3.9
Mercury	0.96	0.92	Silver	0.016	3.8

where ρ is the specific conductance corresponding to the temperature t, ρ_0 is the specific conductance corresponding to $t = 0$ °C, and α is the *resistance temperature coefficient*. The values of ρ_0 and α for some metals are given in Table 15.2.

Resistance of some metals and of some alloys can turn to zero at sufficiently low temperatures. This property of these materials is called *superconductivity*.

15.2.2 Series and parallel connection of resistors

When several resistances are arranged *in series* (Figure 15.9), the total resistance is equal to their sum:

$$R = R_1 + R_2 + \ldots + R_n. \tag{15.34}$$

In all the points of the above circuit the current strength has the same value and the voltages between the ends of the segments of this circuit are proportional to the resistances of these segments:

$$\frac{U_1}{U_2} = \frac{R_1}{R_2}. \tag{15.35}$$

The total voltage between the ends of this circuit is equal to the sum of the voltages between the ends of its segments:

$$U = U_1 + U_2 + \ldots + U_n.$$

The value inverse to the resistance is called *conductivity*. When several resistances are arranged *in parallel* (Figure 15.10), the total conductivity of the circuit is equal to the sum of the conductivities corresponding to R_1, R_2, ..., R_n:

$$\frac{1}{R} = \frac{1}{R_1} + \frac{1}{R_2} + \ldots + \frac{1}{R_n}. \tag{15.36}$$

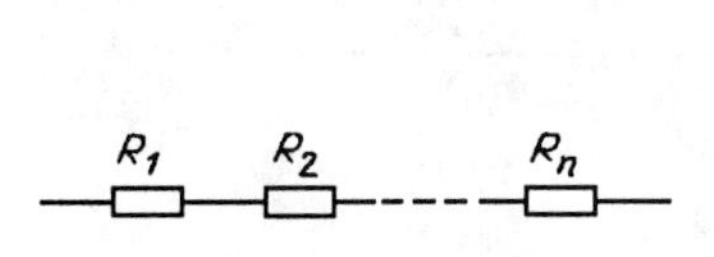

Figure 15.9: Several resistances arranged in series.

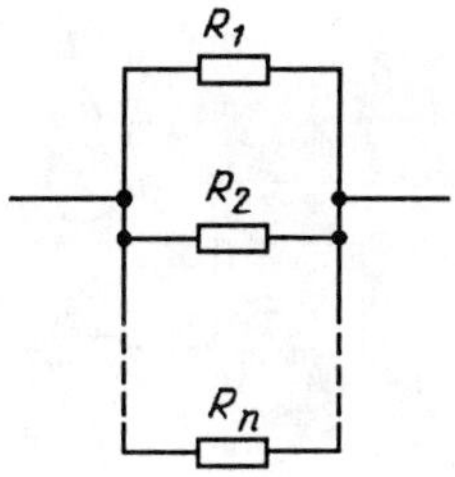

Figure 15.10: Parallel grouping of resistances.

In the case of the circuit shown in Figure 15.10, the voltages between the ends of any of the resistances are equal to each other, and the currents in separate parallel segments are proportional to the corresponding conductivities:

$$\frac{I_1}{I_2} = \frac{R_1}{R_2}. \tag{15.37}$$

The current strengths in non-ramified parts of the above circuit are equal to the sum of the currents in its branches:

$$I = I_1 + I_2 + \ldots + I_n.$$

This formula can be regarded as a special case of Kirghoff's first rule (see Section 15.2.5).

15.2.3 Measurements in direct current circuits

An instrument for measuring a current strength is called an *ammeter*. It is connected *in series* with the circuit where the current strength is measured. The inner resistance R_a of the ammeter should be as small as possible. A *shunt* resistance can be connected in parallel to the ammeter in order to extend its range (Figure 15.11*a*). The shunt with the resistance

$$R_s = \frac{R_a}{n-1} \tag{15.38}$$

is needed to extend n times the range of the ammeter.

An instrument for measuring a voltage is called a *voltmeter*. It is connected *across* a circuit where the voltage is measured. The inner resistance R_v of the voltmeter should be as large as possible. In order to extend the range of the voltmeter a *multiplier* should be switched into it (Figure 15.11*b*). A multiplier with a resistance

$$R_m = R_v(n-1) \tag{15.39}$$

is needed to extend n times the range of the voltmeter.

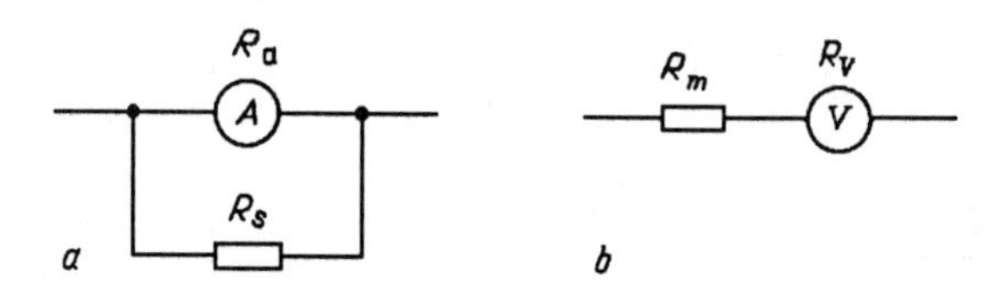

Figure 15.11: A shunt (a) and a multiplier (b).

To measure a resistance the *Wheatstone's bridge* (named after Charles Wheatstone) can be used (Figure 15.12). The bridge is at balance when the current in the galvanometer is absent. The position of the contact arm D in this case corresponds to the relation

$$R_x = R\,\frac{R_1}{R_2}. \tag{15.40}$$

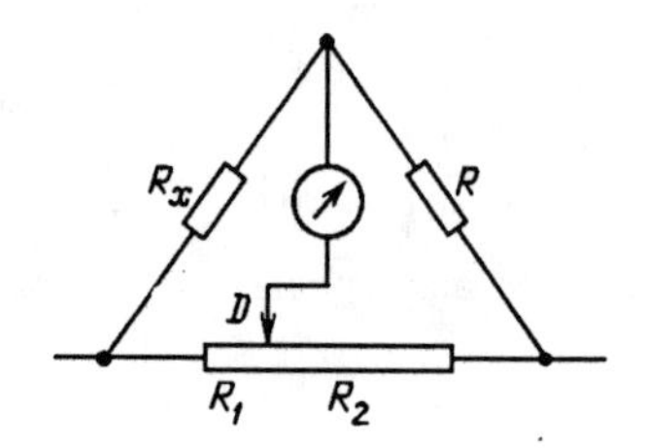

Figure 15.12: Wheatstone's bridge.

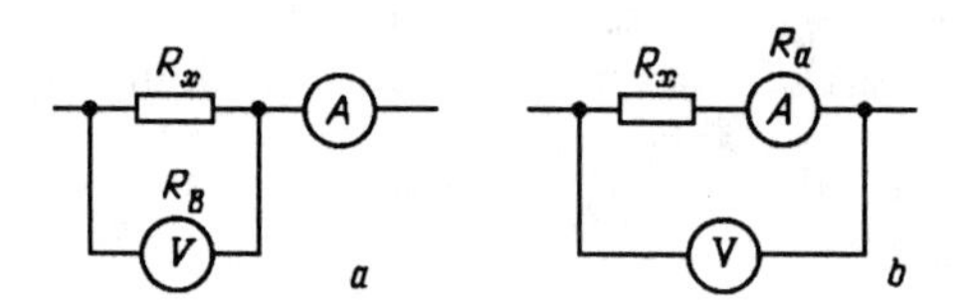

Figure 15.13: The measurements of R_x.

To measure a resistance with the help of a voltmeter and an ammeter one can use any of the circuits shown in Figure 15.13. In the case of the circuit a the unknown resistance is given by the formula

$$R_x = \frac{U}{I - U/R_v}, \tag{15.41}$$

where U and I are the readings of the voltmeter and the ammeter respectively. In the case of the circuit *b* the unknown resistance is given by the formula

$$R_x = \frac{U}{I} - R_a. \tag{15.42}$$

The first circuit is convenient to measure resistances much smaller than the resistance of the voltmeter R_v. The second circuit is convenient to measure resistances much larger than the resistance of the ammeter R_a. In these cases one can write, with sufficient precision,

$$R_x = \frac{U}{I}. \tag{15.43}$$

15.2.4 Circuit with a source

Ohm's law for *the unbranched* closed circuit (Figure 15.14) containing the power source with an electromotive force $\mathcal{E}$ and an internal resistance r is written in the form

$$I = \frac{\mathcal{E}}{R + r}. \tag{15.44}$$

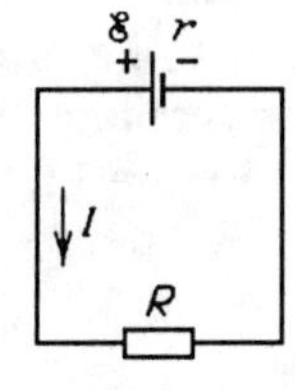

Figure 15.14: An unbranched closed circuit with a source.

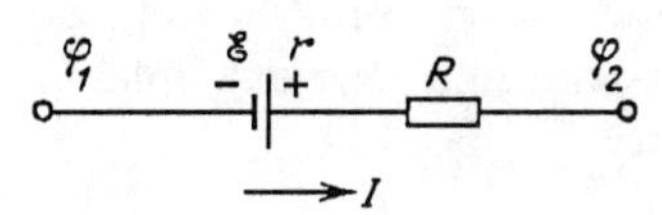

Figure 15.15: A part of an inhomogeneous circuit.

Ohm's law for an *inhomogeneous* part of a circuit (Figure 15.15) having the resistance R and containing the source with an electromotive force $\mathcal{E}$ is written in the form

$$I = \frac{U + \mathcal{E}}{R + r}, \tag{15.45}$$

where $U = \varphi_1 - \varphi_2$, and the electromotive force $\mathcal{E}$ is considered to be positive if the source is placed in the circuit as shown in Figure 15.15, and $\mathcal{E}$ is considered to be negative if the source is placed in the circuit in the opposite polarity. If the current strength calculated by means of Equation (15.45) turns out to be negative, then actually the current has the opposite direction.

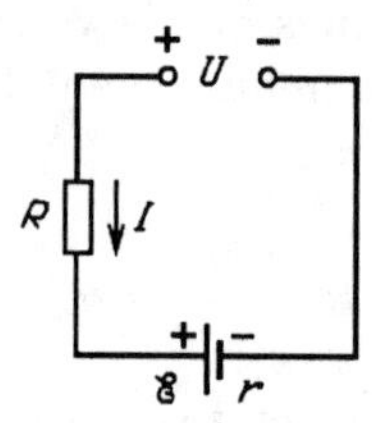

Figure 15.16: Charging of a storage battery.

Example. The electric circuit for the charging of a storage battery is shown in Figure 15.16. The current strength in this case is determined by the expression

$$I = \frac{U - \mathcal{E}}{R + r}. \tag{15.46}$$

15.2.5 Kirchhoff's rules

Kirchhoff's rules are used for calculating the currents in *ramified* (branching) electric circuits.

Kirchhoff's first rule reads: the algebraic sum of the current strengths in all the circuit segments converging in a node is equal to zero. In the sum, the currents flowing into a node are considered to be of one sign, while the currents flowing out of a node are considered to be of the opposite sign. For example, for the circuit shown in Figure 15.17 we can write:

$$I_1 + I_2 - I_3 + I_4 = 0. \tag{15.47}$$

Kirchhoff's first rule follows from the electric charge conservation law.

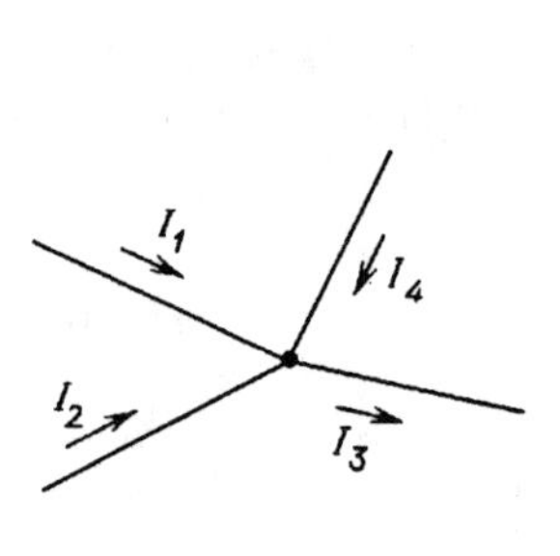

Figure 15.17: Kirchhoff's first rule.

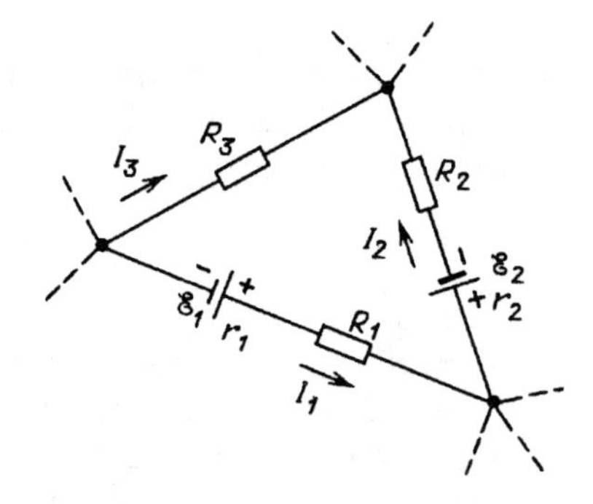

Figure 15.18: Kirchhoff's second rule.

Kirchhoff's second rule reads: for any closed contour in a ramified circuit the algebraic sum of products of the current strengths and resistances of the corresponding parts is equal to the algebraic sum of all the electromotive forces in the contour. The currents with directions coinciding with the direction of the bypassing of the contour are taken with positive signs; the electromotive forces that raise the electric potential in the direction of bypassing of the contour are considered to be positive. For example, for the contour shown in Figure 15.18 one can write:

$$I_1(R_1 + r_1) + I_2(R_2 + r_2) - I_3R_3 = \mathcal{E}_1 - \mathcal{E}_2. \qquad (15.48)$$

Kirchhoff's second rule follows from Ohm's law, Equation (15.45), applied to all the parts of the contour.

The rules of calculating the ramified circuits read:

1. Designate the currents in all nonramified parts and assign arbitrarily the directions of the currents.

2. Write down the equations for sums of the currents in all the nodes except one according to Kirchhoff's first rule.

3. According to Kirchhoff's second rule, write down the equations for all the simple contours that cannot be treated as a superposition of already considered ones. (A contour is called simple if one occurs to be only once at any of its points while bypassing it). In properly chosen system of contours each part of the circuit should appear at least in one of the contours.

4. Currents that would occur to be negative at solving the system of the equations, actually have directions opposite to those assigned to them previously.

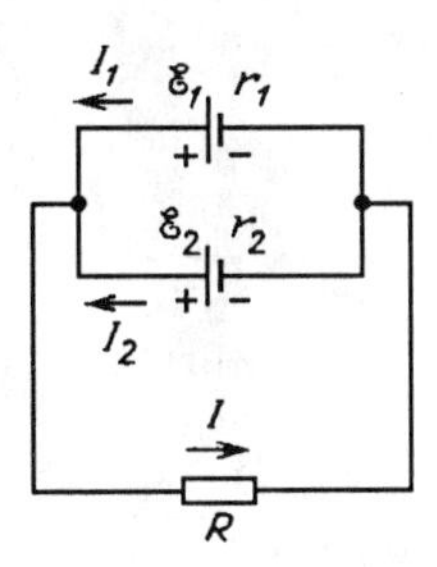

Figure 15.19: A branched circuit.

Example. The system of equations corresponding to the circuit shown in Figure 15.19 can be written in the form:

$$I_1 + I_2 - I = 0, \quad I_1r_1 + IR = \mathcal{E}_1, \quad I_2r_2 + IR = \mathcal{E}_2. \qquad (15.49)$$

Solving this system (see Section 3.5.2) we get:

$$I = \frac{\mathcal{E}_1 r_2 + \mathcal{E}_2 r_1}{R(r_1 + r_2) + r_1 r_2}, \quad I_1 = \frac{(\mathcal{E}_1 - \mathcal{E}_2)R + \mathcal{E}_1 r_2}{R(r_1 + r_2) + r_1 r_2},$$

$$I_2 = \frac{(\mathcal{E}_2 - \mathcal{E}_1)R + \mathcal{E}_2 r_1}{R(r_1 + r_2) + r_1 r_2}.$$

15.2.6 The work of electric current

The *work* A done by a direct current I on the circuit segment with a voltage U during a time interval t is given by the formula:

$$A = IUt. \tag{15.50}$$

The *power* P generated by a current I is given by the formula:

$$P = \frac{A}{t} = IU. \tag{15.51}$$

The *heat* Q released during time interval t in a conductor with resistance R is determined by *Joule's law*

$$Q = I^2 Rt, \tag{15.52}$$

where I is the current in the conductor. Respectively, the heat per unit time (a power of dissipated heat) is given by the formula:

$$P_Q = \frac{Q}{t} = I^2 R. \tag{15.53}$$

For a *homogeneous* part of a circuit, for which $\mathcal{E} = 0$ and Ohm's law $I = U/R$ is valid, we have

$$A = IUt = I^2 Rt = \frac{U^2}{R}t = Q. \tag{15.54}$$

All the work done by electric current in a homogeneous part of a circuit implements in a heat.

For an *inhomogeneous* part of the circuit containing the source with an electromotive force $\mathcal{E}$ and an internal resistance r, where $I = (U + \mathcal{E})/(R + r)$ (see Section 15.2.4), the following formulas describe the work A_c done by the current, the work A_s of the source, and the released heat Q:

$$A = IUt = \frac{U + \mathcal{E}}{R + r}Ut, \quad A_s = I\mathcal{E}t,$$

$$Q = I^2(R + r)t = I(U + \mathcal{E})t = \frac{(U + \mathcal{E})^2}{R + r}t. \tag{15.55}$$

Released heat is equal to the sum of the work done by the current and the work of the source: $Q = A + A_s$.

Example. At charging of a storage battery, the power P, generated by the charging device, is given by the relation

$$P = IU. \tag{15.56}$$

For the heat per unit time released in the storage battery and in the wires, we have

$$P_Q = I^2(R + r). \tag{15.57}$$

The power P accumulated in the battery can be expressed in the form

$$P_a = I\mathcal{E}. \tag{15.58}$$

In accordance with the energy conservation law we have $P = P_Q + P_a$.

15.2.7 A power source in a circuit

A *source* (a generator) of electric current is a device maintaining a voltage on a circuit. The voltage occurs due to the work done by *extraneous* forces, i.e., by forces of a nonelectrostatic nature. The ratio $\mathcal{E}$ of the work A of the extraneous forces, permutating a charge q in the circuit, to the charge q is called an *electromotive force* of the source:

$$\mathcal{E} = \frac{A_{\text{ext}}}{q}.$$

The electric resistance of a source is called the *internal* resistance. Contact points of a source are called *terminals*. The electromotive force $\mathcal{E}$ of a source is equal to the voltage on the interrupted terminals of the source. For a closed circuit the electromotive force of the source is equal to the sum of the voltages on the external and the internal parts of the circuit:

$$\mathcal{E} = U_{\text{ext}} + U_{\text{int}} = IR + Ir.$$

Total power P generated by a source is given by the following formula (Figure 15.14):

$$P = I\mathcal{E} = \frac{\mathcal{E}^2}{R + r}. \tag{15.59}$$

Useful power P_{usf}, i.e., the power released in the external part of a circuit having the resistance R, can be written as follows:

$$P_{\text{usf}} = IU = I^2 R = \frac{\mathcal{E}^2 R}{(R+r)^2}. \tag{15.60}$$

The efficiency of a source is

$$\eta = \frac{P_{\text{usf}}}{P} = \frac{R}{R+r}. \tag{15.61}$$

The conditions of employment of a power source are shown by the graphs in Figure 15.20.

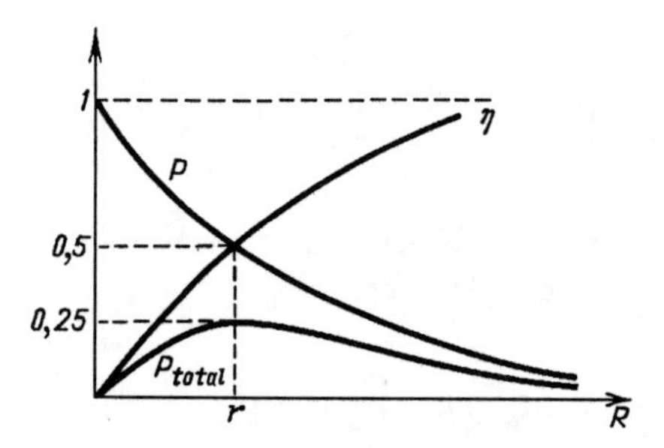

Figure 15.20: The graphs illustrating the work of a current source.

Maximal useful power that can be generated by a source with the electromotive force $\mathcal{E}$ and the internal resistance r is given by the formula

$$P_{\text{usf}}^{\max} = \frac{\mathcal{E}^2}{4r}. \tag{15.62}$$

This power is attained when the resistance R of the load is equal to the internal resistance r; in this case the efficiency is equal to 50%: $\eta = 0.5$.

Any useful power less than the maximum one can be attained at two different values of the load resistance:

$$R_{1,2} = \frac{\mathcal{E}^2}{2P_{\text{usf}}} - r \pm \sqrt{\frac{\mathcal{E}^2}{P_{\text{usf}}}\left(\frac{\mathcal{E}^2}{4P_{\text{usf}}} - r\right)}, \tag{15.63}$$

with $R_1 > r$ and $R_2 < r$. In the first case, $\eta > 0.5$ and in the second case, $\eta < 0.5$.

The quantity $I_0 = \mathcal{E}/r$ is called a *short-circuit current.* Adding one more current source in series in a circuit is advantageous only if its short-circuit current is larger than the current that flows in the circuit before adding the source. In the opposite case the heat released inside the source would be larger than the power generated by the source.

15.2.8 Faraday's laws of electrolysis

The *electrolytes* are conductors of the second kind. These are solutions of acids, alkalis, and salts in water and other solvents. The melt salts are also electric conductors. The positive and negative ions and the electrons are carriers of charge in these cases. The electrode connected with the positive terminal of a source is called an *anode*, the other one is called a *cathode*. Positive ions move towards the cathode, negative ions and electrons move towards the anode.

Electrolysis is a change of chemical composition of an electrolyte (a solution where a dissociation of the molecules has occurred) and a separation of different substances on the electrodes due to electric current.

Faraday's first law: the mass of a material separated on an electrode is proportional to the passed electric charge:

$$m = kq = kIt, \tag{15.64}$$

where k is the *electrochemical equivalent* of the material.

Faraday's second law: the electrochemical equivalent of a material is proportional to its chemical equivalent M/n (M is the molar mass and n is the valence of the material):

$$k = \frac{1}{F}\frac{M}{n}, \tag{15.65}$$

where the *Faraday's constant* F is the charge necessary for separation of one mole of a material on an electrode:

$$F = N_A e = 96,485.309 \text{ C/mol}.$$

The joint form of Faraday's first and second laws is

$$m = \frac{1}{F}\frac{M}{n}q = \frac{1}{F}\frac{M}{n}It. \tag{15.66}$$

15.3 Magnetic Field

15.3.1 Induction of magnetic field

Magnetic fields can be created by permanent magnets and by moving electric charges (electric currents). A magnetic field exhibits itself in acting on a magnet needle, on a wire or on a coil with an electric current, or on a moving charge. A magnetic field is a *solenoidal* (or vortex) field, and the lines of the magnetic induction are always closed: the sources of the field—magnetic charges—do not exist.

Induction $\mathbf{B}$ of a magnetic field is its force characteristic. Magnetic fields satisfy the superposition principle as electric fields do. Only the resultant induction of the magnetic field created by all the elements of an electric current can be measured experimentally.

Biot–Savart–Laplace law: an element Δl of a wire with a current I creates at some point in a vacuum a magnetic field with the induction

$$\Delta B = \frac{\mu_0}{4\pi} \frac{I \Delta l \sin\alpha}{r^2}, \tag{15.67}$$

where $\mu_0 = 4\pi \cdot 10^{-7}$ N/A^2 is the *magnetic constant*, r is the distance between the element Δl of the wire and the point of observation and α is the angle between the direction to the point of observation and the direction of the element Δl of the wire (Figure 15.21).

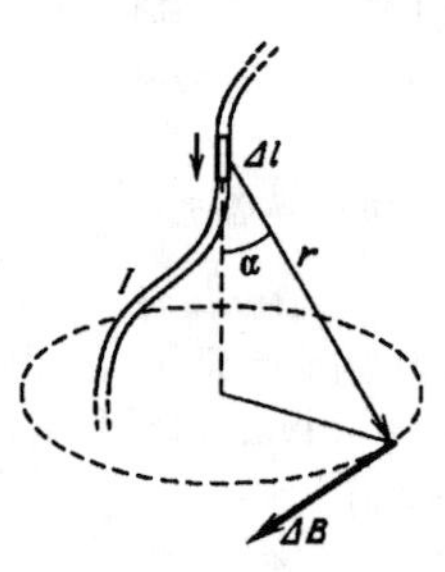

Figure 15.21: Magnetic field of the element $\Delta \mathbf{l}$.

Vector $\Delta \mathbf{B}$ is perpendicular to the plane containing the element $\Delta \mathbf{l}$ and the radius-vector r. The direction of $\Delta \mathbf{B}$ is determined by the *right screw rule*: it coincides with the direction of rotation of a screw during its movement along the current. The expression for $\Delta \mathbf{B}$ can be written in terms of the vector product (see Section 8.2):

$$\Delta \mathbf{B} = \frac{\mu_0}{4\pi} \frac{I}{r^3} (\Delta \mathbf{l} \times \mathbf{r}). \tag{15.68}$$

In a material with a magnetic permeability μ (see Section 15.3.4) a factor μ appears in Equations (15.67) and (15.68) for the magnetic induction.

The direction of the vector $\Delta \mathbf{l}$ coincides with the direction of motion of positive charges along the wire.

Magnetic induction in the center of a circular conducting wrap of the radius R with a current I (Figure 15.22) is equal to

$$B = \mu_0 \frac{I}{2R} \quad \left(B = \mu_0 \mu \frac{I}{2R} \right). \tag{15.69}$$

Magnetic induction at a distance r from a rectilinear wire with a current I (Figure 15.23) is given by the expression

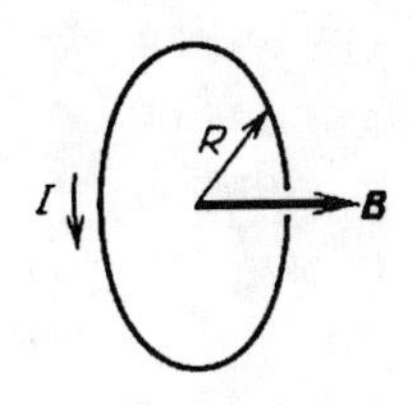

Figure 15.22: Magnetic field in the center of a circular current.

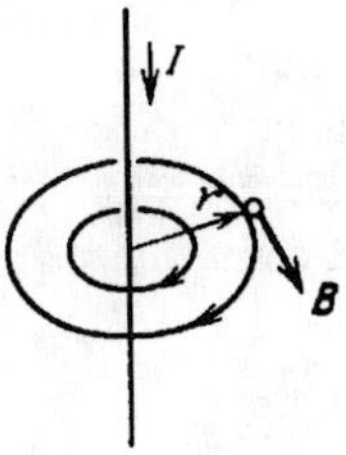

Figure 15.23: Magnetic field of a rectilinear wire.

$$B = \frac{\mu_0}{2\pi}\frac{I}{r} \qquad \left(B = \frac{\mu_0 \mu}{2\pi}\frac{I}{r}\right). \tag{15.70}$$

Inside a relatively long cylindrical coil with the length l, the number of turns N, and the current I magnetic field is homogeneous:

$$B = \mu_0 \frac{IN}{l} \qquad \left(B = \mu_0 \mu \frac{IN}{l}\right). \tag{15.71}$$

15.3.2 Ampere's force and Lorentz' force

The force exerted on an element Δl of a wire bearing a current in a magnetic field is determined by the *Ampere's law*:

$$\Delta \mathbf{F} = I(\Delta \mathbf{l} \times \mathbf{B}). \tag{15.72}$$

The magnitude of the Ampere's force is equal to:

$$F = I \Delta l B \sin \alpha, \tag{15.73}$$

where α is the angle between the directions of the induction $\mathbf{B}$ and Δl of a wire with a current. The direction of the force that acts on the wire with a current is determined by the vector product (see Section 8.2) or by the *left-hand rule* (Fleming's rule): if we place the left hand so that the magnetic induction lines would enter the palm and stretched four fingers would indicate the electric current direction, then the turned back thumb would show the direction of a force, that acts on the conductor (Figure 15.24).

Two long rectilinear wires with currents attract each other if the currents are parallel. The wires repel each other if the currents flow in opposite directions. The force of their interaction is equal to:

$$F = \frac{\mu_0 \mu I_1 I_2 l}{2\pi d}, \tag{15.74}$$

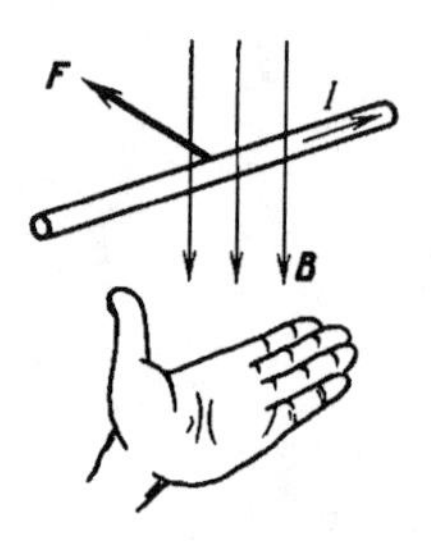

Figure 15.24: The left-hand rule.

where d is the distance between the wires, l is their length, I_1 and I_2 are the currents in the wires, and μ is the *magnetic permeability* of the surrounding medium (see Section 15.3.4).

The *Lorentz force* is the force that acts on an electric charge, moving in a magnetic field:

$$\Delta\mathbf{F} = q(\mathbf{v} \times \mathbf{B}), \quad F = qvB\sin\alpha, \tag{15.75}$$

where α is the angle between the vectors $\mathbf{v}$ and $\mathbf{B}$. Direction of the Lorentz force is determined by the vector product (see Section 8.2) of $\mathbf{v}$ and $\mathbf{B}$ or by the left-hand rule.

Example. A charged particle of mass m moves with a velocity $\mathbf{v}$ perpendicular to the lines of magnetic induction $\mathbf{B}$ in a homogeneous magnetic field along a circular trajectory of the radius $R = mv/(qB)$ with a period T that does not depend on the velocity:

$$T = \frac{2\pi m}{qB}.$$

This property of motion in a magnetic field gives rise to a *cyclotron*—a device for the acceleration of charged particles.

15.3.3 Magnetic field energy

A magnetic field, distributed in space, possesses an energy. The *energy of a magnetic field*, created by a conductor with the inductance L (see Section 15.4.2), is equal to

$$W = \frac{1}{2}LI^2, \tag{15.76}$$

where I is the current in the conductor. The *volume density* of magnetic field energy is determined by the expression

$$w = \frac{1}{2}\frac{B^2}{\mu_0 \mu}. \tag{15.77}$$

15.3.4 Magnetic field in substances

In all bodies placed in a magnetic field, a *magnetization* occurs: they acquire *magnetic moment.* A magnetized body is called magnetic. *Magnetic permeability* μ shows whether magnetic induction in the material is greater ($\mu > 1$) or smaller ($\mu < 1$) in comparison with its value in a vacuum (for the same currents, creating the magnetic field).

Substances with $\mu \gg 1$ (iron, cobalt, nickel, some alloys) are called *ferromagnets*. The magnetic field in such materials increases multiple. A characteristic temperature exists for each ferromagnetic material, which is called the *Curie point*. Above this point a ferromagnetic material becomes paramagnetic. Substances with $\mu \sim 1$ or $\mu > 1$ (aluminum, platinum, air) are called *paramagnets*. Materials with $\mu < 1$ are called *diamagnets* (copper, silver, bismuth). In a non-homogeneous magnetic field paramagnets are attracted to the region of a strong field and diamagnets are forced out of it.

Diamagnetism of a substance is an inductive phenomenon that is determined by orbital currents induced by a magnetic field. Diamagnetism of metals is caused by the quantization of the motion of free electrons in a magnetic field. Diamagnetism is a common property of all materials, but it exhibits itself indicatively in materials whose atoms (or ions) don't possess their own magnetic moments. Paramagnetism and ferromagnetism of a substance are usually connected with intrinsic magnetic moments of electrons, which are not caused by their orbital motion. In crystals of ferromagnetic materials the parallel orientation of electrons' magnetic moments is energetically advantageous and magnetized regions with the size of 10^{-2} to 10^{-4} cm appear. These regions are called *domains.* The magnetic field tends to order the fields of separate domains.

15.4 Electromagnetic Induction

15.4.1 Faraday's law

Magnetic flux through a surface S bounded by a plane loop in a homogeneous magnetic field is a scalar product (see Section 8.1.4) of the vectors **B** and **S**:

$$\Phi = (\mathbf{B} \cdot \mathbf{S}) = BS \cos \alpha. \tag{15.78}$$

Here α is the angle between the direction of magnetic induction vector **B** and the direction of normal to the surface S.

A magnetic field that changes in time induces ***a vortex electric field***—a field with closed electric field lines. An induced vortex electric field exhibits itself in a conductor as an action of *extraneous forces* (see Section 15.2.7). This phenomenon is called ***electromagnetic induction.***

Faraday's law of electromagnetic induction: variation of the magnetic flux through the surface, bounded by a closed loop, induces an electromotive force whose value is determined by the rate of change of the magnetic flux:

$$\mathcal{E} = -k\frac{\Delta\Phi}{\Delta t}. \tag{15.79}$$

The coefficient $k = 1$ in SI (see Section 12.2.5). The electromotive force creates an electric current in a conducting loop. The sign "minus" corresponds to the *Lentz rule*: direction of the induction current is determined by the condition that its magnetic field hinders the changing of the magnetic flux that causes the induction current.

An electromotive force appearing due to the movement of a conductor in a constant magnetic field has a different physical nature. In this case the vortex electric field is absent and an extraneous force is a ***Lorentz force***, acting on electric charges that move with the conductor in a magnetic field. The electromotive force is also given by Equation (15.79), which gives the following expression for its value in a conductor of the length l, moving with the velocity v perpendicularly to the lines of the magnetic induction $\mathbf{B}$:

$$\mathcal{E} = -\frac{\Delta\Phi}{\Delta t} = Blv. \tag{15.80}$$

In this case the direction of the induced current according to the Lentz rule can be determined by the ***right-hand rule***: if we place the right hand so that magnetic induction lines enter the palm and the turned back thumb would indicate direction of the conductor movement, then the stretched four fingers would show the electric current direction.

Thus, Equation (15.79) is only the rule for calculation of an electromotive force. It does not reveal the physical reason of the origin of this force.

There are exclusions to this rule (for example, the unipolar induction, shown in Figure 15.25). The magnetic flux through the loop $ABCD$ does not change, but the electric current exists.

15.4.2 Inductance

Any change of current in a conductor gives rise to an induced electromotive force, which is produced by the change of the magnetic flux created

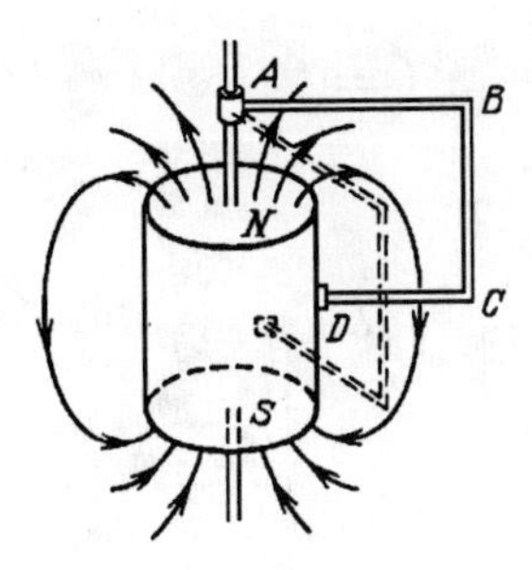

Figure 15.25: An unipolar inductor.

by this current. This phenomenon is called *self-induction*. The electromotive force of self-induction is determined by Equation (15.79), which in this case can be written in the form

$$\mathcal{E} = -L\frac{\Delta I}{\Delta t}, \tag{15.81}$$

where L is called the *inductance* of the circuit. Its value depends on the size and shape of the conductor and on the properties of a media surrounding the conductor. The inductance L relates the current I that is flowing in the loop and creates the magnetic flux Φ through the loop:

$$\Phi = LI.$$

The inductance of a long solenoid with a soft iron core (provided that the ratio of the length of the solenoid to its diameter is more than 10) is given by the expression

$$L = \frac{\mu_0 \mu N^2 S}{l} = \mu_0 \mu n^2 V, \tag{15.82}$$

where N is the number of turns, S is the area of a turn, l is the length of the solenoid, $n = N/l$ is the number of turns per unit of length, and $V = Sl$ is the volume of a solenoid.

15.5 Alternating Electric Current (AC)

15.5.1 AC in circuits with one element

Physical processes that take place in circuits of *sinusoidal alternating current* are stationary (or steady-state) forced *electromagnetic oscillations* (see Section 16.4.2).

Different circuits of alternating current with the applied voltage

$$U = U_0 \cos \omega t \tag{15.83}$$

are investigated below.

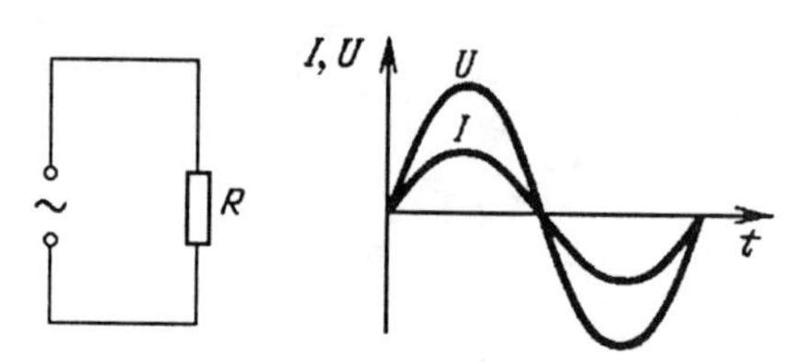

Figure 15.26: The voltage and the current in a circuit with an active resistance.

In the case of a usual *ohmic* (or *active*) resistance R (Figure 15.26) a current is determined by the expressions

$$I(t) = I_0 \cos \omega t, \quad I_0 = U_0/R. \tag{15.84}$$

The current varies in phase with the applied voltage.

In the case of a capacitor with a capacitance C (Figure 15.27) a current in the circuit is determined by the expressions

$$I(t) = I_0 \cos(\omega t + \pi/2), \quad I_0 = U_0 C\omega. \tag{15.85}$$

The current leads the voltage by $\pi/2$ in phase. A capacitor with a capacitance C is associated with *a capacitive reactance* $R_c = 1/(\omega c)$.

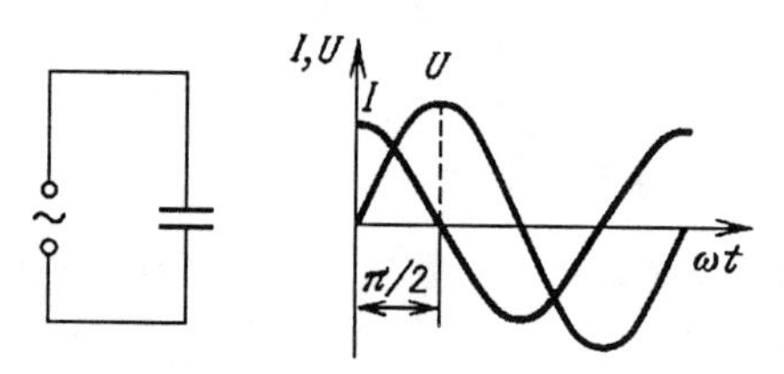

Figure 15.27: The voltage and the current in a circuit with a capacitance.

In the case of a coil with an inductance L (Figure 15.28) a current in the circuit is determined by the expressions

$$I(t) = I_0 \cos(\omega t - \pi/2), \quad I_0 = U_0/(L\omega). \tag{15.86}$$

The current lags behind the voltage by $\pi/2$ in phase. A coil with an inductance L is associated with an *inductive* reactance $R_L = L\omega$.

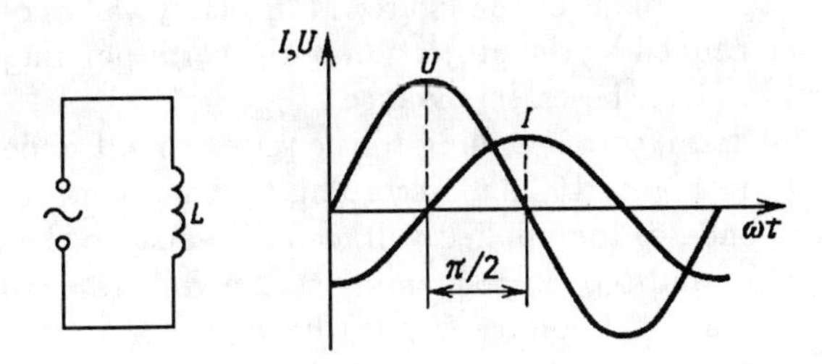

Figure 15.28: The voltage and the current in a circuit with an inductance coil.

15.5.2 Series RLC-circuit

When an active resistance R and reactive resistances R_C and R_L are connected *in series* a current is the same in any part of the circuit at every time instant (Figure 15.29), and the sum of momentary values of voltages U_R, U_C, and U_L is equal to the applied voltage at the same time instant:

$$U = U_R + U_C + U_L. \tag{15.87}$$

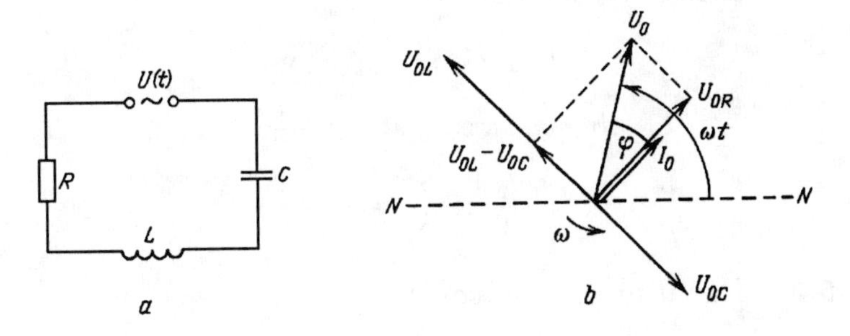

Figure 15.29: A circuit containing a resistance, an inductance, and a capacitance, and the corresponding vector diagram.

It is convenient to find the relations between currents and voltages for different loads with the help of vector diagrams (see Section 16.4.2). Each physical quantity varying in time according to sinusoidal law is associated with a vector (a *phasor*) whose length is equal to the amplitude value of the corresponding quantity, and the angles between different vectors are equal to the phase shifts between corresponding quantities. All system of vectors rotates as a whole with an angular velocity ω around the axis, perpendicular to the plane of the figure. Momentary values of I, U_R, U_L, and U_C are determined by the projection of the corresponding vectors on certain direction NN, chosen in advance.

A vector diagram, corresponding to the series circuit under consideration, is shown in Figure 15.29*b*. According to Equation (15.87), vector U_0 that corresponds to the applied voltage U is equal to the sum of the vectors U_{0R}, U_{0C}, and U_{0L} corresponding to the voltages across the separate loads. The applied voltage U and the current I in the circuit are written in the form

$$U(t) = U_0 \cos \omega t, \quad I(t) = I_0 \cos(\omega t - \varphi). \tag{15.88}$$

The quantities I_0 and φ are determined with the help of the vector diagram on Figure 15.29:

$$I_0 = \frac{U_0}{\sqrt{R^2 + (L\omega - 1/(\omega C))^2}}, \qquad \tan\varphi = \frac{L\omega - 1/(\omega C)}{R}; \tag{15.89}$$

The relation between the amplitude of the current and the amplitudes of the voltages on separate elements of the circuit is as follows:

$$U_{0R} = I_0 R, \quad U_{0C} = I_0/(\omega C), \quad U_{0L} = I_0 L_\omega. \tag{15.90}$$

Momentary values of the voltages on the separate elements of the circuit are given by the formulas

$$\begin{aligned} U_R &= I_0 R \cos(\omega t - \varphi), \\ U_L &= I_0 L\omega \cos\left(\omega t - \varphi + \frac{\pi}{2}\right), \\ U_C &= \frac{I_0}{\omega C} \cos\left(\omega t - \varphi - \frac{\pi}{2}\right). \end{aligned} \tag{15.91}$$

15.5.3 Parallel RLC-circuit

In the circuit with the *parallel* grouping of an active resistance R and reactive resistances R_L and R_C (Figure 15.30), a momentary value of the current in the unbranched part of the circuit is equal to the sum of the currents in all parallel parts:

$$I = I_R + I_C + I_L, \tag{15.92}$$

and a momentary value of the voltage is the same across all the loads. A vector diagram for the case is shown in Figure 15.30. The applied voltage U and the current I in the circuit are determined by the expressions

$$U(t) = U_0 \cos \omega t, \quad I(t) = I_0 \cos(\omega t - \varphi), \tag{15.93}$$

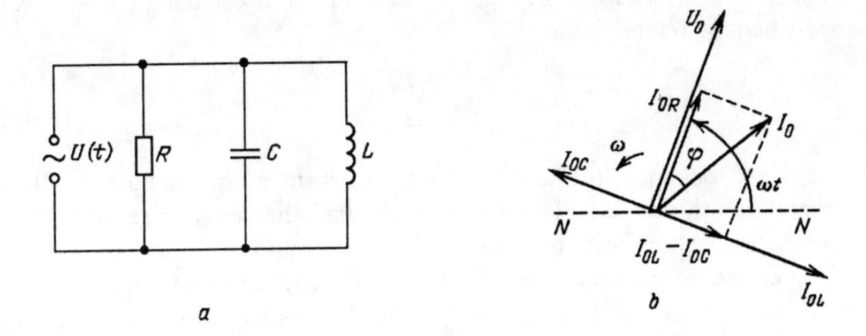

Figure 15.30: A circuit with the parallel grouping of a resistance, an inductance, and a capacitance, and the corresponding vector diagram.

where

$$I_0 = U_0\sqrt{\frac{1}{R^2} + \left(\frac{1}{L\omega} - \omega C\right)^2}, \qquad \tan\varphi = R\left(\frac{1}{L\omega} - \omega C\right). \quad (15.94)$$

The amplitude values of the currents and of the voltage are connected by the relations

$$U_0 = I_{0R}R = I_{0C}\frac{1}{\omega C} = I_{0L}L\omega. \quad (15.95)$$

Momentary values of currents in separate parts of the circuit are given by the formulas:

$$I_R = \frac{U_0}{R}\cos\omega t,$$
$$I_L = \frac{U_0}{L\omega}\cos\left(\omega t - \frac{\pi}{2}\right), \quad (15.96)$$
$$I_C = U_0 C\omega\cos\left(\omega t + \frac{\pi}{2}\right).$$

Vector diagrams can be constructed for all fashions of connecting of the loads R, L, and C in a circuit.

15.5.4 Impedance of a circuit

If one specifies the position of the vectors, associated with currents and voltages on vector diagrams, with the help of complex numbers (see Section 7.3.1),

$$U = U_0(\cos\omega t + i\sin\omega t), \quad \tilde{I} = I_0[\cos(\omega t - \varphi) + i\sin(\omega t - \varphi)], \quad (15.97)$$

then the *Ohm's law* for a segment of a circuit of alternating current can be represented in the form

$$\tilde{I} = \frac{\tilde{U}}{Z}, \tag{15.98}$$

where $Z = Z_0(\cos\varphi + i\sin\varphi)$ is some constant complex number that characterizes the circuit. This number is called the *complex resistance* or *impedance* of the circuit (here $Z_0 = |Z|$, $\varphi = \arg Z$).

To determine the impedance Z of a circuit with an alternating current, we should attribute definite values of complex resistance to the elements R, L, and C of the circuit according to the rule

$$R \to X_R = R, \quad L \to X_L = iL\omega, \quad C \to X_C = \frac{1}{i\omega C}. \tag{15.99}$$

Then these resistances should be combined according to the rules of combining resistances in circuits of a direct current, using the rules of operation with complex numbers (see Section 7.2). Transforming the obtained expression for an impedance to the form

$$Z = \operatorname{Re} Z + i \operatorname{Im} Z, \tag{15.100}$$

we obtain the formulas for the amplitude I_0 of the current in the circuit and for the phase shift φ between the applied voltage and the current in the circuit:

$$I_0 = \frac{U_0}{\sqrt{(\operatorname{Re} Z)^2 + (\operatorname{Im} Z)^2}}, \quad \tan\varphi = \frac{\operatorname{Im} Z}{\operatorname{Re} Z}. \tag{15.101}$$

Examples.

1. For a serial circuit shown in Figure 15.29, we have

$$Z = R + iL\omega + \frac{1}{i\omega C} = R + i\left(L\omega - \frac{1}{\omega C}\right). \tag{15.102}$$

Then with the help of (15.101) we obtain (see Section 7.2)

$$I_0 = \frac{U_0}{\sqrt{R^2 + (L\omega - 1/(\omega C))^2}}, \quad \tan\varphi = \frac{L\omega - 1/(\omega C)}{R}. \tag{15.103}$$

These expressions coincide with the formulas (15.89).

2. For a parallel circuit shown in Figure 15.30, we have

$$\frac{1}{Z} = \frac{1}{R} + \frac{1}{iL\omega} + i\omega C. \tag{15.104}$$

Transforming this expression to the form (15.100), we obtain

$$Z = \frac{L^2\omega^2 R + iL\omega R^2(1 - \omega^2 LC)}{L^2\omega^2 + R^2(1 - \omega^2 LC)^2}. \tag{15.105}$$

Using the expressions (15.101), we get the formulas (15.94).

15.5.5 Resonance of voltages and resonance of currents

Resonance of voltages takes place in a *series* circuit of alternating current (Figure 15.29) if the condition $\omega^2 = 1/(LC)$ is satisfied. In this case the amplitude of the current strength reaches its maximum equal to U_0/R and the phase shift between the voltage and the current is absent. Voltages U_L on an inductance L and U_C on a capacitance C are equal in magnitudes and change with time in opposite phases. The total voltage across the circuit coincides with the voltage U_R across the active resistance R. The amplitudes of the voltages across reactive resistances can significantly exceed the voltage applied to the circuit.

Resonance of currents takes place in a *parallel* circuit of an alternating current containing an inductance L and a capacitance C if $\omega^2 = 1/(LC)$ (this circuit corresponds to the circuit shown in Figure 15.30 if the condition $R \to \infty$ is fulfilled). At resonance of currents, the current strength in the unbranched part of the circuit tends to zero, while the currents I_L and I_C achieve considerably large amplitudes and change in antiphase.

15.5.6 Power of alternating current

Momentary power in a circuit of an alternating current is given by the expression

$$P(t) = IU = U_0 I_0 \cos\omega t \cos(\omega t - \varphi). \tag{15.106}$$

Using the trigonometric formula for a product of cosines of two angles (see Section 4.2.2) we get

$$P(t) = \frac{1}{2} U_0 I[\cos\varphi + \cos(2\omega t - \varphi)]. \tag{15.107}$$

The mean value of the second term in square brackets over the period $T = 2\pi/\omega$ equals zero. Therefore, it is possible to neglect this term while calculating the work performed during an interval of time that is much greater than T ($t \gg T$). In this case we have

$$P = \frac{1}{2} U_0 I_0 \cos\varphi, \tag{15.108}$$

$$A = Pt = \left(\frac{1}{2} U_0 I_0 \cos\varphi\right) t. \tag{15.109}$$

Introducing the *effective values* of the voltage $U = U_0/\sqrt{2}$ and of the current $I = I_0/\sqrt{2}$, we can rewrite the expression for the power in the form

$$P = UI\cos\varphi. \tag{15.110}$$

15.5.7 Transformer

A passive device that is used to change the voltage up or down is called a *transformer* (Figure 15.31). It usually operates with an efficiency close to 99%. A basic transformer consists of two coils, a primary and a secondary, wound on the same iron core. The sinusoidal current in the primary coil produces a sinusoidally changing magnetic field in the core. Due to the phenomenon of electromagnetic induction the electromotive forces are induced in both coils. The ratio of the electromotive forces is equal to the ratio of numbers of turns in the coils if the same magnetic flux penetrates both coils:

$$\frac{\mathcal{E}_1}{\mathcal{E}_2} = \frac{n_1}{n_2}. \tag{15.111}$$

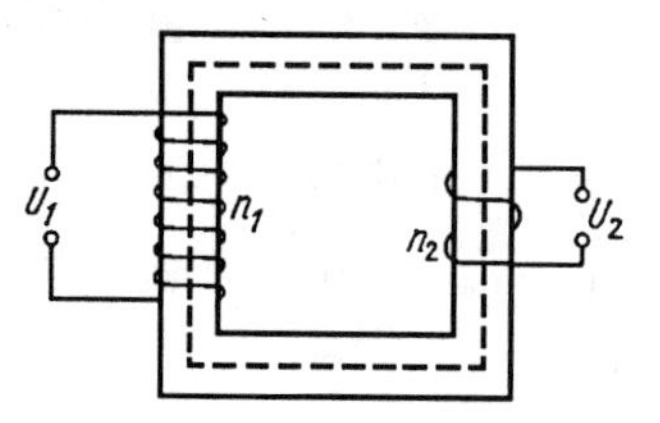

Figure 15.31: A circuit with the parallel grouping of a resistance, an inductance, and a capacitance, and the corresponding vector diagram.

If the secondary coil is not closed, the induced voltage in it is equal to the electromotive force induced in it:

$$U_2 = \mathcal{E}_2.$$

If the inductive reactance $L\omega$ of the primary coil is much larger than its active resistance R ($L\omega \gg R$), the primary voltage U_1 is equal to the voltage on the inductive reactance $U_L = -\mathcal{E}_1$. Hence if the secondary coil is not closed, then

$$U_2 = -\frac{n_2}{n_1} U_1. \qquad (15.112)$$

The sign "minus" means that these voltages are in antiphase. By virtue of a large inductive reactance of the primary coil, the current in it is small if the secondary coil is not closed.

If a current is drawn from the secondary coil, the secondary current tends to produce a change in the magnetic flux in the core. But the magnitude of the flux is maintained by driving electromotive force of the generator and must remain essentially unchanged. The current I_0 increases to counteract the effect of the secondary current: the primary coil has to draw a current from the source which changes in antiphase with the secondary current. Thus, loading the secondary coil is equivalent to reducing the inductive reactance of the primary coil.

If the current in the secondary coil is in a resistive load, the current is in phase with the secondary electromotive force. The extra current drawn by the primary is also in phase with the primary electromotive force, and hence power is drawn from the generator. For an ideal transformer in which all losses are eliminated, primary power drawn equals secondary power consumed. In the general case,

$$\frac{1}{2} U_1 I_1 \cos\varphi_1 = \frac{1}{2} U_2 I_2 \cos\varphi_2, \qquad (15.113)$$

where I_1 and I_2 are the phase shifts in the primary and in the secondary.

15.6 Electromagnetic Field

15.6.1 Relative character of electric and magnetic fields

Electric and magnetic fields are of a relative character. Electric charge stationary in one reference frame occurs in motion with respect to another reference frame. Such a moving charge is similar to an electric current—it creates a magnetic field. Hence if there is only electric field in some reference frame, there would be also a magnetic field in any other one.

According to the *principle of relativity* (see Section 18.1), all inertial reference frames are equivalent and physical laws have the same form in any inertial system. Thus, due to the relative character of electric and magnetic fields, they should be treated together as a unique electromagnetic field. If the relative velocity of two reference frames is small ($v \ll c$), the formulas for transformation of the fields when we change from one reference frame to another one are written in the form

$$\mathbf{E}' = \mathbf{E} + \mathbf{v} \times \mathbf{B}, \quad \mathbf{B}' = \mathbf{B} - \varepsilon_0\mu_0\mathbf{v} \times \mathbf{E}. \tag{15.114}$$

Here $\varepsilon_0\mu_0 = 1/c^2$ (c is the velocity of light in a vacuum).

15.6.2 Invariants of electromagnetic field

At the transition from one inertial system to another, two *invariant* combinations of the vectors $\mathbf{E}$ and $\mathbf{B}$ exist that do not change their values:

$$\mathbf{E} \cdot \mathbf{B} = \mathbf{E}' \cdot \mathbf{B}', \quad E^2 - c^2B^2 = E'^2 - c^2B'^2. \tag{15.115}$$

These relations do not depend on the relative velocity of the reference frames. If electric and magnetic fields are mutually perpendicular in some inertial reference frame, they would be mutually perpendicular in any other one. For such mutually perpendicular fields a reference frame exists where either $\mathbf{B} = 0$ or $\mathbf{E} = 0$, depending on the fact whether the invariant $E^2 - c^2B^2$ is positive or negative.

Both invariants expressed by Equations (15.115) are equal to zero for the mutually perpendicular electric and magnetic fields of a traveling electromagnetic wave (see Section 16.7.12). Hence the ratio $E = cB$, as well as the orthogonality of $\mathbf{E}$ and $\mathbf{B}$ at the same space-time point is valid in all inertial reference frames.

There is a certain symmetry between electric and magnetic phenomena: a magnetic field changing in time generates a solenoidal electric field and (conversely) an electric field, changing in time, generates a magnetic field. This property was predicted theoretically by Maxwell—the law of electric charge conservation can be fulfilled only if this property is valid.

15.6.3 Maxwell's equations

The basic concepts of the theory of electromagnetic field were developed by Maxwell by way of generalization of experimental data. The system of Maxwell's equations contains all the laws of electromagnetism, discovered experimentally.

Analyzing these equations, Maxwell predicted the possibility of the existence of electric and magnetic fields connected with each other and propagating through free space with the velocity of light—*electromagnetic waves* (see Section 16.7.12). These waves were later discovered experimentally by Hertz.

The system of Maxwell's equations contains four principal laws of electromagnetic phenomena.

The first law is the law of interaction of electric charges—Coulomb's law (see Section 15.1.1). This law can also be formulated as the so-called Gauss' theorem. To formulate this theorem the *flux of the field strength vector* $\mathbf{E}$ of an electric field through the element of a surface ΔS

should be introduced. The flux of the field strength vector is introduced as a product $E_n \Delta S$, where E_n is the projection of the vector $\mathbf{E}$ on the direction of a normal n to the surface element ΔS. The flux through the arbitrary surface is determined as an algebraic sum of fluxes through its elements. The physical sense of the flux N is the number of force lines of the electric field intersecting the surface if one considers that the "density" of the lines (i.e., the number of lines intersecting a small surface of unit area) equals the magnitude of the field strength at the point. *Gauss' theorem* (or Gauss' law) reads: the flux N of the electric field strength vector $\mathbf{E}$ in vacuum through any closed surface is proportional to the total electric charge q inside the surface:

$$N = q/\varepsilon_0.$$

This statement can be treated as if every positive charge q generated q/ε_0 force lines, going out of the charge. Similarly, every negative charge attracts $|q|/\varepsilon_0$ force lines, ending on the charge.

The second law is *Gauss' theorem for a magnetic field*, according to which the flux of the magnetic field induction vector through any closed surface equals zero. The theorem reflects the absence of magnetic charges in nature and stresses the vortex character of a magnetic field: the lines of magnetic field induction are always closed.

The third law is *Faraday's law of electromagnetic induction*, according to which a changing magnetic field generates a vortex electric field (see Section 15.4.10).

The *fourth law* is the generalization of Biot–Savart–Laplace law: magnetic field can be created either by moving electric charges (currents) or by changing an electric field.

Maxwell's equations satisfy the principle of relativity (see Section 18.1) and do not change at a transition from one inertial reference frame to another.

Chapter 16

Oscillations and Waves

Alternating processes or movements which, to some extent, repeat themselves in time are usually called *oscillations* or *vibrations.* Oscillations in physical systems are followed by alternate transformations of energy from one kind to another and back. According to the physical nature of the phenomena involved, oscillations are classified as *mechanical* and *electromagnetic* ones and their combinations, e.g., oscillations in plasma. Vibrations of different physical nature have much in common and are the subject of the *theory of oscillations.*

16.1 Classification of Oscillations

1. According to *kinematics*, i.e., to the character of dependence of a certain physical quantity $x(t)$ on time, oscillations are classified as *periodic* and *non-periodic.* For periodic oscillations any value of $x(t)$ repeats itself after identical intervals T called the *period* of oscillation:

$$x(t+T) = x(t).$$

The value inverse to the period T and equal to the number of oscillations (of full cycles) per unit time is called the *frequency* ν (or f) of oscillations:

$$\nu = \frac{1}{T}.$$

The frequency ν is measured in *hertz* (Hz), named after Heinrich Hertz, a German physicist. During oscillations with the frequency of 1 Hz, one complete oscillation (one cycle) per one second is executed.

The most important of periodic oscillations are *sinusoidal* or *harmonic* oscillations (see Figure 16.1, where a – harmonic (sinusoidal) oscillations,

b – square-wave oscillations, c – saw-tooth oscillations, d – damped oscillations, e – amplitude-modulated oscillations, and f – frequency-modulated oscillations).

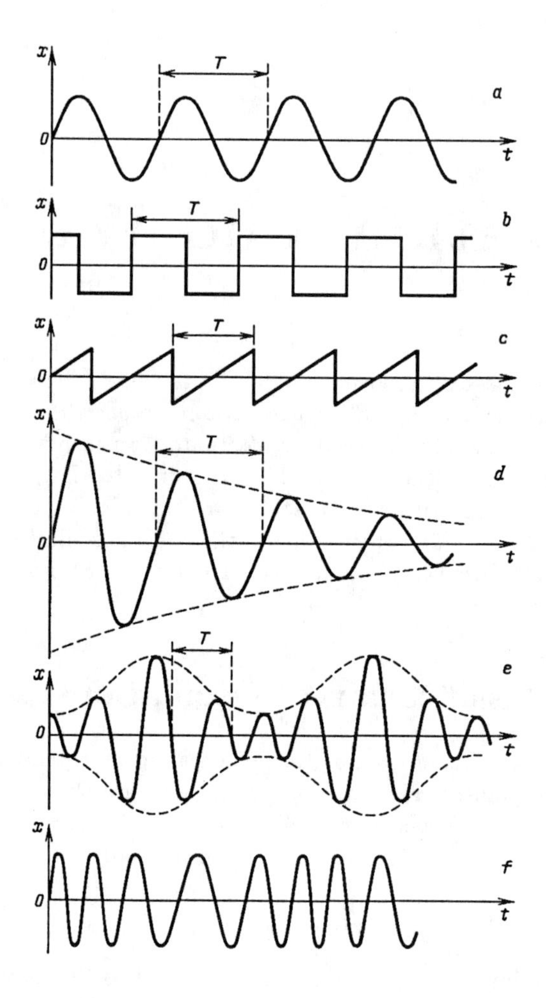

Figure 16.1: Oscillations with different shape of time-dependence.

The most important non-periodic oscillations are almost sinusoidal ones, characterized by a slow variation of the amplitude (Figure 16.1d, e), phase, or frequency (Figure 16.1f). If the amplitude changes slowly, the interval between consecutive passages of the oscillating quantity through zero in the same direction is conventionally assumed to be the period of oscillations. Almost sinusoidal oscillations with slowly varying amplitude are used in radio communication: an electromagnetic wave of a high frequency (a carrier wave) is modulated by oscillations of a low (acoustic) frequency, that transfers the information.

2. According to the *means of excitation*, oscillations are divided into *free* or *natural* ones, excited by some initial action on an isolated system which is left to itself afterwards; *forced oscillations*, occurring under periodic external action; *parametric oscillations*, occurring at a periodic change of some parameter of the oscillating system (pendulum length, capacity, or an inductance of a circuit, etc.); and *self-excited oscillations*, occurring in systems with a feedback which can regulate the supply of energy from a constant source to compensate losses.

3. According to *complexity*, oscillatory systems are divided into *simple* systems (with a single degree of freedom), characterized by a single natural frequency; systems with *concentrated* or *lumped parameters*, which have a finite number of natural frequencies (those of normal oscillations) coinciding with the number of degrees of freedom; and systems with *distributed parameters*, having an infinite number of degrees of freedom and, consequently, an infinite number of natural frequencies.

16.2 Harmonic Oscillations

16.2.1 Kinematics of simple harmonic motion

Harmonic oscillations are described by the function

$$x(t) = A\cos\varphi(t) = A\cos(\omega t + \varphi_0) \tag{16.1}$$

or

$$x(t) = A\cos(2\pi\nu t + \varphi_0) = A\cos(\frac{2\pi}{T}t + \varphi_0), \tag{16.2}$$

where A is the *amplitude*, $\varphi(t)$ is the *phase*, φ_0 is the phase value at $t = 0$ (*initial phase*), $\omega = 2\pi\nu$ is the *angular frequency* measured in the same units as angular velocity ω, i.e., in radians per second (see Section 13.1.5). If $x(t)$ is the coordinate of a body executing harmonic oscillations, then its velocity is $v_x(t) = dx/dt = \dot{x}$ and the acceleration is $a_x(t) = dv_x/dt = d^2x/dt^2 = \ddot{x}$. Both of these quantities also change in time harmonically:

$$\begin{aligned} v_x(t) &= \dot{x}(t) = -A\omega\sin(\omega t + \varphi_0) = A\omega\cos(\omega t + \varphi_0 + \pi/2), \\ a_x(t) &= \ddot{x}(t) = -A\omega^2\cos(\omega t + \varphi_0), \end{aligned} \tag{16.3}$$

i.e., oscillations of velocity lead the phase of oscillations of the coordinate by $\pi/2$ (by a quarter of period), and oscillations of acceleration lead the phase of the coordinate by π (or lag by π), i.e., go on in antiphase with those of the coordinate. Plots $x(t)$, $\dot{x}(t)$, and $\ddot{x}(t)$ are shown in Figure 16.2.

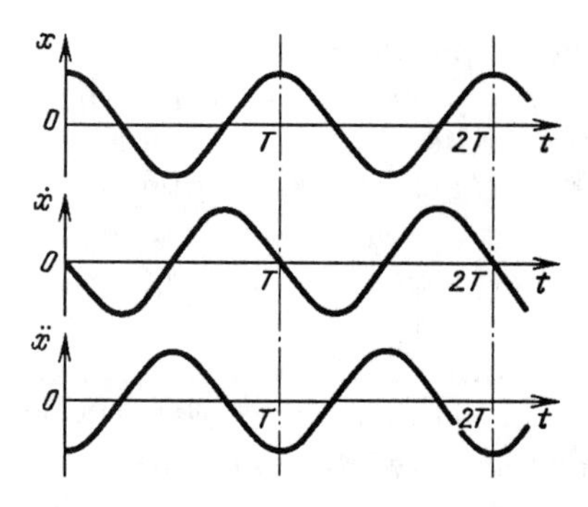

Figure 16.2: The graphs of displacement, velocity, and acceleration for harmonic oscillations.

16.2.2 Vector diagrams for harmonic oscillations

To have a better idea of kinematic characteristics of harmonic oscillations it is convenient to consider them in interconnection with a uniform circular motion of a point (Figure 16.3). Radius-vector $\mathbf{r}$, whose modulus is equal to A, is rotating counter-clockwise with a constant angular velocity ω so that the angle φ formed by the radius with the axis x is growing linearly with time: $\varphi(t) = \omega t + \varphi_0$. Components of radius-vector $\mathbf{r}$, vectors of velocity $\mathbf{v}$, and acceleration $\mathbf{a}$ along the axis x are changing harmonically in accordance with Equations (16.1) through (16.3). Such a graphic method of representation of harmonic oscillations is called a *vector diagram*: any quantity that executes harmonic oscillation is associated with a projection of a uniformly rotating vector called a *phasor.* The modulus of a phasor is equal to the amplitude of the oscillation in question, and the phasor rotates with the angular velocity that is equal to the angular frequency ω of the oscillation. It is worth mentioning that this vector should not be confused with the usual vector physical quantities.

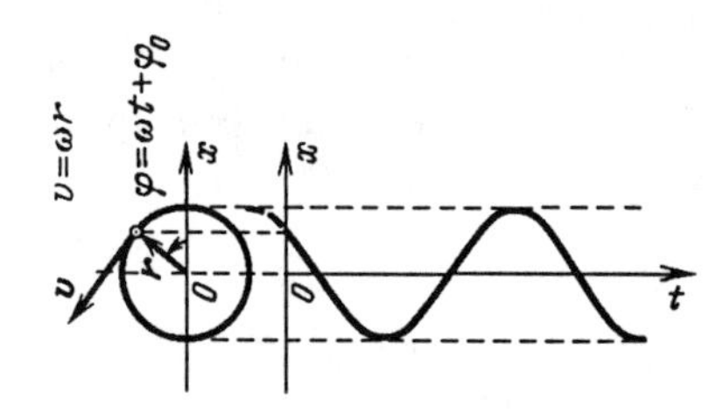

Figure 16.3: Uniform circular motion and harmonic oscillations (a vector diagram for harmonic oscillations).

In all cases where some harmonic quantities are subjected to linear operations (multiplication by a number, addition, differentiation, integration), these operations can be done directly with the corresponding phasors, and then the projection on the x-axis can be taken in the final result.

16.3 Natural Oscillations of Simple Systems

16.3.1 Differential equation of harmonic oscillator

If a physical system is disturbed from the state of stable equilibrium by some external action and then left to itself, oscillations occurring in the system in the vicinity of the stable equilibrium are called *natural*, or *free* oscillations. A system able to execute natural oscillations is called an oscillator. Examples of oscillators are shown in Figure 16.4.

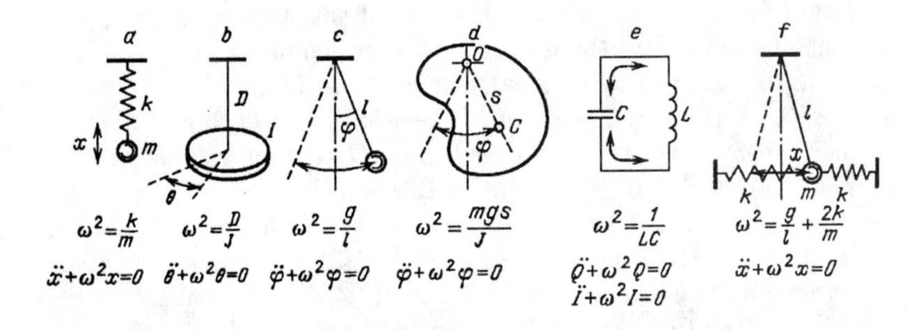

Figure 16.4: Examples of oscillators: a – spring pendulum, b – torsion pendulum (disc on an elastic wire), c – simple (mathematical) pendulum, d – physical pendulum, e – oscillatory circuit, and f – compound pendulum.

Natural oscillations in the absence of friction, that is, in idealized *conservative systems* (see Section 13.4.7), are *non-damped* and exactly *periodic.* If the restoring force, i.e., the force tending to put the conservative mechanical system back into equilibrium, is proportional to the displacement, natural oscillations are *harmonic.* For example, in a spring pendulum (Figure 5.4a) the restoring elastic force F of the strained spring is proportional to the shift x from the equilibrium position:

$$F = -kx, \tag{16.4}$$

where the coefficient of proportionality k is called the *spring constant.*

In the absence of other forces, the differential equation (see Section 16.5.1) of motion of mass m (equation of Newton's second law) is the following:

$$m\frac{d^2x}{dt^2} = -kx, \qquad \text{or} \qquad \ddot{x} + \omega^2 x = 0, \tag{16.5}$$

where the quantity $\omega^2 \equiv k/m$ characterizing the system is introduced. The sense of ω is the frequency of free oscillations.

Equations for other oscillators have the same form if in each of them $x(t)$ is understood as a corresponding quantity characterizing deflection from the equilibrium position: angle θ of twisting of elastic suspension in Figure 16.4*b*; angle φ of deviation from the vertical line of a mathematical or a physical pendulum in Figure 16.4*c*, *d*; charge Q of the capacitor or current I in the oscillatory circuit in Figure 16.4*e*. The factor ω^2 standing together with x in Equation (16.5) equals the square of the angular frequency of natural oscillations (see Section 16.2.1). The angular frequency depends on the characteristics of the oscillator: on the spring rigidity (spring constant) k and mass m of the weight in a spring pendulum ($\omega^2 = k/m$); on the modulus of suspension twisting D and moment of inertia J of the disc in a torsion pendulum ($\omega^2 = D/J$); on the strength g of the gravitational field and length l in a simple pendulum ($\omega^2 = g/l$); on the distance s from suspension point O to mass center C, on the moment of inertia J relative to axis O, and on mass m in a physical pendulum ($\omega^2 = mgs/J$); and on the capacitance C of the capacitor and inductance L of the coil in the oscillatory circuit ($\omega^2 = 1/(LC)$).

For systems where several restoring forces are acting simultaneously (e.g., the compound pendulum in Figure 16.4*f*), coefficient ω^2 in Equation (16.5), i.e., the natural frequency squared, equals the sum of the squares of the natural frequencies which would be under separate restoring forces: $\omega^2 = \omega_1^2 + \omega_2^2 + \ldots$. In the compound pendulum (Figure 16.4*f*) $\omega_1^2 = g/l$, due to contribution of the gravitational force into the restoring force, $\omega_2^2 = 2k/m$ due to the force of spring elasticity.

In the presence of friction in an oscillatory system (or electric resistance in an oscillatory circuit), damping of natural oscillations occurs because of the energy dissipation (see Section 16.3.5).

16.3.2 Initial conditions

The *general solution* of Equation (16.5) is the harmonic oscillation expressed by Equation (16.1) with an arbitrary amplitude A and initial phase φ_0. The values of A and φ_0 are determined by the *initial conditions*, i.e., by way of excitation of oscillations. For example, if the system is deflected from the equilibrium position through the distance x_0 and released without an initial push (the initial conditions $x(0) = x_0$ and $\dot{x}(0) = 0$), then $a = x_0$ and $\varphi_0 = 0$:

$$x(t) = x_0 \cos \omega t. \tag{16.6}$$

If there is an initial push, that is, if the system is excited from the equilibrium position with an initial speed ($x(0) = 0$ and $\dot{x}(0) = v_0$), then $a = v_0/\omega$ and $\varphi_0 = -\pi/2$:

$$x(t) = \frac{v_0}{\omega} \sin \omega t. \tag{16.7}$$

In the general case of initial excitation $x(0) = x_0$ and $x(0) = v_0$. The amplitude and initial phase of the corresponding *particular* solution of Equation (16.5) are expressed by the formulas

$$a = \sqrt{x_0^2 + \left(\frac{v_0}{\omega}\right)^2}, \qquad \tan \varphi_0 = -\frac{v_0}{\omega x_0}. \tag{16.8}$$

The frequency ω of oscillations, unlike the amplitude and the initial phase, does not depend on the way of excitation, but is wholly determined by the properties of the system. Independence of the period and frequency of free oscillations on initial conditions is called the *isochronism* of a harmonic oscillator.

16.3.3 Transformations of energy in oscillations

Natural oscillations are accompanied by *energy transformations* from one kind of energy to another and backwards (Figure 16.5). At the points of maximum deflection of a mechanical oscillator from equilibrium (at the *turning points*), the kinetic energy becomes zero, and the total energy of the oscillator is the *potential* energy of the strained spring (in the spring oscillator), of the twisted suspension (in the torsion pendulum), or of the weight in gravitational field (in mathematical and physical pendulums). In a quarter of a period the oscillator passes through the equilibrium point. At this moment the potential energy becomes zero, and the total energy of the oscillator is the *kinetic* energy of the weight. During the next quarter of a period the reverse energy transformation takes place: the kinetic energy converts into the potential one. Such transformations occur twice during one period. Dependence of the potential energy on time (for a spring pendulum) is expressed by the equation

$$W_{\text{pot}}(t) = \frac{1}{2}kx^2(t) = \frac{1}{2}kA^2 \cos^2(\omega t + \varphi_0) = \frac{1}{4}kA^2[1 + \cos 2(\omega t + \varphi_0)].$$

Oscillations of the potential and kinetic energies occur with the frequency 2ω in opposite phases (see Figure 16.5). The sum of these energies (i.e., the total energy of the oscillator) remains constant and is equal to the maximum values of the kinetic and the potential energies. Their values, averaged over a period, coincide:

$$\langle W_{\text{kin}}(t) \rangle = \langle W_{\text{pot}}(t) \rangle.$$

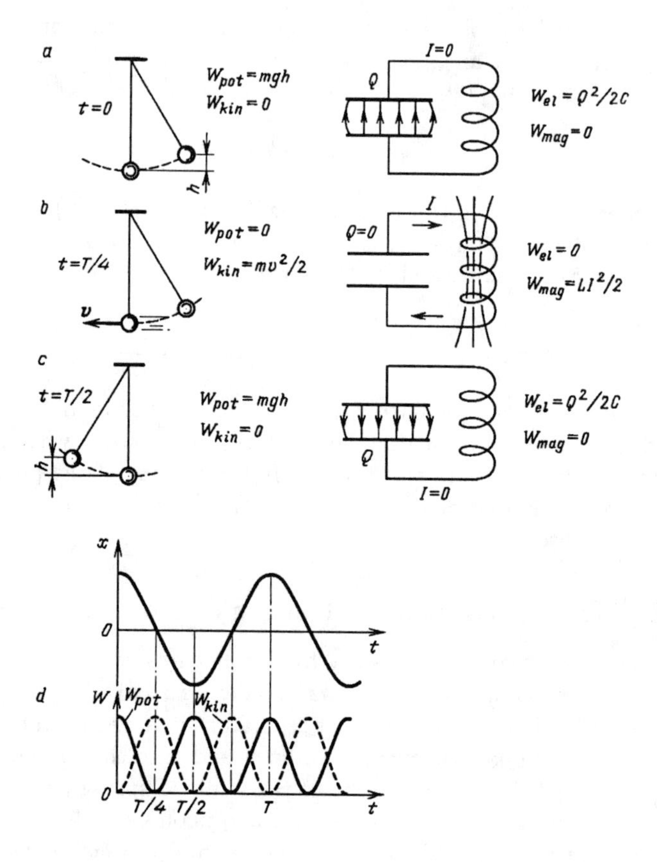

Figure 16.5: Transformations of energy at oscillations (a – c) and the plots of energy time dependence (d).

For an oscillatory circuit, the energy of the electric field in the capacitor is the analogue of the potential energy of a mechanical oscillator, and the energy of the magnetic field in the inductance coil is the analogue of the kinetic energy of the weight (see Figure 16.5).

16.3.4 Nonlinear free oscillations

If natural oscillations are harmonic and their frequency does not depend on the amplitude, the *restoring force* must be *proportional to the elongation* ($F = -kx$ for a spring pendulum), and the *potential energy* must be *proportional to the elongation squared*: $W_{\text{pot}} = kx^2/2$. Such oscillatory systems are called *linear systems* because their behavior is described by linear differential equations (the desired function and its derivatives are to the first power). Mathematical (simple) and physical pendulums can be considered approximately as linear systems only for small angles of deflection from the vertical line (Figure 16.6, see also Section 10.2.1):

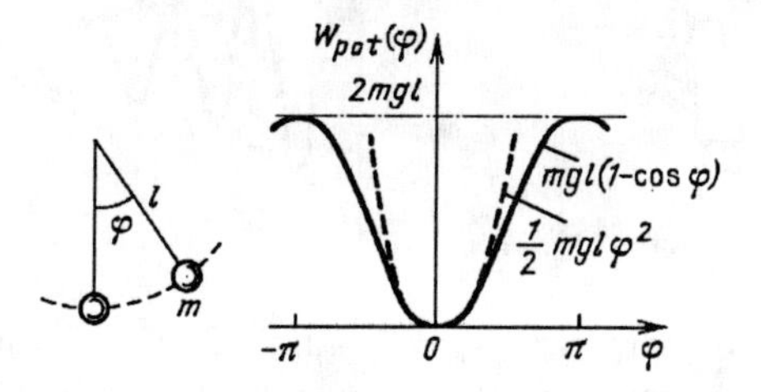

Figure 16.6: Graph of potential energy for a pendulum (solid curve) and its approximation with a parabolic potential well (dotted curve) for a harmonic oscillator.

$$W_{\mathrm{pot}}(\varphi) = mgl(1 - \cos\varphi) \approx \frac{1}{2} mgl\,\varphi^2.$$

For large amplitudes the pendulum is described by the following *nonlinear* equation ($\varphi(t)$ enters the argument of the sine function):

$$\ddot{\varphi} + \omega_0^2 \sin\varphi = 0.$$

Large oscillations of the pendulum are *unharmonic*, and their period depends on the amplitude φ_m. For relatively small amplitudes

$$T \approx T_0(1 + \varphi_m^2/16), \qquad T_0 = 2\pi\sqrt{l/g}.$$

16.3.5 Damped natural oscillations

In real oscillatory systems the dissipation of mechanical (or electromagnetic) energy occurs due to friction (or electric resistance), and natural oscillations inevitably damp. It is a case of common occurrence that the frictional force is proportional to the velocity: $F = -\lambda v$ (viscous friction), and the differential equation of motion (Newton's second law), e.g., for a weight of mass m on a spring of rigidity k is the following:

$$m\ddot{x} = -kx - \lambda\dot{x} \qquad \text{or} \qquad \ddot{x} + 2\gamma\dot{x} + \omega_0^2 x = 0, \tag{16.9}$$

where $\omega_0 = \sqrt{k/m}$ is the angular frequency of natural oscillations in the absence of friction, $\gamma = \lambda/2m$ is the damping constant. The equation for the current in a circuit (Figure 16.7) containing a series connection of a capacitor, an inductance coil, and a resistor (the role of the latter may be played by the resistance of the coil and conductors) is reduced to the same form, Equation (16.9).

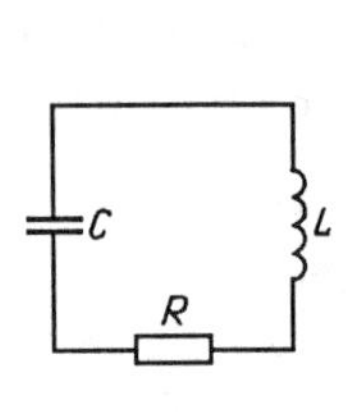

Figure 16.7: Oscillatory circuit with electric resistance.

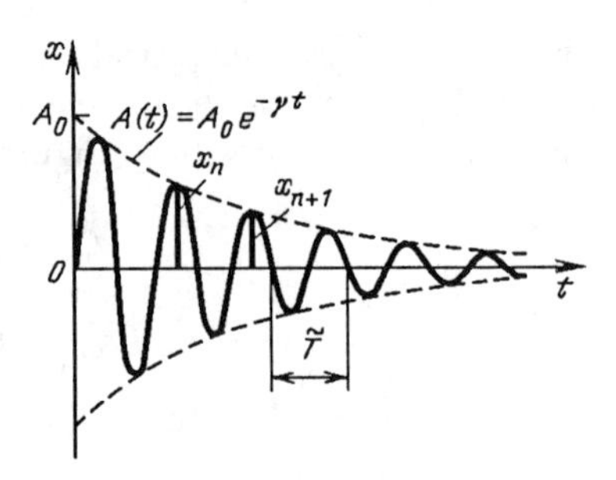

Figure 16.8: Damping of oscillations at viscous friction.

The general solution of Equation (16.9) has the form (Figure 16.8):

$$x(t) = A_0 e^{-\gamma t}\cos(\tilde{\omega} t + \varphi_0), \qquad \tilde{\omega} = \sqrt{\omega_0^2 - \gamma^2}, \tag{16.10}$$

where A_0 and φ_0 are arbitrary constants with their values determined by initial conditions (i.e., they depend on the method of excitation of natural oscillations). In the case of *weak damping* ($\gamma \ll \omega_0$) the frequency $\tilde{\omega}$ is practically the same as ω_0:

$$\tilde{\omega} = \omega_0 \sqrt{1 - \frac{\gamma^2}{\omega_0^2}} \approx \omega_0 \left(1 - \frac{\gamma^2}{2\omega_0^2}\right) = \omega_0 - \frac{\gamma^2}{2\omega_0} \tag{16.11}$$

(see Section 10.2.1). In Equation (16.10) the factor $A_0 e^{-\gamma t}$ standing before the cosine can be considered as the amplitude of oscillations, which slowly changes in time. The amplitude decreases by a factor of $e \approx 2.72$ within the time interval $\tau = 1/\gamma$ (the *decay time* or the time of damping of the amplitude). The condition of weak damping $\gamma \ll \omega_0$ means that $\tau \gg \tilde{T}$, i.e., within the time $\tau = 1/\gamma$ many oscillations occur. The period $\tilde{T}$ of slowly damping oscillations is conventionally assumed to be equal to the period of cosine cofactor in Equation (16.10). Successive maximum deflections in the case of weak damping decrease in a geometric progression (Figure 16.8, see Section 5.1.6):

$$\frac{x_{n+1}}{x_n} = \frac{A_0 e^{-\gamma(t+\tilde{T})}}{A_0 e^{-\gamma t}} = e^{-\gamma \tilde{T}} \approx 1 - \gamma \tilde{T}.$$

To characterize damping, the dimentionless logarithmic decrement (damping factor) is used, defined as the logarithm of the ratio of consecutive maximum elongations:

$$D = \ln \frac{x_n}{x_{n+1}} = \ln e^{\gamma \tilde{T}} = \gamma \tilde{T} = \frac{\tilde{T}}{\tau} = 2\pi \frac{\gamma}{\tilde{\omega}}. \tag{16.12}$$

The inverse quantity $1/D = \tau/\tilde{T}$ gives the number of oscillations executed during the time of damping τ.

Another equivalent characteristic of damping is the *quality factor* Q of the oscillator (see also Section 16.4.2, Equation (16.21)):

$$Q = \frac{\pi}{D} = \frac{\pi\tau}{\tilde{T}} = \frac{\tilde{\omega}}{2\gamma} \approx \frac{\omega_0}{2\gamma}.$$

The energy of oscillations averaged over the period of oscillations is proportional to the amplitude squared. In damped natural oscillations the average energy decreases in time exponentially:

$$W(t) = W_0 e^{-2\gamma t} = W_0 \exp(-t/\tau_{\text{en}}), \tag{16.13}$$

where $\tau_{\text{en}} = 1/2\gamma = \tau/2$ is the time of energy damping, W_0 is the initial energy value.

In the case of *strong damping* ($\gamma \geq \omega_0$) the oscillator asymptotically returns to the equilibrium position after an initial excitation (without oscillations). In particular, at $\gamma = \omega_0$ (the *critical* or the *quickest damping*) the general solution to the differential equation, Equation (16.9), has the form

$$x(t) = (C_1 t + C_2)e^{-\gamma t},$$

where C_1 and C_2 are constants determined by initial conditions. For example, with $x(0) = 0$, $\dot{x}(0) = v_0$ (a push from equilibrium), constants $C_2 = 0$, $C_1 = v_0$, and $x(t) = v_0 t e^{-\gamma t}$. The plot of $x(t)$ is shown in Figure 16.9.

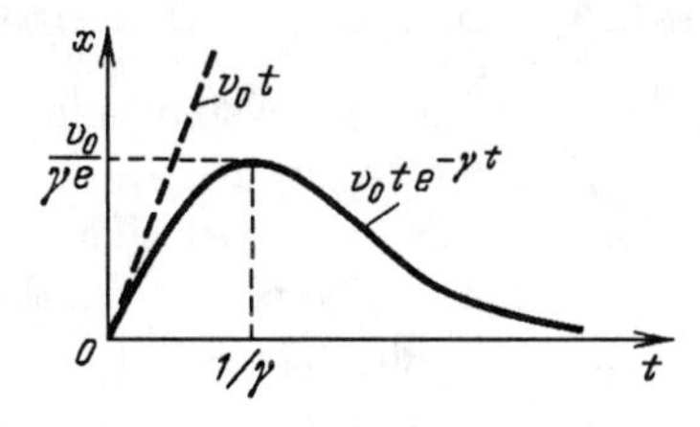

Figure 16.9: Non-oscillatory motion at critical damping ($\gamma = \omega_0$).

Damping is introduced intentionally in measuring instruments of various kinds to overcome the problem of taking a reading from an oscillating needle. A measuring instrument is said to be *critically damped* if the system just fails to oscillate and the system comes to rest in the shortest possible time.

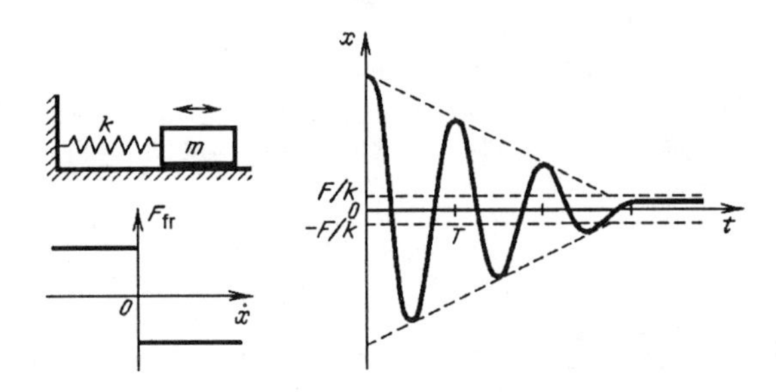

Figure 16.10: Damping of oscillations at dry friction.

16.3.6 Damping by dry friction

At *dry friction* (Coulomb's friction), damping of oscillations goes on in another way (Figure 16.10). Static friction is the reason that equilibrium is possible in any place of the *stagnation zone*, limits of the latter being determined by the ratio of maximum force of static friction F to the spring constant k. Every subsequent maximum displacement is smaller than the preceding one done to the same side by the double width of the stagnation zone. In this case the amplitude decreases in an arithmetic progression (linearly). Oscillations stop after a finite number of cycles, and the halt may occur at any point of the stagnation zone (depending on the initial conditions). Dry friction does not influence the period of oscillations (i.e., the time interval between consecutive maximum deflections to one side): $T = T_0 = 2\pi/\omega_0$, where $\omega_0 = \sqrt{k/m}$.

16.4 Forced oscillations. Resonance

16.4.1 Steady-state forced oscillations

If a damped oscillator is excited by a sinusoidal external force (Figure 16.11), harmonic oscillations are eventually established during some time after the driving force begins to operate. These oscillations have the frequency ω of the external force. The phase and the amplitude of these terminal steady oscillations are independent of the initial conditions. The dependence of the amplitude of *steady-state oscillations* on frequency ω of the external force has a *resonant character*, i.e., it grows sharply as ω approaches the natural frequency ω_0 of the oscillator.

The differential equation (Newton's second law) for forced oscillations with viscous friction (Figure 16.11) has the form

$$m\ddot{x} = -kx - \lambda\dot{x} + F_0 \cos\omega t,$$

or

$$\ddot{x} + 2\gamma\dot{x} + \omega_0^2 x = f_0 \cos\omega t, \qquad (16.14)$$

where $\omega_0^2 = k/m$, $2\gamma = \lambda/m$, and $f_0 = F_0/m$ (for the physical sense of λ, see Section 16.3.5). Steady-state oscillations are described by the *particular solution* of the *inhomogeneous* differential equation, Equation (16.14):

$$x(t) = a\cos(\omega t + \delta). \qquad (16.15)$$

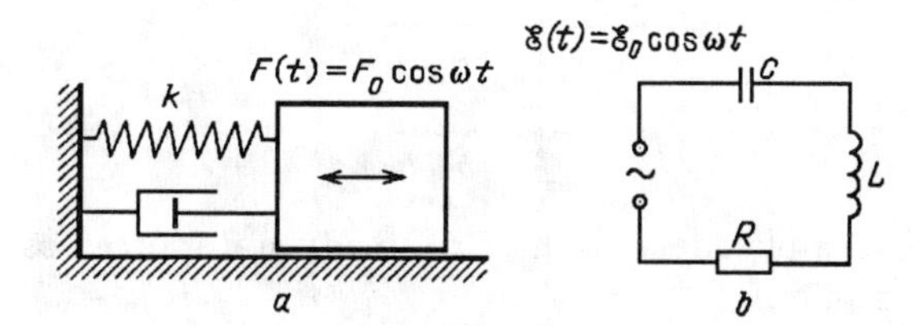

Figure 16.11: Mechanical (a) and electrical (b) damped oscillators excited by sinusoidal external action.

Far from resonance (when $|\omega - \omega_0| \gg \gamma$) the influence of damping on amplitude a and phase δ of steady-state oscillations can be ignored, that is, instead of Equation (16.14) it is possible to consider the equation for a non-damped oscillator:

$$\ddot{x} + \omega_0^2 x = f_0 \cos\omega t. \qquad (16.16)$$

Searching for its solution in the form $x(t) = a\cos\omega t$, we obtain the following expression for the amplitude a (Figure 16.12a):

$$a(\omega) = \frac{f_0}{\omega_0^2 - \omega^2}. \qquad (16.17)$$

The value $a = f_0/\omega_0^2 = F_0/k$ of the amplitude at $\omega = 0$ corresponds to the static deformation of the spring under the influence of constant force F_0. At $\omega \to \omega_0$ the amplitude grows indefinitely. The physical sense of this fact is that close to resonance ($\omega \approx \omega_0$) it is impossible to ignore damping. For $\omega > \omega_0$, amplitude a in Equation (16.17) becomes negative. The negative sign of a means that for $\omega > \omega_0$, steady-state forced oscillations occur in *antiphase* with the external driving force. If an amplitude is supposed to be always non-negative, we can write the solution in the form of Equation (16.15), understanding a in it as the modulus of the right side of Equation (16.17) and considering the phase shift δ as being equal to zero for $\omega < \omega_0$ and equal to $-\pi$ for $\omega > \omega_0$ (Figure 16.12b). For $\omega \gg \omega_0$ the amplitude a tends to zero.

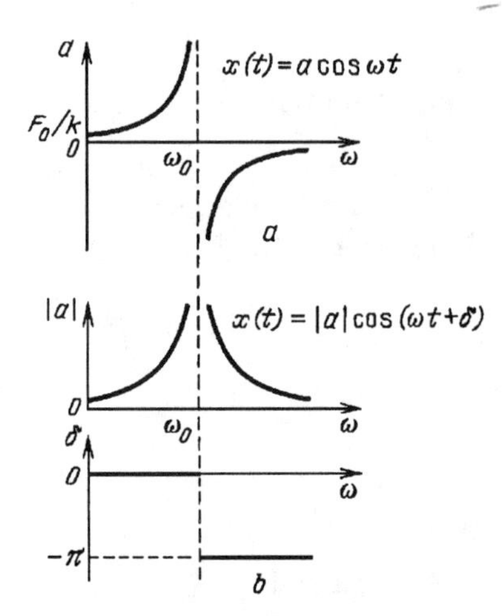

Figure 16.12: Amplitude and phase of steady-state forced oscillations at negligible damping.

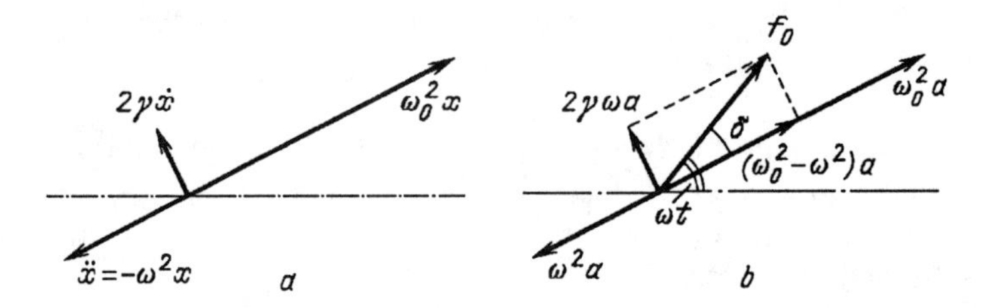

Figure 16.13: Vector diagrams for steady forced oscillations.

16.4.2 Resonance curves of linear oscillator

In the close vicinity of resonance (more precisely, in the interval of frequencies $|\omega - \omega_0| \approx \gamma$), damping should be taken into account in Equation (16.14). The amplitude a and the phase shift δ can be found with the help of a vector diagram (Figure 16.13*a*) by associating each term of the left side of Equation (16.14) with a rotating vector (phasor), whose length equals the amplitude value of this term (see Section 16.2.2). The sum of these vectors forms a vector that must be equal to the rotating vector (phasor) associated with the right side of Equation (16.14), i.e., $f_0 \cos\omega t$ (Figure 16.13*b*). According to Pythagoras' theorem,

$$f_0^2 = (\omega_0^2 - \omega^2)^2 a^2 + 4\gamma^2\omega^2 a^2,$$

whence

$$a(\omega) = \frac{f_0}{\sqrt{(\omega_0^2 - \omega^2)^2 + 4\gamma^2\omega^2}}. \qquad (16.18)$$

The vector corresponding to the steady-state forced oscillation $x(t) = a\cos(\omega t + \delta)$ lags in phase from the vector corresponding to the driving force $f_0 \cos\omega t$, as can be seen in Figure 16.13*b*. Therefore,

$$\tan\delta = -\frac{2\gamma\omega}{\omega_0^2 - \omega^2}. \tag{16.19}$$

Plots of $a(\omega)$ (the resonance curve) and of $\delta(\omega)$ for different values of damping constant γ are shown in Figure 16.14*a*. Resonance of the amplitude (i.e., maximum of the amplitude $a(\omega)$) corresponds to the frequency

$$\omega_{\text{res}} = \sqrt{\omega_0^2 - 2\gamma^2}.$$

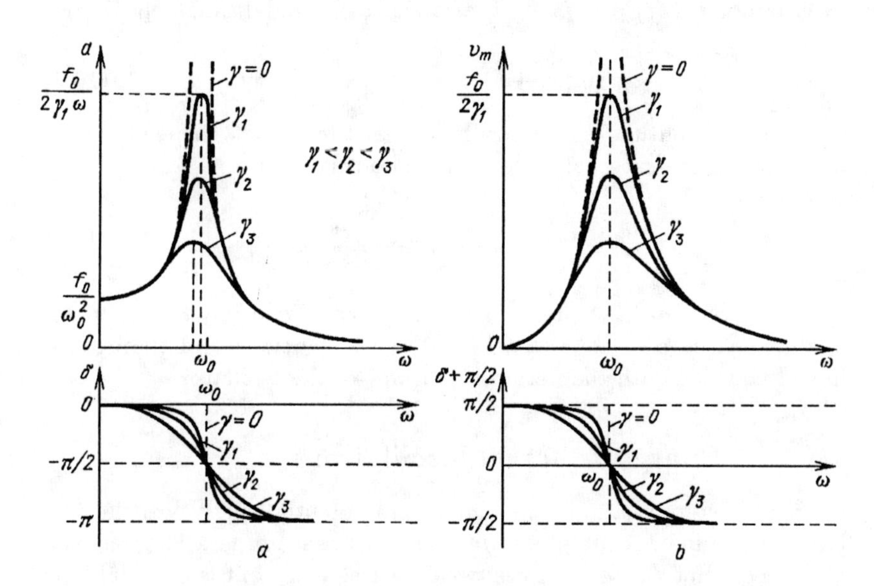

Figure 16.14: Dependence of the amplitude and phase of steady-state forced oscillations on the frequency of external force: *a* – for the displacement $x(t) = a\cos(\omega t + \delta)$; *b* – for the velocity $\dot{x}(t) = v_m \cos(\omega t + \delta + \pi/2)$.

For weak damping ($\gamma \ll \omega_0$) we get $\omega_{\text{res}} \approx \omega_0 - \gamma^2/\omega_0$, i.e., the resonant frequency practically coincides with the natural frequency ω_0. The amplitude of forced oscillations at resonance is

$$a_{\max} = \frac{f_0}{2\gamma\sqrt{\omega_0^2 - \gamma^2}} \approx \frac{f_0}{2\gamma\omega_0}. \tag{16.20}$$

The ratio of $a_{\max}$ to the static displacement $a(0) = f_0/\omega_0^2$ under the action of the constant force F_0 is called the *quality factor* Q of the oscillator (see also Section 16.3.5):

$$Q = \frac{\omega_0}{2\gamma}. \tag{16.21}$$

The quality factor characterizing the amplitude of oscillations at resonance is inversely proportional to the logarithmic decrement of natural oscillations of the same oscillator (see Equation (16.12)):

$$Q = \pi/D.$$

16.4.3 Resonance of velocity

The velocity $v_x(t) = \dot{x}(t)$ at forced oscillations under the influence of the external force $F(t) = mf_0 \cos\omega t$ also has a sinusoidal time dependence:

$$v_x(t) = -a\omega \sin(\omega t + \delta) = v_m \cos(\omega t + \delta + \pi/2), \tag{16.22}$$

where the amplitude v_m of oscillations of the velocity depends on the frequency of the external force in the following way:

$$v_m(\omega) = \omega a(\omega) = \frac{f_0}{\sqrt{(\omega_0^2/\omega - \omega)^2 + 4\gamma^2}}. \tag{16.23}$$

The maximum of the resonance curve for the velocity (Figure 16.14*b*) is always situated at $\omega = \omega_0$, i.e., at exact coincidence of the external force frequency with the natural frequency of the oscillator.

16.4.4 Energy in forced oscillations

During forced oscillations, the kinetic and potential energies of the oscillator change in opposite phase with respect to each other. The frequency of these oscillations of energy is twice the frequency of the external force.

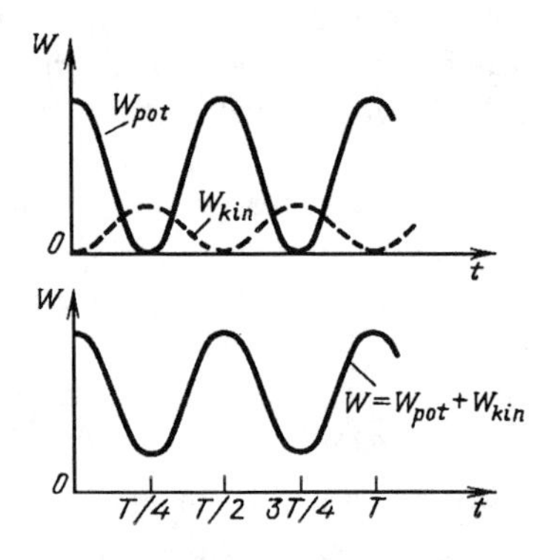

Figure 16.15: Energy of an oscillator at steady-state forced oscillations.

Unlike natural oscillations, average values of $\langle W_{\text{pot}} \rangle$ and $\langle W_{\text{kin}} \rangle$ are not equal (Figure 16.15). For $\omega < \omega_0$, the potential energy prevails (for $\omega \ll \omega_0$ the deformation is quasistatic, and the total energy of the oscillator is practically the potential one); for $\omega > \omega_0$, the kinetic energy prevails. As it is shown in Figure 16.15, the total energy of the oscillator remains constant only on the average: during a quarter of a period energy is delivered to the oscillator by the external source, and during the next quarter period the oscillator returns energy (the work of the external force being negative during that time). Only at resonance (when $\omega = \omega_0$) the total energy of the oscillator is constant. At resonance the external force changes in phase with the velocity (see Figure 16.14*b*), i.e., the driving force is constantly acting in the direction of the motion. This provides for most favorable conditions for energy transfer to the oscillator. The work done by the external force compensates for the energy losses due to friction, keeping the amplitude of oscillations constant.

16.4.5 Transient processes

For high values of the quality factor Q the amplitude of forced oscillations at resonance essentially exceeds (Q times) the static displacement under the action of the same (but constant) force. This large amplitude cannot be established immediately after activating the driving force, since it takes time for the oscillator initially at rest to receive a great energy store from the weak external force. The *transient process* of growth of the amplitude lasts (at $\omega = \omega_0$) approximately Q periods (Figure 16.16*a*).

In the general case, the transient process of stabilization of oscillations is a superposition of steady-state oscillations of a constant amplitude with the driven frequency ω (i.e., the particular solution, given by Equation (16.15), of the inhomogeneous differential equation (16.14)) and damped oscillations with the natural frequency ω_0 (i.e., the general solution, Equation (16.10) of the homogeneous differential equation (16.9)), their amplitude and phase being determined by the initial conditions. The plots of transient processes for the oscillator, initially resting in the equilibrium position and influenced by the external force $F(t) = F_0 \cos \omega t$ after the moment $t = 0$, are shown in Figure 16.16.

16.4.6 Non-sinusoidal external force

Forced oscillations under the action of short periodic external impulses (Figure 16.17) will be nearly sinusoidal when the alteration period of the impulses coincides with the period of natural oscillations of the oscillator or is a multiple to it. The phase of oscillations establishes itself so that the pushes occur at the moments at which the oscillator passes through the equilibrium position. In the case of weak friction the energy received

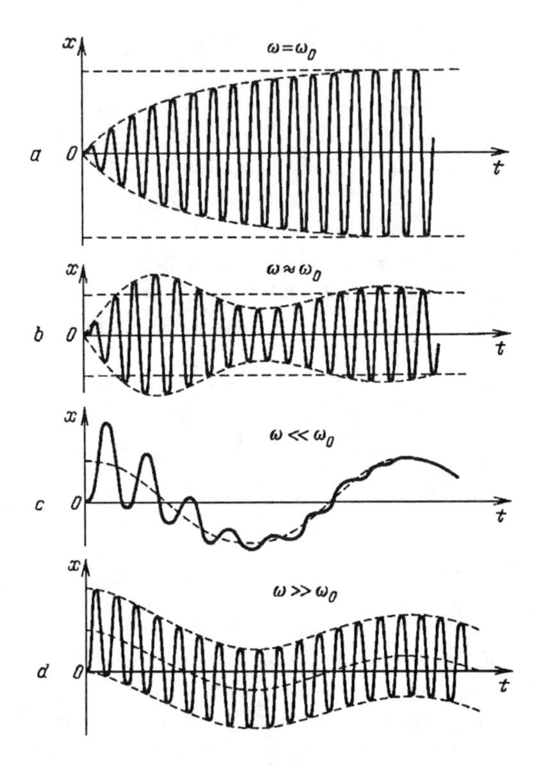

Figure 16.16: Transient processes at forced oscillations.

by the oscillator at every push is much smaller than the energy store and equals the energy dissipated by the oscillator during one period.

16.5 Parametric Resonance. Self-Excited Oscillations

16.5.1 Parametric excitation of oscillations

Non-damped oscillations are possible not only under the action of an external periodic force, but also at periodic changes of some parameter (length of a pendulum, rigidity of a spring, moment of inertia, capacity of a capacitor, inductance of a coil, etc.). Such excitation of oscillations by modulation of some parameter of the system is called *parametric resonance.* A familiar example of parametric resonance is swinging, when somebody on the swing regularly squats and draws himself up, thereby moving periodically the center of mass of the system.

To simulate parametric excitation of oscillations, let us slightly re-

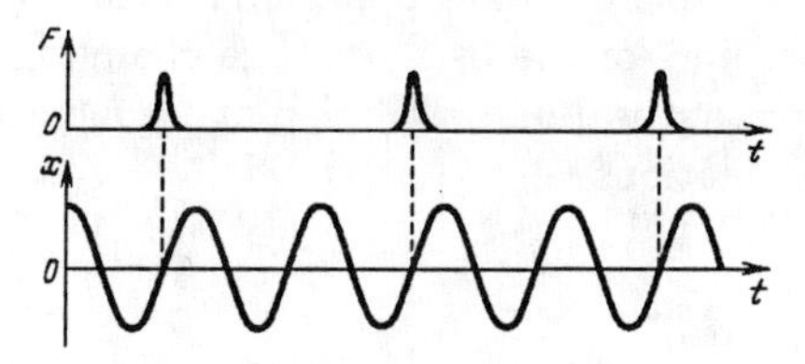

Figure 16.17: Forced oscillations under the action of short periodic external impulses.

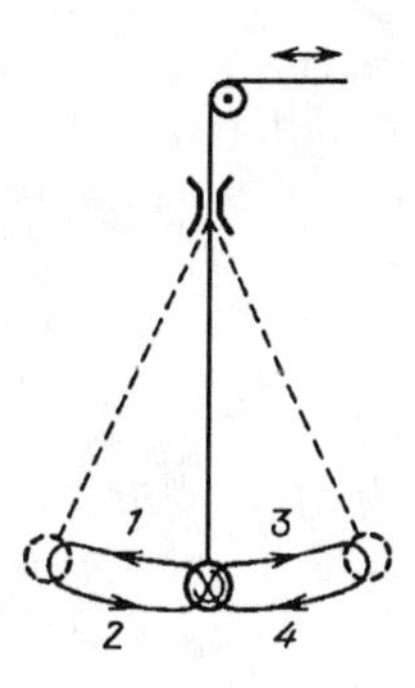

Figure 16.18: Parametric excitation of a pendulum.

duce the suspension length of the pendulum (Figure 16.18) each time it passes through the equilibrium position, pulling up the thread thrown over the pulley, and in every extreme position letting it down by the same length. The tension of the thread at extreme points is smaller than at the equilibrium position, therefore the positive work of the external force when the thread is pulled up is greater than the modulus of the negative work when the thread is let down. On the whole, the work of the external force, averaged over a period, is positive. This work can surpass energy losses due to friction if the amplitude of modulation exceeds some threshold value. The increase in the energy of oscillations is proportional to the energy already stored by the pendulum, and the amplitude of the pendulum increases progressively. Over the threshold, friction cannot restrict the growth of oscillations. The growth of parametrically excited oscillations is limited by nonlinear effects (by the change of the natural period with the amplitude).

Variations of the length of the thread cannot take a resting pendulum out of equilibrium, so for parametric excitation there must already exist (even small) natural oscillations in the system. There are always such initial perturbations due to fluctuations inevitable in any real system.

Unlike ordinary resonance at forced oscillations, parametric resonance can occur if parameters of the oscillator are changing with a frequency equal to the double natural one, and also at frequencies ω_n (or periods T_n) satisfying the relations

$$\omega_n = \frac{2\omega_0}{n}, \qquad T_n = n\frac{T_0}{2}, \tag{16.24}$$

where n is an integer. Growth of oscillations is possible not only at exact satisfaction of the relationship expressed by Equation (16.24), but also in some finite intervals of the modulation frequency ω near the resonant frequencies ω_n (in the *ranges of instability*). The greater the changes of the oscillator parameter (e.g., of the pendulum length), the greater the width of these intervals of instability.

Parametric resonance is used for amplification and generation of electromagnetic oscillations of various frequencies, up to optical ones (parametric oscillators of light).

16.5.2 Self-excited oscillations

The difference between *self-excited oscillations* (or *auto-oscillations*) and other oscillatory processes is that the former occur without any external periodic disturbance. Non-damped oscillations are maintained due to the capability of self-excited oscillatory systems to control the delivery of energy from a constant external source. Unlike forced oscillations, the form, amplitude, and frequency of self-excited oscillations are determined by the system itself. Examples of self-excited oscillations are oscillations of air in an organ pipe, vibrations of a violin string at a uniform motion of a bow, alterations of current in a radio-frequency oscillator, and oscillations of a pendulum in a clock.

Characteristic elements of simple self-excited oscillatory systems are a resonator, a constant source of energy, and a device of feedback between them. A *resonator* is a system in which natural damped oscillations may occur: a clock pendulum, an oscillatory circuit of a radio-frequency (rf) generator, or a string of a bow instrument. The *feedback* is realized by a device helping the resonator to control the delivery of energy from the source. By virtue of the feedback, energy is supplied to the resonator, compensating for the losses so that the amplitude of the oscillations remains constant.

In a *mechanical clock* the feedback is carried out by the pallet mechanism (Figure 16.19). A curved lever (an anchor) is attached to the pendulum. This lever has projections of a special shape (pallets) on its ends. The spring tends to turn the running wheel, but the latter for the most time rests with one of its teeth against the lateral surface of the right or the left pallet sliding along the tooth surface during the pendulum oscillation. Only at the moments when the pendulum is passing through the equilibrium position, the pallets stop barring the teeth's way and the wheel turns by one tooth, pushing the anchor and the pendulum by the tooth, sliding with its point along the beveled end of the pallet. These pushes deliver some energy to the pendulum and compensate for energy losses of the pendulum caused by friction.

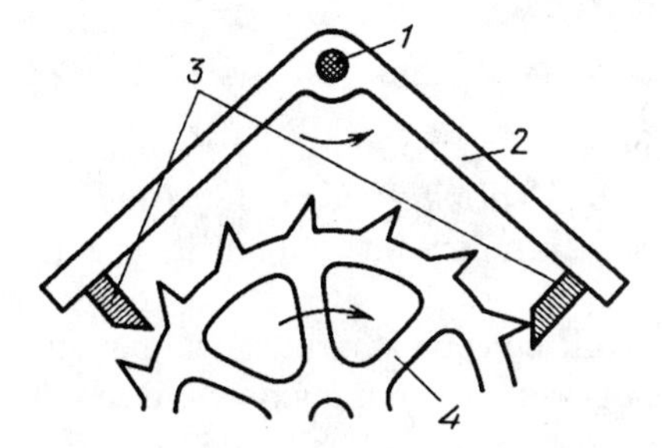

Figure 16.19: Pallet mechanism of a mechanical clock: *1*–axis of the pendulum, *2*–anchor, *3*–pallets, *4*–running wheel.

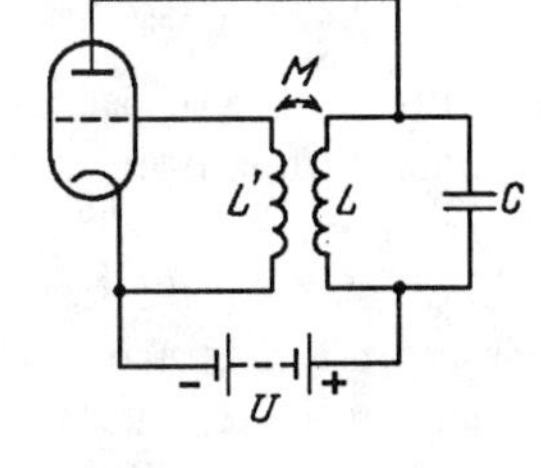

Figure 16.20: The circuit of a valvc rf-oscillator.

In a *valve* rf-*oscillator* (Figure 16.20) the feedback is fulfilled by the coil L' included between the cathode and the grid and coupled inductively with the coil L of the oscillatory circuit. Small natural oscillations (casually occurring in the circuit) control through the coil L' the anode current I of the valve, which amplifies oscillations in the circuit. The amplification occurs only for proper mutual orientation of coils L and L', that is, if the feedback is positive. If losses in the circuit are less than the energy supplied to it, the amplitude grows. Its growth is limited by the nonlinear dependence of the anode current on the grid voltage. As a result, stationary oscillations of a certain amplitude are established, when all energy losses are compensated by the battery included into the anode circuit. Such a way for excitation of self-excited oscillations without an initial push is called *soft excitation*. In systems with *hard excitation*, oscillations increase only beginning with a certain threshold initial amplitude. For transition of such systems from the state of rest to the generation mode, an initial excitation is necessary, exceeding a certain critical value.

It should be noted that the shape, amplitude, and frequency of self-excited oscillations are determined only by the system parameters (unlike forced oscillations whose amplitude, frequency, and phase are determined by the external force).

Self-excited oscillatory systems are described by nonlinear differential equations. In the phase plane, periodic self-excited oscillations are represented by a closed curve called the *limiting cycle*. The phase trajectories of transient processes in a self-excited system, occurring at various initial conditions, are attracted by the limiting cycle.

16.6 Oscillations of Complex Systems. Composition of Oscillations

16.6.1 Degenerate oscillatory systems

A pendulum able to oscillate simultaneously in two mutually perpendicular planes is a simple example of a system with *two degrees of freedom*: to characterize its position it is necessary to indicate two quantities, e.g., coordinates x and y (Figure 16.21*a*).

By a certain choice of the initial conditions one can make the pendulum execute free oscillations either along axis x or along axis y. At small amplitudes these oscillations are almost harmonic and have the same frequency $\omega = \sqrt{g/l}$ (coincidence of frequencies of such *normal oscillations* is called *degeneracy*):

$$x(t) = x_m \cos(\omega t + \varphi_1), \qquad y(t) = y_m \cos(\omega t + \varphi_2). \tag{16.25}$$

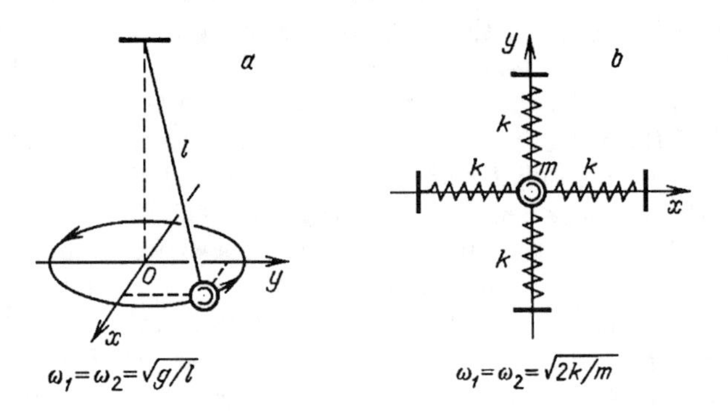

Figure 16.21: Degenerate systems with two degrees of freedom.

The same equations are used for the description of motion of a weight attached to two pairs of identical springs (Figure 16.21*b*).

At arbitrary initial conditions, oscillations along both axes are excited simultaneously, and the motion of the pendulum is their superposition. The result of the composition depends on the relationship between amplitudes and phases of these oscillations.

If the phases are equal or differ by π, oscillations occur in a plane, and the trajectory of the pendulum is a segment of a straight line (one of the diagonals of a rectangle with sides $2x_m$, $2y_m$).

When the phase shift is $\pm\pi/2$, and the amplitudes are equal, the trajectory is a circle $x^2 + y^2 = a^2$. One can easily make sure of it, squaring and adding the following equations:

$$x(t) = a\cos(\omega t + \varphi), \qquad y(t) = \pm a\sin(\omega t + \varphi).$$

In the general case the trajectory is an ellipse inscribed into a rectangle with sides $2x_m$, $2y_m$. The results of the composition of such mutually perpendicular oscillations of the same frequency for various values of difference of phases $\delta = \varphi_1 - \varphi_2$ and equal amplitudes $x_m = y_m = a$ are shown in Figure 16.22.

Thus, any motion of a pendulum can be represented as a sum of normal oscillations with certain amplitudes x_m, y_m and phases φ_1, φ_2. However, in the case of degeneracy (that is, at coincidence of frequencies $\omega_1 = \omega_2$) the choice of normal oscillations is not unique: it is possible to take any two mutually perpendicular directions as the axes x and y in this plane. In particular, for the system shown in Figure 16.21*b* axes x and y can be chosen along directions other than the springs as well.

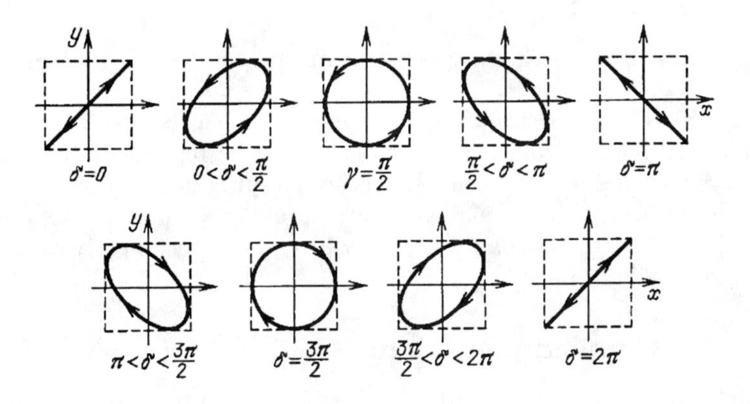

Figure 16.22: Composition of mutually perpendicular oscillations of the same frequency and equal amplitudes at various values of the phase difference.

16.6.2 Normal oscillations (modes)

A pendulum suspended on a loop (Figure 16.23*a*) and a weight attached to the springs of different spring constants (Figure 16.21*b*) are also systems with two degrees of freedom. Here one can also find initial conditions under which the motion of the system will be a pure harmonic oscillation. Any system has as many different kinds of these movements, called *normal oscillations* or *modes*, as it has *degrees of freedom*. In the examples of such systems in Figure 16.23 the normal modes are represented by oscillations along axes x and y. However, now the frequencies corresponding to each of the modes are different:

$$x(t) = x_m \cos(\omega_1 t + \varphi_1), \qquad y(t) = y_m \cos(\omega_2 t + \varphi_2). \tag{16.26}$$

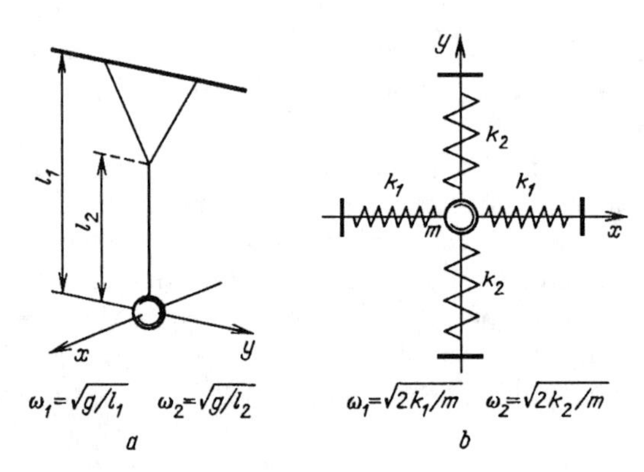

Figure 16.23: Oscillatory systems with two degrees of freedom.

The set of frequencies of normal modes is called the *spectrum* of the system. Any free oscillation of the system can be represented as a sum of normal modes with certain frequencies and phases.

When frequencies ω_1 and ω_2 relate as integers, the trajectory is a closed curve (a *Lissajous figure*) inscribed into a rectangle with sides $2x_m, 2y_m$ (Figure 16.24).

16.6.3 Coupled pendulums

Coupled pendulums (Figure 16.25*a*) give an example of a system with two degrees of freedom, in which oscillations corresponding to different modes occur along one direction. The frequency of the first mode ω_1 is $\sqrt{g/l}$, because during synphase oscillations the spring connecting the pendulums is not strained and does not influence the oscillations (Figure 16.25*b*).

The frequency of the second mode (Figure 16.25*c*) is determined by the expression

$$\omega_2 = \sqrt{g/l + 2k/m}$$

(during oscillations occurring in antiphase the central point of the spring is motionless, and the frequency is equal to that of the combined pendulum with a spring whose constant equals $2k$, see Figure 16.25*d*).

For arbitrary initial conditions, the motion of each pendulum is a sum of harmonic oscillations with frequencies ω_1 and ω_2 (the spectrum of the

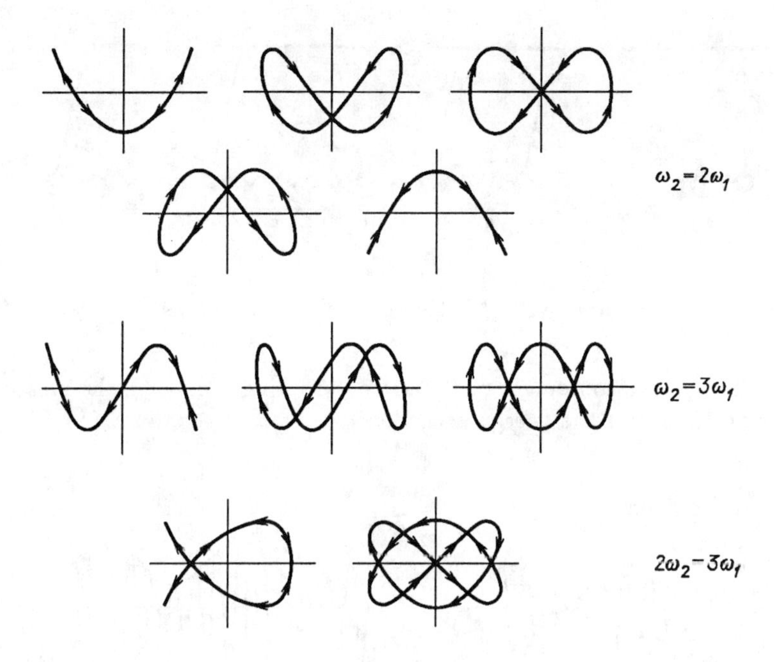

Figure 16.24: Lissajous figures.

system consists of two frequencies ω_1 and ω_2). There is an interesting case of *weak coupling*, in which $k/m \ll g/l$ and the difference between the frequencies of normal modes $\Delta\omega = \omega_2 - \omega_1$ is small compared with each of them. For example, let the left pendulum be deflected through the distance $2a$ at the initial moment, let the right one be in the equilibrium position, and let the system be released without an initial push. Such initial conditions correspond to the following solution (see Section 4.2.2):

$$x_1(t) = a(\cos\omega_1 t + \cos\omega_2 t) = 2a\cos\left(\frac{\Delta\omega}{2}t\right)\cos\omega t,$$

$$x_2(t) = a(\cos\omega_1 t - \cos\omega_2 t) = 2a\sin\left(\frac{\Delta\omega}{2}t\right)\sin\omega t, \qquad (16.27)$$

where $\Delta\omega = \omega_2 - \omega_1$, $\omega = (\omega_2 + \omega_1)/2$ (the average frequency), and $\Delta\omega \ll \omega$. These expressions, Equations (5.27), can be considered as almost harmonic oscillations with a frequency ω and slowly pulsing amplitudes (Figure 16.26*a*). This type of motion is called *beats*. In the case of such initial excitation, the energy store is imparted to one pendulum, but then the energy gradually passes to the other one and then back to the first one. The cycle of beats is completed during the time interval $T_0 = 2\pi/\Delta\omega$.

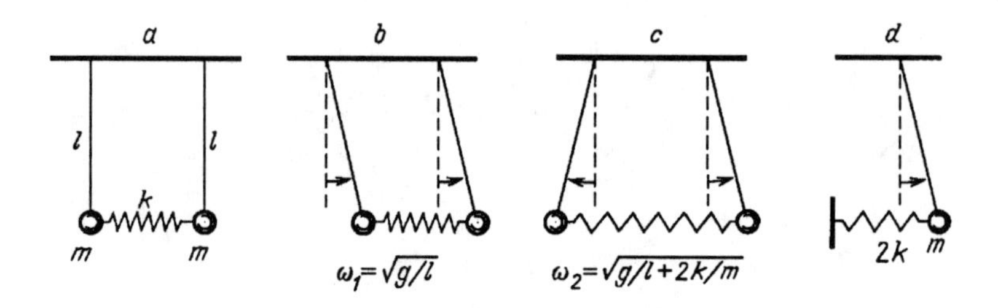

Figure 16.25: Modes of coupled pendulums.

The stronger the coupling (i.e., the more the spring constant), the greater $\Delta\omega$ and the faster the energy exchange.

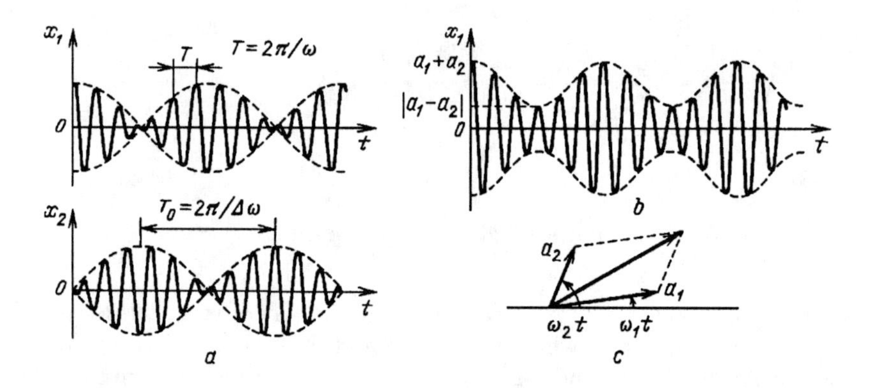

Figure 16.26: Beats (a), modulated oscillations (b), and the vector diagram (c) for the add on of oscillations with frequencies ω_1 and ω_2, whose amplitudes are a_1 and a_2.

If both modes of different amplitudes ($a_1 \neq a_2$) are excited simultaneously, oscillations of each pendulum are *modulated* (Figure 16.26b), i.e., the amplitude is slowly changing between the values a_1+a_2 and $|a_1-a_2|$. We can easily see this by means of the vector diagram in Figure 16.26c, where vectors of length a_1 and a_2 rotate with slightly different angular velocities ω_1 and ω_2.

16.6.4 Forced oscillations of coupled pendulums

In the case of *forced oscillations* in systems with several degrees of freedom under the action of a sinusoidal external force with a frequency ω, resonance occurs at coincidence of ω with each of the natural frequencies. For example, under the action of an external force $F(t) = F_0 \cos\omega t$

applied to the first of the coupled pendulums (Figure 16.25*a*), stationary oscillations of each of them are executed with a frequency ω of the external action, and in the absence of damping they have the form

$$x_1 = b_1 \cos \omega t, \qquad x_2 = b_2 \cos \omega t. \tag{16.28}$$

The dependence of amplitudes b_1 and b_2 on ω is shown in Figure 16.27*a*. When ω approaches ω_1 or ω_2, amplitudes b_1 and b_2 grow (indefinitely if damping is ignored), and close to resonance at a frequency ω_1 oscillations of the pendulums occur in phase (b_1 and b_2 have the same sign), i.e., in the same way as at free oscillations in the first mode. At a frequency ω_2 the pendulums swing in antiphase (b_1 and b_2 are opposite in signs), i.e., like at free oscillations in the second mode.

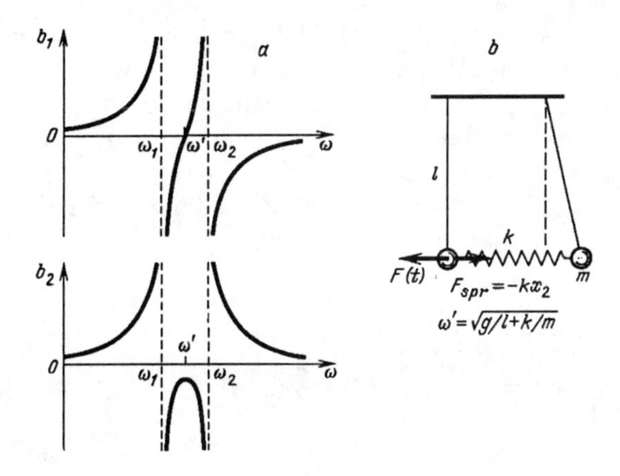

Figure 16.27: Resonance curves for steady forced oscillations of coupled pendulums (*a*), and dynamic damping of oscillations (*b*).

It is interesting to note that at some frequency ω' lying between ω_1 and ω_2, the amplitude b_1 of the first pendulum, which is subjected to the action of an external force, becomes zero. The second pendulum at this time oscillates in antiphase with the external force ($b_2 < 0$), and the elastic force of the connecting spring compensates for the external force acting upon the first pendulum, ensuring its immobility (Figure 16.27*b*). The frequency ω' is equal to that of free oscillations of the second pendulum while the first one is fixed (that is, to *the partial frequency*): $\omega' = \sqrt{g/l + k/m}$. There is a method of eliminating harmful vibrations based on this phenomenon. The damping of such vibrations of machine foundations (by attachment of an additional device with harmless oscillations) is called *dynamic damping* and is widely used in different engineering devices in which the frequency of harmful vibrations is constant.

16.6.5 Coupled electromagnetic circuits

Free and forced electromagnetic oscillations in circuits connected by capacitive coupling (Figure 16.28*a*) are characterized by the same properties as oscillations of coupled pendulums. Oscillations with phase coincidence, when the net current in the tap containing a coupling capacitor $C_{\rm cpl}$ is equal to zero, correspond to the first mode. Since the capacity $C_{\rm cpl}$ in this case does not take part in oscillations, the frequency of this mode is $\omega_1 = 1/\sqrt{LC}$. In the second mode, oscillations occur in antiphase, their frequency being $\omega_2 = \sqrt{1/LC + 2/LC_{\rm cpl}}$ (each circuit uses only "one half" of the capacitor $C_{\rm cpl}$, like in the mechanical analogy in Figure 16.28*b* in which "one half" of connecting spring $k_{\rm c}$ takes part in antiphase oscillations of the weights, which gives the frequency $\omega_2 = \sqrt{k/m + 2k_{\rm c}/m}$). In this circuit the condition $1/C_{\rm cpl} \ll 1/C$ corresponds to the condition of weak coupling ($k_{\rm c} \ll k$) of the mechanical analogue. Under the action of sinusoidal external voltage $U(t) = U_0 \cos\omega t$, resonance occurs at frequencies $\omega = \omega_1$ and $\omega = \omega_2$.

Oscillations in circuits with inductive coupling (Figure 16.28*c*) are characterized by similar properties.

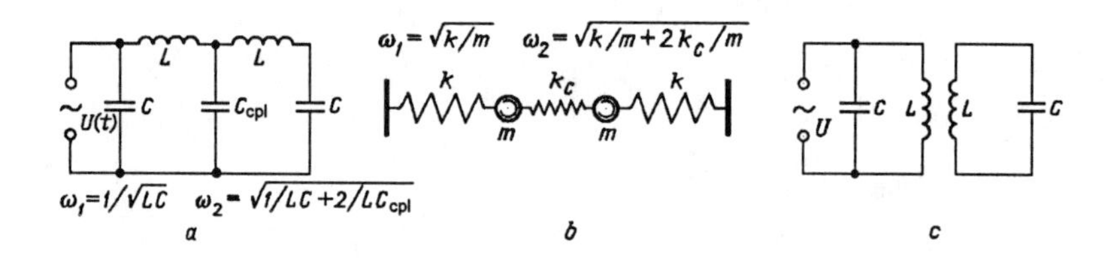

Figure 16.28: Electrical circuits with capacitive coupling (*a*), their mechanical analogue (*b*), and circuits with inductive coupling (*c*).

16.6.6 Standing waves as normal oscillations

Systems with *distributed parameters* have an infinite number of degrees of freedom and are therefore characterized by an infinite set of modes and discrete frequencies ω_n ($n = 1, 2, 3, \ldots$) corresponding to these modes. The simplest of such systems are: an elastic stretched string whose elementary parts have both inertia (mass) and elasticity; a double-wire (or coaxial) transmission line whose every length element has both inductance and capacity.

Some transverse modes of a string with fixed ends are shown in Figure 16.29*a*. Any two points of the string, lying between two neighboring *nodes*, that is, points of zero amplitude, execute harmonic oscillations in the same phase. Amplitude of oscillations depends on the position of the point sinusoidally, achieving its maximum half-way between the nodes—in *antinodes*. Points of the string, located on the opposite sides of a node, oscillate in antiphase.

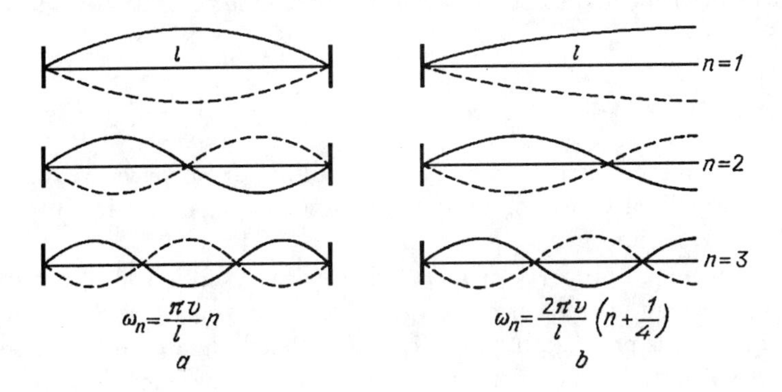

Figure 16.29: The first three modes of a string with fixed ends (a), and of a string with one free end (b).

Normal oscillations of a string can be considered as *standing waves* (see Section 16.7.9) produced by a superposition of running waves of equal frequency and amplitude traveling in opposite directions. The distance between adjacent nodes is equal to one half of the wavelength λ. Frequencies of normal oscillations ω_n can be found from the condition that for the mode with a number n there are n half-waves on the length of the string: $l = n\lambda/2$. Since $\lambda = vT = 2\pi v/\omega$ (v is the velocity of waves in the string, see Section 16.7.4), we obtain

$$\omega_n = \frac{\pi v}{l}\, n. \tag{16.29}$$

Any natural oscillation of the string can be represented as a sum (in the general case with an infinite number of terms) of normal oscillations with discreet frequencies ω_n (5.29) and certain amplitudes and phases. Frequencies ω_n form an equidistant spectrum, i.e., they are separated by equal intervals $\Delta\omega = \pi v/l$ between them. The frequency $\omega_1 = \pi v/l$ is called the *fundamental* one, frequencies ω_n ($n > 1$) are called *higher harmonics* (or *overtones*).

In the string fixed at one end, modes are characterized by configurations with an antinode at the free end (Figure 16.29b). The length of the string is equal to an integral number of half-waves and a quarter of the wavelength: $l = n\lambda/2 + \lambda/4$. The same configurations are typical for oscillations of the air in an organ pipe closed at one end, or of electromagnetic oscillations in a line short-circuited at one end.

Normal modes of more complex systems (e.g., of elastic plates of a definite shape, of a glass or a bell, etc.) also form an infinite number of discrete frequencies. However, relationships between the spectrum of frequencies and the form and location of nodal lines appear quite complicated even for bodies of a comparatively simple geometric shape.

16.7 Waves

Waves are disturbances, *traveling* in a medium or in space. Waves in a medium are disturbances of its equilibrium state (i.e., excitations) spreading in this medium.

The principal property of a traveling wave of any physical nature is that in a wave the transfer of energy goes without any transfer of mass. When a wave is propagating, particles of the medium are executing some oscillatory motion near their average equilibrium positions.

16.7.1 Waves of different physical nature

Wave processes are extremely diverse and occur almost in every field of physical phenomena. According to their physical nature we discern *mechanical waves* (particular cases of them being elastic waves, including acoustic and seismic waves, waves on a liquid surface), and *electromagnetic waves* in wave guides, in various media, and in free space (including radio waves, light, X-radiation). Oscillations of electromagnetic field in certain types of waves may couple with mechanical oscillations of particles of a medium (e.g., in waves in plasma).

16.7.2 Polarization of waves

Depending on the orientation of disturbance relative to the direction of wave propagation, waves may be *longitudinal* (e.g., acoustic waves in gas, where particles are displaced along the direction of wave propagation, see Figure 16.30*a*) or *transverse* (e.g., waves in a string, whose particles are displaced perpendicularly to the direction of propagation, see Figure 16.30*b*). In liquids and gases, only the longitudinal elastic waves can propagate: transverse waves are impossible because elastic forces occur in such media only at compression and expansion. On the other hand, in solid bodies transverse waves can propagate as well, because elastic forces occur there also under shears. In anisotropic solid media and in solid bodies of limited size (in rods and plates), as well as in plasma in a magnetic field, other (more complicated) types of waves are possible.

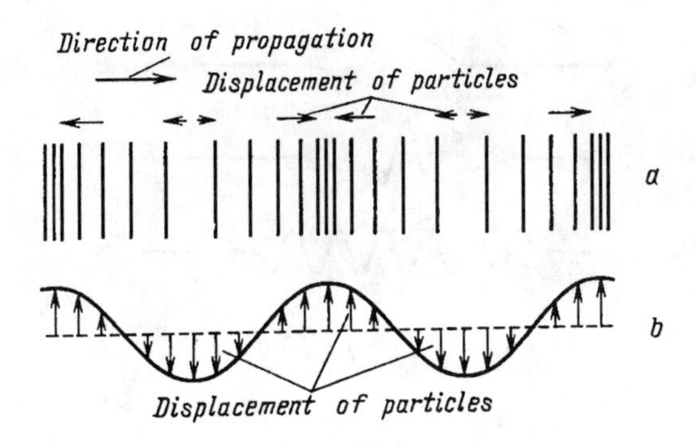

Figure 16.30: Longitudinal (*a*) and transverse (*b*) waves.

Electromagnetic waves in free space are transverse, because the directions of oscillating electric and magnetic fields are perpendicular to the direction of propagation (however, in wave guides, vectors of the fields may also have longitudinal components). In waves on a liquid surface (see Section 16.7.12), particles execute circular or elliptical movements, i.e., such waves are neither longitudinal nor transverse. In plasma, we can discern wave polarization in the electrodynamical and hydrodynamical sense. This classification takes into account the orientation of the electric field strength of the wave and the direction of the displacement of plasma particles in the wave.

16.7.3 Kinematics of wave motion

According to *kinematics*, i.e., to the character of dependence of disturbance on time at a fixed spatial point (or spatial dependence at a fixed moment of time), in other words, according to their *shape*, waves are divided into *single* waves, or *wave pulses*—comparatively short disturbances of irregular character; *wave packets* (Figure 16.31*b*)—limited sets of regularly repeating disturbances; and *wave trains* (Figure 16.31*c*)—limited disturbances of a sinusoidal shape. The idea of a *harmonic* or *monochromatic* wave (Figure 16.31*d*)—a wave of an infinite extent with sinusoidal dependence of disturbance on coordinates and time—is very important for the theory of wave processes.

In a monochromatic wave, disturbance at some spatial point (e.g., $z = 0$) changes in time harmonically:

$$x(t) = a \cos \omega t.$$

Here x can be understood as the displacement of particles of the medium from their equilibrium positions along the direction of propagation (for

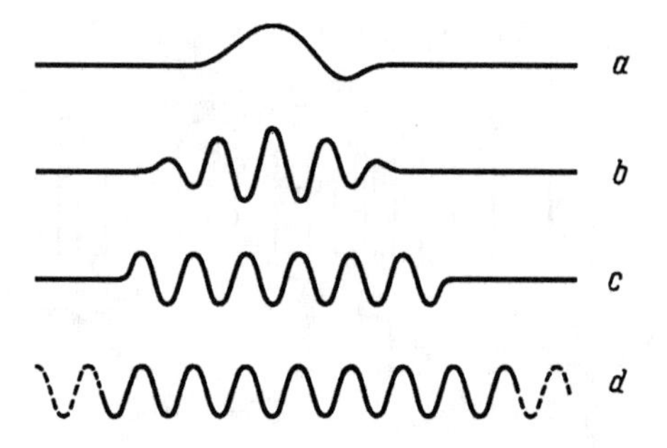

Figure 16.31: Different kinds of waves.

longitudinal elastic waves) or across it (for transverse waves), as a deviation of medium pressure from its average value (for acoustic waves), as a component of the electric field strength (for electromagnetic waves), etc. Time dependence of the disturbance at a point located at a distance z along the direction of the wave propagation is the same, but with a time delay $t = z/u$, where u is the *speed of the wave propagation* (more precisely, the *phase velocity* of the wave):

$$x(z,t) = a\cos\omega(t - z/u) = a\cos 2\pi(t/T - z/\lambda), \qquad (16.30)$$

i.e., disturbances at all points execute harmonic oscillations of the same frequency ω and amplitude a, but of a different phase.

If t in Equation (16.30) is fixed, then the dependence of $x(z,t)$ on z gives us an instantaneous picture (a snapshot) of the distribution of disturbances along the direction of propagation (Figure 16.32). The spatial period of a sine curve, i.e., the distance between successive points oscillating in the same phase, is called the *wavelength* λ. One can imagine the picture of the wave propagation, moving this sine curve with the velocity u along the axis z. At the next moment the spatial distribution of disturbances is represented by a sine curve shown by the dotted line in Figure 16.32. The length of the wave equals the distance the sinusoid (i.e., any hump of it) moves during the oscillation period T:

$$\lambda = uT = u/\nu = 2\pi u/\omega, \qquad u = \nu\lambda. \qquad (16.31)$$

For a wave propagating in the negative direction of the z-axis:

$$x(z,t) = a\cos\omega(t + z/u) = a\cos 2\pi(t/T + z/\lambda).$$

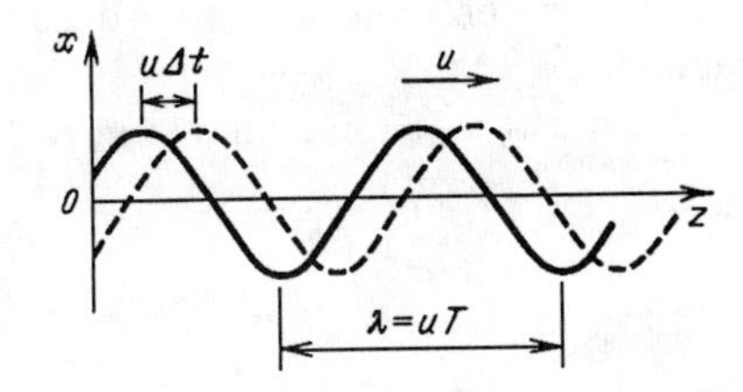

Figure 16.32: Monochromatic wave.

16.7.4 The speed of waves

The greater the restoring force tending to return the disturbed medium into the equilibrium position, the greater the velocity of wave propagation, and the greater the inertia of that medium, the smaller the velocity. In the simple cases the speed of propagation of a disturbance in a medium depends only on characteristics of the medium, but not on the form or size of the disturbance. For monochromatic waves this means that the phase velocity is independent of the wavelength. Such independence is called the absence of *dispersion*. This property is characteristic of waves in a stretched string, elastic waves (sound) in a homogeneous medium, and electromagnetic waves (light) in a vacuum. In more complicated cases the speed of wave may also depend on the shape of the wave, its amplitude, and wavelength. The dependence of the phase velocity of monochromatic waves on the wavelength (or frequency) is called the *dispersion*. Dispersion is inherent to waves on water surface, electromagnetic waves (light) in substances and transmission lines. Propagation of wave packets in a dispersive medium has some peculiarities (see Section 16.7.13).

The character of dependence of wave speed on parameters of the medium is sometimes easily found by the method of dimensional analysis (see Section 12.3.4). For example, in the case of a thin string it is plausible from physical considerations that the speed of a wave depends on the tension force F and on the linear density (mass of a unit length) $\rho_l = \rho S$ (S – cross-sectional area). These quantities can be combined in a unique way to get a combination having velocity dimensions $u \sim \sqrt{F/\rho_l}$. Similarly, for compression waves in a rod, the speed depends on Young's modulus of elasticity E and density ρ, and dimensional analyses gives $u \sim \sqrt{E/\rho}$. To obtain exact numerical factors in these formulas, a dynamic investigation is necessary.

Below are some formulas for the speed of waves.

1. Transverse waves in a stretched string:

$$u = \sqrt{\frac{F}{\rho S}} = \sqrt{\frac{F}{\rho_l}}$$

(F – tension force, S – cross-sectional area, ρ – density of the material, $\rho_l = \rho S$ – linear density).

2. Longitudinal elastic waves (sound) in a thin rod:

$$u = \sqrt{\frac{E}{\rho}}$$

(E – Young modulus, ρ – density).

3. Longitudinal elastic waves in an isotropic solid:

$$u = \sqrt{\frac{E}{\rho}\frac{1-\mu}{(1+\mu)(1-2\mu)}}$$

(E – Young modulus, ρ – density, μ – Poisson's ratio).

4. Transverse (shear) waves in an isotropic solid:

$$u = \sqrt{\frac{G}{\rho}}$$

(G – shear modulus, ρ – density).

5. Longitudinal waves (sound) in an ideal gas:

$$u = \sqrt{\frac{\gamma p_0}{\rho}} = \sqrt{\gamma \frac{RT}{M}}$$

($\gamma = C_p/C_V$ – ratio of specific heat capacities of the gas at constant pressure and at constant volume, p_0 – average pressure, R – ideal gas constant, M – molar mass of the gas). In particular case of the speed of sound waves in air,

$$u = 20.1\sqrt{T}\ \mathrm{m/s} \approx (331 + 0.6\,t)\ \mathrm{m/s}$$

(T – absolute temperature in kelvins, t – temperature in Celsius scale). Values of the speed of sound in some substances are shown in Table 16.1.

A human ear perceives sounds of frequencies ranging from 20 Hz to 20 kHz (wave length in air from 17 m to 17 mm), the maximum sensitivity corresponding to the frequency of about 3.5 kHz. The minimum pressure amplitude at which sound can be discerned by a human being is about $3 \cdot 10^{-5}$ Pa, and the threshold of pain is about 30 Pa (atmospheric pressure being 10^5 Pa). The spectral range of audible sounds becomes narrower with age. The sensation of a *sound pitch* (tone) is related with the frequency of vibrations, and the sensation of its *loudness* depends on their amplitude. Sound *timbre* is determined by the set of harmonics in the sound spectrum, i.e., by the components with frequencies multiple of the main tone (fundamental) frequency. Inaudible vibrations of a frequency lower than 20 Hz are called *infrasounds*, and those of a frequency higher than 20 kHz are called *ultrasounds*.

Table 16.1: Speed of Sound in some Substances (m/s)

Gas	0 °C	20 °C	Liquid	20 °C
Air	331	343	Water	1490
Nitrogen	334	346	Sea water	1530
Oxygen	316	327	Alcohol	1180
Helium	965	981	Mercury	1453
Hydrogen	1284	1328	Glycerol	1923

Solid substance	Longitudinal waves	Transverse waves	Waves in the rod
Melted quartz	5970	3762	5760
Glass	3760–4800	2380–2560	3490–4550
Gold	3220	1200	2030
Brass	4600	2080	3450

16.7.5 Energy transferred by waves

The *energy of a wave* in an elastic medium consists of kinetic energy of oscillating particles and potential energy of elastic deformation. The velocity of a particle in a monochromatic wave that propagates along the z-axis, Equation (16.30), depends on equilibrium coordinate z of the particle and on time:

$$v(z,t) = \dot{x} = -\omega a \sin \omega(t - z/u). \tag{16.32}$$

The density of kinetic energy (energy per unit volume of a medium)

$$w_{\text{kin}}(z,t) = \frac{1}{2}\rho v^2 = \frac{1}{2}\rho\omega^2 a^2 \sin^2 \omega(t - \frac{z}{u}) \tag{16.33}$$

is proportional to the wave amplitude squared. Dependence of $w_{\text{kin}}(z,t)$ on z for a fixed moment of time is shown in Figure 16.33. Kinetic energy reaches its maximum for the particles passing through the equilibrium point at the moment. The elementary portion of the medium around this point, being distorted more than others, also has maximal potential energy.

The density of potential energy in a monochromatic traveling wave, Equation (16.30), is equal to that of potential energy at every point at every moment: $w_{\text{pot}}(z,t) = w_{\text{kin}}(z,t)$, that is, oscillations of kinetic and potential energy in the wave have the same amplitude and phase (unlike the oscillator, where these changes occur in antiphase). Averaged over the period or over the wave length, density of energy in a monochromatic wave is proportional to the frequency squared and amplitude squared:

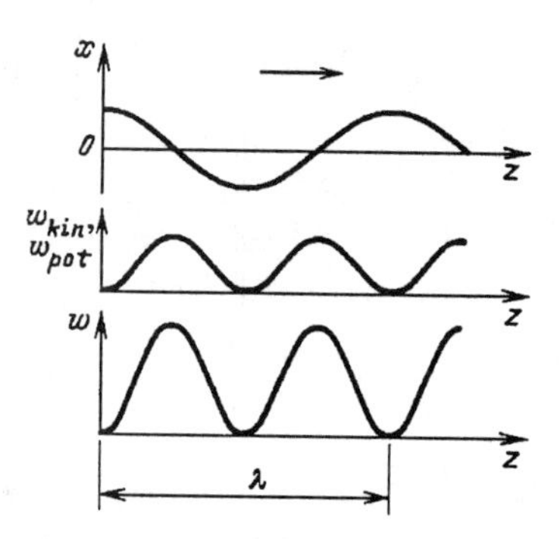

Figure 16.33: Dependence of the disturbance x and energy density w on coordinate z for a fixed instant t in a monochromatic traveling wave.

$$\langle w \rangle = \langle w_{\text{kin}} \rangle + \langle w_{\text{pot}} \rangle = 2\langle w_{\text{kin}} \rangle = \frac{1}{2}\rho\omega^2 a^2. \qquad (16.34)$$

There is no transfer of substance in a traveling wave, because the particles of the medium oscillate about their middle positions. However, the energy of these oscillations does not remain localized and travels along with the wave with a velocity u. The *intensity*, or surface density of energy flow, i.e., the average power carried by the wave through the unit area oriented perpendicularly to the direction of wave propagation, is measured by the product of the energy density $\langle w \rangle$, Equation (16.34), and the velocity u of the wave propagation:

$$\langle j \rangle = \frac{1}{2}\rho\omega^2 a^2 u. \qquad (16.35)$$

16.7.6 Plane, spherical, and cylindrical waves

Depending on the conditions of their emission and propagation in a three-dimensional medium (or on a boundary surface between two media), waves can have different shapes of *surfaces* (or *lines*) of equal phases, also called *wave surfaces* or *wavefronts*. For single waves, the wavefront is the leading edge of the disturbance, contiguous to the undisturbed medium. Lines whose directions at each spatial point coincide with the direction of wave energy propagation are called *rays*. In an isotropic medium, rays are orthogonal to wave surfaces (to wavefronts).

In a *plane wave*, surfaces of equal phases are planes traveling in the direction of wave propagation with the velocity u. Such a wave corresponds to a *parallel beam* of straight rays. The amplitude of a plane wave in the absence of energy dissipation is the same at any distance from the source.

In a *spherical wave* (e.g., the wave from a point source in a homogeneous isotropic medium), wave surfaces are concentric spheres spreading with the velocity u, and the rays form a radial beam diverging from the source. The amplitude of a spherical wave decreases with the distance r from the source as $1/r$, because in the absence of absorption the same energy flow is distributed over a larger area (the latter is proportional to r^2, and consequently, $\langle j \rangle \sim 1/r^2$ and $a \sim 1/r$). The equation of a spherical wave is as follows:

$$x(r,t) = \frac{a_0 r_0}{r} \cos\omega(t - \frac{r}{u}), \tag{16.36}$$

where a_0 is the amplitude at a distance r_0 from the wave center (from the point source).

In a *cylindrical wave* (e.g., in a wave from a linear source), wave surfaces are coaxial circular cylinders. The amplitude of such a wave (and of a circular wave on water surface) decreases with the increasing distance r from the axis as $1/\sqrt{r}$ (in the absence of absorption).

16.7.7 Reflection and refraction of waves

On a boundary between two media a wave is partially *reflected* and partially penetrates into the second media, changing its direction (the wave is *refracted*). The *laws of reflection* and *refraction* can be deduced from Huygens' principle, according to which every point on a wavefront may itself be regarded as a source of secondary spherical waves. Position and form of the wave surface at any subsequent time is given by the *envelope* (i.e., the common tangent surface) of the secondary waves. As follows from Huygens' construction for a plane wave on the boundary (Figure 16.34), the reflected and the refracted rays lie in the same plane with the incident ray and the normal to the boundary surface at the point of incidence. The angle of reflection is equal to the angle of incidence ($\gamma = \alpha$—the *law of reflection*), and the ratio of sines of incident and refraction angles is independent of the angle of incidence and equals the ratio of phase velocities of waves in the first and the second media ($\sin\alpha/\sin\beta = u_1/u_2$—the *law of refraction*). As the wave passes into the second medium, the frequency ν remains unchanged, but the wavelength changes: according to Equation (16.31), $\lambda_1 = u_1/\nu$, $\lambda_2 = u_2/\nu$ and hence $\lambda_2/\lambda_1 = u_2/u_1$.

16.7.8 Interference of waves

When several waves are propagating simultaneously, each of them travels just as in the absence of other waves, and the resulting disturbance at every point is equal to the sum of disturbances created by every wave. This statement (*principle of superposition*) is usually fulfilled with great

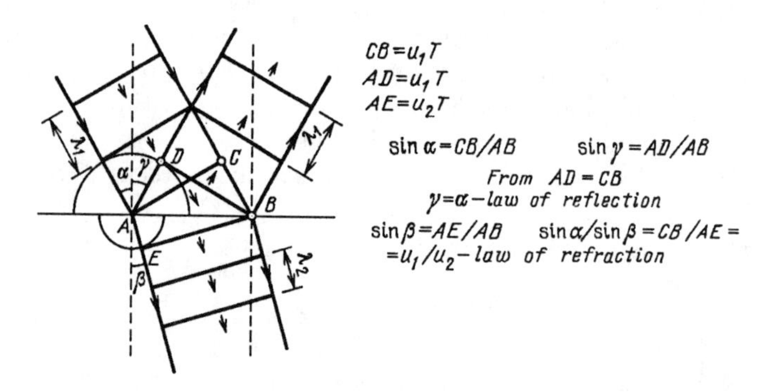

Figure 16.34: Huygens' construction for reflected and refracted waves.

accuracy and is violated only for waves of large amplitudes, because their propagation is described by nonlinear differential equations (nonlinear acoustics, nonlinear optics).

Interference of waves is the superposition of two or more waves, causing the spatial distribution of amplitudes of resulting oscillations, constant in time (a stationary *interference pattern*). Interference of waves is possible for the superposition of waves from *coherent* (matched) sources. The condition of coherence is fulfilled if oscillations of the sources are executed at the same frequency and with a constant phase difference (for details see Section 17.3.3).

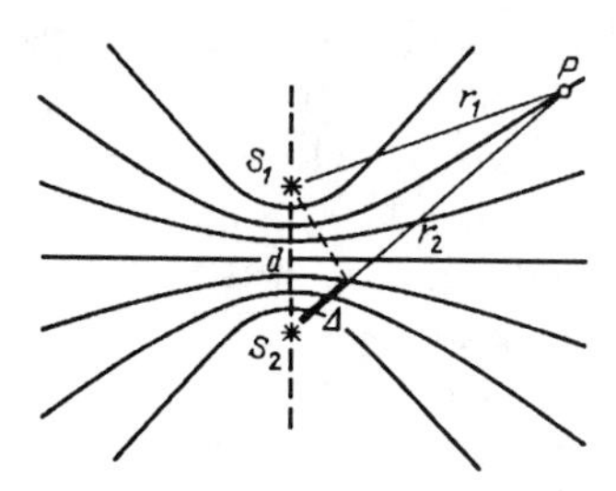

Figure 16.35: Position of maximums in the interference pattern produced by two point sources.

The simple case of interference is the superposition of two monochromatic waves of the same frequency and direction (polarization) of vibrations (Figure 16.35). For the observation point P located at a great distance from two identical sources (compared with the distance d between them), amplitudes of both waves are practically the same. The result of interference at the point P depends on the phase difference of coming oscillations: if *path-length difference* $\Delta = r_2 - r_1$ equals an integral number of wavelengths λ $(r_2 - r_1 = k\lambda;\ k = 0,\ \pm1,\ \pm2,\ \dots)$, the waves come in phase and their superposition causes an oscillation of a double amplitude (compared to the amplitude produced by a single source); if the path-length difference is equal to an odd number of half-waves $(r_2 - r_1 = (2k+1)\lambda/2)$, the waves come to the point P in antiphase and suppress each other.

In the plane of Figure 16.35, maximums of the interference pattern are located on hyperbolas (see Section 9.3.3). Damping of oscillations in some places and their amplification in others testify that at interference of waves the energy flow from the sources does not change as a whole, but is redistributed in space.

16.7.9 Standing waves

Standing waves result from the interference of two monochromatic waves of the same frequency, amplitude, and polarization, propagating in opposite directions. the distribution of amplitudes in a standing wave can be easily found by the addition of disturbances $x_1(z,t)$ and $x_2(z,t)$, created by the direct wave and the reflected one (which propagate along the z-axis):

$$x(z,t) = a\cos\omega\left(t - \frac{z}{u}\right) - a\cos\omega\left(t + \frac{z}{u}\right) = 2a\sin\frac{\omega z}{u}\ \sin\omega t \qquad (16.37)$$

(the initial phase of the reflected wave is chosen so that at $z = 0$ the displacements $x_1(0,t)$ and $x_2(0,t)$ were equal and opposite: $x_1 = -x_2$). At every point a harmonic vibration of the frequency ω is executed, with the amplitude depending periodically on the position of the point:

$$a(z) = 2a\left|\sin\frac{\omega z}{u}\right|. \qquad (16.38)$$

The points $z = n\pi u/\omega = nuT/2 = n\lambda/2$, in which the amplitude is zero are called *nodes* of a standing wave. The distance between neighboring nodes is equal to $\lambda/2$ (Fig. 16.36). Points of maximum amplitude, located between nodes, are called *antinodes*. All points between neighboring nodes oscillate in the same phase, and any two points located on opposite sides of a node oscillate in antiphase. Unlike the traveling wave, oscillations of kinetic and potential energy at every point here occur in antiphase: at the moment when disturbance becomes zero simultaneously at all points, kinetic energy reaches its maximum, and potential energy has its minimum (turns to zero).

For a string fixed at its ends, standing waves are *normal oscillations*, or *modes* (see Section 16.6.3). The length of a string here is equal to

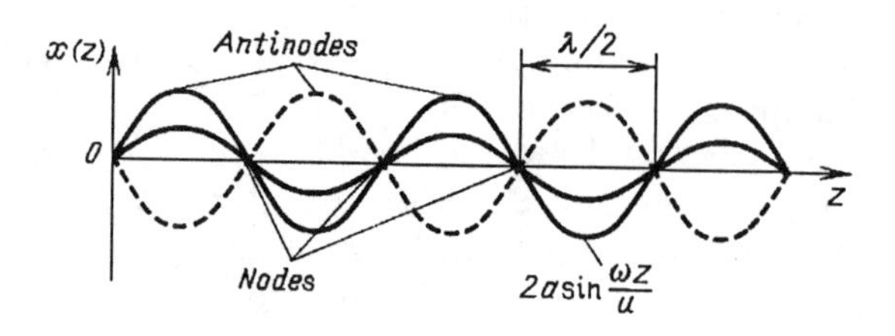

Figure 16.36: Standing wave.

an integral number of half-waves. For a stretched string, frequencies ω_n of these modes form an equidistant spectrum, described by Equation (16.29). (For other systems, the spectrum can be more complex.) Each time the frequency of external sinusoidal action is close to one of the natural frequencies ω_n, a *resonance* occurs in the system (i.e., a sharp growth of the amplitude of forced oscillations).

16.7.10 Diffraction of waves

Obstacles on the way of waves cast a *shadow*, i.e., an area free of wave motion (Figure 16.37). However, the boundaries of this area are diffuse due to diffraction. *Diffraction of waves* is a violation of their rectilinear propagation, when they bend around the obstacles and partially spread to the areas of geometric shadow. A qualitative explanation of this phenomenon is given by Huygens' principle (see Section 16.7.7), but to find the wave field behind an obstacle it is necessary to take into account the interference of secondary waves entering the area of geometric shadow (principle of Huygens–Fresnel, see Section 17.4.1). Just behind an obstacle (at $\lambda \ll d$, where d is the size of the obstacle) the amplitude of oscillations is very small, but at large distances l of the order of d^2/λ the penetration of oscillations into the shadow area becomes noticeable. Diffraction of sound and of radio waves and *diffraction of light* are particularly important in applications.

16.7.11 Doppler effect

When the source of waves or the receiver moves with respect to the medium, the frequency of oscillations registered by the receiver differs from that of the source. This phenomenon is called the *Doppler effect*. Let the source of waves move along the line AB with a velocity v_s (Figure 16.38). Surfaces of equal phases (crests) of the waves emitted by the source condense in front of the source and become rarefied behind it. The wavelength λ' of the wave emitted forward is smaller than the wavelength

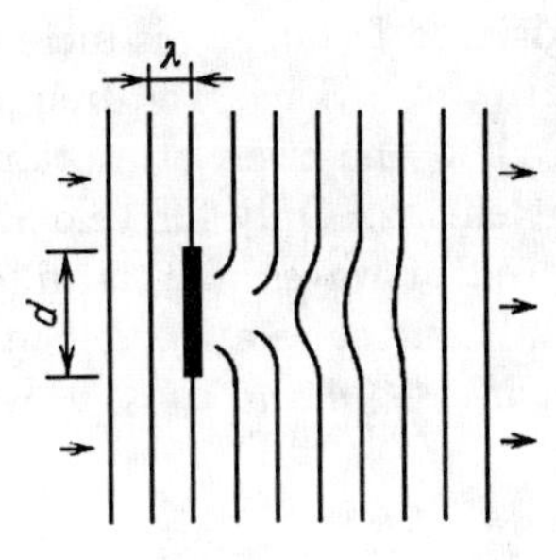

Figure 16.37: Diffraction of a plane wave on an absorbing obstacle.

$\lambda = uT$ of the wave emitted by an immovable source by the distance v_sT passed by the source during one period of oscillations:

$$\lambda' = \lambda - v_sT = (u - v_s)T = \frac{u - v_s}{\nu_s}.$$

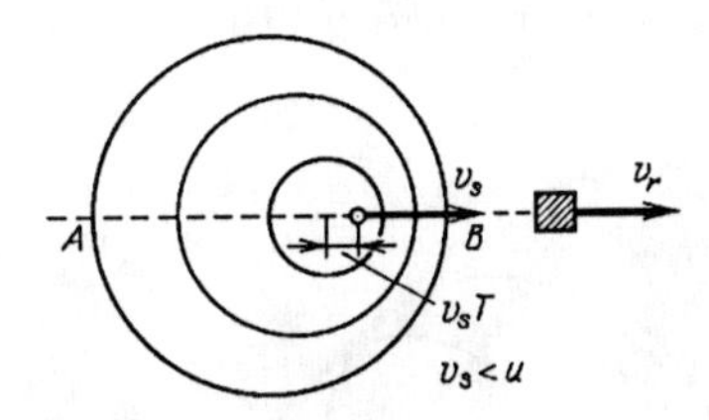

Figure 16.38: Doppler effect.

If the receiver located in front of the source is moving in the same direction with the velocity v_r with respect to the medium, the wave velocity relative to the receiver is equal to $u - v_r$. The frequency ν_r of the registered oscillations is equal to the ratio of this velocity to the wavelength λ':

$$\nu_r = \frac{u - v_r}{u - v_s}\nu_s. \tag{16.39}$$

Here the source and the receiver are supposed to move in the same direction with respect to the medium, the receiver being ahead.

If the source is moving away from the receiver (to the left in Figure 16.38), one must consider $v_s < 0$ in Equation (16.39). If the receiver is moving towards the source, one must consider $v_r < 0$. In any case, when the distance between the receiver and the source is decreasing, we get $\nu_r > \nu_s$, i.e., the registered frequency is increased, and when the distance is growing, we get $\nu_r < \nu_s$, i.e., the frequency is reduced. From the qualitative point of view, the cases of motion of the source and the receiver lead to different effects. Only for velocities v_r and v_s, small in comparison with the speed of waves u, this difference disappears, and the frequency depends only on the relative velocity of the source and the receiver $v = v_s - v_r$: when $v_s \ll u$ and $v_r \ll u$, we have

$$\nu_r = \frac{1 - v_r/u}{1 - v_s/u}\nu_s \approx \left(1 - \frac{v_r}{u}\right)\left(1 + \frac{v_s}{u}\right)\nu_s \approx \left(1 + \frac{v}{u}\right)\nu_s. \qquad (16.40)$$

For electromagnetic waves (light) in a vacuum the wave speed in all frames of reference is the same, and the frequency shift is determined only by the relative velocity of the source and the receiver.

When the velocity of the source exceeds that of the waves ($v_s > u$), i.e., when the source outstrips the waves, superposition of waves emitted at different moments forms a cone-shaped wave front (Figure 16.39) moving at the velocity u together with the source located on its vertex. This wave front is called *Mach's cone*. The angle ϕ between the generator of the cone and its axis is determined by the ratio $v_s/u = M$, called the *Mach number*:

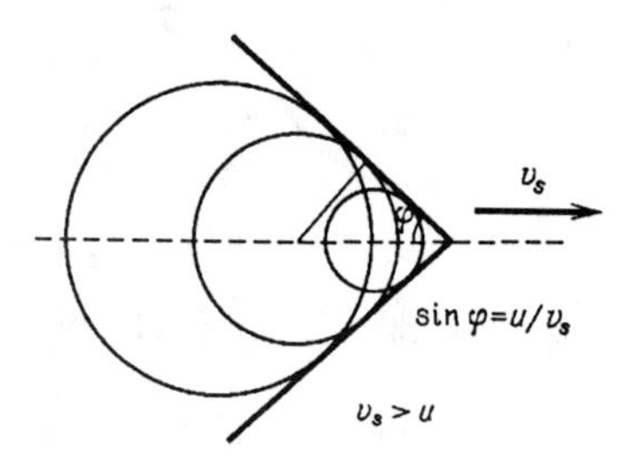

Figure 16.39: Mach cone.

$$\sin\phi = u/v_s = 1/M. \qquad (16.41)$$

Mach's cone in the form of a shock wave (a sonic boom) is formed at the motion of any body (not necessarily the source of periodic waves) in a medium with a velocity exceeding that of the waves.

An analog to Mach's cone in optics is the Cerenkov radiation emitted at the motion of charged particles in a medium with a speed exceeding the speed of light in this medium.

16.7.12 Electromagnetic waves

The principal laws of wave phenomena are valid for waves of any physical nature, including electromagnetic waves, i.e., for the process of propagation of oscillating electric and magnetic fields coupled with each other. Unlike all other kinds of waves needing some medium for their existence, electromagnetic waves can travel in the vacuum. The possibility of the existence of an electromagnetic field in a vacuum without any sources (charges and currents), in the form of waves propagating with the speed of light,

$$c = 1/\sqrt{\varepsilon_0 \mu_0} = 299,792,458 \text{ m/s} \approx 3 \cdot 10^8 \text{ m/s}$$

follows from Maxwell's equations—the basic laws of electrodynamics (see Section 15.6.3).

In a traveling electromagnetic wave, vectors of electric and magnetic fields $\mathbf{E}$ and $\mathbf{B}$ at every spatial point at a given moment of time are perpendicular to each other and to the direction of propagation $\mathbf{s}$ (Figure 16.40). Due to this property electromagnetic waves are called *transverse.* Vectors $\mathbf{E}$, $\mathbf{B}$, and $\mathbf{s}$ at every spatial point form a right-hand screw (like unit vectors $\mathbf{i}$, $\mathbf{j}$, and $\mathbf{k}$ of the right coordinate system, see Section 8.3.3). Modules of vectors $\mathbf{E}$ and $\mathbf{B}$ at the same point at any time moment are proportional to each other:

$$B = E/c. \tag{16.42}$$

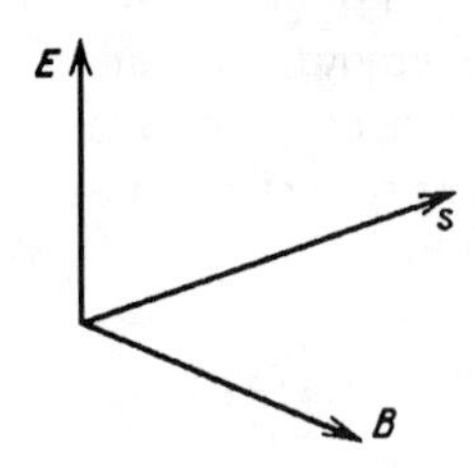

Figure 16.40: Mutual orientation of **E** and **B** vectors and vector **s** of the direction of propagation in a traveling electromagnetic wave.

Orthogonality of vectors $\mathbf{E}$ and $\mathbf{B}$ of an electromagnetic wave, and the relationship between their instantaneous values at every space-time point, Equation (16.42), are valid simultaneously in all inertial frames of reference. This follows from the expressions for the invariants of electromagnetic field (see Equations (15.115) in Section 15.6.2).

The instantaneous picture of vectors of electric and magnetic fields at different points along the line of propagation of a monochromatic plane wave is shown in Figure 16.41. If vector **E** at a given point oscillates along a certain direction, the passing wave is called *linearly polarized* wave (Figure 16.41*a*); if vector **E**, being constant in modulus, rotates around the direction of propagation, the wave is called *circularly polarized* (Figure 16.41*b*). A wave of circular polarization can be represented as a superposition of two waves of the same amplitude, propagating in the same direction, linearly polarized in mutually perpendicular directions and with the phase difference $\pi/2$ (i.e., a quarter of the wavelength).

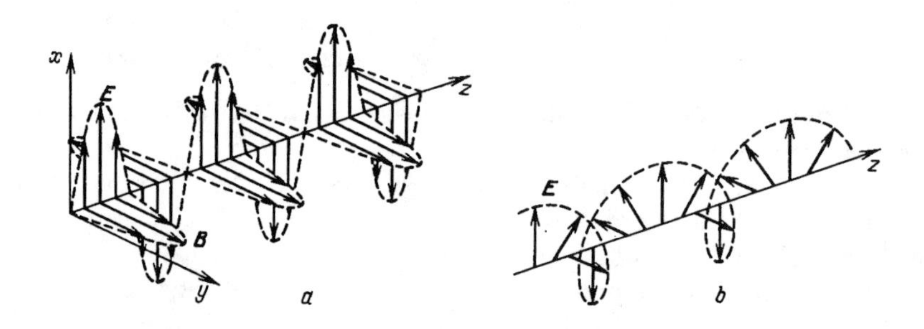

Figure 16.41: Electromagnetic waves of linear polarization (*a*) and of circular polarization (*b*).

The density of energy of electric field $w_e = \varepsilon_0 E^2/2$ in a traveling wave equals that of magnetic field $w_m = B^2/(2\mu_0)$, because **B** and **E** satisfy the relationship expressed by Equation (16.42). *Intensity*, or the *density of energy flow* S (i.e., the amount of energy carried by the wave per unit time through unit area oriented perpendicularly to the direction of propagation) is equal to the product of the energy density $w = w_e + w_m$ and the wave velocity c:

$$S = c(w_e + w_m) = 2cw_e = c\varepsilon_0 E^2. \qquad (16.43)$$

In a monochromatic wave $E(z,t) = E_0 \cos\omega(t - z/c)$, and the average value of S for a period equals $\langle S\rangle = c\varepsilon_0 E^2/2$ (because $\langle\cos^2\omega(t - z/c)\rangle = 1/2$). The electromagnetic field of a wave transfers not only the energy, but also the momentum. The *volume density of the field momentum* is equal to w/c. The momentum transferred by the wave to an absorbing surface of unit area is the *pressure* p of the electromagnetic wave. The pressure is equal to the volume density of energy in an incident wave: $p = w$. At a normal incidence on an ideal reflecting surface, the transferred momentum and the pressure of electromagnetic wave is twice as much: $p = 2w$.

Emission of electromagnetic waves is caused by an *accelerated motion* of electric charges. If a charge Q oscillates along the axis z so that its coordinate z changes harmonically, $z(t) = z_0 \cos\omega t$ (Figure 16.42), the strength of the electric field of the emitted wave at the point P located at the distance r from the charge, is proportional to the acceleration a of the charge in the previous moment of time $t - r/c$:

$$E(r,t) = \frac{1}{4\pi\varepsilon_0}\frac{Qa(t-r/c)}{c^2 r}\sin\theta =$$

$$= \frac{1}{4\pi\varepsilon_0}\frac{Q\omega^2 z_0 \cos\omega(t-r/c)}{c^2 r}\sin\theta. \tag{16.44}$$

Here θ is the angle between the direction to this point and the axis z, along which the charge is oscillating. The wave emitted by the oscillator is spherical, and the amplitude of the field strength diminishes in inverse proportion to the distance r from the source. The wave is linearly polarized, with the directions of oscillations of vectors **E** and **B** depending on the position of the point P (Figure 16.42). The time-average density of energy flow in such a wave is

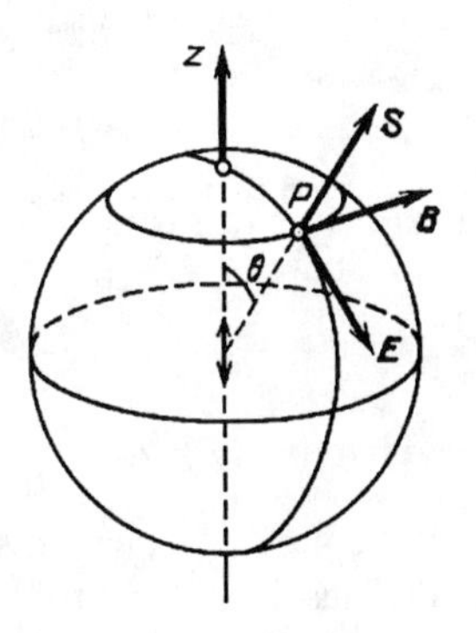

Figure 16.42: Vectors **E** and **B** at the point P in the wave emitted by the electric charge that oscillates along z-axis.

$$\langle S\rangle = \frac{1}{32\pi^2\varepsilon_0}\frac{Q^2\omega^4 z_0^2}{c^3 r^2}\sin^2\theta. \tag{16.45}$$

The energy emitted by a charged oscillator is proportional to the *square of the amplitude* z_0 of its oscillations and to the *fourth power of the frequency* ω. The surface density of the energy flow (intensity) is inversely proportional to r^2, so the total energy flow through the sphere surrounding the source does not depend on the radius of the sphere. This fact is in accordance with the conservation of energy: all the energy emitted by the source is transferred by the wave away from the source.

The density of the energy flow depends on its direction (on the angle θ, see Figure 16.42): no energy is emitted in the direction along which the charge is oscillating ($\sin\theta = 0$ for $\theta = 0$ or $\theta = \pi$). The energy flow is maximal in the plane perpendicular to the direction of oscillations of the charge ($\sin\theta = 1$ for $\theta = \pi/2$). The angular distribution of the energy emitted from the source is shown in Figure 16.43 (the length of radial line segment to the intersection with the curve is proportional to $\sin^2\theta$, i.e., to the energy emitted in the given direction). In space, the angular distribution of the energy is characterized by the surface formed by rotation of the curve around the axis z.

Equations (16.44) and (5.45) are valid for distances r that are large in comparison with the wavelength λ (in the so-called *wave zone*). Every small area of a spherical wave can be considered as a plane wave of a certain amplitude, direction of propagation, and polarization.

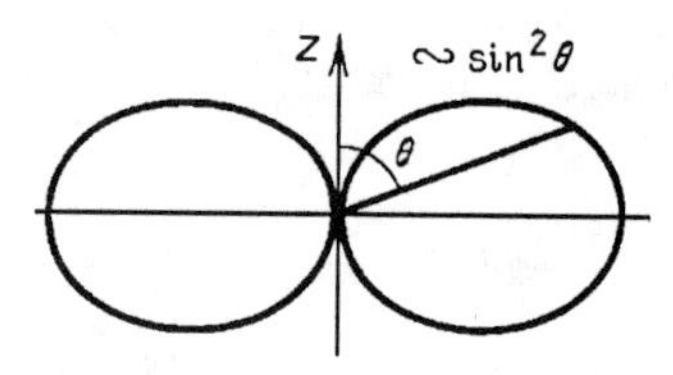

Figure 16.43: Angular distribution of the energy emitted by an oscillating electric charge.

16.7.13 Waves on the water

Waves on a water surface have narrow sharp crests and wide flat troughs. Their shape (Figure 16.44*a*) clearly differs from a sinusoidal one. Such a shape is due to the circular paths of the liquid particles (Figure 16.44*b*). That is, waves on the surface are neither transverse nor longitudinal. The diameters of the circular paths rapidly (exponentially) diminish with depth, so that only the surface layer of liquid (its thickness being approximately equal to the wavelength) is involved into the motion.

When the shape of a surface deviates from the equilibrium one (from a plane), restoring forces occur due to the gravitational force and the surface tension. The relative importance of these factors depends on the wavelength: for short waves the forces of surface tension prevail (*ripples*, or *capillary* waves); in the case of rather long waves (more than 5 cm) the forces of gravity prevail (*gravitational* or *heavy waves*).

The speed of waves on the water surface depends on the wavelength λ, i.e., these waves are characterized by a dispersion. The speed of capillary waves is inversely proportional to the square root of the medium density and of the wavelength:

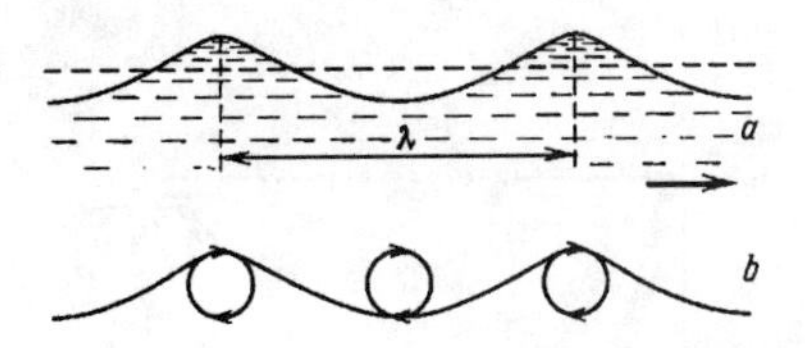

Figure 16.44: Waves on water surface: the shape of the crests (a) and trajectories of the water particles (b).

$$u_c = \sqrt{\frac{2\pi\sigma}{\rho\lambda}}, \tag{16.46}$$

where σ is the surface tension, and ρ is the liquid density. The dependence of u_c on σ, ρ, and λ, can be found by the method of dimensional analysis (see Section 12.3.5). The speed of gravitational waves is

$$u_g = \sqrt{\frac{g\lambda}{2\pi}} \qquad (h > \lambda). \tag{16.47}$$

This expression is valid if the depth h exceeds the wavelength λ. In the limiting case of a small depth ($h \ll \lambda$) the wave speed does not depend on the wavelength:

$$u_g = \sqrt{gh} \qquad (h \ll \lambda), \tag{16.48}$$

i.e., the speed of very long waves increases as the depth of the water in the reservoir is increased. For waves with $\lambda \approx 1$–2 cm, both the forces of surface tension and the forces of gravitation are equally important. The speed of these waves is determined by the expression

$$u = \sqrt{u_c^2 + u_g^2} = \sqrt{\frac{2\pi\sigma}{\rho\lambda} + \frac{g\lambda}{2\pi}}. \tag{16.49}$$

In Figure 16.45 the dependencies of speed of capillary and gravitational waves on $\sqrt{\lambda}$ are shown. The minimum speed corresponds to the wavelength $\lambda_0 = 1.73$ cm. Waves of such length propagate with the speed $u = 23.2$ cm/c. At $\lambda < \lambda_0$ the waves are mainly capillary, at $\lambda < \lambda_0$ they are mainly gravitational. With the growth of wavelength the speed of gravitational waves increases in proportion with $\sqrt{\lambda}$. This growth slows down when the wavelength becomes more than the depth h of the reservoir. Therefore, the maximal possible speed of gravitational waves is determined by the depth h : $u_{\text{max}} = \sqrt{gh}$. In an ocean, at the depths of $h \approx 5$ km, the speed of *tsunami waves* (with the wavelength of $\lambda > h$) is 200 m/s (i.e., approximately 700 km per hour).

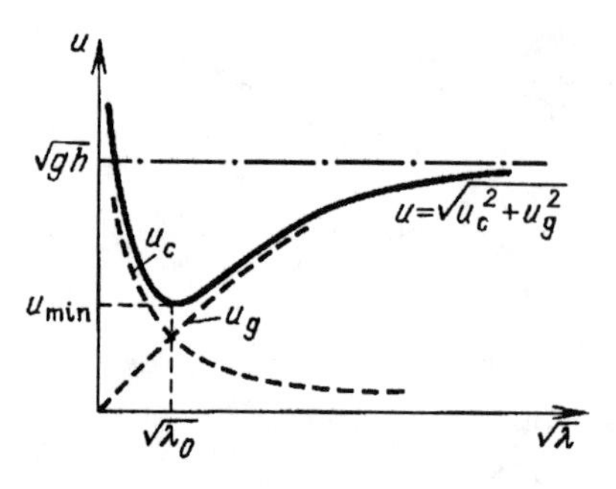

Figure 16.45: Speed of waves on water surface.

16.7.14 Speed of wave packets

When the speed of monochromatic waves does not depend on the wavelength, i.e., in the absence of dispersion (elastic waves, electromagnetic waves in vacuum), disturbances of any type propagate with the same speed without changing their shape. Generally speaking, in the presence of dispersion, single pulses and wave packets (see Figure 16.31*a, b*) change their shape during propagation. However, under certain conditions (weak dispersion of the medium, sufficiently long wave packet) the envelope of a wave packet keeps its shape (while the packet travels to a limited distance), and the *group velocity of waves* may be introduced as the velocity of the envelope.

One can get the idea of the motion of a wave packet by considering the wave formed by a superposition of two monochromatic waves with slightly different wavelengths (Figure 16.46). In the place where a hump of one wave coincides with that of the other one, the resulting wave has a hump of double height. In the place where a hump of one wave coincides with the trough of the other one, the resulting displacement is zero. The whole wave is a train of wave packets, or *wave groups* (Figure 16.47). In the absence of dispersion the velocities of superimposed monochromatic waves are the same, and the resulting wave moves with the same velocity without changing its shape. In a medium with dispersion the velocities of superposed waves differ, and the mutual positions of their humps change with time. Therefore, the centers of wave groups move with a velocity different from that of separate humps and troughs. If at some moment the maximum of the envelope, i.e., the center of the group, is located at a point P (Figure 16.46), after the time interval $\tau = \Delta\lambda/\Delta u$ the hump Q' reaches Q, and the center of the group shifts backward by the distance equal to the wavelength λ. So the velocity of the group center is less than the velocity u of individual humps or troughs approximately by λ/τ:

$$u_{gr} = u - \frac{\lambda}{\tau} = u - \lambda \frac{du}{d\lambda} \tag{16.50}$$

(*Rayleigh formula*). For gravitational waves on water surface $u_g(\lambda) = C\sqrt{\lambda}$ (see Equation (16.47)), and Equation (16.50) gives $u_{gr} = u_g/2$ — the center of the group moves with the velocity which is one half of the velocity of separate humps and troughs. For capillary waves $u_c = C'/\sqrt{\lambda}$ (see Equation (16.46)), and the computation by the Equation (16.50) gives $u_{gr} = 3u_c/2$ — the group center moves with the velocity one and a half times greater than that of separate humps.

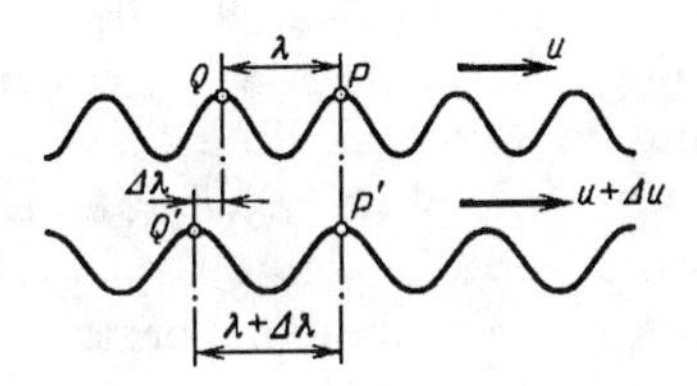

Figure 16.46: For the calculation of the group velocity of waves.

Rayleigh formula, Equation (16.50), is applicable not only to wave groups (Figure 16.47) formed by a superposition of two monochromatic waves, but also to single wave packets (Figure 16.31*b*), if such packets keep the form of the envelope while the wave is propagating. Any packet can be expanded into a sum of an infinite number of monochromatic components whose frequencies fill a certain interval $\Delta\nu$. The duration Δt of the wave packet is connected with this interval of frequencies $\Delta\nu$ by the relationship

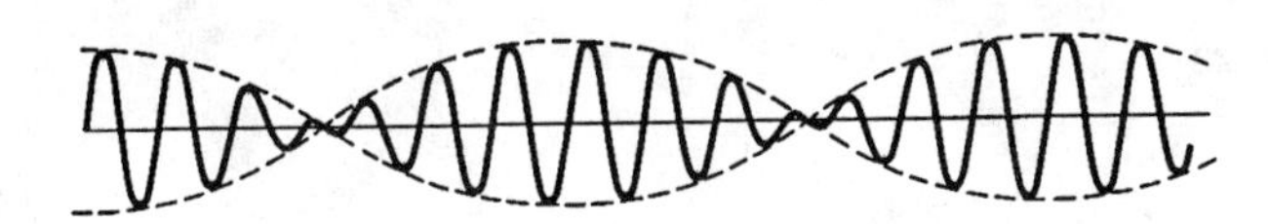

Figure 16.47: Wave groups.

$$\Delta\nu \cdot \Delta t \approx 1, \tag{16.51}$$

i.e., the narrower the interval $\Delta\nu$ of frequencies (or the corresponding interval $\Delta\lambda$ of wavelengths) of monochromatic components, the greater the duration and spatial extension of the wave packet (see Section 18.4.1). The packet practically does not spread during its propagation, if the group velocity, Equation (16.50), is approximately constant within the limits of the wavelength interval $\Delta\lambda$ corresponding to this packet.

A non-periodic external action (e.g., a stone thrown into the water) causes a single wave pulse. In a medium with dispersion, monochromatic components of the wave pulse have essentially different velocities, i.e., in some sense they seem to get "out of step," and the single pulse eventually turns into an extended wave process. Long waves on water surface get farther forward, leaving the shorter waves behind. For such a wave process with a constantly changing shape, the concept of group velocity is not applicable.

In a nonlinear medium, the spreading of a wave packet due to dispersion can be compensated under certain conditions by the non-linearity effects leading to the dependence of the wave velocity and shape on the amplitude. As a result, in such a medium the propagation of solitary waves of a certain shape, i.e., of so-called *solitons*, becomes possible. Solitons can exist in various media—on water, in crystals, magnetic materials, superconductors, and even in living organisms. In many respects solitons behave themselves like particles.

Chapter 17

Optics

The subject of *optics* envelopes a wide region of physical phenomena associated with light—its emission, transmission, propagation, detection, and interaction with substance. *Light* is the kind of electromagnetic radiation with wavelengths in the range from approximately 400 nm (violet light) up to 700 nm (red light) to which the human eye is sensitive. In a broader sense, optics includes also all the phenomena associated with *ultraviolet* and *infrared* radiation. The boundaries of the optical spectrum are rather conventional and are determined mainly by existing means and experimental methods of generation and detection of radiation. The optical band of electromagnetic spectrum is adjacent at one end to the *X-ray radiation*, and at the other end—to the *microwave radiation.* In a vacuum the light travels with a constant speed of $c = 2.997,924,58 \cdot 10^8$ m/s, as well as electromagnetic radiation of any other wavelength.

17.1 Geometrical Optics

17.1.1 The principal laws

Geometrical or *ray optics* assumes that light travels in straight lines in transparent optically homogeneous media. It uses the concept of *light rays,* which are rectilinear in a homogeneous medium, are partially reflected, and partially refracted on a boundary of two media with different optical properties. Ray optics is concerned with the laws controlling the reflection and refraction of light rays. The energy of electromagnetic oscillations is transferred along the light rays. In an isotropic medium, light rays are orthogonal to the wave surfaces (see Section 16.7.6).

Optical properties of a transparent medium are characterized by *refractive index n*. The ***absolute refractive index*** of a medium is the ratio of the speed of electromagnetic radiation in free space to the phase velocity of light in the medium: $n = c/v$. The refractive index depends on the

Table 17.1: Refractive index of some transparent media

Substance	Wavelength, m			Δn (blue – red)*
	$4.4 \cdot 10^{-7}$ (blue)	$5.89 \cdot 10^{-7}$ (yellow)	$6.6 \cdot 10^{-7}$ (red)	
Air	1.000296	1.000293	1.000291	0.000005
Water	1.340	1.333	1.331	0.009
Ethyl alcohol	1.370	1.362	1.360	0.010
Melted quartz	1.470	1.458	1.455	0.015
Crown-glass	1.528	1.517	1.513	0.015
Flint-glass	1.594	1.575	1.570	0.024
Heavy flint	1.945	1.890	1.875	0.070
Diamond	2.465		2.407	0.058

* The quantity $\Delta n = n_{\text{blue}} - n_{\text{red}}$ (blue – red) serves as a measure of *dispersion* of a substance. Refraction of the white light in diamond or in heavy flint (in crystal) produces wide colored spectra. A hollow prism filled with water does not produce a wide spectrum.

wavelength λ (or on frequency ω): $n = n(\lambda)$. This dependence is specific for a medium and is called the *law of dispersion* (see Section 16.7.13). The absolute refractive index is connected with the permittivity (dielectric constant) ε of the medium (for a given wavelength) by the relation $n = \sqrt{\varepsilon}$. The *relative refractive index* is the ratio of the absolute refractive indexes of two media. The inverse quantity gives the ratio of the speed of light in one medium to that in an adjacent medium. A medium of larger refractive index is called having greater *optical density*. The values of refractive index for some transparent media are given in Table 17.1.

The rays of light obey the following *principal laws of ray optics*:

- In a homogeneous medium the rays of light are straight lines (the law of rectilinear propagation of light).

- On a boundary surface of two media (or on a boundary of a medium and a vacuum) a reflected ray arises which lies in the same plane with the incident ray and the normal to the reflecting surface at the point of incidence, and *the angle of reflection* φ_1 *equals the angle of incidence* φ (Figure 17.1):

$$\varphi_1 = \varphi \tag{17.1}$$

—the law of reflection of light.

- The ray of light changes its direction as it passes obliquely from one medium to another in which the speed of light is different. The refracted ray lies in the same plane with the incident ray and the normal to the reflecting surface at the point of incidence, and *the angle of refraction* φ_2 obeys *Snell's law*:

$$n_1 \sin\varphi_1 = n_2 \sin\varphi_2, \qquad \text{or} \qquad \frac{\sin\varphi_1}{\sin\varphi_2} = \frac{n_2}{n_1} = \frac{v_1}{v_2} \tag{17.2}$$

—the law of refraction of light on a boundary of a transparent medium.

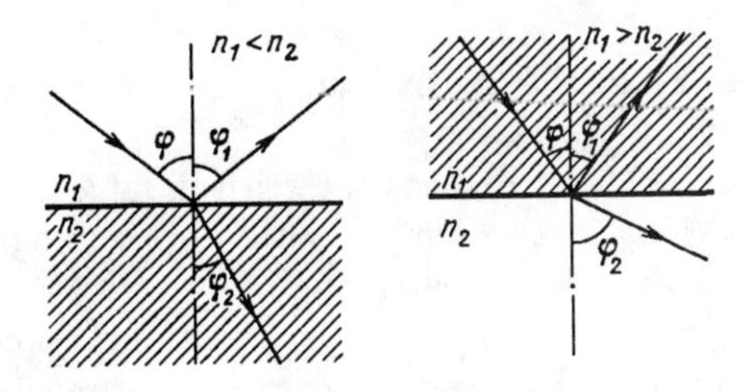

Figure 17.1: Reflection and refraction of light on a flat boundary of two media: φ – the angle of incidence, φ_1 – the angle of reflection, φ_2 – the angle of refraction.

The laws of reflection and refraction of light are valid not only for plane boundary surfaces but also for curved ones, e.g., for spherical surfaces of lenses. In this case the normal to the interface between the two media has different directions at various points of the boundary surface.

The refracted ray of light is bent toward the normal ($\varphi_2 < \varphi_1$) while it passes into the medium with greater optical density ($n_2 > n_1$), and the ray deflects farther from the normal ($\varphi_2 > \varphi_1$) if the relative refractive index $n_{12} = n_2/n_1$ is less than unity (Figure 17.1). When light passes from a vacuum into a transparent medium with the refractive index n, Snell's law of refraction, Equation (17.2), takes the form:

$$\frac{\sin\varphi}{\sin\varphi_2} = n. \tag{17.3}$$

In the air the absolute refractive index is very close to unity ($n_{\text{air}} = 1.0003$). So it is possible to use Equation (17.3) for light passing into some medium from the air.

When light is transmitted into a medium with smaller optical density ($n_2 < n_1$), the angle of incidence can not exceed some critical value $\varphi_{\rm cr}$ because the angle of reflection φ_2 cannot exceed $\pi/2$ (Figure 17.2):

$$\sin\varphi_{\rm cr} = n_2/n_1, \qquad (n_2 < n_1). \tag{17.4}$$

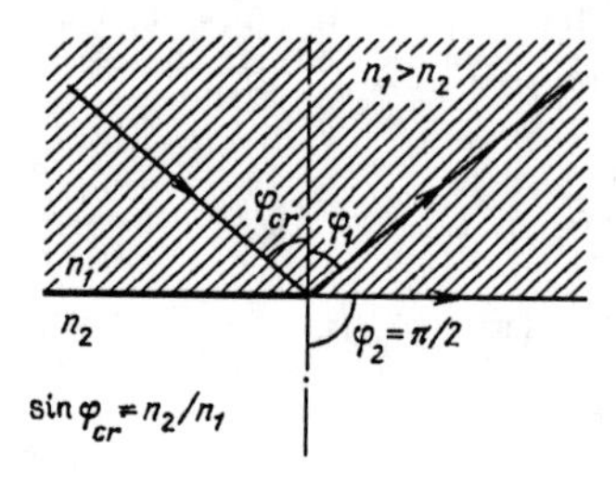

Figure 17.2: The limiting angle of total reflection.

If the angle of incidence exceeds the critical value $\varphi > \varphi_{\rm cr}$, the *total reflection* occurs, i.e., all the energy of the incident light returns into the first medium, which has greater optical density. For the boundary glass—air ($n_1/n_2 = 1.5$) the critical angle of total reflection equals $41°50'$.

The laws of ray optics had been discovered experimentally long before the nature of light was discovered. These laws can be derived in the wave theory of light by means of Huygens' principle (see Section 16.5.7). The scope of validity of these laws is restricted by the *diffraction phenomena* (see Section 17.4).

The rectilinear propagation of light causes the formation of shadows. A *shadow* is the region into which light does not penetrate. In the case of a *point source* (a light source of negligible size) an obstacle casts a shadow with sharp edges (Figure 17.3*a*). These edges are slightly blurred only by diffraction phenomena (see Section 17.4). In the case of two or more sources of light, and also in the case of an *extended source* that has an appreciable size, the shadow has two different regions: one of full shadow, called the *umbra*, and the other of half-shadow, called the *penumbra* (Figure 17.3*b*). This transition region of half-shadow between full darkness and illuminated area increases with the size of light source, and in the case of a very extended source (e.g., a cloudy sky) the full shadow may disappear completely.

A pencil of light rays all of which pass through the same point is called a *homocentric bundle* (Figure 17.4). In *astigmatic bundles* (Figure 17.5) the light rays lying in the two mutually perpendicular axial sections (in the *main sections*) intersect in different places—along the two segments, displaced from each other by some distance along the bundle. The wave

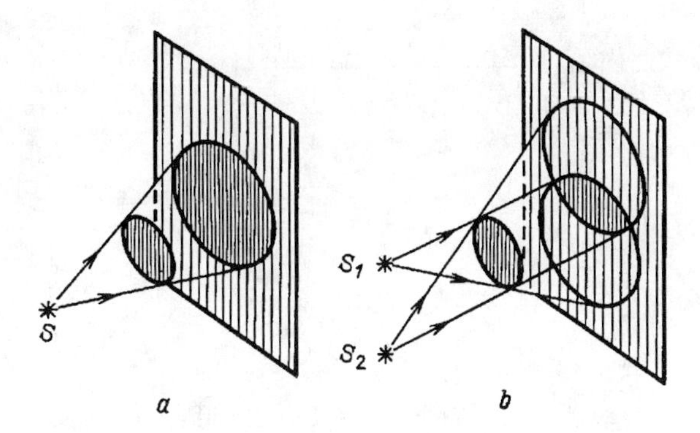

Figure 17.3: Formation of full shadow (the *umbra*) and half-shadow (the *penumbra*).

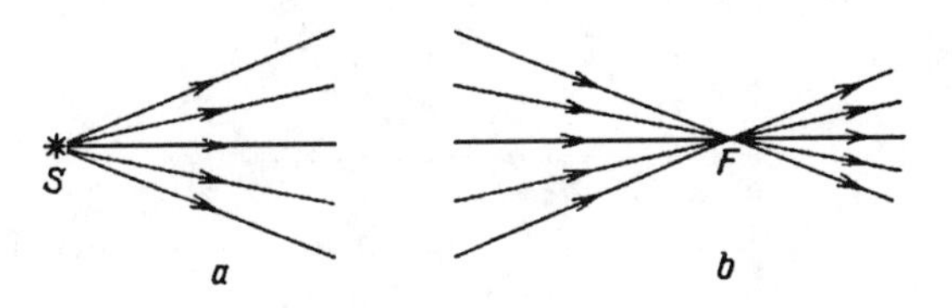

Figure 17.4: Divergent (*a*) and convergent (*b*) homocentric bundles of light rays.

surfaces of an astigmatic bundle, which are orthogonal to the rays, are the surfaces of double curvature (see Section 9.7) characterized by the radii R_1 and R_2 in Figure 17.5, unlike the wave surfaces of a homocentric ray bundle which are concentric spheres.

17.1.2 Plane mirrors

A smooth surface that reflects most of the incident light is called a *mirror*. A homocentric bundle of light reflected by a *plane mirror* changes its direction of propagation, but remains homocentric (Figure 17.6*a*). The observer perceives the reflected rays as if they were emitted from the point S' located behind the mirror symmetrically to the point S, which is the real source of the ray bundle. The image of a real object produced by a plane mirror is called *virtual,* because any point S' of this image is not a point where the reflected rays themselves intersect: in the virtual image the *imaginary* straight lines intersect, which are the backward continuation of the reflected rays. The virtual image of a real object that

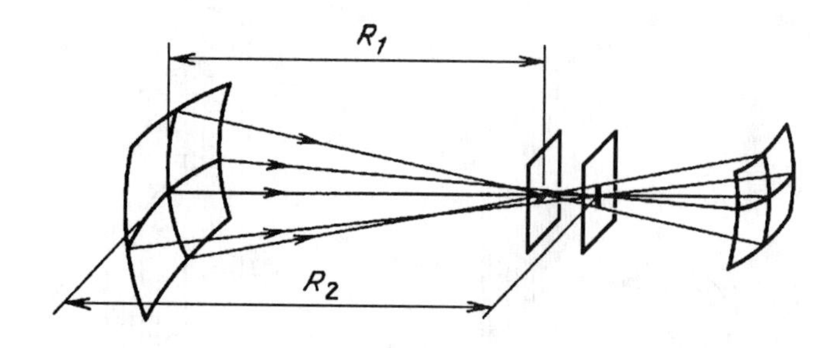

Figure 17.5: Astigmatic bundle of light rays.

we see in a plane mirror is *erect* with *reversed front and back*. This image is sharp and all its sizes are equal to the sizes of the object.

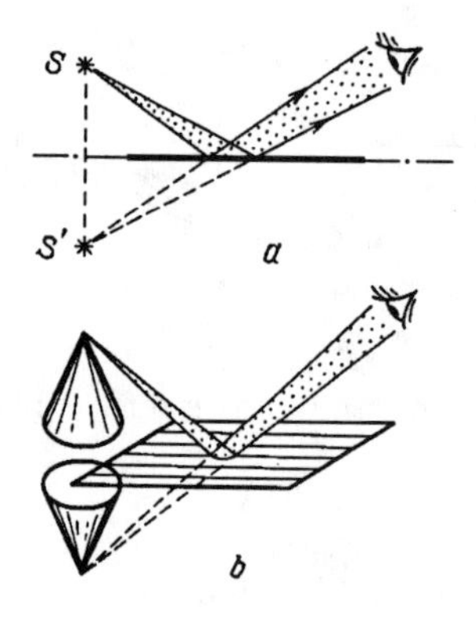

Figure 17.6: Plane mirror.

17.1.3 Paraxial approximation and optical images

Homocentric bundles of rays originating from individual points of an object lose, as a rule, this quality after reflection or refraction and become astigmatic (reflection by a plane surface is an exception). But in practically very important case of rather narrow paraxial bundles of rays in *coaxial optical systems* like lenses, the homocentricity of bundles is approximately preserved.

An optical system is called coaxial if it consists of spherical refracting and reflecting surfaces whose centers are located on the same straight line called the *optical axis*. A ray of light is called *paraxial* if it falls on a reflecting or a refracting surface close to the optical axis (the distance must be small compared to the radius of the surface curvature) and almost parallel to the axis. For such paraxial rays a simple lens theory can be developed by means of using the small angle approximation.

After passing through an optical system, bundles of paraxial rays originating from different points of an object form the *optical image* of the object. A certain point S' of the image corresponds to every point S of the object (Figure 17.7). If the rays of light actually pass through the image, it is called a *real image*. When a screen is placed in the plane of a real image, the image becomes visible. If the image is located at a point from which the rays only appear to come to the observer, but do not actually do so, the image is called a *virtual image*.

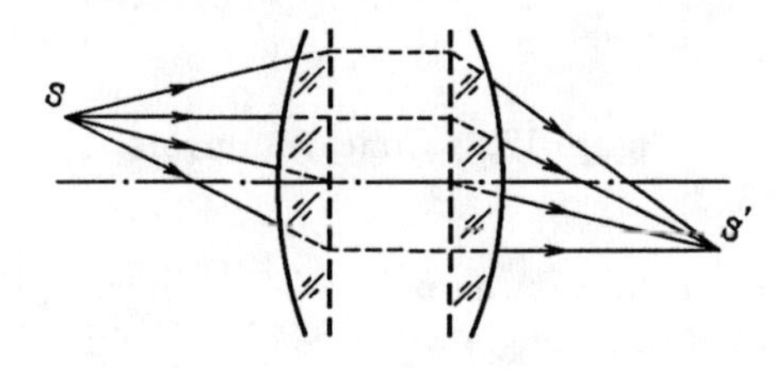

Figure 17.7: Formation of an image by an optical system.

17.1.4 Spherical mirrors

When a parallel bundle of light rays hits the surface of a *concave spherical mirror*, reflected rays are converged and come to a focus at the *focal point* F of the mirror (Figure 17.8*a*). The focal point is located in the middle of the segment OP connecting the center O of the spherical mirror surface and the top P (the *pole*, or the *optical center*). The *focal length* f of a spherical mirror is $R/2$, where R is the radius of the surface ($f = R/2$).

A method of locating the image of an arbitrary point of an object is based on drawing *ray diagrams*. It is convenient to draw several specific rays which are easy to trace (Figure 17.8*b*):

- The ray AOB through the center O of the sphere; the reflected ray is directed along the same line backwards.
- The ray AFD through the focal point F; the reflected ray DA' is parallel to the optical axis PO.
- The ray AC, which is parallel to the optical axis OP; the reflected ray CFA' is aimed toward the focal point F.

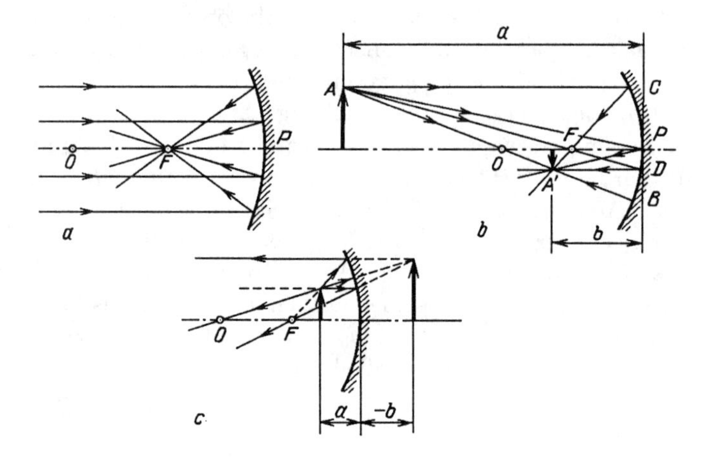

Figure 17.8: Concave mirrors.

- The ray AP which hits the pole P of the mirror; the reflected ray PA' makes the same angle with the optical axis PO.

Any two of the rays are sufficient to localize the image point A', because in paraxial approximation all reflected rays are aimed toward A'. If an object is located apart from the mirror at a distance exceeding the focal length, a *diverging* pencil of rays coming from any point of the object becomes *convergent* after reflection. To produce a sharp image, all the rays coming from one point of an object must converge to one point of the image. Then one point of the image (not a fuzzy region) corresponds to each point of the object. This is approximately valid for the image formed by paraxial bundles of rays.

The *mirror equation* has the form:

$$\frac{1}{a} + \frac{1}{b} = \frac{1}{f}, \tag{17.5}$$

where a and b are the distances of the object and of the image (measured from the mirror pole P), and $f = R/2$ is the focal length of the mirror.

When the object distance a is greater than $2f$, the image is real inverted diminished. If $f < a < 2f$, the image is real inverted magnified. If the object distance is smaller than the focal length ($a < f$), the image is virtual upright diminished and is located behind the mirror (Figure 17.8c). The mirror equation, Equation (17.5), is valid in this case also, if the image distance b is considered to be negative ($b < 0$).

A parallel pencil of rays becomes divergent after reflection from a *convex* spherical mirror (Figure 17.9). The reflected rays appear to come from the focal point F, located behind the mirror at a distance $R/2$ from the mirror. The image of an object in a convex mirror is always virtual erect diminished and is located behind the mirror at a distance shorter than the focal length f.

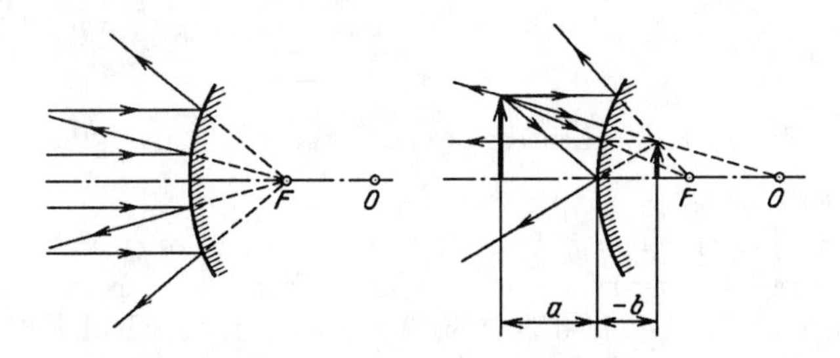

Figure 17.9: Convex mirrors.

The tracing of an image formed by a convex mirror is fulfilled by drawing the specific rays similar to those mentioned above for a concave mirror. Equation (17.5) is also valid for a convex mirror if the focal length f in it is considered to be negative ($f = -R/2$).

17.1.5 Lenses

The straight line passing through the centers of curvature of the spherical surfaces forming the lens faces is called the ***principal optical axis.*** In the case of a lens with one flat surface it is the line passing through the center of curvature of the spherical surface perpendicularly to the plane surface. ***Converging lenses*** are thicker in the middle than at the edges; ***diverging lenses***, on the contrary, are thinner in the middle (Figure 17.10). This is true for the case when the transparent material of the lens has a greater optical density (greater refraction index) than the surrounding medium. Converging lenses include biconvex, planoconvex, and meniscus lenses (Figure 17.10*a*). Diverging lenses include biconcave, planoconcave, and diverging meniscus lenses (Figure 17.10*b*). A lens is called ***thin*** if its thickness is negligible compared with the radii of curvature of its faces and with the object distance from the lens. In this case the points of intersection of the lens faces with the optical axis are located so close to each other that they are assumed to be the one point O (Figure 17.11), which is called the ***optical center*** of the lens. Any ray passes through this point of a lens without deviation.

A converging lens brings the rays, incident close to and parallel to the optical axis of the lens, to a real ***principal focus*** F (Figure 17.11*a*). The distance between the optical center and the principal focus of the lens

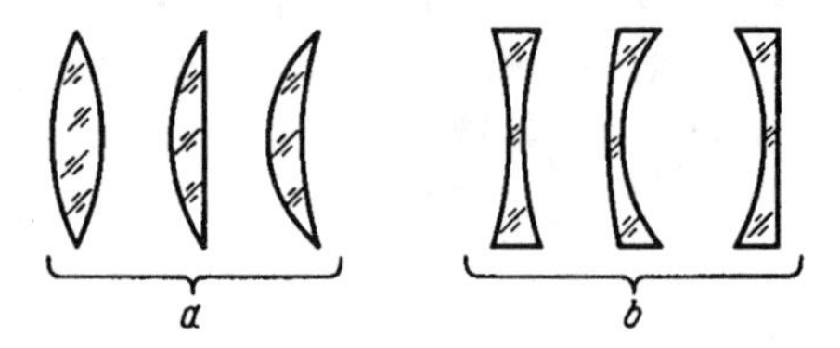

Figure 17.10: Lenses: *a* – converging, *b* – diverging.

is called the *focal length f*. It depends on the radii of curvature R_1 and R_2 of the refracting surfaces and on the refractive index n of the lens' material. For a convex lens the focal length f can be calculated with the help of the *lens maker's formula*:

$$\frac{1}{f} = (n-1)\left(\frac{1}{R_1} + \frac{1}{R_2}\right). \tag{17.6}$$

It is assumed that the lens is surrounded by a medium whose refractive index is equal to unity (a vacuum, an air). The values of refractive index for some transparent media are given in Section 17.1.1.

The lens maker's formula, Equation (17.6), is valid for different kinds of lenses if the definite sign convention is adopted. If one of the surfaces is plane, its radius of curvature $R = \infty$. For a convexo-concave lens the radius R_2 of a concave surface should be assumed as negative ($R_2 < 0$). The quantity inverse to the focal length is called the *optical power D*:

$$D = 1/f.$$

Optical power D is measured in reciprocal meters. In this case the $(\text{meter})^{-1}$ is called the *diopter*. The focal length of a lens with optical power of 1 diopter equals 1 meter.

If a pencil of paraxial rays parallel to the optical axis hits a lens from the opposite site, it converges towards the other principal focus F'. Left and right principal focuses of a lens are located at equal distances from the lens provided that the medium is the same on both sides.

It is convenient to use the following rays while drawing ray diagrams to trace the image formation (Figure 17.11*b*):

- The ray AO passing through the optical center O of the lens without deflection.
- The ray AB parallel to the principal optical axis; the refracted ray BFA' is aimed toward the focal point F.
- The ray $AF'C$ passing through the front focal point F'; the refracted ray is parallel to the principal optical axis OF.

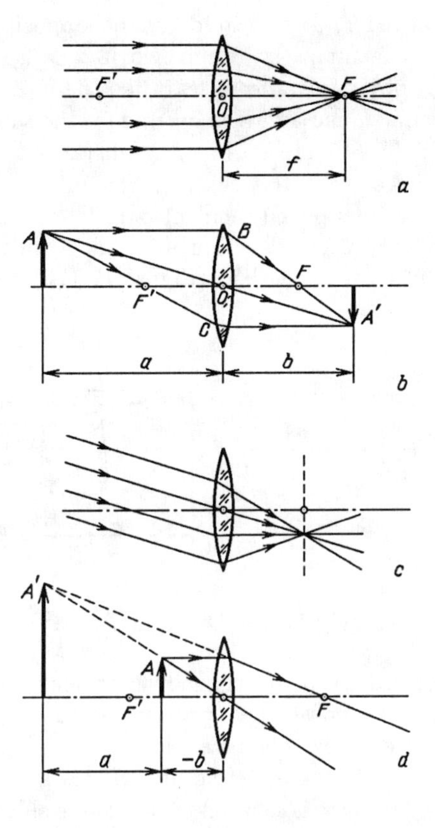

Figure 17.11: Converging lenses.

A parallel pencil of paraxial rays hitting a convergent lens at some (small) angle with the principal axis becomes convergent after refraction and comes to a focus lying in the *focal plane* of the lens—the plane passing through the focal point perpendicular to the optical axis (Figure 17.11*c*).

The *lens formula*

$$\frac{1}{a} + \frac{1}{b} = \frac{1}{f} \tag{17.7}$$

expresses the relation which connects the object distance a, the image distance b, and the focal length f. It is valid for any kind of lens provided that the "real-is-positive" sign convention is used. It takes distances to real objects, images, and foci as positive; those to virtual objects, images, and foci as negative. The focal length f in Equation (17.7) is assumed to be negative for diverging ("negative") lenses.

If an object distance is greater than the focal length ($a > f$), the image is real inverted and is located on the opposite side of the lens (Figure 17.11*b*). The image is diminished if $a > 2f$ and magnified if $2f > a > f$. If the object distance is less than the focal length, the image is virtual upright magnified and it is located on the same side of the lens with the object (Figure 17.11*d*). The image distance b in Equation (17.7) is assumed to be negative in this case.

The rays of a parallel pencil incident parallel to the optical axis of a *divergent* lens appear, after refraction, to come from the principal focal point F, located in front of the lens (Figure 17.12*a*).

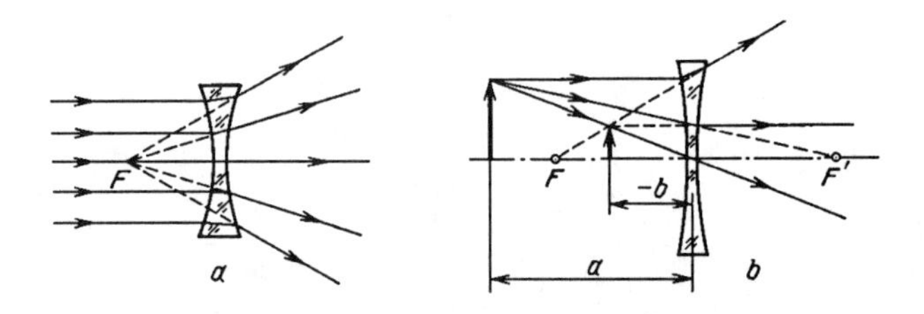

Figure 17.12: Diverging lenses.

The image formed by a diverging lens (a lens with negative focal length f and negative power D) is always virtual upright diminished independently of the object distance. For a divirging lens $f < 0$, and from Equation (17.7) in this case we always get a negative value of b. This means that the image is located on the same side of the lens as the object is.

17.2 Optical Instruments

17.2.1 Camera

In a *camera* for obtaining still photographs or movie films the images of objects (real inverted and usually diminished) are produced by a *camera lens* (Figure 17.13). A single lens has chromatic and spherical aberrations, astigmatism, and other deficiencies. That is why a perfect camera lens is a multi-lens converging optical system, in which some or other aberrations are corrected or minimized. The glass surfaces of individual lenses are covered with special thin layers in order to reduce the losses of light due to harmful reflections. The function of these layers is based on the interference phenomena (see Section 17.3.6).

The lens forms sharp images in the plane of a light-sensitive plate or film only if the objects are located at a definite distance from the camera (the point A in Figure 17.13). In order to get sharp, "focused" images of desired objects, one can change the film distance by shifting the lens

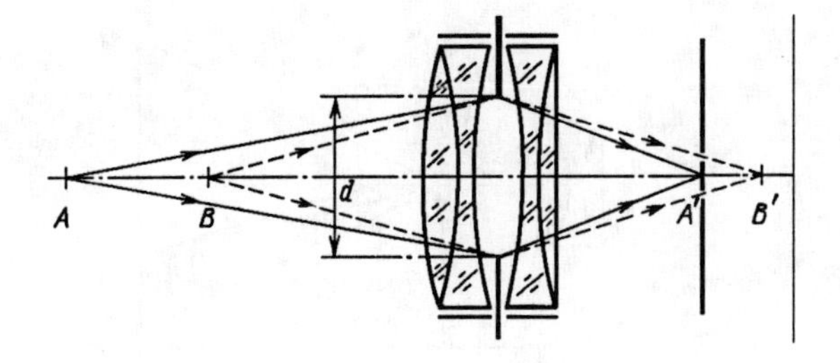

Figure 17.13: Camera.

along its axis. The images of objects' points that are located at other distances (the point B on Figure 17.13) are fuzzy. They have the form of diffuse circles, whose size diminishes while the aperture (the "opening") of the camera lens is decreased (i.e., when the ratio d/f, called the *relative aperture,* is reduced) by means of an iris diaphragm to let light pass only through the central part of the lens. In this way it is possible to get a greater *depth of field*: in order to get objects at different distances "in good focus", you should confine the light rays to a narrower aperture. However, this method gives reduced image brightness because the light flux forming the image is restricted, and hence a protracted exposition is required.

The maximum value of the relative aperture d/F corresponds to the completely opened diaphragm. The ratio d/F is an important characteristic of the *light-gathering power* of a lens and is usually expressed by the reciprocal quantity F/d, called the *focal ratio.* The numerical value of the focal ratio is known as the *f-number* of a lens. A camera lens with a 50 mm focal length and a 20 mm aperture has a relative aperture of 0.4 and a focal ratio of 2.5. Its f-number is $f/2.5$, which is commonly written as $f2.5$.

17.2.2 Diascope

In slide or movie *projector*, or a *diascope*, a flat, transparent object O—a slide or a film—is placed somewhere at a distance a between f and $2f$ from the lens L, so that a real magnified image is formed on the screen S (Figure 17.14). The *linear magnification* γ of a projector, i.e., the ratio of the image and the object sizes increases with the distance b to the screen:

$$\gamma = \frac{b}{f} - 1 \approx \frac{b}{f}. \tag{17.8}$$

A *condenser* C (a common form consists of two plano-convex lenses with the plane faces pointing outward) and a spherical mirror M serve

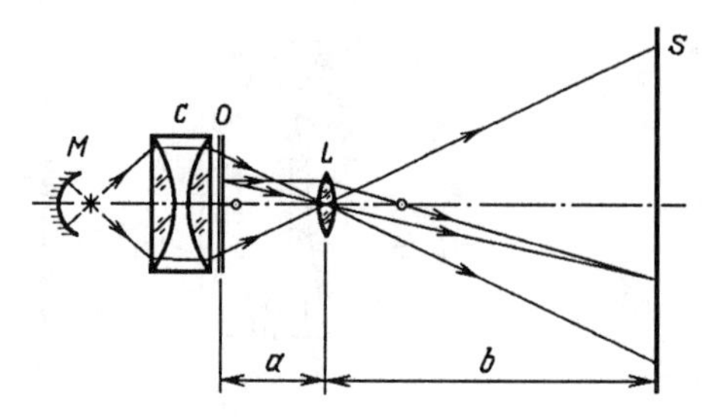

Figure 17.14: Diascope.

to concentrate the light diverging from a compact source into the lens. The optical system of a condenser is designed so that the real image of the radiating body of the light source is located inside the aperture of the lens. The light source is located in the center of curvature of the spherical mirror.

17.2.3 Magnifying glass

The size of an image on the eye's retina depends on the angle formed by the rays from the edges of the object when the object is viewed directly (Figure 17.15). In order to enlarge the *angle of vision* and hence the size of an image, we try to bring the eye nearer to the object, but it is possible to approach the eye only up to the distance of comfortable vision l_0 (approximately 25 cm for a normal eye), because at smaller distances it is difficult to achieve the eye's *accomodation* necessary to get sharp images on the retina.

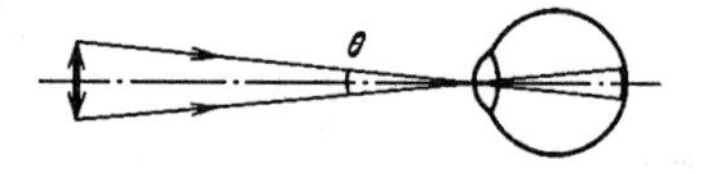

Figure 17.15: Angle of vision.

By placing a converging lens—a *magnifying glass* or an *eyeloupe*—in front of the eye, we can considerably approach the object and thus increase the angle of vision. The ratio of the angle of vision while using a magnifying glass (or any other optical system) to the angle of vision when the object is viewed directly (by the naked eye) in the most favorable available position is called the *angular magnification* (or the *magnifying power*) of the optical system.

An object is placed in front of the magnifying glass in its focal point or a bit nearer (the point A in Figure 17.16). The image A' is virtual upright and is located on the same side with the object A. The angular magnification Γ approximately equals the ratio of the distance l_0 of comfortable vision to the focal length f:

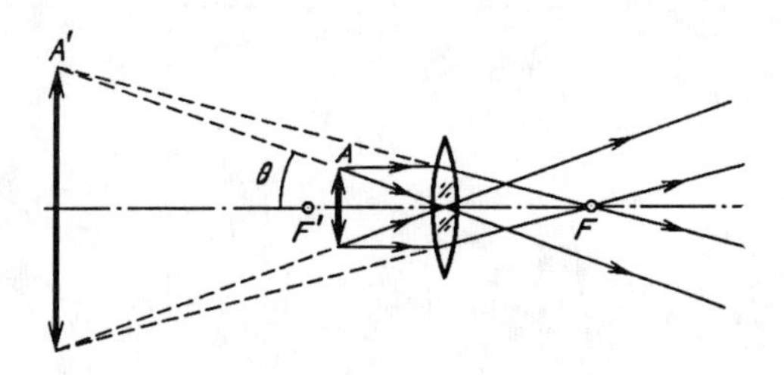

Figure 17.16: Eyeloupe.

$$\Gamma = \frac{l_0}{f}. \tag{17.9}$$

A magnifying glass with a focal length of 10 cm provides the magnification of 2.5, a jeweler's eye loupe with focal length of 5 cm—the magnification of 5.

17.2.4 Microscope

The optical system of a *microscope* (Figure 17.17) consists of a compound multi-lens *objective* with a short (a few millimeters) focal length f_1, and an *eyepiece* with a focal length f_2 of about several centimeters. The lenses are mounted at the opposite ends of a tube that can be moved in relation to the object by means of a special screw. The objective forms a real inverted magnified image $A'B'$ of an object AB, which is placed directly before the front focal point of the objective. The intermediate magnified real image $A'B'$ is observed through the eyepiece serving as a loupe and is additionally magnified by it. The position of the eyepiece is fitted so that the image $A'B'$ lies in its focal plane (or a bit nearer).

The magnification of the objective $\Gamma_1 = b/a$ approximately equals l/f_1, where l is the length of the tube. This can be seen from Figure 17.17, where $b \approx l$ because the intermediate image $A'B'$ is located inside the tube immediately before the eyepiece, and $a \approx f_1$. The magnification of the eyepiece Γ_2 equals the ratio l_0/f_2, like for a magnifying glass. The total magnification of the microscope is given by the formula:

$$\Gamma = \Gamma_1 \Gamma_2 = \frac{l\, l_0}{f_1 f_2}. \tag{17.10}$$

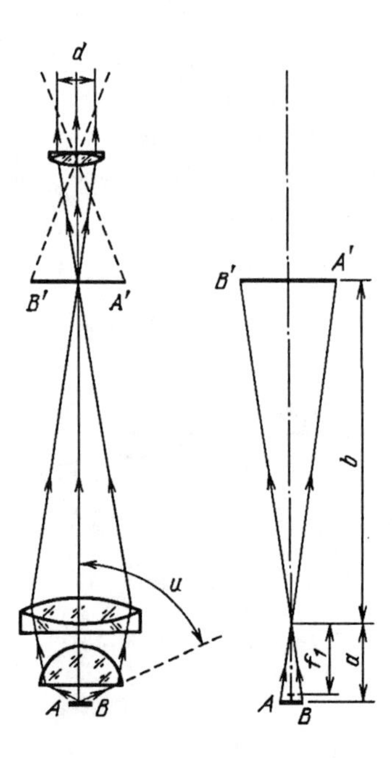

Figure 17.17: Microscope.

To achieve compatibility of the optical system of a microscope with the eye of an observer, the diameter d of the parallel light beam emerging from the eyepiece, emitted by some point of an object, should be equal to the diameter d_0 of the eye's pupil (or from two to four times smaller for bright objects). Therefore, the focal length f_2 of the eyepiece (for a given focal length f_1 of the objective) should have a quite definite value. This condition limits the admissible magnification of a microscope by a value between 250 and 1000. At greater magnifications d becomes smaller than d_0, and the illuminance of the image on the retina diminishes.

The *resolving power*, i.e., the minimal size $l_{\rm min}$ of an object's details, or the minimum distance between two points of an object that can be seen separately (can be *resolved*), is determined by the wave nature of light. The image of a shining point, as a result of diffraction by the lens aperture, consists of a bright central blob surrounded by dark and light rings. The resolving power of a microscope depends on the *numerical aperture* $n \sin u$ of the objective, where $2u$ is the opening angle of the cone of rays from an object, intercepted by the objective, and n is the refractive index of the medium between the object and the objective:

$$l_{\rm min} \approx \frac{\lambda}{n \sin u}, \tag{17.11}$$

where λ is the wavelength of light. In perfect objectives the value of aperture is close to the theoretical limit ($2u = \pi$), and for $n = 1$ the value $l_{\min} \approx \lambda$. Thus, the resolving power of the best optical microscopes approaches the wavelength of visible light $\lambda \approx 5 \cdot 10^5$ cm. That is why at magnifications exceeding 1000, no new details of the object can be resolved.

17.2.5 Telescope

The objective lens of a refracting optical *telescope* produces a real diminished inverted image of a distant object (Figure 17.18). This image is located in the focal plane of the objective lens, and is observed through the eyepiece, which serves as a loupe. The focal plane of the eyepiece is usually matched to the focal plane of the objective lens; in this case a parallel incident beam of rays from some point of a distant object emerges from the eyepiece also parallel, and the observation is performed with an unstrained eye (accomodated to infinity).

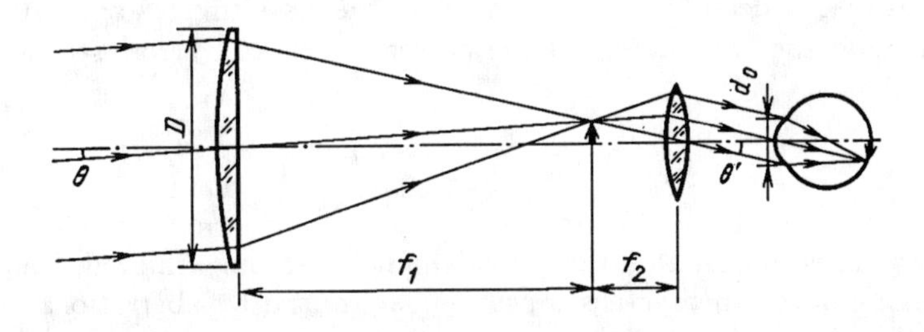

Figure 17.18: Telescope.

The *magnification* Γ of a telescope is the ratio of the angle θ' formed by the final image, to the angle θ formed by the object when viewed directly with the naked eye:

$$\Gamma = \frac{\theta'}{\theta} = \frac{f_1}{f_2} \tag{17.12}$$

—the magnification equals the ratio of the focal length f_1 of the objective lens to the focal length f_2 of the eyepiece. In order to get maximal possible illuminance of the images of extended objects in visual observations, diameter D of the objective lens must be Γ times greater than the diameter d_0 of the eye's pupil. Therefore, for a given objective lens (with definite diameter) the eyepiece ought to be chosen so that the total magnification Γ of the whole optical system is equal to the ratio D/d_0 (the *normal magnification*).

In large astronomical telescopes, concave spherical or paraboloid mirrors are used as objectives, where the chromatic aberration is suppressed because light of all wavelengths is reflected in the same way. Such an optical instrument is called a *reflecting telescope*, in contrast to a *refracting telescope*, in which lenses are used. For terrestrial telescopes an additional lens is usually inserted in order to provide an upright image. In *binocular field glasses*, in each tube a pair of prisms of internal reflection is used to increase the effective length and to produce an upright image. There is a variety of a telescope, in which a diverging lens is used as an eyepiece. Such a *Galilean telescope* provides upright images originally and is used for the compact telescopes in opera glasses.

The theoretical limit of the *angular resolving power* of a telescope, i.e., the smallest angular separation of the images, is imposed by diffraction phenomena. As a result of diffraction by the lens aperture, the image of a distant star in the focal plane of an objective is fuzzy. It consists of a bright round central blob surrounded by alternating dark and light rings. According to the *Rayleigh criterion* for resolution, two near images are perceived separately if the central bright circle of one image falls on the first dark ring of the other. The angular resolving power $\theta_{\min}$ (in radians) in this case is determined by the ratio of the wavelength $\lambda \approx 5 \cdot 10^{-7}$ m of light to the diameter D of the objective lens or mirror (see Section 17.4.4):

$$\theta_{\min} \approx 1.22\frac{\lambda}{D}. \tag{17.13}$$

In real instruments this theoretical limit is seldom achieved, and the resolving power in practice is usually determined by aberrations or atmospheric conditions.

17.3 Interference of Light

17.3.1 Interference and coherent light

In the phenomena of interference the *wave properties of light* are demonstrated in the most distinct manner. Interference exhibits itself when two or more coherent wave processes superimpose to produce a stationary spatial distribution of resulting instantaneous disturbances. Interference is a characteristic property of waves of any physical nature. It is easily observed in experiment for sound waves or waves on a water surface. In order to observe the interference of light, it is necessary to provide some special conditions.

The interference of light was first described by Thomas Young in 1801. It provided strong evidence for the wave theory of light. Interference of light leads to formation of stationary *interference fringes*—bright places where resulting illuminance is greater than the sum of illuminances produced by separate light beams (*constructive interference*), which alternate with dark places where illuminance is smaller than this sum (*destructive interference*).

In an ordinary light source, excited atoms or ions *spontaneously* emit individual trains of electromagnetic waves with random phase relationships between separate wave trains. Duration of these wave trains does not exceed 10^{-8} s (spatial extent less than 1 m), even in the case of narrow spectral lines of radiation from low pressure gaseous discharge light sources. Therefore, the radiation of usual (non-laser) light sources is a superposition of a huge number of incoherent wave trains. It may be considered as a "light noise"—random, chaotic, *noncoherent* oscillations of electromagnetic field. Only in a *laser,* where a *stimulated emission* of radiation is used, do all excited atoms emit radiation coherently. In this case the wave trains emitted by individual atoms are matched in all characteristics—wavelength, phase, direction of propagation. The radiation emerges from a laser as a powerful coherent nearly monochromatic and parallel beam of light.

Coherent radiation produced by lasers is often used in order to get interference effects. One important practical application of coherent radiation is the holography (see Section 17.4.7).

It is possible to observe interference of light from non-laser sources only if this light is divided into two or more beams and then these beams are superimposed again. In each of the beams the phase relationships between individual wave trains change randomly during the time of observation, but these changes are nearly the same in every beam. The interference can be observed, if the *path difference* (see Section 16.7.8) between the beams does not exceed the length of individual wave trains.

17.3.2 Interference fringes

The methods of dividing a light wave from a primary source into two mutually coherent light waves can be classified as *division of the wave front* (Young and Fresnel experiments) and *division of the amplitude* (interference in thin films and plates, Newton's rings, fringes of equal slope). In any case instead of a primary point source it is possible to consider two coherent sources S_1 and S_2, separated by some distance d (Figure 17.19). Their radiation can be considered to be monochromatic, if our aim is to determine only the spatial position and the form of the interference fringes. Two coherent waves from these sources come to every point of observation, and the amplitude of resulting oscillations is maximal at the points where the path difference from the sources equals an integral number of wavelengths λ:

$$l_1 - l_2 = n\lambda, \qquad n = 0,\ \pm 1,\ \pm 2,\ \ldots. \tag{17.14}$$

The number n in Equation (17.14) is called the *order of interference.*

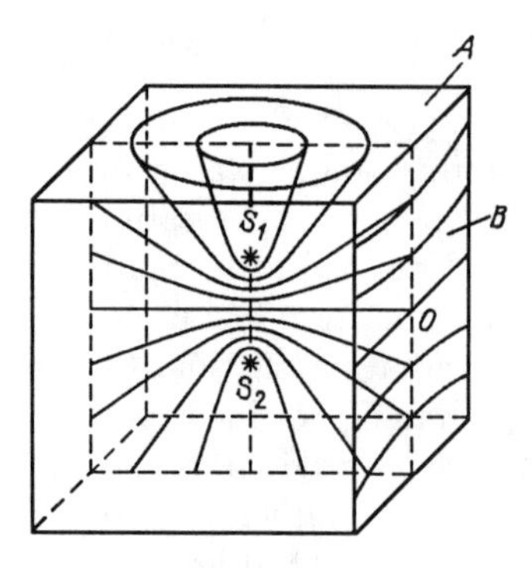

Figure 17.19: The shape of interference fringes formed by two point sources.

In a plane containing the sources the locus of points of equal path differences from the sources S_1 and S_2 is a hyperbola (a family of hyperbolas for various values of the path difference). In a three-dimentional space the set of points of equal path differences $l_1 - l_2 = n\lambda$ is the surface formed by rotation of a hyperbola about the axis passing through S_1 and S_2 (hyperboloid). The form of interference fringes on some screen is determined by the lines of intersection of such hyperboloids with the plane of the screen (Figure 17.19). In particular, on screen A these lines produce a family of concentric circles; on screen B, in the vicinity of point O, the interference fringes can be considered as nearly equidistant parallel straight lines.

17.3.3 Young's double-slit experiment

In *Young's experiment* the role of secondary coherent sources S_1 and S_2 is performed by two tiny adjacent holes (or two narrow long parallel slits) illuminated by one source of small angular size, or by the light from an extended source after this light passed through a single tiny hole S (Figure 17.20). In the light beam from S (which is spreading due to diffraction, see Section 17.4.2), electromagnetic oscillations at different points of the wave surface (including the points S_1 and S_2) are coherent, and a stationary *interference pattern* is observed on the screen B, where the light beams spreading from the secondary coherent sources S_1 and S_2 superimpose. This pattern is a series of straight parallel equidistant interference fringes, oriented perpendicularly to the plane of the diagram in Figure 17.20. In point P, the direction to which makes the angle θ with the direction toward the center O of the interference pattern, the path difference has the value $l_1 - l_2 \approx d \sin\theta \approx d\theta$, where d is the distance between the sources S_1 and S_2. Assuming here the path difference $l_1 - l_2$ to be equal to an integral number of wavelengths $n\lambda$, we get the direction θ_n, indicating to the interference maximum (bright stripe) of the order n:

$$\theta_n = n\lambda/d, \qquad n = 0, \pm 1, \pm 2, \ldots . \tag{17.15}$$

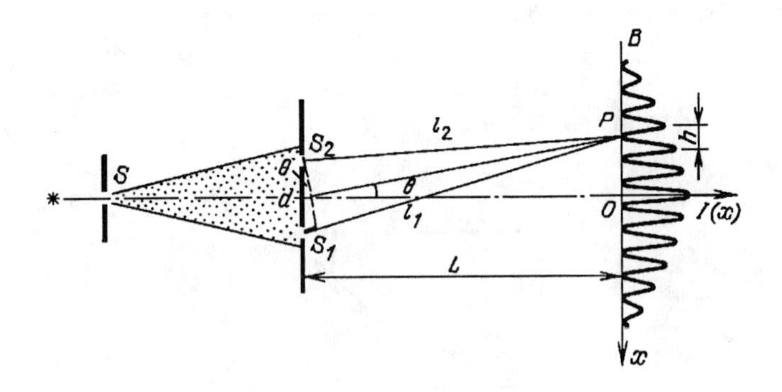

Figure 17.20: Young's double-slit experiment.

The angular distance between neighboring stripes $\Delta\theta = \theta_{n+1} - \theta_n$ equals λ/d, and the separation h of the stripes on screen B is found by multiplying $\Delta\theta$ and the distance L:

$$h = L\Delta\theta = \lambda L/d. \tag{17.16}$$

For instance, at $L = 1$ m, $d = 0.5$ mm, and $\lambda = 5 \cdot 10^{-7}$ m, the separation of the fringes $h \approx 1$ mm. The fringes the wider, the closer to each other the holes S_1 and S_2 and the greater the distance L to screen B. The plot of illuminance $I(x)$ of the screen is shown in the right side of Figure 17.20. Near the center of the interference pattern illuminance $I(x)$ changes sinusoidally:

$$I(x) = 2I_0 \left(1 + \cos \frac{2\pi}{h} x\right). \tag{17.17}$$

The illuminance at minimums equals zero, and at maximums it is four times greater than the illuminance produced by a single source. Reduction of light oscillations in some places and their reinforcement in other places is caused by redistribution of the energy flow in space without any other energy transformations: according to Equation (17.17), the average over the screen value of illuminance is $2I_0$, i.e., just double the illuminance produced by one source.

In order to increase the brightness of the observed pattern, two long, narrow adjacent parallel slits are used instead of the holes S_1 and S_2, and a long, narrow slit is used as a primary source instead of the hole S.

It is possible to create two coherent light waves by means of a *reflection of light* from the two surfaces of a thin plane-parallel transparent plate (or film), see Figure 17.21. To every observation point P two light waves come with the path difference just though as these waves originated from the coherent sources S_1 and S_2—the virtual images of the real light source S formed by reflection from the top and the bottom surfaces of the plate. The interference fringes on screen B have the form of concentric rings with the center at point O. The separation between adjacent rings shrinks as we go farther from the center.

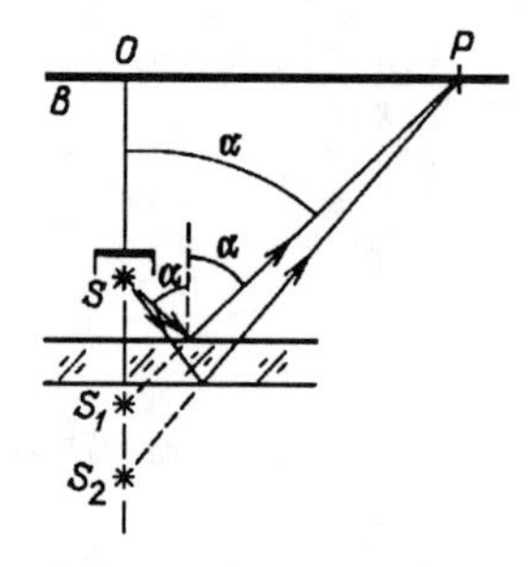

Figure 17.21: Interference of light reflected by two parallel planes.

17.3.4 Localized interference patterns

Any radiating small part of an extended light source can be considered as a point source whose radiation is incoherent with light waves emitted by other parts of the source. Interference patterns produced by different parts of an extended light source simply superimpose (add without any further interference), and if the fringes of these individual patterns are displaced with respect to each other, the resulting pattern is blurred, producing nearly uniform illuminance up to full disappearance of fringes. Coincidence of fringes in individual interference patterns is possible only under specific conditions of observation (or specific disposition of the screen of observation). That is why these interference fringes in the case of extensive light source are called *localized.*

Fringes of equal slope

The *fringes of equal slope* are formed in the focal plane of a converging lens gathering the light from an extended source after reflection by two parallel plane surfaces (Figure 17.22).

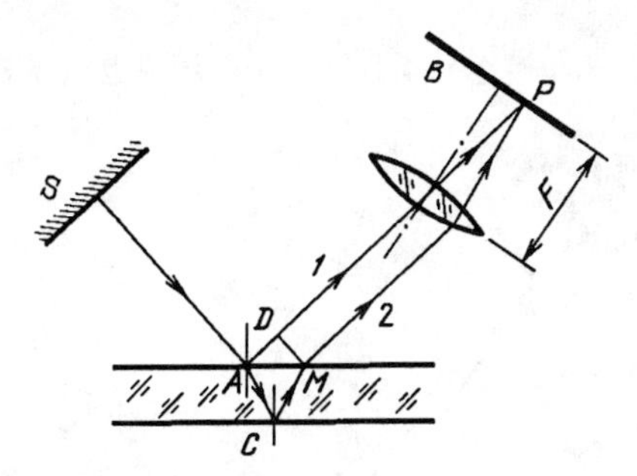

Figure 17.22: Observation of the fringes of equal slope.

The rays that arrive at any point P of screen B were *parallel* to each other before the refraction in the lens and therefore had a quite definite *slope* with respect to the planes of reflection. From point A, where the incident ray is divided into two mutually coherent rays *1* and *2*, ray *2* passes additionally the distance $|AC| + |CM|$ inside the plate, and in the air its path is shorter than the path of ray *1* by the length of the segment $|AD|$. It is obvious that the path difference of rays *1* and *2* does not depend on the position of point A, in which the primary incident ray hits the top surface of the plate, so that the result of interference in the point of observation A (bright or dark stripe) will be the same for all radiating elements of an extended light source. The interference fringes have the form of concentric rings with the center in the focal point of the lens. The fringes of equal slope are used in *interferometers*—optical instruments designed for many different scientific and practical purposes (measuring wavelengths, small distances, and displacements with high precision, testing the form and quality of optical surfaces, etc.).

Fringes of equal thickness

Fringes of equal thickness can be observed near the surface of a thin film when it is illuminated by an extended light source (e.g., by a cloudy sky). In particular, familiar iridescent patterns formed by oil or gasoline on a puddle of water, as well as by soap bubbles, are also interference fringes of equal thickness. Light reflected at the air-oil interface interferes with the light reflected at the oil-water interface. Assume that ray *1* forms the angle θ with the normal to the film surface (Figure 17.23). The other ray *2*, coherent with ray *1*, arrives to the point P from the same element S of the extended light source after reflection from the bottom face of the film at the point B. The path difference of the interfering rays *2* and *1* can be estimated neglecting refraction in the film, i.e., assuming the refractive index $n = 1$:

$$l_2 - l_1 \approx |AB| + |BP| - |CP| = 2h\cos\theta.$$

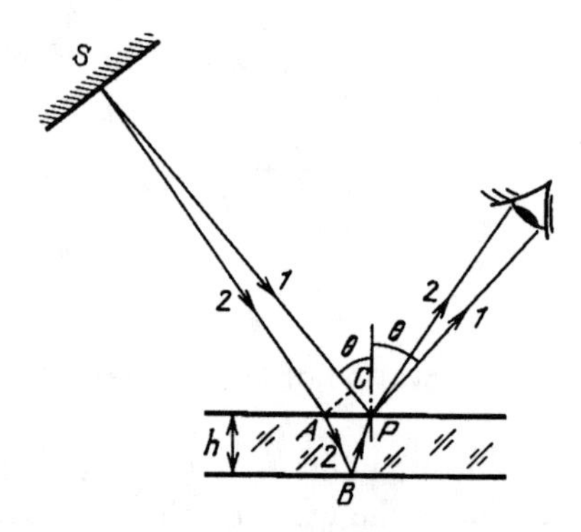

Figure 17.23: Observation of the fringes of equal thickness.

Due to the small size of the eye's pupil the image of P on the retina is produced only by the rays from P that have nearly the same value of θ (see Figure 17.23). Therefore, all pairs of mutually coherent rays that penetrate into the eye from point P have the same path difference determined by the film thickness h at point P. The contour of each interference fringe passes through the places where the thickness h of the film has the same value (if $\cos\theta$ is nearly the same for the whole region of observation). What we really see is the variation in thickness of the film.

The fringes of equal thickness can be observed in a thin layer of air between two glass surfaces, e.g., a layer between a slightly convex lens placed on a flat glass plate (*Newton's rings*, Figure 17.24). If the direction of observation is close to the normal ($\theta \approx 0$), transition from one bright fringe to the next one corresponds to the change in the thickness of the layer that equals one half of the wavelength. In reflected light the whole picture looks like a dark round spot surrounded by a series of bright and dark rings. The center of the picture (the place of contact of the surfaces) is dark due to different reflection conditions: the phase of the wave reflected from the surface of the medium with greater optical density is opposite to the phase of the incident wave. The radius r_n of the n-th dark ring is given by the formula

$$r_n = \sqrt{n\lambda R}, \tag{17.18}$$

where R is the radius of curvature of the lens' spherical surface. Due to the dependence of r_n on λ the rings in white light are colored. The phenomenon is used in the interference quality testing of lens surfaces. The irregular pattern like "zebra stripes", often produced by thin film interference in the air layer between a projected slide and its cover glass, is a variety of Newton's rings.

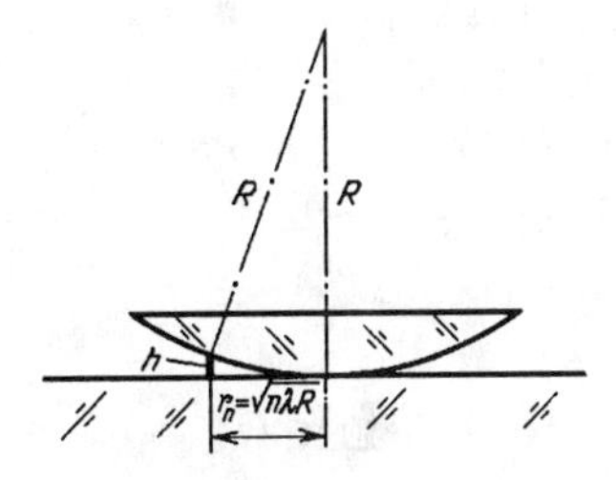

Figure 17.24: Newton's rings.

17.3.5 Multiple-ray interference

When only *two coherent waves* interfere, interference fringes are produced in which the illuminance changes according to Equation (17.17). In this case bright and dark stripes have the same width, and transition from light and dark places is smooth (sinusoidal).

Multiple-ray interference is produced by superposition of a large number of coherent light waves originating from one primary wave as a result of wave front division by many slits (*diffraction grating*, see Section 17.4.5) or of amplitude division by multiple reflections on parallel surfaces (*Fabry–Perot interferometer*).

The principal distinctive property of a multiple-ray interference pattern is the formation of very sharp and narrow bright stripes whose width is much smaller than the separation between adjacent stripes. The greater the number of rays taking part in interference, the narrower the bright stripes. This feature of a multiple-ray interference pattern can be easily explained on the basis of light energy redistribution: as the number of interfering beams increases n times, total amount of light energy also increases n times, but intensity in maxima of the interference pattern (which is proportional to the square of the amplitude) increases n^2 times. Such a situation is possible only if the width of the maxima simultaneously decreases n times. The width of bright stripes is inversely proportional to the number n of interfering beams.

17.3.6 The enlightenment of optical systems

In order to reduce the losses of light caused by harmful reflections in optical instruments, the surfaces of lenses are covered by a thin layer of transparent substance whose refracting index n' is smaller than that of glass, $n \approx 1.5$ (the *enlightenment of lenses*). The double thickness of such a layer must be equal to one half of the light wavelength in its substance (or to an odd number of half-wavelengths). In this case the light waves, reflected at the surfaces of the layer and the lens, are half a period out of phase (for normal incidence) and interfere destructively. For total suppression of reflected light the amplitudes of these two waves (reflected by the layer and by the lens surface) must be equal. This condition is fulfilled if $n' = \sqrt{n}$.

17.4 Diffraction of Light

17.4.1 The Huygens–Fresnel principle

The phenomenon of *diffraction* is common to wave processes of any physical nature (see Section 16.7.10), but for light waves the phenomenon has some peculiarities caused by the fact that the wavelength λ of light is, as a rule, much smaller than the size d of an obstacle (or of an aperture). Therefore it is possible to observe diffraction of light, i.e., the spreading and bending of light round the edge of an obstacle into the region of geometric shadow, only at rather large distances l from the barrier ($l > d^2/\lambda$).

An explanation of diffraction phenomena is given by the *principle of Huygens–Fresnel,* according to which any point of a wave surface can be considered as a source of secondary spherical waves, and the light oscillations at some point of observation P (Figure 17.25) are found by adding the oscillations produced by all secondary waves reaching this point. The resulting oscillation in P depends on amplitudes and phases of these secondary waves. As a result of this interference of the secondary waves, a dark region can be observed where a straight path lies from the light source, and in the limits of a geometric shadow, a bright region can be produced, as if light rays were bending round the obstacles.

17.4.2 Diffraction spreading of a parallel light beam

A *plane light wave* is characterized by a definite direction of propagation. This means that a parallel bundle of light rays corresponds to a plane wave. In the case of an unlimited wave surface the secondary waves fully cancel each other in interference for any direction of propagation (the forward direction with $\theta = 0$ being the only exception). Indeed, for any arbitrary element ΔS_1 of the unrestricted wave surface, we can

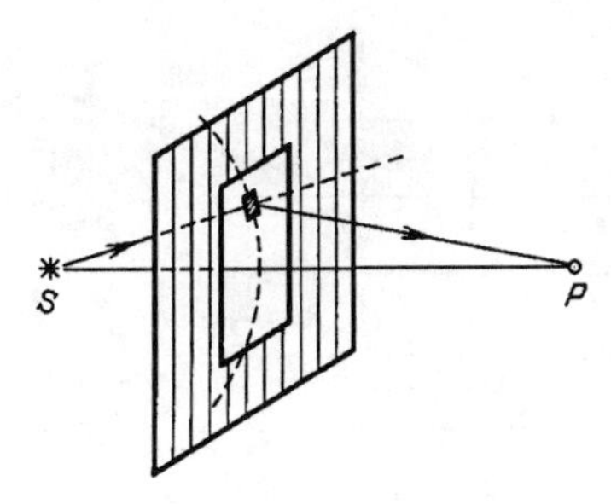

Figure 17.25: The Huygens–Fresnel principle.

always find the corresponding element ΔS_2 of equal area, the secondary wave from which traveling in a considered direction lags by $\lambda/2$ and produces oscillations, at a distant point of observation, which have the same amplitude and opposite phase with the oscillations coming from ΔS_1 (Figure 17.26).

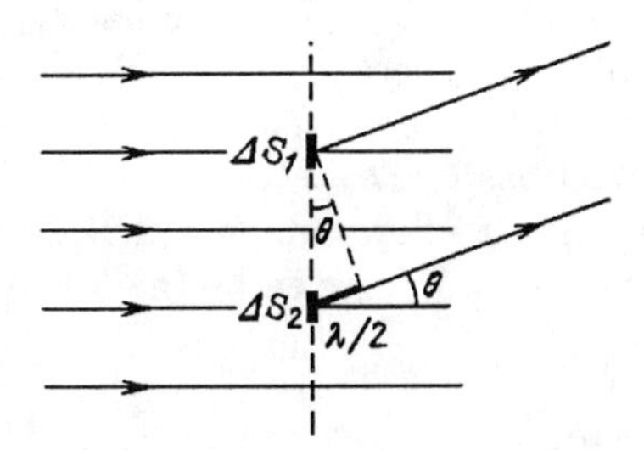

Figure 17.26: Mutual cancellation of secondary waves due to interference for any direction (except $\theta = 0$) in a plane wave with unrestricted wave surface.

Restriction of a wave surface by the edges of a slit (Figure 17.27) causes the passing light to form *diffracted rays* bending round the edges: a beam of light of a restricted cross-section inevitably spreads in the process of propagation. The minimal angle $\theta_{\min}$ that determines the direction of complete interference cancellation of the secondary waves can be determined in the following way. Let us consider the secondary waves traveling in this direction that originate from a pair of elements S_1 and S_2 of the primary wave surface. Let the elements be apart at a distance $a/2$, where a is the width of the slit (see Figure 17.27). The path difference Δ of these secondary waves for a distant point of observation must be equal to $\lambda/2$:

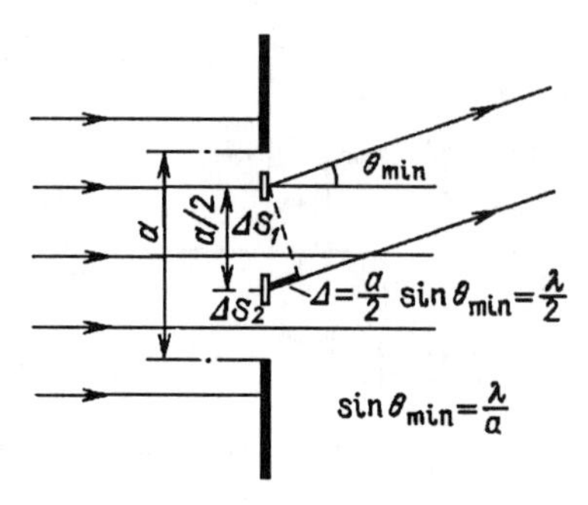

Figure 17.27: Diffraction spreading of a light beam with a restricted cross-section.

$$\Delta = \frac{a}{2}\sin\theta_{\min} \approx \frac{a}{2}\theta_{\min} = \lambda/2, \qquad \text{i.e.,} \quad \theta_{\min} \approx \frac{\lambda}{a}. \tag{17.19}$$

Full darkness will be observed in the direction $\theta_{\min}$ determined by the condition of Equation (17.19), because the whole slit aperture consists of such pairs of elements. Total destructive interference of secondary waves occurs also in the directions $\theta_n = n\lambda/a \quad (n = \pm 1,\ \pm 2,\ \pm 3,\ \dots)$. The angular distribution of energy flow behind a slit of the width a, through which a plane wave has passed, is given by the following expression:

$$I(\theta) = I_0\left(\frac{\sin u}{u}\right)^2, \quad \text{where} \quad u = \frac{\pi a \sin\theta}{\lambda} \approx \frac{\pi a \theta}{\lambda}. \tag{17.20}$$

The plot of this function is shown in Figure 17.28. The central maximum takes over about 85% of the incident energy, so $\theta_{\min} \approx \lambda/a$ determined by Equation (17.19) can be regarded as the angular measure of *diffraction spreading* of a light beam with a restricted cross-section. The smaller the diameter of a light beam, the greater its angular diffraction spreading. The spreading of a laser beam is caused by diffraction, but the spreading of light beams emitted by projectors with filament lamps is determined by the size of the radiating body.

The distance l_0 from a slit, at which the diffraction widening of a light beam reaches the magnitude a of its original width, is determined by the condition $l_0\theta_{\min} \approx a$. Since $\theta_{\min} = \lambda/a$ (see Equation (17.19)), we obtain:

$$l_0 \approx a^2/\lambda. \tag{17.21}$$

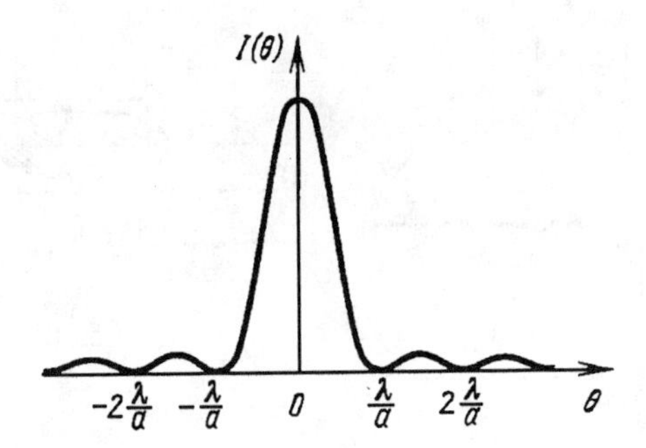

Figure 17.28: Angular distribution of energy flow produced by diffraction of a plane wave falling on a slit.

When the distance l from the hole or a slit is small compared with $l_0 = a^2/\lambda$, the distribution of light flow is satisfactorily described by the *geometric optics*, and shadows have sharp outlines. If a distance l from a hole or a slit to the point of observation is of an order of l_0, the phenomenon is called *Fresnel diffraction.* If l is significantly greater then l_0 and the wavefronts of secondary waves at an observation point are practically plane, the phenomenon is usually referred as *Fraunhofer diffraction.*

17.4.3 Fresnel diffraction

The distribution of illuminance in a *Fresnel diffraction pattern* produced by a *rectilinear edge* of a screen is shown in Figure 17.29*a*. At the boundary of geometric shadow the amplitude of light oscillations is one half of the amplitude which would be in the absence of a screen, and, consequently, the intensity equals one fourth of the value corresponding to the illuminated area. The intensity of light diminishes monotonically to zero as we advance into the region of geometric shadow, and the light in the illuminated area is split into alternating bright and dark diffraction fringes. These fringes become closer and less contrasting with the distance from the boundary of geometric shadow and finally they vanish, producing homogeneous illumination. The width of the first fringes is of the order of $\sqrt{\lambda l}$, where l is the distance from the screen to the point of observation.

The shadow outline produced by a *circular screen* (a disc) also becomes diffuse and surrounded by alternating bright and dark diffraction circles (Figure 17.29*b*). In the center of the geometric shadow the secondary waves reinforce each other due to the constructive interference and produce a bright round small spot called the *Arago–Poisson spot.*

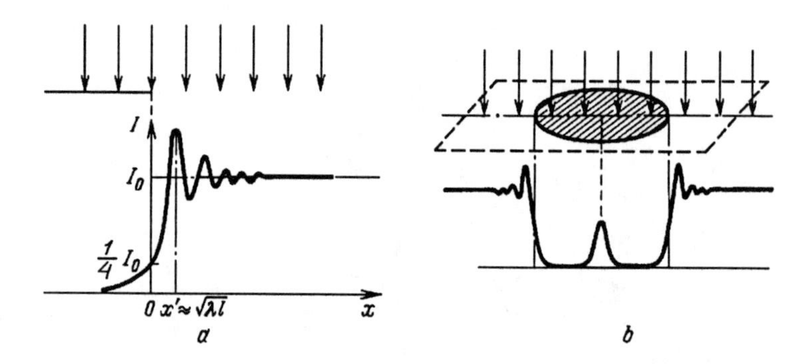

Figure 17.29: Fresnel diffraction pattern produced by an edge of a screen.

17.4.4 Fraunhofer diffraction

The *Fraunhofer diffraction* can be observed in the focal plane of a converging lens (Figure 17.30). At a point x of the focal plane the lens gathers the diffracted rays, which formed a definite angle θ with the optical axis: $x = F\tan\theta \approx F\theta$. These rays were parallel to each other and the corresponding diffracted wave had a planar wave surface. That is why the intensity distribution $I(x)$ in the focal plane is equivalent to the angular energy distribution observed at "infinity" ($l \gg l_0$) without the lens (in "parallel rays"). When the width of light beam is restricted by a slit, the diffraction pattern in the focal plane is given by the function $I(\theta)$ from Equation (17.20) and by the plot shown in Figure 17.28, if we substitute there $\theta = x/F$.

In the case of a circular diaphragm with a diameter a a system of diffraction rings is formed whose center is located at the focal point of the lens. The radial distribution of intensity is similar to that shown by the plot in Figure 17.28. The first dark ring surrounding the central maximum is characterized by the angular radius $\theta_{\min} = 1.22\lambda/a$.

The real image of a point light source S (Figure 17.30) formed by an ideal optical system (a system with entirely eliminated aberrations) in reality is not a point (as it should be, according to geometrical optics), but rather a tiny Fraunhofer diffraction pattern. Two adjacent points of an image are conventionally assumed to be *resolved* if the center of the diffraction pattern formed by one of the points is displaced from the center of the other by a distance which is not smaller than the separation between the center and the first minimum of the diffraction pattern (*Rayleigh criterion*, see Figure 17.31). In this case the "trough" between the maxima of resulting picture is about 20 percent. This is not a real limit of resolution, because with some skill you might be able to do better—but not much better. And so this criterion gives a useful means

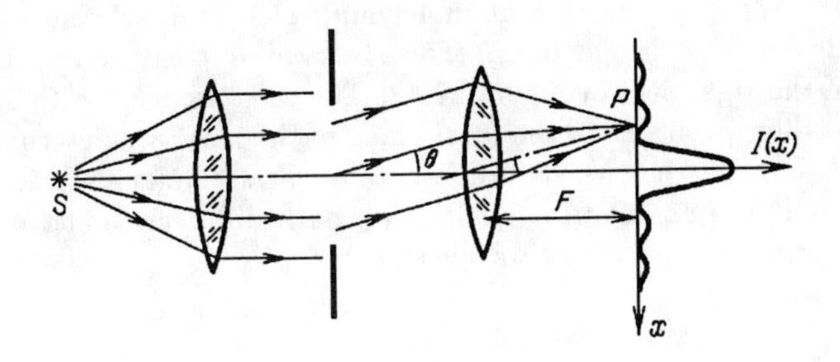

Figure 17.30: Fraunhofer diffraction.

of comparing the *resolving power* of different optical instruments.

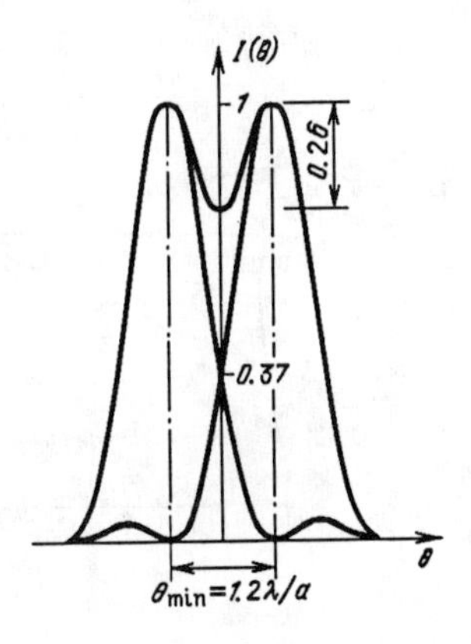

Figure 17.31: Rayleigh criterion.

The angular resolving power θ_{min} of a telescope (or of a binocular) the better the greater the diameter of the light beam forming the image:

$$\theta_{\text{min}} = 1.22\lambda/a, \tag{17.22}$$

where a is the objective lens (or the main mirror) diameter.

17.4.5 Diffraction grating

When a plane light wave hits a system of many evenly spaced parallel slits (a *diffraction grating*), the secondary beams of light originating from individual slits are coherent with each other. Broadened due to diffraction, these beams overlap and interfere. In a focal plane of a lens

(Figure 17.32) they form a Fraunhofer multiple-ray interference pattern with sharp narrow bright lines. The observed intensity distribution is shown in the right side of Figure 17.32. The directions θ_n to these maxima (and their positions in the focal plane of the lens) are determined by the condition of constructive interference of the secondary waves, taking the origin at the neighboring slits—their path difference must be equal to an integral number of the wavelength λ:

$$d \sin\theta_n = n\lambda, \qquad n = 0,\ \pm 1,\ \pm 2,\ \ldots. \tag{17.23}$$

Here d is the separation between the centers of adjacent slits (the *period* of diffraction grating), and an integral number n is the "order" of the principal maximum. The greater the number of slits (or scratches) in the grating, the narrower the bright stripes. The *sharpness* of the principal maxima (i.e., the ratio of their separation and their width) is equal to the total number N of the slits in the grating (though their position θ_n does not depend on N).

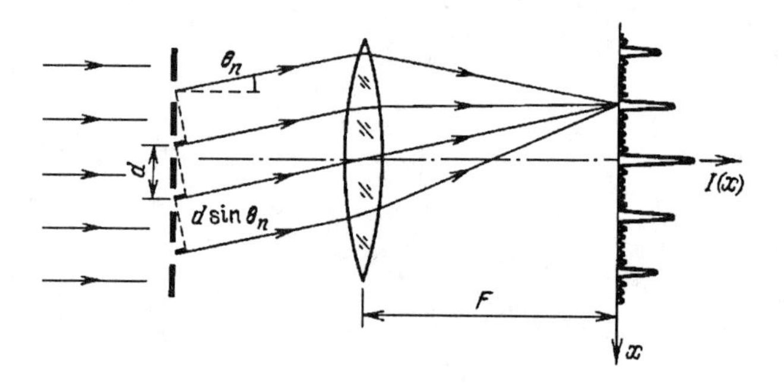

Figure 17.32: Diffraction grating.

Positions of the principal maxima depend on the wavelength λ (central maximum with $n = 0$ being the only exception). That's why a diffraction grating can be used as a dispersing device of a *spectroscopic instrument* intended for decomposition of investigated radiation into pure spectral components (for examination of the spectral composition of non-monochromatic radiation). A spectrum corresponds to every principal interference maximum (spectrum of n-th order). The ability of a spectroscopic instrument to separate close wavelengths of radiation is characterized by its *chromatic resolving power*, which is equal to $\lambda/\Delta\lambda$, where $\Delta\lambda$ is the smallest difference in wavelength of two equally strong spectral lines that the instrument can separate, and λ is the average wavelength of the two lines. For a diffraction grating, the resolving power equals the product of the total number N of scratches and the order n of the spectrum:

$$\lambda/\Delta\lambda = Nn. \tag{17.24}$$

The maximal possible order $n_{\max}$ for a given period d does not exceed d/λ. Therefore, the maximal resolving power of a given diffraction grating (in the spectrum of maximal order)

$$Nn_{\max} = Nd/\lambda = L/\lambda$$

is determined only by the total size $L = Nd$ of the grating. The advantage of gratings with small periods (with large numbers of scratches per unit length) becomes apparent in the investigation of radiation with a wide spectrum: the neighboring spectra of high orders begin to overlap earlier than the spectra of low orders ($n = 1, 2$), which are used in gratings with a small period. To concentrate in the spectrum of a certain order as much of the diffracted light as possible, diffraction gratings with a special profile of the scratches are used.

17.4.6 Dispersion spectrometer

The phenomenon of ***dispersion***, i.e., the dependence of the refractive index $n(\lambda)$ of a substance (or of the phase velocity of light) on the wavelength λ is also used for spectral decomposition of radiation. If a ray of white light strikes one face of a glass prism and passes out of another face, the white light is split into its spectral components. A scheme of simple spectrometer with a prism as dispersing device is shown in Figure 17.33.

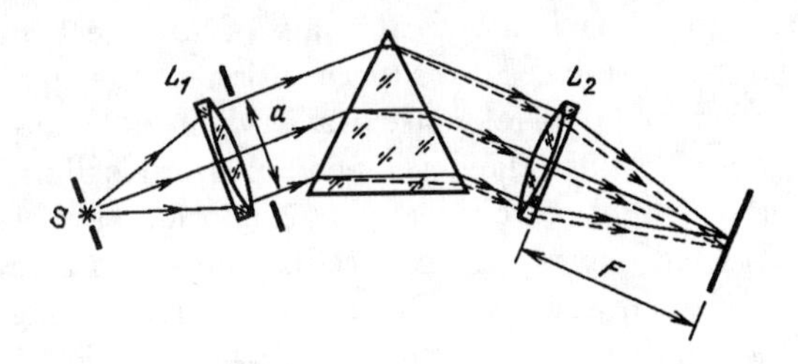

Figure 17.33: Dispersion spectrometer.

The *collimator*, consisting of the input slit S, illuminated by the investigated radiation, and the convergent achromatic lens L_1, the slit being in the principal focus of the lens, forms a parallel beam of radiation falling on the face of the prism. Due to the dispersion in the prism, the parallel beams of different wavelengths are bent through different angles, and in the focal plane of the camera lens L_2 the images of the input slit produced by these beams are located at different places, and thus the spectrum of radiation is formed.

The chromatic resolving power of such a spectrometer is restricted finally by the fact that a monochromatic radiation gives in the instrument not an indefinitely narrow line, but rather a spectral contour of finite width—an image of the input slit. This image converts into a Fraunhofer diffraction pattern as the slit becomes narrower. The minimal width of the spectral line is determined therefore by the width a (Figure 17.33) of the parallel beam of light which forms the image (the width of the central Fraunhofer maximum is inversely proportional to a).

17.4.7 Holography

The phenomena of interference and diffraction of light lie in the basis of *holography*—a method of recording and subsequent reconstruction of light waves. Like photography, it provides the possibility of forming a permanent record and conservation of visual images of objects, but, unlike photography, holography ensures recording and reconstruction not of simple two-dimensional distribution of illuminance in the plane of a photographic plate or a film, but rather of the *light waves scattered by an object* with all their characteristics—the direction of propagation, amplitude, phase, and wavelength. These waves, reaching the eye of an observer, produce a full impression of reality of the observed objects—their images are truly three-dimensional and, what is more, you can see the objects from different angles if you move your head to change the point of view or if you turn the hologram. It is possible to "look around the corner": one object in the field of view may block another at one angle. Changing the angle of viewing, you can see behind it.

A hologram records not an optical image of an object, but rather an *interference pattern,* produced by superposition of the light wave scattered by an object and the reference wave which is coherent with the wave from the object. When the developed hologram is illuminated afterwards by a coherent light wave (ideally, by the original reference beam), two sets of diffracted waves are produced. One set is a rather precise reproduction of the subject wave and forms a virtual image coinciding with the original object position. This reconstructed object wave and the virtual image are just what you need for naked-eye viewing. The other set forms a real image on the other side of the hologram.

The easiest for explanation is the case of holographic recording and reconstruction of a plane wave. Let such a wave *1* coming from an object strike a photosensitive plate at some angle θ' with the normal (Figure 17.34*a*). Let simultaneously with this *object wave 1* another plane wave *2* coherent with the object wave fall normally at the plate. These two plane waves form an interference pattern on the photographic plate. The distribution of illuminance $I(x)$ in this pattern has the form of parallel equidistant fringes oriented perpendicularly to the plane of Figure 17.34. The distance d between adjacent fringes equals $\lambda/\sin\theta'$.

After the photosensitive plate is developed, we get the hologram looking like a diffraction grating with smooth (sinusoidal) transition from dark to transparent stripes.

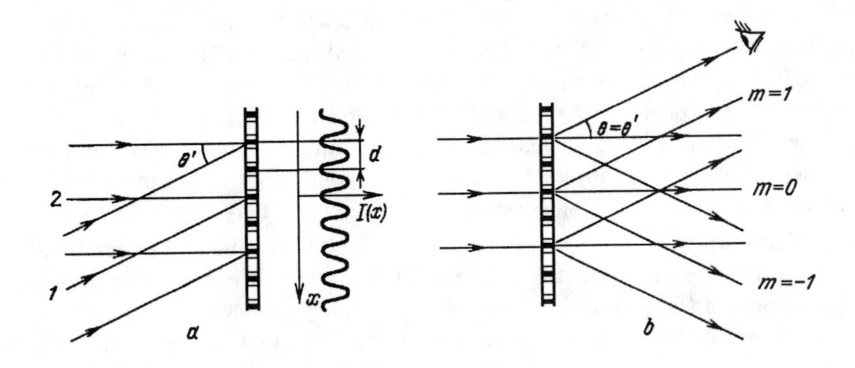

Figure 17.34: Holographic registration (a) and reconstruction (b) of a plane wave.

To reproduce the object wave which was recorded in such way, we can illuminate the hologram by a *reconstructing wave* that is similar to the reference wave *2* used during registration (Figure 17.34*b*). Diffraction of this reconstructing wave on the grating with sinusoidal transmission produces three waves. One of them corresponds to the principal maximum of the order $m = 0$ and travels in the direction of the original incident wave; the other two diffracted waves correspond to the principal maxima of the order $m = \pm 1$.

The direction of propagation of the wave with $m = 1$ is determined by the condition $d \sin\theta = \lambda$ (see Section 17.4.5), and since $d = \lambda / \sin\theta'$, we obtain $\theta = \theta'$—the direction of this wave, as well as all other characteristics, are exactly the same as those of the object wave *1* used in the process of registration. Coming to the eye of an observer, this reconstructed wave produces the same visual impression as would produce the object wave in direct observation.

In the case of a complex object, the coherent laser light scattered by the object can be represented as a set of elementary plane waves. Each of the plane waves from this set interferes with the referent coherent wave from the same laser source and produces an individual interference pattern (a diffraction grating) with definite separation and orientation of fringes. On the stage of reconstruction a beam of coherent light similar to the reference beam that was used in registration passes through this grating and produces a diffracted wave similar to the corresponding partial plane wave from the object. The whole set of these reconstructed plane waves produces the same visual images that would have been created by direct observation of the object. Besides this (virtual) image of an object, the set of diffracted waves with $m = -1$ forms one more (real) image of the object.

In contrast to usual photography where information about some point of an object is registered at one definite point of the film, each part of a hologram contains, in a coded form, information about all points of the object, because in the process of registration the light scattered by any point of an object usually falls on the whole surface of the photographic plate. That is why it is possible to reconstruct an object wave and an image of the whole object by means of a small part (a fragment) of a hologram. However, in this case the field of vision becomes restricted.

Holography was invented by Dennis Gabor in 1948, but practical holography became possible only after 1961, due to the invention of new light sources (lasers), producing coherent highly monochromatic parallel light beams. More recent technique of registration (developed by Yuri Denissjuk) uses special thick-emulsion plates. By this method it is possible to produce holograms that make the image visible in the ordinary white light.

17.4.8 Photometry

Photometry deals with the study of visual radiation, especially with the measurements and calculations of luminous intensity, luminous flux, illuminance of a surface, etc. The sensitivity of the human eye is different for radiation of various wavelengths: it reaches its maximum somewhere near the middle of visible spectrum (for green light) and gradually diminishes, tending to zero when we move to infrared ($\lambda > 760$ nm) and to ultraviolet ($\lambda < 400$ nm) radiation (Figure 17.35). Therefore, in photometry, two types of measurements are used: those that rely on perception by the human eye (corresponding quantities are called *luminous*), and those that rely on the use of photoelectric devices to measure electromagnetic energy (corresponding quantities are called *radiant*).

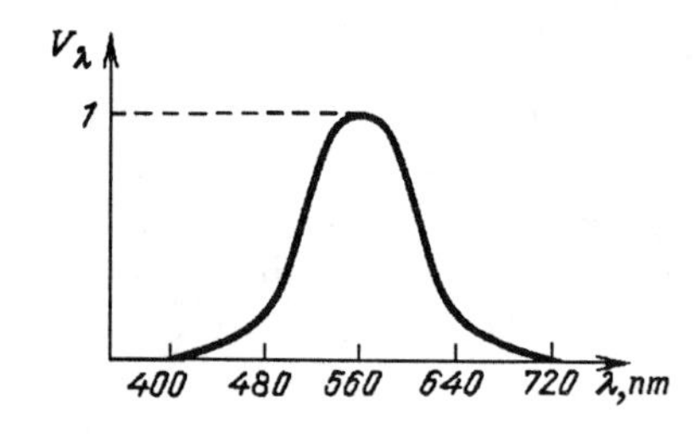

Figure 17.35: Spectral luminous efficiency.

Luminous flux Φ is a measure of the rate of flow of light in the wavelength range 400–760 nm, estimated by visual perception. It is measured by reference to emission from a standard source (usually by comparing the illuminance of two surfaces).

A source of light is regarded as a *point source* if it radiates uniformly in all directions and its size is much less than the distance at which the illuminance produced by it is estimated. *Luminous intensity* I is a measure of light-emitting ability of a light source. It equals the luminous flux radiated by the source in the limits of a solid angle (see 9.6.6) of 1 steradian: $I = \Phi/\Omega$. The whole luminous flux spreading from the source in all directions (i.e., in the solid angle $\Omega = 4\pi$ steradians) is expressed by the relation

$$\Phi = 4\pi I. \tag{17.25}$$

The base unit of luminous (photometric) quantities is the unit of luminous intensity *candela* (cd). Candela equals the luminous intensity in a given direction of a source that emits monochromatic radiation of frequency $5.4 \cdot 10^{14}$ Hz and has a radiant intensity in this direction of 1/683 watt per steradian. The unit of luminous flux is a *lumen*. It equals the flux from a source of 1 candela spreading in the solid angle of 1 steradian.

Illuminance E of a surface is the ratio of the luminous flux Φ hitting some part of the surface to the area S of this part: $E = \Phi/S$. The unit of illuminance is *lux*. Illuminance equals 1 lux if there is a luminous flux of 1 lumen per 1 square meter of uniformly illuminated surface. Illuminance of a surface oriented perpendicularly to rays coming from a source (point A in Figure 17.36) is inversely proportional to the square of the distance from the source:

$$E = \frac{I}{h^2}. \tag{17.26}$$

Illuminance of a surface by oblique light rays (the point B in Figure 17.36) depends on the angle of incidence α:

$$E = \frac{I\cos\alpha}{r^2} = \frac{I}{h^2}\cos^3\alpha. \tag{17.27}$$

Here $r = h/\cos\alpha$ is the distance from the source to the point of observation B, and h is the height of the source over the flat illuminated surface. In the case of several independent (incoherent) light sources, illuminance of some surface equals the sum of illuminances produced by the separate sources.

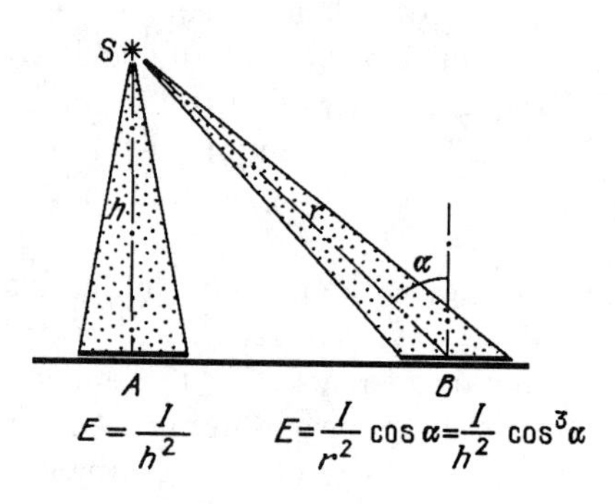

Figure 17.36: Illuminance of the surface produced by the point source S.

Chapter 18

Relativistic and Quantum Physics

The progress of physics in the twentieth century is deeply associated with the discovery of the *theory of relativity* and the *quantum theory*. Development of these fundamental physical theories led to the profound changes in our comprehension of all principal notions of classical physics. Relativistic and quantum theories form the basis of modern understanding of the surrounding world.

18.1 The Theory of Relativity

The *theory of relativity* is the physical theory of space and time. In other words, it is the theory of the most general space–time relationships valid for all physical and, more widely, for all natural processes. According to the *general theory of relativity*, or relativistic theory of gravitation, the properties of space and time are determined by gravitational fields acting in the space region under consideration. The *special theory of relativity* was proposed by Albert Einstein in 1905. It describes the properties of space and time in weak gravitational fields when the influence of the gravitation on these properties can be ignored.

The phenomena described by the special theory of relativity (such phenomena are called *relativistic*) reveal themselves at the motions of objects with velocities approaching to the speed of light in a vacuum $c = 299792458$ m/s $\approx 3 \cdot 10^8$ m/s. The quanta of light called *photons* are characterized by the rest mass of zero and move with this speed c in a vacuum. The speed of particles having finite rest mass is always less than c (though it may approach arbitrarily close to c). The value c is the general *limit of speed* accessible for transmission signals or any interactions from one point of the space to another. This speed c is the highest speed attainable in the universe. Its value is a universal constant and is independent of the speed of the observer. The existence of the universal limiting speed requires a deep revision of familiar classical concepts concerning space and time which are based on our common sense and everyday experience.

All phenomena in an isolated physical system will develop just in the same way if we displace the system as a whole through some distance in space or if we turn it as a whole through some angle. In this fact the *properties of symmetry* of the laws of nature display themselves, namely the properties of symmetry that express the *homogeneity of the physical space* (i.e., the equivalence of all its points), and the *isotropy of space* (i.e., the equivalence of all directions). Independence of the laws of physics of time (invariability of the laws during the course of time) expresses the *homogeneity of time.* Along with this invariance of physical laws and the laws of nature with respect to translations and rotations in space and with respect to translation in time, another important invariance is established experimentally—the invariance of laws with respect to *transformations of motion*, i.e., with respect to transitions from one inertial frame of reference to another: all phenomena in a closed physical system obey the same laws independently of its rectilinear uniform motion. This statement declaring the *equivalence of all inertial frames of reference* is the essence of the *principle of relativity*: experiments performed in different inertial reference frames give the same results.

18.2 Relativistic Kinematics

18.2.1 Galilean transformation

The central notion of the theory of relativity is the concept of an *event*—something that occurs at a definite space point in a definite time instant (e.g., a decay of some elementary particle). Any physical process can be treated as a consequence of events. Some event is characterized by the space coordinates x, y, z, and by the time instant t in some inertial frame of reference K. In another inertial frame of reference K', which moves with respect to K with a constant speed v in the direction of the x-axis (Figure 18.1), coordinates and time of the same event will be expressed by different values x', y', z', and t'.

According to the concepts of *classical physics* that were elaborated on the experimental basis of observations of relatively slow motions ($v \ll c$) of macroscopic bodies, *time intervals* between events and *space distances* between the points of their occurrence are *absolute*—their values do not depend on the frame of reference. Therefore, the classical formulas relating the coordinates and time of some event in the two reference frames (the *Galilean transformations*) have the form

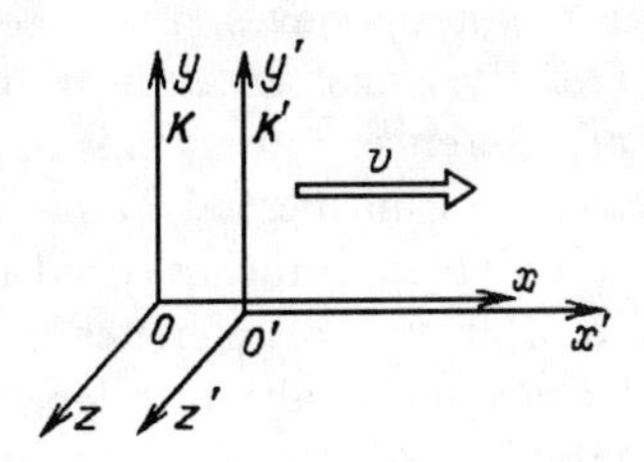

Figure 18.1: Frames of reference K and K'.

$$x = x' + vt, \qquad y = y', \qquad z = z', \qquad t = t'. \tag{18.1}$$

It is supposed here that at the time instant $t = t' = 0$, the coordinate systems K and K' coincide.

From the Galilean transformations, Equation (18.1), it follows the classical low of *velocity transformation* for a moving particle in transition from one frame of reference to another. If $\mathbf{u} = d\mathbf{r}/dt$ is the velocity of some particle in K, and $\mathbf{u}' = d\mathbf{r}'/dt$ is the velocity of the same particle in K', then

$$u_x = u'_x + v, \qquad u_y = u'_y, \qquad u_z = u'_z. \tag{18.2}$$

Equations (18.2) mean that classical transformation of velocity in transition from K to K' is simply the vectorial addition of the relative velocity $\mathbf{u}'$ and the transfer velocity $\mathbf{v}$: $\mathbf{u} = \mathbf{u}' + \mathbf{v}$. Acceleration of a particle $\mathbf{a} = d\mathbf{u}/dt$ is the same in K and K', as can be seen from Equations (18.2): $\mathbf{a} = \mathbf{a}'$. The equation of motion of classical mechanics $\mathbf{a} = \mathbf{F}/m$ (Newton's law of motion) retains its form in transition from K to K'. The invariance of this equation means that Galilean transformations satisfy the principle of relativity with respect to the laws of mechanics.

18.2.2 Insufficiency of classical concepts

Numerous experiments give the evidence that the principle of relativity is valid not only for mechanical phenomena, but for all other physical phenomena as well. In particular, all attempts to detect the influence of the earth's orbital motion on electromagnetic and optical phenomena gave the negative result (the famous Michelson–Morley experiment and other attempts to measure the velocity of the earth through the ether). Nevertheless, the laws of electrodynamics (Maxwell's equations) do not retain their form in transition to another inertial frame of reference by the Galilean transformations, Equations (18.1). This is obvious from the fact that the speed of electromagnetic waves (the speed of light), according to the classical law of velocity transformation, Equations (18.2), cannot have the same value in all reference frames, as is required by the principle of relativity (the equivalence of all inertial frames of reference). That is why the existence of the universal limiting speed c (the speed of light), having the same value in all frames of reference, requires the fundamental revision of classical concepts of absolute length and absolute time. The Galilean transformations, Equations (18.1), based on these classical concepts must be replaced by other transformations (Lorentz transformations, see Section 18.2.5). The laws of classical dynamics must be revised also.

18.2.3 Main principles of the theory of relativity

The following two principles (postulates) lie in the basis of the theory of relativity:

- The principle of relativity, which states the equivalence of all inertial frames of reference for all physical phenomena.
- The principle of the universal speed limit, which states that the speed of light in a vacuum c is constant throughout the universe and is independent of the speed of the observer. Its value gives the upper limit of propagation speed for any signals and interactions.

The procedures of time and space measurements ought to be defined in agreement with these principles. The definition of *simultaneity of events* or the *procedure of clock synchronization* is based on the statement that the speed of a signal traveling with the maximal speed is independent of the signal direction. Let a signal be sent from the point A (Figure 18.2) at the instant t_1 measured by the clock A. This signal gets to the space point B at the instant t' measured by the clock B. The reflected signal returns to point A at the instant t_2. By definition, the clocks in space points A and B are considered as synchronized if $t' = (t_1 + t_2)/2$.

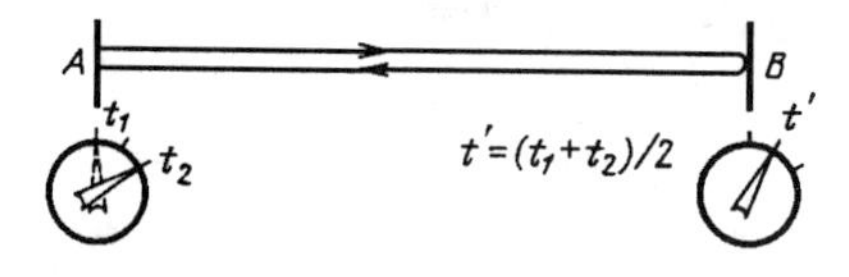

Figure 18.2: Procedure of clock synchronization.

As a proper procedure for measuring *space distances* a "radar method" of ranging can be adopted, in which the time spent by a signal to travel to the object and backwards is measured. The desired distance l to the target can be calculated by multiplying the universal speed c of the signal by half of the full time $t_2 - t_1$ of its propagation: $l = c(t_2 - t_1)/2$.

18.2.4 The relativity of simultaneity. Time dilation and Lorentz contraction

Two events occurring at some distant space points are regarded by an observer as simultaneous if the synchronized clocks of his reference frame located at the points where the two events occur read the same time. But the simultaneity of events depends on the frame of reference.

The relativity of simultaneity is easily seen from the following example in which the argument relies on the fact that light travels at the same speed in all directions in all reference frames. The sketch in Figure 18.3 represents the sending of a light signal from the midpoint A of a segment BC. These signals reach points B and C simultaneously if we judge by the readings of the clocks in K' reference frame, where the segment BC is at rest. But the same events of receiving the signals are not simultaneous in the K reference frame: point B in this reference frame is moving towards the signal, and the light on its way to B travels a lesser path than on the way to C. As the speed of light c is independent of the direction of propagation, in K reference frame the signal will be received in B earlier than in C.

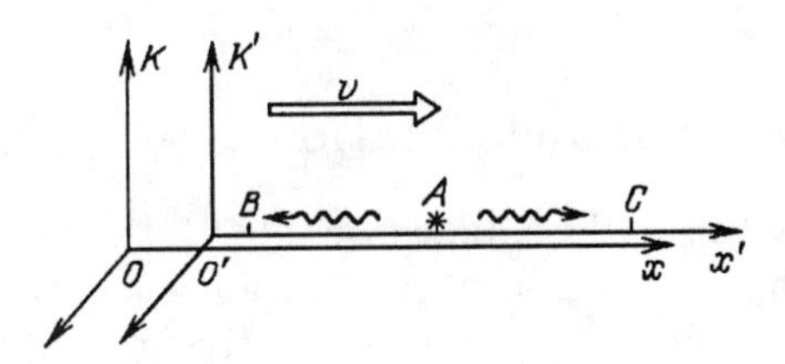

Figure 18.3: Relativity of simultaneity of events.

The time intervals and space distances between events, measured by the procedures mentioned above, are also *relative* (depending on the reference frame). The time interval between two events is called the *proper time* τ_0 if it is measured in the reference frame in which the two events occur at the same place (at the same spatial point). The proper time is the shortest time that any observer measures for the time interval between those two events. In the reference frame moving at the *relative speed* v, the time interval τ between the same events is dilated ($\tau > \tau_0$):

$$\tau = \frac{\tau_0}{\sqrt{1 - v^2/c^2}}. \tag{18.3}$$

The dependence of space distances between events on the frame of reference is usually called the *relativistic length contraction* or *Lorentz–Fitzgerald contraction* of a moving body. The length l_0 of an object measured in the reference frame where the object is at rest is called the *proper length*. In any other reference frame moving with relative velocity v with respect to the first frame, the length l of the same object in the direction of motion is shorter ($l < l_0$) by the Lorentz square-root factor:

$$l = l_0\sqrt{1 - \frac{v^2}{c^2}}. \tag{18.4}$$

This relativistic kinematic effect of length contraction refers not only to the length of a moving body, but also to the space distance between any two events as well, if this distance is measured by two observers in the direction of their relative motion.

In agreement with the principle of relativity, the kinematic effects of time dilation and length contraction expressed by Equations (18.3)–(18.4), are mutual: if there are two identical clocks in reference frames K and K', the observer in K finds that the moving clock (i.e., the clock in K') is slow, and the observer in K' finds that the clock that moves with respect to K' (i.e., the clock in K) is slow. Similarly, if there are two identical rigid rods in K and K', oriented along the relative velocity of the two frames of reference, each of the observers regards the rod moving with respect to his reference frame to be contracted.

18.2.5 Lorentz transformations

The Galilean transformations, Equations (18.1), that are valid in classical Newtonian mechanics, in relativistic mechanics are replaced by *Lorentz transformations*, which arise from the relativistic main principles. The relativistic set of equations expressing the space coordinates x, y, z and the time t of an arbitrary event in the K reference frame in terms of the coordinates x', y', z' and the time t' of the same event in K' has the following form:

$$x = \frac{x' + vt'}{\sqrt{1 - v^2/c^2}}, \quad y = y', \quad z = z', \quad t = \frac{t' + (v/c^2)x'}{\sqrt{1 - v^2/c^2}}. \tag{18.5}$$

The inverse transformations from K to K' can be obtained from Equations (18.5) if we replace there $v \to -v$.

The relativistic formulas for transformation of the velocity in transition from one reference frame to another follow from the Lorentz transformations, Equations (18.5):

$$u_x = \frac{u'_x + v}{1 + vu'_x/c^2}, \quad u_y = \frac{u'_y\sqrt{1 - v^2/c^2}}{1 + vu'_x/c^2}, \quad u_z = \frac{u'_z\sqrt{1 - v^2/c^2}}{1 + vu'_x/c^2}. \quad (18.6)$$

These formulas manifest the limiting character of the speed of light: if a short light impulse moves along the x'-axis with the speed $u'_x = c$ with respect to K', then its speed with respect to K, as it follows from Equations (18.6), is also equal to c:

$$u_x = \frac{c + v}{1 + v/c} = c.$$

In the limiting case of slow relative motion $v \ll c$ Lorentz transformations, Equations (18.5), coincide with Galilean transformations, Equations (18.1), and relativistic velocity addition formulas, Equations (18.6), coincide with the corresponding classical law, Equations (18.2). This coincidence means that classical concepts concerning space and time are valid only for rather slow ($v \ll c$) motions.

18.2.6 Relativistic interval

In transition from one reference frame to another by Galilean transformations, the time interval between any two events remains the same as well as the distance between the space points where the events occurred. Thus, both these quantities are invariant under the classical Galilean transformations. Relativistic Lorentz transformations change time intervals (time dilation) and space distances (length contraction), but they leave invariant a quantity s_{12} called the *relativistic interval* between events, which has the same value in all inertial reference frames:

$$s_{12} = \sqrt{c^2 t_{12}^2 - l_{12}^2}. \quad (18.7)$$

In this definition $t_{12} = t_1 - t_2$ is the time interval between the events and l_{12} is the space distance between the points where the events occurred:

$$l_{12}^2 = \sqrt{(x_1 - x_2)^2 + (y_1 - y_2)^2 + (z_1 - z_2)^2}.$$

The concept of the space–time relativistic interval is a generalization of the concepts of the time interval and space distance between events. Depending on what part of the interval—spatial or temporal—predominates, intervals are classified as space-like ($l_{12}^2 > c^2 t^2$) and time-like ($c^2 t^2 > l_{12}^2$).

In the case of a time-like interval between events it is always possible to find a reference frame K' in which the two events occurred at the same place ($l'_{12} = 0$) (at the same spatial point), and the interval s_{12} coincides with the proper time τ_0 between the events (to within the constant factor c). The order of such events in the time is the same in all reference frames and the notions "earlier" and "later" have absolute meaning (independent of the reference frame). This means that one of these events can be the cause of the other—the reason precedes the consequence for all possible observers.

In the case of a space-like interval between two events it is always possible to find the reference frame K'' in which the two events occur simultaneously ($l'_{12} = 0$), and the modulus of the interval is equal to the proper distance l_0. For such events casual relationship is impossible, because no signal or interaction can propagate with a speed greater than c. The notions "simultaneously", "earlier", and "later" for such events have relative meaning (their sequence in the time is different depending on the frame of reference).

18.3 Relativistic Dynamics

18.3.1 Relativistic momentum and energy

The laws of classical dynamics are invariant under the Galilean transformations. That is why the denial of classical concepts of space and time and of the Galilean transformations based on these concepts requires the revision and correction of the laws of dynamics. The equations of relativistic dynamics turn into classical ones in the case of slow movements ($v \ll c$) where the validity of classical dynamics is proved experimentally by all our experience.

In the theory of relativity, as well as in classical mechanics, for an isolated physical system the total momentum $\mathbf{p}$ and the total energy E are conserved. But the relativistic expressions for the momentum and energy differ from the corresponding classical expressions:

$$\mathbf{p} = \frac{m_0 \mathbf{v}}{\sqrt{1 - v^2/c^2}}, \qquad E = \frac{m_0 c^2}{\sqrt{1 - v^2/c^2}}. \tag{18.8}$$

Here m_0 is the *rest mass* which characterizes the inertia of an object at slow ($v \ll c$) motions. Relativistic momentum and energy satisfy the equations that are analogous to the corresponding classical equations (see Section 2.2.2):

$$\frac{d\mathbf{p}}{dt} = \sum \mathbf{F} \quad \text{(the Newton's second law)}, \qquad \frac{dE}{dt} = \sum \mathbf{F} \cdot \mathbf{v}. \tag{18.9}$$

18.3.2 Mass and energy

The rest mass m_0 is not a conserving physical quantity. In particular, in the events of nuclear decay and transmutation of elementary particles, the sum of relativistic energies and relativistic momenta of all involved particles conserves, but the sum of their rest masses changes. For example, in the act of annihilation (see Section 18.7) of an electron and a positron, the energy and momentum are carried away by two photons of γ-radiation, but the rest mass is diminished by $2m_e$ (the rest mass of a photon moving with the highest possible speed c equals zero).

In the reference frame in which a body is at rest, its energy E_0 according to Equation (18.8) is given by the relation:

$$E_0 = m_0 c^2. \tag{18.10}$$

The energy E_0 is called the *rest energy*. It is proportional to the rest mass m_0 of the body. The mass–energy relationship, Equation (18.10), is called *Einstein's equation*. It expresses the *law of proportionality of mass and energy*. If a body at rest receives (or gives back) some amount of energy ΔE in the form of radiation or heat, then its rest mass changes in proportion to the amount of energy received (or given back):

$$\Delta m_0 = \Delta E / c^2.$$

The existence of the rest energy is demonstrated in the most obvious way in the act of annihilation of a particle and its antiparticle, when the whole rest energy of the particles transforms into the energy of γ-radiation. However, for many physical processes the major part of the rest energy (and of the corresponding rest mass) does not take part in transformations. That is why the mass determined by weighing is practically conserved in spite of the fact that the body absorbs or emits energy. For example, in the chemical reaction of burning, when 1 kg of hydrogen is combined with 8 kg of oxygen, the energy ΔE released in the form of light and heat amounts to 10^8 J. The rest mass of the water produced in this reaction is only by $\Delta m_0 = \Delta E/c^2 \approx 10^{-9}$ kg smaller than the mass of the initial substances, so the relative change of rest mass $\Delta m_0/m_0 \approx 10^{-10}$. To detect such a small change of mass the weighing would have to be done to better than 1 part in 10 billion. The best laboratory balances provide weighings up to 7 significant figures, but not to 10. In nuclear reactions (see Section 18.6.2) the rest energy of all the nucleons (protons and neutrons) taking part in reaction remains the same, but in this case the active part of the rest energy (i.e., the changing part of the rest energy)—the energy of interaction of the nucleons—makes a significant part of the rest energy.

In some books the *relativistic mass* m of a particle is introduced, which depends on the velocity v of the particle. The relativistic mass is connected with the rest mass m_0 and the velocity v by the relation

$$m = \frac{m_0}{\sqrt{1 - v^2/c^2}}.$$

Using this quantity, it is possible to express the relativistic energy E, Equation (18.8), in the following way:

$$E = mc^2.$$

Relativistic mass of a particle differs from its relativistic energy (which includes the kinetic energy) only by the constant factor c^2, so the relation $E = mc^2$ expresses the *law of equivalence of relativistic mass and relativistic energy*. By means of the relativistic mass m, the relation between momentum $\mathbf{p}$ and velocity $\mathbf{v}$ of a particle $\mathbf{p} = m\mathbf{v}$ can be written in the same form as in non-relativistic mechanics.

18.3.3 Relativistic kinetic energy

For a moving particle, the difference between relativistic energy, Equation (18.8), and the rest energy, Equation (18.10), determines the *kinetic* energy of the particle:

$$E_{\text{kin}} = E - E_0 = m_0 c^2 \left(\frac{1}{\sqrt{1 - v^2/c^2}} - 1 \right). \qquad (18.11)$$

At $v/c \ll 1$ we can use the expansion

$$\frac{1}{\sqrt{1 - v^2/c^2}} \approx 1 + \frac{1}{2}\frac{v^2}{c^2}$$

(see Section 10.2.1), and Equation (18.11) transforms to the usual non-relativistic expression (see Section 13.4.5) for kinetic energy:

$$E_{\text{kin}} \approx \frac{m_0 v^2}{2}.$$

From Equation (18.11) we see that relativistic kinetic energy increases without limit as the speed v approaches the speed limit c. In order to accelerate a body of a finite rest mass up to the speed c it would require an infinitely large amount of energy. This fact is in agreement with the statement that c is the limit of speed for particles of non-zero rest mass: it is impossible to reach this speed, though we can approach as close to it as desired.

18.3.4 Relativistic transformation of energy and momentum

The relativistic formula expressing the relation between energy E and momentum $\mathbf{p}$ can be derived from Equations (18.8) by excluding the velocity v:

$$E^2 - p^2c^2 = m_0^2c^4. \tag{18.12}$$

The energy E and momentum $\mathbf{p}$ of a particle depend on the frame of reference, but the right side of Equation (18.12) is a relativistic invariant. We can see here an analogy with the behavior of time and coordinates of an event, expressed by Equation (18.7): the time t and radius-vector $\mathbf{r}$ depend on the frame of reference, but their combination,

$$c^2t^2 - r^2 = c^2\tau_0^2,$$

has the same value in all reference frames (it is the square of the space–time interval between the origin $t = 0$, $\mathbf{r} = 0$ and the considered event with $c^2t^2 > r^2$). This analogy permits us to suppose that the energy and momentum of a particle transform in transition from one inertial reference frame to another in the same way as the time and spatial coordinates of an event, i.e., the transformation is expressed by the same formulas, Equations (18.5), of Lorentz' transformations, in which the following substitution is made:

$$t \rightarrow E/c, \qquad x \rightarrow p_x \qquad y \rightarrow p_y \qquad z \rightarrow p_z,$$

and similar substitution is made for the primed quantities (i.e., in K' reference frame). The rest energy of a particle on the right side of Equation (18.12) is the analog of the time-like relativistic interval (of the proper time between the origin and the event).

For ultrarelativistic particles whose energy E is much greater than the rest energy m_0c^2, Equation (18.12) can be approximately represented in the form:

$$E \approx pc \qquad (\text{at} E \gg m_0c^2). \tag{18.13}$$

From this equation it follows, in particular, that the momentum of a photon with the energy $E = \hbar\omega$ is proportional to the frequency ω of corresponding electromagnetic oscillations: $p = \hbar\omega/c$. The same relation $p = E/c$ is valid for the momentum and energy of any traveling electromagnetic wave.

18.3.5 Example: acceleration by a constant force

Let a particle move under the action of a constant force. From Equation (18.9) it follows that at $\mathbf{F} = \text{const}$, the momentum $\mathbf{p}$ of the particle changes in time linearly, and if at the initial moment $t = 0$ the particle was at rest ($\mathbf{p} = 0$), then $\mathbf{p}(t) = \mathbf{F}t$. Substituting $\mathbf{p}$ in Equation (18.12), we obtain the dependence of energy E on time:

$$E^2 = (Ft)^2c^2 + m_0^2c^4. \tag{18.14}$$

In order to get an explicit dependence of the particle's velocity on time, let us divide term by term the first of Equations (18.8) by the second:

$$\mathbf{v} = \frac{c^2}{E}\mathbf{p} = \frac{c\mathbf{F}}{\sqrt{(Ft)^2 + m_0^2c^2}}\,t. \tag{18.15}$$

At the beginning the motion of the particle is non-relativistic. While $Ft \ll m_0c$, we can ignore the first term in the square root in Equation (18.15). Then we get the ordinary non-relativistic expression for the motion under the action of a constant force: $\mathbf{v} = (\mathbf{F}/m_0)t$.

At large t values the growth of velocity v diminishes (Figure 18.4), and its value tends to the finite limit c. This limit can be seen from Equation (18.15) in the case of $t \to \infty$ (see Section 5.2.5).

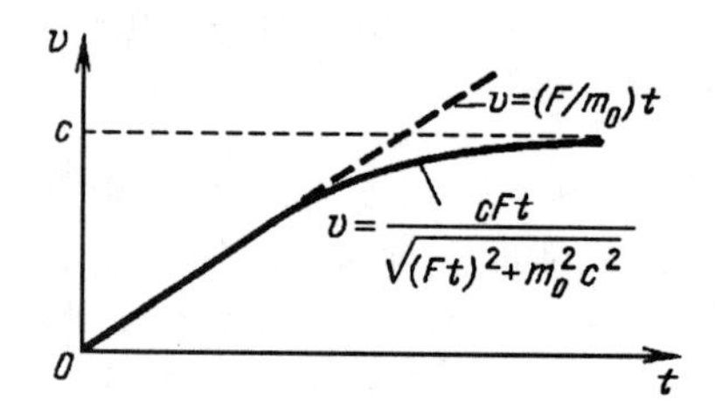

Figure 18.4: The graph of velocity versus time in the motion under the action of constant force.

18.3.6 Example: relativistic particle in magnetic field

The Lorentz force $\mathbf{F}$ (see Section 15.3.2) experienced by a moving charged particle traveling with a velocity $\mathbf{v}$ in a magnetic field of magnetic induction $\mathbf{B}$ (magnetic flux density) is directed perpendicularly to the velocity $\mathbf{v}$. Consequently, this force changes only the direction of $\mathbf{v}$, but not its magnitude v. This means that the relativistic mass m also remains constant. In this case the main law of dynamics, expressed by the first of Equations (18.9), can be written as follows:

$$m\frac{d\mathbf{v}}{dt} = q\mathbf{v} \times \mathbf{B}. \tag{18.16}$$

If velocity $\mathbf{v}$ is perpendicular to magnetic induction $\mathbf{B}$, the particle moves along a circle with the centripetal acceleration v^2/R, where R is the radius of the circle. In this case Equation (18.16) gives

$$m\frac{v^2}{R} = qvB,$$

and we get

$$R = \frac{mv}{qB}, \qquad \omega = \frac{2\pi}{T} = \frac{v}{R} = \frac{qB}{m}.$$

The above expressions for R and ω have the same form as they do in the non-relativistic case (see Section 15.3.2), but the relativistic mass m in the relativistic expressions depends on the velocity v of the particle (see Section 18.3.2). Velocities of electrons in modern cyclic accelerators used for research in nuclear and particle physics are so close to the speed of light c that their relativistic mass becomes greater than the rest mass of protons. To hold the relativistic electrons on a circular path of the same radius as at the beginning of the acceleration cycle, very strong magnetic fields are required.

18.3.7 Example: transmutations of elementary particles

Let us show that according to relativistic conservation laws the emission of light by a free electron is impossible. In the frame of reference in which the electron is at rest, its energy before emission of a photon (of a quantum of gamma radiation) is $E_0 = m_0c^2$. After such a hypothetical emission the electron, according to the law of the momentum conservation, would inevitably acquire some speed v by recoil, and so the system including the electron and the emitted photon would have the following energy:

$$E = \frac{m_0c^2}{\sqrt{1 - v^2/c^2}} + h\nu,$$

where $h\nu$ is the energy of the emitted photon (see Section 18.4.4). It is clear from this expression that it is impossible to fulfill the requirement of the law of the energy conservation, since E is always greater than E_0. The above considerations concerning the possibility of light emission apply to particles that are unable to change their inner state. Atoms and ions have intrinsic structure and are able to emit or to absorb light in the state of free motion by virtue of a transition from one inner energy level to another.

The laws of conservation prohibit *annihilation* of a particle and its antiparticle (see Section 18.7), e.g., annihilation of an electron and positron, with formation of only *one* photon. We can easily see this if we use the reference frame in which the center of mass of the particle and its antiparticle is at rest and the full momentum of the system equals zero. The zero total momentum can retain its zero value after annihilation only if at least two photons are produced in the act.

One photon of a sufficient energy can generate an electron-positron pair only in the presence of some other particle that would provide the fulfillment of the law of momentum conservation. The minimal energy of the photon that is necessary to generate an electron-positron pair the less, the greater the mass M of the particle that receives and takes away the momentum of the photon. The smallest possible energy of the photon is determined by the relation $h\nu = 2m_0c^2$ and equals 1.02 MeV for an electron-positron pair. This threshold energy corresponds to the generation of an electron and positron at rest in the presence of a particle of a mass so large ($M \gg m_0$) that its kinetic energy of recoil can be neglected.

18.4 The Principles of Quantum Physics

18.4.1 Uncertainty relations

Heisenberg's uncertainty relation determines the boundaries of validity of the laws of classical physics in the description of the properties of atomic and subatomic microparticles:

$$\Delta x \cdot \Delta p_x \geq h, \tag{18.1}$$

where Δx is the uncertainty of the coordinate x of a particle, Δp_x is the uncertainty of the corresponding component p of the momentum at the same time instant, and $h = 6.6260755 \cdot 10^{-4}$ J·s is the Planck's constant. It follows from the uncertainty relation that the more precisely the position is determined, i.e., the smaller Δx, the larger the uncertainty Δp_x of the momentum p_x after the measurement. If the position is determined exactly ($\Delta x \to 0$), the momentum p_x after the measurement would be completely uncertain ($\Delta p_x \to \infty$).

The relation connecting the uncertainty in the change of energy of a particle and the uncertainty of the time instant of this change is called the *Bohr–Heisenberg inequality:*

$$\Delta E \cdot \Delta t \geq h.$$

This relation means that the determination of energy with accuracy ΔE demands a period of time not less than $\Delta t = h/\Delta E$.

A similar relation connects the duration Δt of a wave packet with the interval $\Delta\nu$ of frequencies of monochromatic waves constituting this wave packet: $\Delta\nu \cdot \Delta t \geq 1$ (see Section 16.7.14). This relationship becomes evident from the formula $E = h\nu$ (see Section 18.4.2). The uncertainty relation is a fundamental law of nature.

Examples

1. The uncertainty Δx of an electron coordinate in an atom should not exceed the size of the atom $d \sim 10^{-8}$ cm. Thus the uncertainty Δp of the momentum is of the order

$$\Delta p \sim \frac{h}{d} \approx 6.6 \cdot 10^{-24} \text{ kg} \cdot \text{m/s}.$$

For the planetary model of an atom with an electron rotating around a nucleus, we derive using Newton's second law:

$$\frac{mv^2}{r} = \frac{1}{4\pi\varepsilon_0}\frac{e^2}{r^2}.$$

Hence

$$p = mv = e\sqrt{\frac{1}{4\pi\varepsilon_0}\frac{m}{r}} \approx 2 \cdot 10^{-24} \text{ kg} \cdot \text{m/s}.$$

The uncertainty of the momentum turned out to be larger than the value of the momentum itself, calculated on the basis of the laws of classical physics. Thus the classical description is not valid for an electron in an atom.

2. The accelerating voltage in a TV cathode-ray tube is of the order $U \approx 15$ kV. The momentum of an electron accelerated in the tube is

$$p = \sqrt{2meU} \approx 6.6 \cdot 10^{-23} \text{ kg} \cdot \text{m/s}.$$

This momentum is directed along the tube axis. The diameter of an electron beam in a modern TV set is of the order $d \geq 10^{-5}$ m. It is just the accuracy for an electron coordinate in a perpendicular direction to be fixed: $\Delta x \sim d$. Due to the uncertainty relation the electron momentum perpendicular to the beam axis is

$$\Delta p \approx h/d \approx 6.6 \cdot 10^{-29} \text{ kg} \cdot \text{m/s}.$$

The uncertainty of the direction of electron motion is given by the expression:

$$\Delta\theta = \Delta p/p = 10^{-6} \text{ rad}.$$

The length of an electron path in a cathode-ray tube does not exceed $l \leq 1$ m and so an unpredictable shift of this path, determined by quantum effects, does not exceed $s = l\Delta\theta \leq 10^{-6}$ m, i.e., it is smaller than the diameter of the beam. Thus an electron in a TV set can be treated adequately on the basis of classical considerations.

18.4.2 Wave–particle dualty

Real physical subatomic microparticles (photons, electrons, etc.) have the potential possibility to reveal both corpuscular and wave properties (*wave–particle dualty*). A wave picture or a particle picture is needed for the description of a physical system (the wave or the particle properties would be revealed) depends on conditions of the experiment and on the measurement to be performed. For every particle variable (like position and momentum) there exists a complementary wave variable (like wavelength and frequency). The frequency ν (or the circular frequency $\omega = 2\pi\nu$), the wavelength $\lambda = c/\nu$, and the wave vector $\mathbf{k}$ that determines the direction of propagation are the wave parameters of a microparticle. The modulus of the wave vector k is inversely proportional to the wavelength λ:

$$k = 2\pi/\lambda = 2\pi\nu/c = \omega/c.$$

The energy E and the momentum $\mathbf{p}$ of a microparticle are its corpuscular parameters. The wave and the corpuscular properties of microparticles are connected by the following relations ($\hbar = h/2\pi$):

$$E = \hbar\omega \quad \text{or} \quad E = h\nu,$$

(the Planck relation) and

$$\mathbf{p} = \hbar\mathbf{k} \quad \text{or} \quad \mathbf{p} = \frac{h}{2\pi}\mathbf{k};$$

hence,

$$p = h/\lambda$$

(the de Broglie relation).

The wave associated with the mechanical motion of a particle is called a *matter wave* or *de Broglie wave*. The phenomenon of electron diffraction gives an experimental evidence for the de Broglie relation.

The Planck and the de Broglie relations constitute a bridge between the world of corpuscles (classical particles) characterized by the momentum $\mathbf{p}$ and the energy E, and the world of waves characterized by the wavelength λ and the frequency ν: the right-hand sides of these relations contain the quantities ω and $\mathbf{k}$ that can be determined from the interference phenomena, while the left-hand sides characterize a microparticle as a corpuscle. It follows from these relations that there is a strict correspondence (or equivalence) between the wave and the corpuscular properties—the measures of these properties are always proportional to each other.

The wave and the corpuscular properties are complementary: only both of them together give an adequate full description of subatomic microparticles in different experiments. Heisenberg's uncertainty relation, Equation (18.17), was used by Niels Bohr as the basis for his *complementarity principle,* which is of decisive importance for the understanding of quantum phenomena. According to Bohr, it should be assumed that in nature both light and elementary particles can appear as waves or corpuscles, depending on the type of experiment performed on them. However, only one possibility can be realized in a certain experiment: either only wave or only corpuscular properties can be observed. According to Bohr, wave or corpuscular aspects of physical phenomena are not incompatible but, on the contrary, they are complementary. It does not make sense to consider wave and corpuscular properties simultaneously.

18.4.3 Range of validity of classical theory

The concept of the *de Broglie matter waves* along with Heisenberg's uncertainty relation can be used for clearing up the question whether classical or quantum theory should be used for the description of a certain phenomenon. In a problem under consideration one should compare the de Broglie wavelength $\lambda = h/p$ with a characteristic linear size: wave properties do not reveal themselves until this wavelength becomes comparable with a typical size. Thus, having chosen for an electron beam such a momentum p that the corresponding wavelength would turn out to be of the same order as a lattice constant of a crystal, we can state that a diffraction will occur if we transmit the beam through such a crystal. The experiment confirms this conclusion.

Using the concept of wave properties of matter it is possible to derive Bohr's rules of quantization (see Section 18.5.1) from the requirement that the length of a stationary orbit should contain an integral number of the de Broglie waves: $n\lambda = nh/p = 2\pi r$. Hence, $mvr = nh/2\pi$.

18.4.4 Quanta of light—photons

The concept of *light quanta*, or *photons*, was introduced in order to explain the character of the black body radiation (see Section 18.5.4). The formula $E = h\nu$ was suggested first just for photons. Some experimental data get a theoretical explanation only on the basis of representation of light as an ideal gas of photons.

Since energy E is always connected with mass m by the relation $E = mc^2$ (see Section 18.3.2), we can write for a photon mass:

$$m = h\nu/c^2.$$

Photons do not exist in a state of rest, so their rest-mass m_0 is equal to zero. The mass m determined by this formula is the mass of a photon moving with the light velocity c (see Section 18.3.2). The momentum of a photon is equal to a product of its mass and velocity:

$$p = mc = \frac{h\nu}{c} = \frac{E}{c} = \frac{h}{\lambda}.$$

18.4.5 Photoelectric effect

A *photoelectric effect* (external) is a phenomenon of the liberation of electrons from a substance by light.

The photoelectric effect is characterized by the following laws:

1. The current of saturation, i.e., the number of electrons ejected by a light beam from the surface of a metal per unit time, depends on the intensity of radiation and is proportional to the light flux falling on the surface.

2. The maximal kinetic energy of photoelectrons is a linear function of a light frequency and is independent of the light flux falling on the surface.

3. There exists for each material the threshold value of light frequency called the red boundary of photoelectric effect.

4. The emission of electrons begins immediately after switching on the electromagnetic field.

The phenomenon of photoelectric effect illustrates the corpuscular properties of light. Each photon is absorbed independently and the energy transformations accompanying the process are described by *Einstein's equation* for the photoelectric effect. This equation corresponds to the energy conservation law in an elementary act of interaction of light with the substance:

$$\frac{mv^2}{2} = h\nu - A,$$

where A is the *work function* of the substance equal to the minimal energy required for the removal of an electron from the material.

The red boundary of photoelectric effect is determined by Einstein's equation with $v = 0$:

$$h\nu_{\min} = A.$$

18.4.6 Light pressure

The *light pressure* can be explained either with the help of the wave theory (see Section 16.7.10) or on the basis of the corpuscular point of view. In the latter case the light pressure can be explained by the transfer of the photon momentum to the bodies the photons interact with. The photon momentum p (see Section 18.4.4) is proportional to the light frequency ν:

$$p = \frac{h\nu}{c}.$$

18.4.7 Doppler effect

The *Doppler effect* can be explained either by the wave or by the corpuscular points of view. In the latter case the energy and the momentum conservation laws for an act of a photon emission by a free moving atom (Figure 18.5) give the equations:

$$h\nu' - h\nu = \frac{p^2}{2m} - \frac{p'^2}{2m}, \quad \mathbf{p} = \mathbf{p}' + \mathbf{p}_{\rm ph}.$$

These equations lead to the following expression for the relative shift of radiation frequency:

$$\frac{\Delta\nu}{\nu} = \frac{v}{c}\cos\theta,$$

where $v = p/m$ is the velocity of the radiating atom.

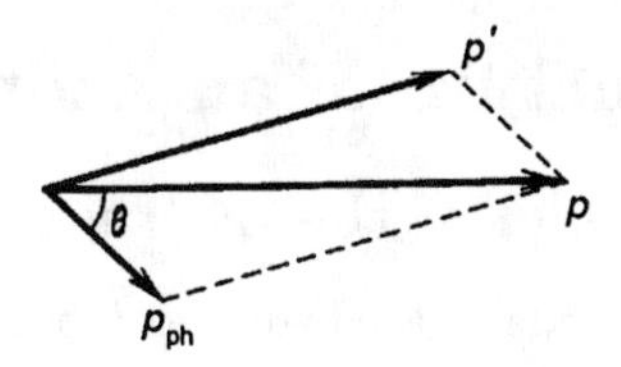

Figure 18.5: The conservation of the momentum for the Doppler effect.

18.4.8 Compton effect

The *Compton effect* is a phenomenon of reduction in the energy of X-rays or gamma-ray photons as a result of elastic scattering of photons by electrons of a substance with light atoms (carbon, paraffin, etc.). Some part of the photon's energy is transferred to the electron and consequently the photon loses energy. The wavelength of scattered radiation is increased in the *Compton scattering*. The observed change $\Delta\lambda$ of wavelength depends on an angle θ between the direction of an initial beam and the direction of a scattered radiation in the following way:

$$\Delta\lambda = 2k\sin^2(\theta/2), \tag{18.2}$$

where the experimentally determined constant k equals 0.0024 nm.

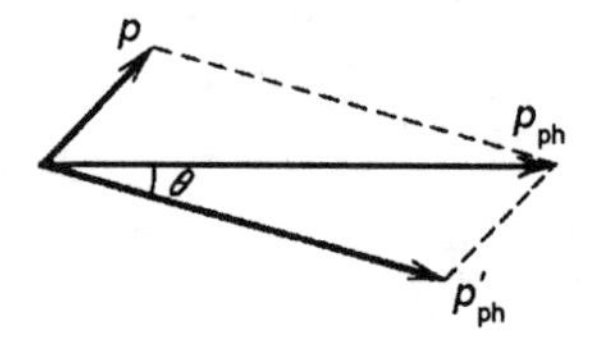

Figure 18.6: The momentum conservation for the Compton effect.

The Compton effect can be explained only on the basis of the corpuscular concept of light. Taking into account the energy and the momentum conservation laws for a collision of a photon with an electron we derive the equations (see Figure 18.6):

$$h\nu = h\nu' + \frac{p^2}{2m}, \quad \mathbf{p}_{\mathrm{ph}} = \mathbf{p}'_{\mathrm{ph}} + \mathbf{p},$$

where $\mathbf{p}$ is the electron momentum after the collision with a photon, and $p_{\mathrm{ph}} = h\nu/c$, $p'_{\mathrm{ph}} = h\nu'/c$ are the photon momenta before and after the scattering. Taking into account that $\lambda = c/\nu$, we obtain Equation (18.18) for $\Delta\lambda$, where $k = 2h/mc = 0.0024$ nm.

18.5 The Structure of an Atom

18.5.1 Bohr's model of the hydrogen atom

Rutherford's planetary model of the hydrogen atom with the only electron orbiting charged nucleus (proton) reveals certain contradictions when treated on the basis of classical theory. The model predicts a continuous spectrum of energies of an atom. Orbits of electrons should look like spirals twisting around the nucleus, due to the emission of light by the electron that moves with an acceleration. Thus, this model cannot explain the stability of atoms. To avoid the contradictions, Niels Bohr introduced in 1913 the following *quantum postulates* for an explanation of the observed stability of atoms and the spectral regularities:

1. An atom can exist only in stationary quantum states with a specific discrete value E_n of energy in each state. An atom in a stationary state does not radiate.

2. Emission of light occurs at a quantum jump-like transition of an atom from a stationary state with a larger value E_m of energy to a stationary state with a lower value E_n of energy. The energy of a radiated photon is equal to the difference of the energies of the stationary states:

$$h\nu = E_m - E_n.$$

Bohr's postulates are in a contradiction with classical mechanics and electrodynamics. They received a theoretical justification in quantum mechanics.

The expressions for the energy values of a hydrogen atom in the stationary states (the energy levels) and for the radii of the corresponding circular Bohr's orbits are determined by the *rules of quantization* (see Section 18.4.3):

$$mvr = n\hbar, \quad n = 1, 2, 3, \ldots, \tag{18.3}$$

where $\hbar = h/(2\pi)$. Using Newton's second law for an electron orbiting around a nucleus:

$$\frac{mv^2}{r} = \frac{1}{4\pi\varepsilon_0}\frac{e^2}{r^2}, \tag{18.4}$$

and the expression for the total energy of the electron in an atom:

$$E = \frac{mv^2}{2} - \frac{1}{4\pi\varepsilon_0}\frac{e^2}{r},$$

we get with the help of Equation (18.19)

$$r_n = 4\pi\varepsilon_0 \frac{\hbar^2}{me^2} n^2 = a_0 n^2, \tag{18.5}$$

$$E_n = -\frac{1}{(4\pi\varepsilon_0)^2}\frac{me^4}{2\hbar^2 n^2} = -\frac{R}{n^2}.$$

Here

$$R = \frac{1}{(4\pi\varepsilon_0)^2}\frac{me^4}{2\hbar^2} = 13.6056981 \text{ eV}$$

is the *ionization energy* of a hydrogen atom, i.e., the minimal energy to be delivered to a hydrogen atom in the ground state in order to remove the electron. The quantity

$$a_0 = 4\pi\varepsilon_0 \frac{\hbar^2}{me^2} = 0.529177249 \cdot 10^{-10} \text{ m}$$

is called the *Bohr's radius*. It is the radius of the nearest electron orbit to the nucleus.

In the lowest energy state with

$$E_1 = -\frac{1}{(4\pi\varepsilon_0)^2}\frac{me^4}{2\hbar^2} = -R = -13.6056981 \text{ eV}$$

an atom can stay during any period of time. The states with $n =$ 2, 3, 4, ... correspond to an excited atom. The lifetime of a free atom in these states is of the order of 10^{-8} s (see Section 18.5.3).

Energies of light quanta emitted by hydrogen atoms during transitions to lower energy states are determined by the energy difference of the initial and the final states:

$$\hbar\omega_{kn} = E_k - E_n = R\left(\frac{1}{n^2} - \frac{1}{k^2}\right).$$

The transitions to the level with $n = 2$ form the Balmer's series of spectral lines in the visible range of spectrum.

The above formulas have a simpler form in the Gaussian system of units (see Section 12.2). This system is most often used in atomic physics (see Table 18.1).

18.5.2 Electron shells

According to quantum mechanics, it does not make sense to speak about the motion of an electron in an atom along a definite orbit. Only the *probability* to discover the electron at some definite spatial point has a physical sense (see Section 18.4.1). However, the radii of Bohr's orbits retain a certain physical sense in quantum mechanics. For the quantum stationary states which are characterized by a spherically symmetrical probability of finding the electron at some distance from the nucleus, the maximums of the probability coincide with the positions of the Bohr's circular orbits.

An electron on the orbit nearest to the nucleus possesses the maximal value of the velocity. This follows from Equations (18.20)–(18.21) at $n = 1$:

$$v^2 = \frac{1}{(4\pi\varepsilon_0)^2}\frac{e^4}{\hbar^2},$$

whence

$$\frac{v}{c} = \frac{1}{4\pi\varepsilon_0}\frac{e^2}{\hbar c} \equiv \alpha \approx \frac{1}{137}.$$

The quantity α is called the *constant of fine structure*. It determines the structure of an atom. The ratio of the absolute value of Coulomb's

Table 18.1: The principal formulas of atomic physics in the Gaussian system of units and in SI

	Gaussian system	SI
The electron orbit radius for the hydrogen atom	$r_n = \frac{\hbar^2}{me^2}n^2$	$r_n = 4\pi\varepsilon_0 \frac{\hbar^2}{me^2}n^2$
The energy levels for the hydrogen atom	$E_n = -\frac{me^4}{2\hbar^2 n^2}$	$E_n = -\frac{1}{(4\pi\varepsilon_0)^2}\frac{me^4}{2\hbar^2 n^2}$
The dimensionless constant of fine structure	$\alpha = \frac{e^2}{\hbar c}$	$\alpha = \frac{1}{4\pi\varepsilon_0}\frac{e^2}{\hbar c}$

interaction energy of an electron on the lowest orbit with the nucleus to the rest energy of the electron mc^2 (see Section 18.3.2) is equal to

$$\frac{1}{4\pi\varepsilon_0}\frac{e^2}{a_0 mc^2} = \frac{1}{(4\pi\varepsilon_0)^2}\frac{e^4}{(\hbar c)^2} = \alpha^2 \ll 1.$$

The binding energy of an electron in an atom is much smaller than the rest energy of the electron. Hence an atom is a rather weakly bounded system.

Two principles determine the occupation by electrons of the allowed stationary states in an atom—the principle of *minimal value of the energy* of the whole electron shell and the *Pauli's principle* that reads: only one electron can occupy each allowed state in any quantum system.

In heavy atoms with the charge of the nucleus Ze, the binding energy of inner electrons is Z^2 times larger than in a hydrogen atom, and the distance of inner electrons from the nucleus is Z times smaller than the radius of the first Bohr's orbit in a hydrogen atom. The binding energy of outer electrons is of the same order as a binding energy of an electron in a hydrogen atom ($\sim$ 10 eV) and the total size of a many-electron atom is about the size of a hydrogen atom.

18.5.3 Light radiation of an atom

Radiation that accompanies transitions of an electron between the stationary states of the outer shell (of the "optical" electron) lies in the *optical region*. Transitions of an electron between the states of inner shells that are close to the nucleus produce the *X-ray emission*.

The wavelength of light emitted by an atom due to a transition of an optical electron is of the order of several hundreds of nanometers. This wavelength is approximately 10^3 times greater than the size of an atom. Actually, assuming the change of the atom energy in accordance with Equation (18.21) to be of the same order as the ionization energy of $1R$, we get the following estimate for the wavelength of the radiation:

$$\lambda = 2\pi\frac{c}{\omega} \approx 2\pi\frac{c\hbar}{R} \approx \frac{4\pi}{\alpha}a_0 \approx 10^3\, a_0.$$

The large wavelength of the radiation compared with the size of an atom leads to a large value for the lifetime of an electron in an excited state by comparison with the period $T = 2\pi/\omega$ of oscillations, i.e., $\omega\tau \gg 1$. Indeed, assuming the amplitude of oscillations of the optical electron to be equal to the size of the atom, we can estimate the radiated power using Eq. (5.45):

$$P \approx \frac{1}{4\pi\varepsilon_0}\frac{e^2\omega^4 {a_0}^2}{c^3}.$$

Thus for the radiation time τ of a photon with the energy $\hbar\omega$ we get:

$$\tau \approx \frac{\hbar\omega}{P} = 4\pi\varepsilon_0\frac{\hbar c^3}{e^2\omega^3 {a_0}^2}.$$

Taking into account that ωa_0 equals the electron velocity at the atom orbit, we get

$$\omega\tau \approx 4\pi\varepsilon_0\frac{\hbar c}{e^2}\left(\frac{c}{v}\right)^2 = \frac{1}{\alpha^3}.$$

For optical frequencies $\omega \approx 10^{15}$ 1/s we get the estimate of the duration of radiation (of the life time of an atom in an excited state): $\tau \approx 10^{-8}$ s.

18.5.4 Black-body radiation

The spectral composition of radiation emitted by isolated exited atoms is a set of relatively narrow lines typical for each chemical element. The spectrum of radiation of *heated* solids and liquids is *continuous*, i.e., it contains a continuous set of frequencies in a wide spectral range. Each body radiates the stronger in any spectral interval, the more it absorbs in this interval at the same temperature. A body is called the *black body* if it absorbs all radiation falling on it.

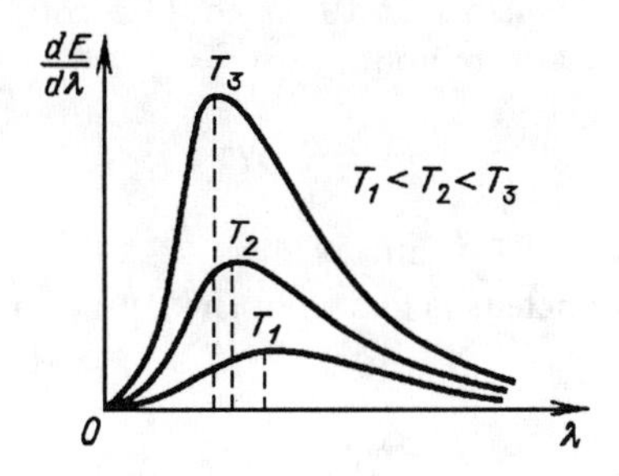

Figure 18.7: Spectral density of the equilibrium thermal radiation.

The radiation which is in thermal equilibrium with bodies having a definite temperature is called the *equilibrium thermal radiation* or the *black-body radiation.* The spectral composition of this radiation coincides with that emitted by a black body at the same temperature. Spectral density (monochromatic specific intensity) of the black-body radiation is expressed by *Planck's formula* (Figure 18.7):

$$\frac{dE}{d\omega} = \frac{\hbar\omega^3}{\pi^2 c^3} \frac{1}{e^{\hbar\omega/kT} - 1}, \tag{18.6}$$

where E is a volume density of the energy of the equilibrium radiation. The peak of the distribution given by Equation (18.22) is displaced toward shorter wavelengths as the temperature is increased, and the total radiation energy increases proportionally to the fourth power of the absolute temperature (see also Section 12.3.9).

18.6 Atomic Nucleus

18.6.1 Composition of atomic nuclei

The *atomic nucleus* of any element consists of *protons* and *neutrons* (see Section 18.7) also called *nucleons.* The principal characteristics of a stable nucleus are its *mass number* A (equal to the number of nucleons—protons and neutrons), the *electric charge* Ze (equal to the number of protons Z multiplied by the proton charge e), the mass M, the binding energy ΔE_b, and the radius R.

The number of protons Z in a nucleus is the ordinal number of the chemical element in Mendeleev's periodic system of elements. Nuclei with the same number of protons Z but with different numbers of neutrons $N = A - Z$ are called *isotopes*. The most stable are the nuclei with the number of neutrons N equal to one of the *magic numbers:* $N =$ 2, 8, 20, 28, 50, 82, 120.

The size of a nucleus is about 10^{-13} cm (1 fermi). The radius R of a nucleus is given by an approximate formula

$$R = R_0 A^{1/3},$$

where $R = (1.2 - 1.3) \cdot 10^{-13}$ cm.

The mass M of a nucleus is less than the sum of masses of its protons and neutrons:

$$M < Zm_p + Nm_n,$$

and the difference $\Delta m = Zm_p + Nm_n - M$ is called the *mass excess*. Thus the rest energy of free protons and neutrons is larger than the total energy of a nucleus formed of them. The quantity

$$\Delta E_b = \Delta m \cdot c^2$$

is called the *nucleus binding energy*. The nucleus binding energy is equal to the work necessary for nucleus fission to separate nucleons.

The nucleus binding energy per nucleon (the *specific binding energy*) can be estimated with the help of the uncertainty relation (see Section 18.4.1): for the modulus of a nucleon's momentum we have

$$p = h/R,$$

where R is the radius of the nucleus. The corresponding value of speed of a nucleus with the mass $m = 1.7 \cdot 10^{-24}$ g equals some tenths of the light velocity. Thus its kinetic energy is given by a non-relativistic expression

$$E_k = \frac{p^2}{2m} \approx \frac{h^2}{2mR^2} \approx 10\ \text{MeV}.$$

A nucleon is in a bound state inside a nucleus and the depth of its potential well is of the same order of magnitude. Typical experimental value of the binding energy per nucleon for the majority of nuclei is 8 MeV/nucleon.

However, the binding energy per nucleon depends on the mass number A. Starting from light nucleons, the binding energy increases with the mass number A from the value 1.1 MeV/nucleon for the deuterium nucleus to the value 8.8 MeV/nucleon for the iron isotope $^{56}_{26}$Fe. Then the binding energy gradually decreases up to the value 7.6 MeV/nucleon for the uranium isotope $^{238}_{92}$U. The dependence of the nucleus binding energy per one nucleus on the mass number A is shown in Figure 18.8. The character of this dependence explains the energy release during the uranium nucleus fission into fragments of the middle part of the Periodic System of Elements (see Section 18.6.3), and during the nuclear fusion of light nuclei with the formation of a heavier nucleus.

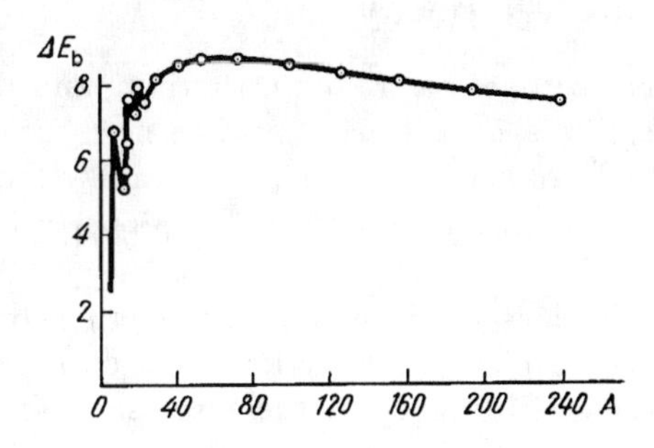

Figure 18.8: The nucleus binding energy per one nucleon.

The energy of 10 MeV constitutes only 1% of a nucleon rest energy $m_0c^2 = 1$ GeV. Thus we may really consider a nucleus to consist of separate nucleons that preserve their individual features also inside a nucleus.

The same conclusions based on the uncertainty relation (see Section 18.4.1) show that a nucleus can not contain electrons: if an electron were localized in a region of a radius $R = 10^{-13}$ cm, then it follows from Equation (18.17) that it would be a relativistic particle with an energy $E_k \approx pc \approx 0.2$ GeV. Such an immense value is incompatible with the value of a specific nucleus binding energy.

Inside a nucleus, the nucleons are bound by nuclear forces appearing due to the *strong interaction*. A proton and a neutron do not differ from each other in this respect.

Nuclear forces depend essentially on the distance between nucleons. At the distance of approximately 10^{-13} cm nuclear forces are 135 times greater than the electric repulsion forces between protons and 10^{38} times greater than the forces of gravitational interaction between nucleons. At distances smaller than $0.7 \cdot 10^{-13}$ cm nuclear forces exhibit themselves as repulsion forces, and at distances greater than $0.7 \cdot 10^{-13}$ cm they turn into attraction forces. Short-range nuclear forces do not exhibit themselves at distances greater than $2 \cdot 10^{-11}$ cm.

Some properties of nuclear forces can be explained by a suggestion that nucleons inside a nucleus exchange the particles—*mesons*. The meson mass can be estimated by using the Bohr–Heisenberg uncertainty relation: the uncertainty ΔE of a nucleus energy when it is radiating a particle with a mass m equals $\Delta E = mc^2$. This uncertainty exists during a time interval Δt that is necessary for a meson to pass through a nucleus ($\Delta t \leq R/c$). Since $\Delta E \cdot \Delta t \geq h$, then $m \approx \hbar/(Rc) = 3 \cdot 10^{-25}\text{g} = 300\ m_e$, where m_e is the electron mass. These particles were discovered experimentally in 1947 and were called *π-mesons*.

18.6.2 Radioactive decay

Some heavy nuclei (isotopes of uranium, thorium, and radium) are unstable and disintegrate spontaneously forming new nuclei and emitting α-particles, i.e., nuclei of helium-4 atoms (α-decay), electrons (β-decay), and photons of high energy (γ-decay). This property is called ***natural radioactivity***.

Radioactive decay does not depend on the external conditions such as temperature, pressure, chemical interactions. The decrease of the number of radioactive nuclei is determined by the expression (see Section 5.5.2)

$$N = N_0 e^{-\lambda t} = N_0 2^{-t/T},$$

where N_0 is the initial number of nuclei at $t = 0$, N is the residual number of nuclei at the time instant t, T is the ***half-life time*** (the time necessary for the half of nuclei to decay), and λ is the probability of decay of one nuclei per unit time (the ***decay constant***). The quantity $\tau = 1/\lambda$ is called the *mean lifetime*. Appearing due to radioactive decay, α- and γ-particles have discrete values of energy, and β-particles have a continuous spectrum of energies. β-decay is accompanied by the emission of neutrino and antineutrino (see Section 18.7).

New nuclei appearing as the result of radioactive decay occupy new places in the Periodic System of Elements: in the case of α-decay the Mendeleev table ordinal number decreases by two, in the case of β-decay it increases by one, and emission of γ-quanta is not being accompanied by the ordinal number change. This rule is called the ***shift law***.

18.6.3 Nuclear reactions

Nuclear reaction is the interaction of an atomic nucleus with another one or with an elementary particle (see Section 18.7), resulting in conversion of the nuclei. Nuclear reactions occur at distances between particles of the order of 10^{-13} cm. When nuclear reactions take place, the laws of conservation of energy, of momentum, of electric charge and of the number of nucleons are fulfilled and new isotopes not existing on the earth at natural conditions can be formed.

The first artificial nuclear reaction was performed by E. Rutherford:

$${}^{14}_{7}\mathrm{N} + {}^{4}_{2}\mathrm{He} \rightarrow {}^{17}_{8}\mathrm{O} + {}^{1}_{1}\mathrm{H}.$$

The upper indexes indicate the mass of a particle expressed in the atomic mass units (see Section 14.2.1), the lower ones indicate the electric charge expressed in units of the elementary charge.

Artificial radioactivity is a radioactivity of a nucleus caused by nuclear reactions. It was discovered in the nuclear reaction

$$^{27}_{13}Al + ^{4}_{2}He \rightarrow ^{30}_{15}P^* + ^{1}_{0}n.$$

The radioactive isotope of phosphorus $^{30}_{15}P^*$ (star is added to the symbol to indicate radioactivity of the isotope) is a source of β^+-radiation—its nucleus emits a *positron* (an antiparticle of an electron, see Section 18.7). The neutron was discovered at the nuclear reaction

$$^{9}_{4}Be + ^{4}_{2}He \rightarrow ^{12}_{6}C + ^{1}_{0}n.$$

The first splitting of a nucleus was carried out with the help of artificially accelerated protons bombarding a target with lithium nuclei:

$$^{7}_{3}Li + ^{1}_{1}H \rightarrow ^{4}_{2}He + ^{4}_{2}He.$$

Fission of uranium: the nucleus of uranium-235 being exposed to slow neutrons decomposes into radioactive fragments. This process is accompanied by emission of several neutrons and some energy:

$$^{235}_{92}U + ^{1}_{0}n \rightarrow ^{145}_{56}Ba^* + ^{88}_{36}Kr^* + 3^{1}_{0}n,$$

$$^{235}_{92}U + ^{1}_{0}n \rightarrow ^{139}_{54}Xe^* + ^{95}_{38}Sr^* + 2^{1}_{0}n,$$

$$^{235}_{92}U + ^{1}_{0}n \rightarrow ^{140}_{55}Ca^* + ^{94}_{37}Rb^* + 2^{1}_{0}n,$$

$$^{235}_{92}U + ^{1}_{0}n \rightarrow ^{145}_{57}La^* + ^{87}_{35}Br^* + 4^{1}_{0}n.$$

When uranium-235 fission takes place *a chain reaction* can occur. It is a process in which a certain reaction causes subsequent reactions of the same type. The minimal uranium mass making possible the chain reaction is called the *critical mass*. Chain nuclear reactions are used in nuclear weapon, and in nuclear power plants.

Reaction of fusion of light nuclei is possible at their collision, so the nuclei should be accelerated up to very high velocities in order to overcome their Coulomb repulsion (see Section 15.1.1). Much more energy is released at fusion of light nuclei than at fission of heavy ones (see Section 18.6.1). Here are some reactions of fusion:

$$^{2}_{1}D + ^{2}_{1}D \rightarrow ^{3}_{1}T + ^{1}_{1}H + 4.0 \text{ MeV},$$

$$^{2}_{1}D + ^{2}_{1}D \rightarrow ^{3}_{2}He + ^{1}_{0}H + 3.25 \text{ MeV},$$

$$^{2}_{1}D + ^{3}_{1}T \rightarrow ^{4}_{2}He + ^{1}_{0}H + 17.7 \text{ MeV}.$$

To realize the synthesis of helium from the heavy hydrogen isotopes—deuterium and tritium—a high temperature of the order of $5 \cdot 10^7$ K is necessary, while synthesis of helium from the light hydrogen isotope takes place at a temperature of about 10^8 K.

The fusion of light nuclei is utilized in the thermonuclear weapon. Realization of a controlled reaction of nuclear fusion may solve the problem of energy supply on the earth.

18.7 Elementary Particles

Besides elementary particles that form atoms—*protons, neutrons,* and *electrons*—a large number of other elementary particles is known at present. Not all of them are stable. For example, electrons escaping a nucleus at β-decay are formed during a neutron decay:

$$\mathrm{n} \rightarrow \mathrm{p} + \mathrm{e}^- + \tilde{\nu},$$

neutron → proton + electron + antineutrino.

The beta-decay is an example of the *weak interaction*.

A simplified modern classification of known elementary particles is given in Supplement VII.

Most of the elementary particles have their *antiparticles* possessing opposite charges (electric, baryon, etc.). The particles coinciding with their antiparticles, such as photons, are called true neutral particles. Antiparticles are denoted by a tilde over their symbol. When a particle collides with its antiparticle *annihilation* occurs: both of them transform to γ-radiation or to lighter particles (see Section 18.3.1). The opposite process is called *pair production*: for example, a γ-quantum passing a nucleus can form an electron–positron pair if its energy exceeds 1.02 MeV.

Besides particles listed in Supplement VII, a great many particles with extremely small lifetimes (of about 10^{-27} s) were discovered later. They are called *resonances*.

Presently the theory (quantum chromodymamics) is developed which considers all particles involved in the strong interaction and called *hadrons* (mesons, barions) to be built of more fundamental particles called *quarks*. Quarks possess fractional electric charges equal to $+2/3$ and $-1/3$ of the elementary charge. Quarks do not exist in a free state, though their presence inside nucleons is confirmed by direct probing of nucleons with high-energy electrons.

APPENDIX

I Fundamental Physical Constants

Universal Constants	
Speed of light in a vacuum	$c = 299\,792\,458$ m/s
Magnetic constant	$\mu_0 = 4\pi \cdot 10^{-7}$ N/A^2
Electric constant	$\varepsilon_0 = 1/(\mu_0 c^2) = 8.854\,188 \cdot 10^{-12}$ F/m
Gravitational constant	$G = 6.672\,59 \cdot 10^{-11}$ m^3/(kg·s^2)
Planck constant	$h = 6.626\,0755 \cdot 10^{-34}$ J·s
Electromagnetic Constants	
Elementary electric charge	$e = 1.602\,177\,33 \cdot 10^{19}$ C
Magnetic flux quantum $h/2e$	$\Phi_0 = 2.067\,834\,61 \cdot 10^{-15}$ Wb
Bohr's magneton $e\hbar/(2m_e)$	$\mu_B = 9.274\,0154 \cdot 10^{-24}$ J/T
Nuclear magneton $e\hbar/(2m_p)$	$\mu_N = 5.050\,7866 \cdot 10^{-24}$ J/T
Atomic Constants	
Fine structure constant $\mu_0 ce^2/2h$	$\alpha = 7.297\,353\,08 \cdot 10^{-3}$ $1/\alpha = 137.035\,9895 \approx 137$
Rydberg's constant $m_e c\alpha^2/2h$	$R_\infty = 10\,973\,731.534$ m^{-1}
Rydberg in electron-volts $R_\infty hc$	13.605 6981 eV
Bohr's radius $\alpha/(4\pi R_\infty)$	$a_0 = 0.529\,177\,249 \cdot 10^{-10}$ m
Electron rest mass	$m_e = 9.109\,3897 \cdot 10^{-31}$ kg
Classical radius of electron $\alpha^2 a_0$	$r_e = 2.817\,940\,92 \cdot 10^{-15}$ m
Proton rest mass	$m_p = 1.672\,6231 \cdot 10^{-27}$ kg
Neutron rest mass	$m_n = 1.674\,9286 \cdot 10^{-27}$ kg
Deutron rest mass	$m_d = 3.343\,5860 \cdot 10^{-27}$ kg
Physico-chemical Constants	
Avogadro's constant	$N_A = 6.022\,1367 \cdot 10^{23}$ 1/mol
Atomic mass unit	1 a.m.u. $= 1.660\,5402 \cdot 10^{-27}$ kg
Faraday's constant	$F = 96\,485.309$ C/mol
Universal gas constant	$R = 8.314\,510$ J/(mol·K)
Boltzmann's constant R/N_A	$k = 1.380\,658 \cdot 10^{-23}$ J/K
Loschmidt's constant	$n_0 = 2.686\,763 \cdot 10^{25}$ m^{-3}

II Physical Quantities and their SI Units

Physical quantity	*Dimensions*	*Name of SI unit*	*Symbol*	*Definition*
Base units				
Length	L	meter	m	The length of the path traveled by light in a vacuum during a time interval of 1/299 792 458 s
Mass	M	kilogram	kg	Mass equal to that of the international prototype kept at Sevres (France)
Time	T	second	s	Time equal to the duration of 9 192 631 770 periods of radiation corresponding to transition between two hyperfine levels of the ground state of the cesium-133 atom
Electric current	I	ampere	A	The constant current that, maintained in two straight parallel infinite conductors of negligible cross-section placed 1 m apart in a vacuum, would produce a force between the conductors of $2 \cdot 10^{-7}$ N per 1 m of length
Thermodynamic temperature	θ	kelvin	K	The fraction 1/273.16 of the thermodynamic temperature of the triple point of water
Amount of substance	N	mole	mol	The amount of substance that contains as many elementary units as there are atoms in 0.012 kg of carbon-12
Luminous intensity	J	candela	cd	The luminous intensity in a given direction of a source that emits monochromatic radiation of frequency 540×10^{12} Hz and has radiant intensity in that direction of 1/683 W per steradian
Supplementary units				
Plane angle	–	radian	rad	The angle subtended at the center of a circle by an arc of equal length to the radius
Solid angle	–	steradian	sr	The solid angle that encloses a surface on a sphere equal to the square of the radius of the sphere

II. Physical Quantities and their SI Units (continued)

Physical quantity	*Dimensions*	*Name of SI unit*	*Symbol*	*Definition*
Derived SI units				
Acceleration	LT^{-2}	–	m/s^2	Acceleration of a uniformly accelerated motion, in which velocity changes by 1 m/s during 1 s
Angular velocity	T^{-1}	–	rad/s	Angular velocity of a uniform rotation, at which during 1 s the angle changes by 1 rad
Frequency	T^{-1}	hertz	Hz	Frequency of an oscillation, in which one cycle is performed during 1 s
Density	ML^{-3}	–	kg/m^3	Density of a substance containing 1 kg of mass in a volume of 1 m^3
Momentum	MLT^{-1}	–	kg·m/s	Momentum of a body with mass 1 kg moving with a velocity of 1 m/s
Force	MLT^{-2}	newton	N	The force that gives a mass of 1 kg an acceleration of 1 m/s^2
Pressure	$ML^{-1}T^{-2}$	pascal	Pa	The pressure equal to 1 N per 1 m^2
Work, Energy	ML^2T^{-2}	joule	J	The work performed by a force of 1 N when the body moves in the direction of the force a distance of 1 m
Power	ML^2T^{-3}	watt	W	The power at which 1 J of work is performed during 1 s
Electric charge	TI	coulomb	C	The charge transferred through a cross-section of a conductor in 1 s by a current of 1 A

II Physical Quantities and their SI Units (continued)

Physical quantity	*Dimensions*	*Name of SI unit*	*Symbol*	*Definition*
		Derived SI units		
Voltage, e.m.f.	$ML^2t^{-3}I^{-1}$	volt	V	The difference of potential between two points of a conductor carrying a constant current of 1 A when the power dissipated between the points is 1 W
Capacitance	$M^{-1}L^{-2}T^4I^2$	farad	F	The capacitance of a capacitor that, if charged with 1 C, has a potential difference of 1 V between its plates
Resistance	$ML^2T^{-3}I^{-2}$	ohm	Ω	The resistance between two points on a conductor that, when a constant voltage of 1 V is applied between these points, it produces a current of 1 A in the conductor
Magnetic induction	$MT^{-2}I^{-1}$	tesla	T	The magnetic flux density equal to 1 Wb of magnetic flux per square meter of cross-section
Magnetic flux	$ML^2T^{-2}I^{-1}$	weber	Wb	The magnetic flux that, linking a circuit with resistance of 1 Ω, produces in it a transfer of 1 C of charge as the flux is reduced to zero
Inductance	$ML^2T^{-2}I^{-2}$	henry	H	The inductance of a closed circuit linked with a magnetic flux of 1 Wb when the circuit carries a current of 1 A
Luminous flux	J	lumen	lm	The luminous flux emitted by a uniform light source of luminous intensity 1 cd in a solid angle of 1 sr
Illuminance	$L^{-2}J$	lux	lx	The illuminance produced by a luminous flux of 1 lm distributed uniformly over an area of 1 m^2

III Conversion of Gaussian Units into SI Units

Physical quantity	*Unit*		*Ratio of the units*
	SI	Gauss	
Length	m	cm	10^2
Time	s	s	1
Velocity	m/s	cm/s	10^2
Acceleration	m/s^2	cm/s^2	10^2
Frequency	Hz	Hz	1
Circular frequency	1/s	1/s	1
Angular velocity	rad/s	rad/s	1
Angular acceleration	rad/s^2	rad/s^2	1
Mass	kg	g	10^3
Density	kg/m^3	g/cm^3	10^{-3}
Force	N	dyne	10^5
Pressure, tension	Pa	$dyne/cm^2$	10
Momentum	kg·m/s	g·sm/s	10^5
Torque	N·m	dyne·cm	10^7
Energy, work	J	erg	10^7
Power	W	erg/s	10^7
Angular momentum	$kg{\cdot}m^2/s$	$g{\cdot}cm^2/s$	10^7
Moment of inertia	$kg{\cdot}m^2$	$g{\cdot}cm^2$	10^7
Viscosity	Pa·s	poise	10
Temperature	K	K	1
Heat capacity, entropy	J/K	erg/K	10^7
Electric charge	C	c.g.s.	$3 \cdot 10^9$
Voltage	V	c.g.s.	1/300
Electric field strength	V/m	c.g.s.	$1/(3 \cdot 10^4)$
Electric capacitance	F	cm	$9 \cdot 10^{11}$
Electric current	A	c.g.s.	$3 \cdot 10^9$
Current density	A/m^2	c.g.s.	$3 \cdot 10^5$
Electric resistance	Ω	c.g.s.	$1/(9 \cdot 10^{11})$
Resistivity	Ω·m	c.g.s.	$1/(9 \cdot 10^9)$
Electric conductance	S	c.g.s.	$9 \cdot 10^{11}$
Conductivity	S/m	c.g.s.	$9 \cdot 10^9$
Magnetic induction (magnetic flux density)	T	G	10^4
Magnetic flux	Wb	Mx	10^8
Inductance	H	cm	10^9
Luminous flux	lm	lm	1
Luminous intensity	cd	cd	1

IV Conversion of Non-system Units into SI Units

Physical quantity	*From unit*	*To* SI *unit*
Length	angstrom (Å)	10^{-10} m
	inch (in)	$2.54 \cdot 10^{-2}$ m
	foot (ft)	0.3048 m
	astronomical unit (AU)	$1.495\,978\,70 \cdot 10^{11}$ m
	light year	$9.460\,530 \cdot 10^{15}$ m
	parsec	$3.085\,678 \cdot 10^{16}$ m
Mass	atomic mass unit (a.m.u.)	$1.660\,5402 \cdot 10^{-27}$ kg
	pound (lb)	0.453 592 37 kg
Area	barn	10^{-28} m^2
	square inch (sq. in)	$6.4516 \cdot 10^{-4}$ m^2
	square foot (sq. ft)	$9.2903 \cdot 10^{-2}$ m^2
Volume	liter (l)	10^{-3} m^3
	cubic inch (cu. in)	$1.63871 \cdot 10^{-5}$ m^3
	cubic foot (cu. ft)	$2.83168 \cdot 10^{-2}$ m^3
	gallon (gal)	$4.54609 \cdot 10^{-3}$ m^3
Velocity	km/hr	0.277 78 m/s
	miles/hr	0.477 04 m/s
Density	lb/in^3	$2.767\,99 \cdot 10^{4}$ kg/m^3
Force	dyne	10^{-5} N
	kgf	9.806 65 N
Energy, work	electronvolt (eV)	$1.602\,177\,33 \cdot 10^{-19}$ J
	kilowatt-hour (kWh)	$3.6 \cdot 10^{6}$ J
	calorie (cal)	4.1868 J
Power	horsepower (hp)	735.489 W
Pressure	mm Hg (torr)	133.322 Pa
	atmosphere (atm)	101 325 Pa
Optical power	diopter	1 m^{-1}

V Main Formulas of Electrodymamics in Gaussian Units and in SI Units

	Gaussian	SI
Coulomb's law	$F = \frac{q_1 q_2}{r^2}$	$F = \frac{1}{4\pi\varepsilon_0}\frac{q_1 q_2}{r^2}$
Electric field strength (definition)	$E = \frac{F}{q}$	$E = \frac{F}{q}$
Strength of the electric field created by a point charge q	$E = \frac{q}{r^2}$	$E = \frac{1}{4\pi\varepsilon_0}\frac{q}{r^2}$
Strength of the electric field near the surface of a conductor	$E = 4\pi\sigma$	$E = \frac{\sigma}{\varepsilon_0}$
Electric flux through a surface (definition)	$\Phi_E = \sum E_n \Delta S$	$\Phi_E = \sum E_n \Delta S$
Gauss' law	$\Phi_E = 4\pi q$	$\Phi_E = \frac{q}{\varepsilon_0}$
Electric potential (definition)	$\varphi = \frac{A}{q}$	$\varphi = \frac{A}{q}$
Electric potential of the field created by a point charge	$\varphi = \frac{q}{r}$	$\varphi = \frac{1}{4\pi\varepsilon_0}\frac{q}{r}$
Relation between the electric field strength and electric potential	$E_l = \frac{\Delta\varphi}{\Delta l}$	$E_l = \frac{\Delta\varphi}{\Delta l}$
Capacitance (definition)	$C = \frac{q}{U}$	$C = \frac{q}{U}$
Capacitance of a plane capacitor with a dielectric	$C = \frac{\varepsilon S}{4\pi d}$	$C = \frac{\varepsilon_0 \varepsilon S}{d}$
Energy of a system of electric charges	$W_e = \frac{1}{2}\sum_i q_i\varphi_i$	$W_e = \frac{1}{2}\sum_i q_i\varphi_i$
Energy of a capacitor	$W_e = \frac{1}{2}CU^2$	$W_e = \frac{1}{2}CU^2$
Density of the energy of electric field	$w_e = \frac{E^2}{8\pi}$	$w_e = \frac{\varepsilon_0 E^2}{2}$
Current (definition)	$I = \frac{dq}{dt}$	$I = \frac{dq}{dt}$
Ohm's law	$I = \frac{U}{R}$	$I = \frac{U}{R}$

V Main Formulas of Electrodymamics in Gaussian Units and in SI Units (continued)

	Gaussian	SI
Joule's law	$Q = I^2 Rt$	
Force of interaction of two parallel currents in a vacuum	$F = \frac{1}{c^2}\frac{2I_1I_2l}{r}$	$F = \frac{\mu_0}{4\pi}\frac{2I_1I_2l}{r}$
Law of Biot–Savart–Laplace	$\Delta B = \frac{1}{c}\frac{I\Delta l \sin\alpha}{r^2}$	$\Delta B = \frac{\mu_0}{4\pi}\frac{1}{c}\frac{I\Delta l \sin\alpha}{r^2}$
Magnetic field of a long straight wire	$B = \frac{1}{c}\frac{2I}{r}$	$\frac{\mu_0}{4\pi}\frac{2I}{r}$
Magnetic field at the center of a circular loop	$B = \frac{1}{c}\frac{2\pi I}{R}$	$B = \frac{\mu_0 I}{2R}$
Magnetic field inside a long coil	$B = \frac{4\pi}{c}nI$	$B = \mu_0 nI$
Magnetic field circulation along a closed loop	$\sum B_l\Delta l = \frac{4\pi}{c}I$	$\sum B_l\Delta l = \mu_0 I$
Ampere's law	$F = \frac{1}{c}IBl\sin\alpha$	$F = IBl\sin\alpha$
Lorentz force	$\mathbf{F} = \frac{1}{c}q\mathbf{v}\times\mathbf{B}$	$\mathbf{F} = q\mathbf{v}\times\mathbf{B}$
Magnetic flux	$\Phi = \sum B_n\Delta S$	
Faraday's law of electromagnetic induction	$\mathcal{E} = -\frac{1}{c}\frac{d\Phi}{dt}$	$\mathcal{E} = -\frac{d\Phi}{dt}$
Inductance (definition)	$\Phi = \frac{1}{c}LI$	$\Phi = LI$
Inductance of a long coil	$L = 4\pi n^2 V$	$L = \mu_0 n^2 V$
Magnetic energy of a current (energy of a magnetic field)	$W_m = \frac{1}{c^2}\frac{LI^2}{2}$	$W_m = \frac{LI^2}{2}$
Density of magnetic field energy	$w_m = \frac{B^2}{8\pi}$	$w_m = \frac{B^2}{2\mu_0}$

VI Atomic Elements and their Masses

Name	*Smb*	N°	*Mass*	*Name*	*Smb*	N°	*Mass*
Actinium	Ac	89	[227]	Mercury	Hg	80	200.59
Aluminum	Al	13	26.98154	Molybdenum	Mo	42	95.94
Americium	Am	95	[243]	Neodymium	Nd	60	144.24
Antimony	Sb	51	121.75	Neon	Ne	10	20.179
Argon	Ar	18	39.948	Neptunium	Np	93	237.0482
Arsenic	As	33	74.9216	Nickel	Ni	28	58.71
Astatine	At	85	[210]	Niobium	Nb	41	92.9064
Barium	Ba	56	137.33	Nitrogen	N	7	14.0067
Berkelium	Bk	97	[247]	Nobelium	No	102	[259]
Beryllium	Be	4	9.01218	Osmium	Os	76	190.2
Bismuth	Bi	83	208.9804	Oxygen	O	8	15.9994
Boron	B	5	10.81	Palladium	Pa	46	106.4
Bromine	Br	35	79.904	Phosphorus	P	15	30.97376
Cadmium	Cd	48	112.41	Platinum	Pt	78	195.09
Calcium	Ca	20	40.08	Plutonium	Pu	94	[244]
Californium	Cf	98	[251]	Polonium	Po	84	[209]
Carbon	C	6	12.011	Potassium	K	19	39.0983
Cerium	Ce	58	140.12	Praseodymium	Pr	59	140.9077
Cesium	Cs	55	132.9054	Promethium	Pm	61	[145]
Chlorine	Cl	17	35.453	Protactinium	Pa	91	231.0359
Chromium	Cr	24	51.996	Radium	Ra	88	226.0254
Cobalt	Co	27	58.9332	Radon	Rn	86	[222]
Copper	Cu	29	63.546	Rhenium	Re	75	186.207
Curium	Cm	96	[247]	Rhodium	Rh	45	102.9055
Dysprosium	Dy	66	162.50	Rubidium	Rb	37	85.467
Einsteinium	Es	99	[254]	Ruthenium	Ru	44	101.07
Erbium	Er	68	167.26	Rutherfordium	Rf	104	[260]
Europium	Eu	63	151.96	Samarium	Sm	62	150.4
Fermium	Fm	100	[257]	Scandium	Sc	21	44.9559
Fluorine	F	9	18.998403	Selenium	Se	34	78.96
Francium	Fr	87	[223]	Silicon	Si	14	28.0855
Gadolinium	Gd	64	157.25	Silver	Ag	47	107.868
Gallium	Ga	31	69.735	Sodium	Na	11	22.98977
Germanium	Ge	32	72.59	Strontium	Sr	38	87.62
Gold	Au	79	196.9665	Sulfur	S	16	32.06
Hafnium	Hf	72	178.49	Tantalum	Ta	73	180.947
Helium	He	2	4.00260	Technetium	Tc	43	98.9062
Holmium	Ho	67	164.9304	Tellurium	Te	52	127.60
Hydrogen	H	1	1.0079	Terbium	Tb	65	158.9254
Indium	In	49	114.82	Thallium	Tl	81	204.37
Iodine	I	53	126.9045	Thorium	Th	90	231.0381
Iridium	Ir	77	192.22	Thulium	Tm	69	168.9342
Iron	Fe	26	55.847	Tin	Sn	50	118.69
Krypton	Kr	36	83.80	Titanium	Ti	22	47.90
Lanthanum	La	57	138.9055	Tungsten	W	74	183.85
Lawrencium	Lr	103	[260]	Uranium	U	92	238.029
Lead	Pb	82	207.2	Vanadium	V	23	50.9415
Lithium	Li	3	6.941	Xenon	Xe	54	131.30
Lutetium	Lu	71	174.967	Ytterbium	Yb	70	173.04
Magnesium	Mg	12	24.305	Yttrium	Y	39	88.9059
Manganese	Mn	25	54.9380	Zinc	Zn	30	65.38
Mendelevium	Md	101	[258]	Zirconium	Zr	40	91.22

VII Table of Elementary Particles

Name	*Symbol*	*Anti-particle*	*Mass* (in m_e)	*Elect. charge*	*Decay time* (c)
Photon	γ	Self	0	0	Stable
			Leptons		
Electron neutrino	ν_e	$\tilde{\nu}_e$	$0\,(< 6 \cdot 10^{-5})$	0	Stable
Muon neutrino	ν_μ	$\tilde{\nu}_\mu$	$0\,(< 1)$	0	Stable
Tau neutrino	ν_τ	$\tilde{\nu}_\tau$	$0\,(< 500)$	0	Stable
Electron	e^-	e^+	1	-1	Stable
Muon	μ^-	μ^+	207	-1	$2.2 \cdot 10^{-6}$
Tau-lepton	τ^-	τ^+	3492	-1	$3 \cdot 10^{-13}$
			Mesons		
Pion	π^0	Self	264.1	0	$0.83 \cdot 10^{-16}$
	π^+	π^-	273.1	1	$2.6 \cdot 10^{-8}$
Kaon	K^+	K^-	966.4	1	$1.2 \cdot 10^{-8}$
	K^0	Self	974.1	0	$8.9 \cdot 10^{-11}\,(K_s^0)$ $5.2 \cdot 10^{-8}\,(K_L^0)$
Eta	η^0	Self	1074	0	$7 \cdot 10^{-19}$
			Barions		
Proton	p	$\tilde{p}$	1836.1	1	Stable (?)
Neutron	n	$\tilde{n}$	1838.6	0	10^3
Lambda-zero	Λ^0	$\tilde{\Lambda}^0$	2183.1	0	$2.63 \cdot 10^{-10}$
Sigma-plus	Σ^+	$\tilde{\Sigma}^+$	2327.6	1	$8 \cdot 10^{-11}$
Sigma-zero	Σ^0	$\tilde{\Sigma}^0$	2333.6	0	$5.8 \cdot 10^{-20}$
Sigma-minus	Σ^-	$\tilde{\Sigma}^-$	2343.1	-1	$1.48 \cdot 10^{-10}$
Cascade-zero	Ξ^0	$\tilde{\Xi}^0$	2572.8	0	$2.9 \cdot 10^{-10}$
Cascade-minus	Ξ^-	$\tilde{\Xi}^-$	2585.6	-1	$1.64 \cdot 10^{-10}$
Omega-minus	Ω^-	$\tilde{\Omega}^-$	3273	-1	$8.2 \cdot 10^{-11}$

VIII Decimal Multiples to be Used with SI Units

Multiple	*Prefix*	*Symbol*	*Multiple*	*Prefix*	*Symbol*
10^{-1}	deci	d	10	deca	da
10^{-2}	centi	c	10^2	hecto	h
10^{-3}	milli	m	10^3	kilo	k
10^{-6}	micro	μ	10^6	mega	M
10^{-9}	nano	n	10^9	giga	G
10^{-12}	pico	p	10^{12}	tera	T
10^{-15}	femto	f	10^{15}	peta	P
10^{-18}	atto	a	10^{18}	exa	E

IX Relations between Fundamental Constants (in Gaussian system of units)

Electric constant	$\varepsilon_0 = 1/(\mu_0 c^2)$
Bohr's magneton	$\mu_B = e\hbar/(2m_e c)$
Fine structure constant	$\alpha = e^2/(\hbar c)$
Rydberg's constant	$R_\infty = m_e e^4/(2\hbar^2)$
Bohr's radius	$a_0 = \hbar^2/(m_e e^2)$
Compton's wavelength for electron	$\lambda_e = h/(m_e c)$
Classical radius of electron	$r_e = e^2/(m_e c^2)$
Faraday's constant	$F = eN_A$
Universal gas constant	$R = kN_A$

Expressions for α, R_∞, a_0, λ_e, r_e in terms of fundamental constants in SI units can be obtained from the above formulas by the substitution $e^2 \to e^2/(4\pi\varepsilon_0)$.

X Table of Mathematical Symbols

N means the set of natural numbers	$\in$ means "belongs" $\notin$ means "does not belong"
Z means the set of integral numbers	$\emptyset$ means empty set $\cup$ means union of sets
Q means the set of rational numbers	$\cap$ means intersection of sets $\subseteq$ means subset of a set
R means the set of real numbers	$\Rightarrow$ means implies $\Leftrightarrow$ means equivalent
C means the set of complex numbers	lim means the limit of a function or a sequence
max means the maximum of a function	$n!$ is the factorial $n!!$ is the semi-factorial
min means the minimum of a function	C_n^m means the number of combinations
$f'(x)$ is the derivative of $f(x)$	$\text{Re}\, z$ is the real part of complex number z
$\int$ means the indefinite integral	$\text{Im}\, z$ is the imaginary part of complex number z
$\int_a^b$ means the definite integral	$\|z\|$ is the modulus of complex number z
$\sum_{k=1}^n$ means the sum from the 1st to the n-th term	$\arg z$ is the argument of complex number z
$\langle A \rangle$ is the mean value of random quantity A	$(\mathbf{a}; \mathbf{b})$ means the angle between vectors **a** and **b**
$\pi = 3.14159265358979...$	$e = 2.718281828459045...$

Index